Translation
Generalized
Quadrangles

SERIES IN PURE MATHEMATICS

Editor: C C Hsiung
Associate Editors: S Kobayashi, I Satake, Y-T Siu, W-T Wu and M Yamaguti

Series in Pure Mathematics – Volume 26

Translation Generalized Quadrangles

J A Thas
K Thas
H Van Maldeghem

Ghent University, Belgium

World Scientific

NEW JERSEY • LONDON • SINGAPORE • BEIJING • SHANGHAI • HONG KONG • TAIPEI • CHENNAI

Published by

World Scientific Publishing Co. Pte. Ltd.

5 Toh Tuck Link, Singapore 596224

USA office: 27 Warren Street, Suite 401-402, Hackensack, NJ 07601

UK office: 57 Shelton Street, Covent Garden, London WC2H 9HE

British Library Cataloguing-in-Publication Data
A catalogue record for this book is available from the British Library.

TRANSLATION GENERALIZED QUADRANGLES
Series in Pure Mathematics — Vol. 26

Copyright © 2006 by World Scientific Publishing Co. Pte. Ltd.

ISBN 981-256-951-0

Printed in Singapore by World Scientific Printers (S) Pte Ltd

To Our Families

Contents

Preface

This book grew out of sets of notes that were provided for students attending the 45-hours elective graduate course *Capita Selecta in Geometry* in 2003 and in 2005 at Ghent University, Belgium. In 2003, the main object of the course was to discuss the Moufang Property for generalized quadrangles, and local versions of it such as being translation with respect to a point or a line. In 2005, the main object was the theory of finite translation generalized quadrangles. In both years, the lecturers were the authors of this book. The audience consisted of undergraduate students, Ph.D. students and post-docs of the research group *Incidence Geometry* at Ghent University. We thank them for their interest and their questions that forced us, once in a while, to review some mathematical details and arguments.

The research needed to write a monograph like the present one is financed by a number of organizations. First of all, we are indebted to the Research Council of Ghent University for awarding a GOA grant *Incidence Geometry*. We also would like to thank the Fund for Scientific Research — Flanders (Belgium) FWO for financial support (research money, Ph.D.-grants, post-doc positions, travel support) and the Institute for Science and Technology IWT for providing Ph.D.-grants. We thank our secretary Sonia Surmont for typing large parts of the manuscript. Finally, a warm "thank you" to all our colleagues and friends in the research group *Incidence Geometry* for the wonderful atmosphere.

J. A. Thas, K. Thas and H. Van Maldeghem
Ghent, March 2006

Foreword

Generalized polygons were introduced by Tits [217] in the appendix of his celebrated work "Sur la trialité et certains groupes qui s'en déduisent". In the latter, Jacques Tits discovers and constructs the triality groups $^3\mathbf{D}_4(\mathbb{K})$, for appropriate fields $\mathbb{K}$. To that aim, he investigates a certain configuration $\mathcal{H}$ of points and lines in a seven-dimensional projective space $\mathbf{PG}(7,\mathbb{K})$ on a quadric of maximal Witt index. Along the course of his argumentation, he verifies in fact the axioms of a generalized hexagon for $\mathcal{H}$, a notion that he then introduces in the appendix in full generality, i.e., replacing 6 by n, he introduces generalized n-gons.

Generalized polygons help to better understand the structure of the semi-simple algebraic groups (including the groups of Lie-type, the (twisted) Chevalley groups and the groups of mixed type) of relative rank 2. Generalized polygons are the building bricks of the more general notion of a Tits building and play a central role in the modern theory of Incidence Geometry.

For finite thick generalized n-gons Feit and Higman [56] show that $n \in \{2,3,4,6,8\}$. Also, several restrictions on the parameters of finite generalized n-gons hold. In the infinite case we do not have a Feit-Higman theorem (although in other natural situations, similar restrictions emerge, such as in the compact connected case [88], the Moufang case [229], the valuation case [230]) and here free generalized n-gons may be constructed in much the same way as free projective planes. Generalized 2-gons are trivial structures (all points are incident with all lines), and the generalized 3-gons are the projective planes. Up to duality just two classes of finite thick generalized hexagons are known, and just one class of finite thick generalized octagons; see e.g. Van Maldeghem [232]. So not surprisingly most of the current literature on generalized n-gons, with $n > 3$, concerns generalized quadrangles, not at least due to the many interconnections with strongly regular graphs, projective planes, hyperovals, circle geometries, codes, designs, classical varieties over finite fields, finite groups.

The standard work on finite generalized quadrangles is due to Payne and Thas, and appeared in 1984. This book, "Finite Generalized Quadrangles" [128], is a systematic treatment of the topic and contains most of the results known up to 1983. The standard work "Generalized Polygons" is authored by Van Maldeghem [232]. Here we find the general theory for finite and infinite generalized polygons, with just a small overlap with the book of Payne and Thas. We also mention the chapter "Generalized Polygons" by Thas in the monumental "Handbook of Incidence Geometry: Buildings and Foundations" [31], edited by Buekenhout in 1995, which collects the most important results — however without proofs — on finite generalized polygons; also, a short section is devoted to the infinite case. Finally, the monograph "Symmetry in Finite Generalized Quadrangles" [206] on automorphisms of finite generalized quadrangles by Koen Thas was published in 2004. This work contains many new results on symmetry, elations and translations of finite generalized quadrangles, all in the bigger framework of a classification of all finite quadrangles based on the notion of "symmetry".

The main examples of finite generalized quadrangles essentially belong to five types:

(1) they are fully embedded in a projective space $\mathbf{PG}(n,q)$ over the Galois field $\mathbf{GF}(q)$, and these are the so-called "classical examples";

(2) they are the point-line duals of the classical examples (the dual $\mathbf{H}(4,q^2)^D$ of the classical generalized quadrangle $\mathbf{H}(4,q^2)$ is never embedded (in the usual sense) in a projective space);

(3) they are of order $(q-1,q+1)$ or, dually, of order $(q+1,q-1)$, and the examples of this type all are in some way connected to ovals or hyperovals of $\mathbf{PG}(2,q)$;

(4) they arise as translation generalized quadrangles (arising from generalized ovals or generalized ovoids);

(5) they arise from flocks of the quadratic cone in $\mathbf{PG}(3,q)$.

The examples which got the most attention in the past fifteen years are of Type (4) and (5), and in particular, many classification results were obtained for the amalgamations of (4) and (5), namely the translation generalized quadrangles which arise from a flock or of which the translation dual arises from a flock. In fact, there is a tight connection between Type (4) and Type (5), due to the the first author's sequence of papers [176], [180] and especially [183], in which he shows that each generalized quadrangle of order (q, q^2), q odd and $q \neq 1$, satisfying Property (G) [118] at some flag is the point-line dual of a flock generalized quadrangle, i.e. is of Type (5). For translation generalized quadrangles, this asserted that a translation generalized quadrangle $\mathcal{S} = \mathbf{T}(\mathcal{O})$ of order (q, q^2), q odd and $q \neq 1$, with $\mathcal{O}$ good, see below, is the translation dual of the point-line dual of a flock generalized quadrangle.

Many of the known generalized quadrangles are of Type (4) and many papers are focussed on the classification of translation generalized quadrangles, on characterizations of certain classes of generalized quadrangles in terms of translation generalized quadrangles, and on the interplay of these generalized quadrangles and other incidence geometries. Without any doubt translation generalized quadrangles play a key role in the general theory of generalized quadrangles, comparable to the role of translation planes in the theory of projective and affine planes. In fact, the notion of translation generalized quadrangle is a local analogue of the more global "Moufang Condition". The finite generalized quadrangles satisfying the Moufang Condition are those of Type (1) and Type (2). Recently, there has been renewed interest in Moufang quadrangles because of the classification of Moufang polygons, a job completed by Tits and Weiss in 1997 after the discovery of a final and unexpected class of Moufang quadrangles, see [229]. We give some attention to the recent results in this direction, too.

Let us now sketch some history of these topics. We begin with translation generalized quadrangles.

In 1969 Thas wrote the monograph "Een studie Betreffende de Projectieve Rechte over de Totale Matrix Algebra $M_3(K)$ der 3×3-Matrices met Elementen in een Algebraïsch Afgesloten Veld" [155] (with English Summary) on the projective line over the matrix algebra $M_3(K)$, where K is algebraically closed. This was the topic of his Ph.D., following the tradition of Ghent University where several Ph.D.s were written on ring geometries, starting with the pioneering work by Julien Bilo on geometries over the quaternions. Thas' Master Thesis [154] was on arcs and caps in finite projective spaces, based mainly on the work of Beniamino Segre. In his Habilitation Thas combined his work on projective spaces over matrix algebras and results on arcs and caps in projective spaces over finite fields, resulting in a paper of 73 pages titled "The m-Dimensional Projective Space $S_m(M_n(GF(q)))$ over the Total Matrix Algebra $M_n(GF(q))$ of the $n \times n$-Matrices with Elements in the Galois Field $GF(q)$", which appeared in 1971 [156]. Here generalizations of arcs and caps of $\mathbf{PG}(m,q)$ appeared, where now the points of the arcs and caps were replaced by $(n-1)$-dimensional subspaces. Around 1972 he read the chapter "Generalized Polygons" in the seminal work "Finite Geometries" by Dembowski [45]. In this chapter the generalized quadrangles $\mathbf{T}_2(\mathcal{O})$ and $\mathbf{T}_3(\mathcal{O})$ of Tits were described, and Thas realized that his generalized arcs and caps could be used to obtain a generalization of the construction of Tits. Along the same lines the construction of 1971 by Hall Jr. in "Affine generalized quadrilaterals" [66] could be extended. These generalizations appeared in the two papers entitled "On 4-gonal configurations" [160, 161]. One of these papers appeared in "Finite Geometric Structures and Their Applications", the proceedings of a Summer school held in Bressanone (Italy) in 1972. In these proceedings one finds another paper by Thas, on flocks of ovoids, and here we have a remarkable coincidence since fifteen years later it would appear that flocks play a central role in the theory of (elation) generalized quadrangles. The aforementioned papers by Thas got published into the proceedings because he was asked to deliver a lecture at the Summer school after he announced, during the lecture series of the celebrated R. H. Bruck, the solution of an open problem on flocks of ovoids posed by Bruck. It was a key problem as the solution implied the classification of all subregular translation planes. The lecture was Thas' first lecture on generalized quadrangles.

In 1974 Thas wrote his paper "Translation 4-gonal configurations" [163] in which he develops a theory for generalized quadrangles similar to the theory of translation planes. It appeared that these translation generalized quadrangles are exactly his quadrangles arising from generalized ovals. In the same year he proved in "Geometric characterization of the $[n-1]$-ovaloids of the projective space $PG(4n-1,q)$" [162] that his generalized ovoids are essentially the common ovoids, but interpreted over a smaller field; hence the corresponding generalized quadrangles are isomorphic to quadrangles of Tits. The appropriate definition of generalized ovoid and its detailed study eventually appeared in "Finite Generalized Quadrangles" [128]; the essentially equivalent term "egg" was introduced by Kantor.

Payne was the first to define and study finite translation generalized quadrangles with any parameters in his paper "Generalized quadrangles as group coset geometries" [112]. A more elaborate treatment of the theory of finite translation generalized quadrangles, with any parameters, is described in "Finite Generalized Quadrangles" [128]. Too early, however, to contain the first family of finite translation generalized quadrangles not isomorphic to a quadrangle of Tits, in 1986 discovered by Kantor in "Some generalized quadrangles with parameters q^2, q" [83]. Hence the book by Payne and Thas only contains Tits' examples, but it provides a strong basis for the vast theory of translation generalized quadrangles.

The year 1987 is another landmark in the theory of generalized quadrangles. By a combination of results of Payne [112, 117] and Kantor [81, 83], with any so-called "q-clan" corresponds a generalized quadrangle of order (q^2, q); a q-clan is a set of q distinct upper triangular 2×2-matrices over $\mathbf{GF}(q)$ such that the difference of any two of them is anisotropic. In 1987 Thas noticed that the algebraic conditions to be satisfied by a q-clan are exactly the algebraic conditions to be satisfied to have a flock of a quadratic cone, that is, a set of q conics of the cone that partition it (up to its vertex) [172]. But already in 1976 Thas and Walker showed that with each such flock corresponds a translation plane of order q^2 [167, 237]. Hence by the observation of 1987 there was now a link between certain classes of generalized quadrangles and certain classes of translation planes. So specialists in one of these topics got interested in both fields. This led to an explosion of the literature on translation planes and generalized quadrangles. Translation generalized quadrangles of order (q, q^2) arising from flocks appeared to be in some sense equivalent to the important class of semifield flocks; these are the flocks for which the corresponding planes are semifield planes.

Another major reason for the fact that flock quadrangles received a lot of attention in the past two decades is the connection with ovals. Every flock quadrangle of even order contains subquadrangles of Tits related to some set of ovals of a projective plane. Especially this link makes the theory available for computer searches, and several people have put a lot of effort in this aspect. This resulted in many "small" new examples, but almost all of them were generalized to infinite series of new flock quadrangles, new hyperovals and new translation planes.

Many papers on generalized quadrangles concern the construction of particular flocks, the determination of their automorphism group, the study of the corresponding planes, characterizations of these flocks and planes, the determination of the so-called "derived flocks", and in the even case the analysis of the corresponding hyperovals. Our book has a small intersection with this specific research area, just mentioning some fundamental results, in particular those which are related to translation generalized quadrangles.

The main aim of the present book is to describe the theory of finite translation generalized quadrangles, but in some chapters also the infinite case is included. It contains several new results and new approaches to known theories and results. Most of the proofs are given, with exception of proofs which are too long, too technical and little instructive for the remainder.

As soon as a generalized quadrangle contains two different translation points, the group generated by the translation groups contains a factor group isomorphic to a so-called "Moufang set", and we also devote some space to present the elementary theory of these objects. Furthermore, generalized quadrangles with two opposite translation points satisfy the Moufang Condition, and we present some recent results in that area, not restricting ourselves necessarily to the finite case here.

We now continue with some history on Moufang quadrangles.

As already mentioned, the concept of a generalized quadrangle was formally introduced in the literature by Jacques Tits in his famous paper on trialities [217], as part of the more general concept of a generalized polygon. Later on, Tits introduced structures called "buildings", an important class being the spherical buildings (they include in particular all finite buildings!). These have a certain rank (dimension) and precisely when the rank is 2, the concept of a thick spherical building coincides with that of a thick generalized polygon. In fact, all other rank 2 buildings are trees without vertices of valency 1.

In 1974, Jacques Tits published a book containing the classification of all thick spherical buildings of rank at least 3, see [222]. In an addendum, he introduces the Moufang Condition for spherical buildings — and thus also for generalized polygons and quadrangles — motivated by his claim that the classification of all Moufang polygons would considerably simplify the classification of spherical buildings of higher rank. Tits started this programme himself already in the sixties, and soon he had a classification of all Moufang hexagons, although he never published this.

In the meantime, John R. Faulkner, from the University of Virginia, Charlottesville, studied certain simple groups — Chevalley groups of rank 2 — by means of their Steinberg representation [55]. This essentially amounts to the commutation relations between the U_i obtained in Section 11.5, including the facts that the U_i are abelian or nilpotent of class 2, etc. He obtained in [55] a wealth of classification results and examples of Moufang hexagons and quadrangles under the ostensibly stronger hypotheses of these commutation relations (but we proved that they follow from the Moufang Condition anyway; this was not yet known to Faulkner, who only derived these relations if there are no involutory root elations). In fact, Faulkner also showed how one can classify certain types of spherical buildings using his results. But since these results were not complete (characteristic 2 was missing for the quadrangles, for instance), they were not popular.

Independently, Jacques Tits worked on his programme, and he was able to classify the thick Moufang octagons (1976) [226], and to prove that no thick Moufang n-gons exist unless $n = 3, 4, 6, 8$ (see [223, 225]). The latter result was also proved by Richard Weiss [238], who derived it in a more general context in a simpler way. The case $n = 3$ amounts to classifying Moufang projective planes and was treated long before (the terminology stems from this case). Hence only the case of Moufang quadrangles was missing. However, Tits knew how to derive the Steinberg relations from the Moufang Property, and he considered this as a first step in the classification. This result was published in 1994, see [228].

In the finite case, Fong and Seitz published two papers [59, 60] in which they classify finite split Tits systems of Type $\mathbf{B}_2$. The geometric interpretation of their result is precisely the classification of finite Fong-Seitz generalized quadrangles. One of the corollaries shown by Fong and Seitz implies the classification of finite thick Moufang generalized quadrangles. This was remarked by Tits.

Until 1993, Jacques Tits refers to the list of Moufang quadrangles as "conjectural" (see for instance [224]). He emphasized the fact that no proof was written down, and that everything was still only in his head. In the meantime, some more results became available. Thas, Payne and Van Maldeghem proved in [196] that every finite half Moufang generalized quadrangle is a Moufang generalized quadrangle, using typically finite arguments. A similar approach was used by Van Maldeghem, Thas and Payne in [236] to show that any finite 3-Moufang generalized quadrangle is a Moufang generalized quadrangle (a result that was generalized by Van Maldeghem and Weiss to arbitrary finite polygons in [235]). In 1998, Van

Maldeghem [232] proved that the 2-Moufang Condition for thick generalized quadrangles is equivalent with the 3-Moufang Condition. Hence, at this moment, all equivalences are proved in the finite case (without invoking the classification of finite simple groups).

In 1993, Jacques Tits started to lecture again about Moufang polygons, with the idea to finish the classification, write it down (only the case of projective planes and octagons [226] was published back then) and publish it. Richard Weiss, who was in the audience, proposed Tits to do this jointly, and the result was that in 1997, the classification of Moufang quadrangles was completed — with an additional new class [229]! At first, the conjecture of Tits was that all Moufang quadrangles were related to classical groups, algebraic groups or mixed groups of relative rank 2, and he had an explicit list. The new example was not only missing in the list, it did not seem to be related to any classical, algebraic or mixed group. Not much later Mühlherr and Van Maldeghem [104] showed that it arises as a certain fixed point structure in a certain mixed building of Type F_4. This was somehow overlooked by Tits since the construction process of [104] is not captured by the theory of algebraic groups, but it is a "mixed analogue" of it. The explicit list of Tits turned out to be incomplete, but his conjecture that all Moufang quadrangles arise from classical groups, algebraic groups or mixed groups of relative rank 2 was still true; he simply overlooked that there also exist mixed groups of *relative* rank 2 (he only lists mixed groups of absolute rank 2).

In September 2002, the full classification of Moufang generalized polygons appeared in a book [229]. The part about quadrangles takes a special place, not only because it is the lengthiest part, but also because it is the most complicated. Tits and Weiss have to consider six different classes of algebraic nature, while in all other cases ($n = 3, 6, 8$), there is only one algebraic structure to consider. In his Ph.D. thesis [46], see also [47], Tom De Medts constructs such a structure for generalized quadrangles, thus unifying, streamlining and shortening the proof of Tits and Weiss. We now have a most beautiful and satisfying proof for the classification (much more "elementary" than the proof for the finite case of Fong and Seitz!).

Before the book of Tits and Weiss was published, Katrin Tent and Hendrik Van Maldeghem considered the question whether every Fong-Seitz generalized quadrangle is a Moufang generalized quadrangle. Together with the result of Tits and Weiss, this would yield a new, shorter, more elegant and simpler proof of the result by Fong and Seitz, and at the same time also a result valid in the infinite case. This was accomplished in 2000, see [153].

Tent and Van Maldeghem proved that every thick Fong-Seitz n-gon, $n \neq 8$, is a Moufang polygon. The case $n = 8$ was done by Tent later [152]. Tent also proved that half Moufang generalized quadrangles are always Moufang generalized quadrangles without finiteness assumptions (see [151]). Not much later, Fabienne Haot and Hendrik Van Maldeghem proved the equivalence of Moufang generalized quadrangles with 3-Moufang generalized quadrangles, and hence also with 2-Moufang generalized quadrangles, in the general case, see [67]. Both results originally used a certain lemma that turned out to be wrong (as noticed by Richard Weiss in a private communication). The proofs were repaired on the occasion of "Moufang Quadrangles: Characterizations, Classification, Generalizations", a course consisting of 27 talks lectured by the authors of this book at Ghent University in the Spring of 2003. Later on, Haot and Van Maldeghem proved that half 3-Moufang generalized quadrangles are Moufang, see [68]. The proof presented in the present book is a further simplification of the proof in [68] by using the general results on Moufang sets noted by Van Maldeghem later on (which were not yet available to Haot and Van Maldeghem).

Even more recently Koen Thas and Hendrik Van Maldeghem [215] finished the classification of finite half 2-Moufang generalized quadrangles: they are all Moufang generalized quadrangles. This is another important step, and one might wonder how long it will take before one can prove this in general without finiteness. Looking at the current proof, there is not much hope, but that is what we said about the half Moufang and 3-Moufang result back in the 90s, too.

We also mention the series of papers [209, 210, 211] of K. Thas on the classification of finite BN-pairs of Type $\mathbf{B}_2$ (without using the classification of finite simple groups).

Finally, Payne and Thas produced several results in the mid 80s that, put together, provide an almost complete, but entirely geometric, proof of the classification of finite Moufang generalized quadrangles (see Chapter 9 of [128]). Payne and Thas came very close to obtaining a proof; their only obstacle was classifying thick generalized quadrangles of order (s, s^2) all lines of which are axes of symmetry. In other words, classifying thick translation generalized quadrangles of order (s, s^2) each point of which is a translation point. William M. Kantor showed in [85] that such quadrangles are always isomorphic to the classical example $\mathbf{Q}(5, s)$, using the classification of finite split BN-pairs of rank 1, and 4B,C of Fong and Seitz [59]. In [199] — see also [202] and Chapter 7 of [206] — Koen Thas gave a proof without the use of any result of [59, 60], but still having to invoke the list of finite split

BN-pairs of rank 1. However, he did not need that result "intrinsically"; at some stage only the order of some group is needed (and it is there that the geometrical proof fails). On the other hand, the result(s) obtained in [206, Chapter 7] are much more general. Recent results on the programme can be found in the paper [188]. These proofs were also partly the motivation for the above mentioned lecture series, which was, on its turn, one of the starting points of this book.

Let us now overview the contents of this monograph chapter by chapter.

Chapter 1

We assume the reader to be familiar with the basic theory of finite generalized quadrangles. Nevertheless, we repeat in Chapter 1 some basic facts on finite generalized quadrangles, including constructions of important classes of generalized quadrangles, the known results on generalized quadrangles with small parameters, and some important theorems that will often be used in this monograph.

Chapter 2

Chapter 2 is on regularity, antiregularity and 3-regularity. The key theorems are proved and examples are given. Section 2.2 is on the connection between regularity and dual nets, with emphasis on the Axiom of Veblen. Section 2.4 is on the connection between antiregularity and Laguerre planes, and Sections 2.6 and 2.7 are about the connections between 3-regularity, subquadrangles and inversive planes. In this chapter also some fundamental characterization theorems, without proofs, are stated.

Chapter 3

Chapter 3 contains the foundations of the theory of elation generalized quadrangles and translation generalized quadrangles. First some notions of (permutation) group theory are introduced. Section 3.2 contains the necessary background on elation generalized quadrangles (EGQs), in particular the equivalence between EGQs and 4-gonal families. In Section 3.3 the background on translation generalized quadrangles (TGQs) is given, including some theory on symmetries of generalized quadrangles (in the more general context of generalized polygons called "axial collineations"). Then the kernel of a TGQ is introduced, generalized ovals and ovoids are defined, and the generalized quadrangles $\mathbf{T}(n, m, q)$ are constructed. Next, the key theorem stating the equivalence between the quadrangles $\mathbf{T}(n, m, q)$ and TGQs is proved. In Section 3.6 regular pseudo-ovals and regular pseudo-ovoids are defined and constructed in detail. Then, in Section 3.7, we prove that any automorphism of a TGQ that fixes the basepoint is induced by an automorphism of the projective space in which the TGQ can be represented. Sections 3.8, 3.9 and 3.10 contain a study of generalized ovals and of generalized ovoids, including several characterization theorems for the regular case. In these sections, the approach taken differs from the one in "Finite Generalized Quadrangles" [128], and many proofs are simplified. Finally, in Sections 3.11 and 3.12, we study the structure of the automorphism group of a TGQ, and explain the connection with the automorphism group of the translation dual (if defined).

Chapter 4

During the last two decades many new classes of finite generalized quadrangles were constructed, but they are all related to flocks of the quadratic cone in $\mathbf{PG}(3, q)$. this relation between flocks and generalized quadrangles is the theme of Chapter 4. As flocks obviously play a key role, several sections are on flocks and their applications. However, we do neither provide a description of all known (finite) flocks, nor discuss their automorphism groups, nor list all planes and hyperovals emerging from them; this is a separate specific research area with an extensive literature. What we do treat, however, is first some preliminaries on flocks of quadratic cones, of ovoids and of hyperbolic quadrics (all in $\mathbf{PG}(3, q)$). In Section 4.2 the Thas-Walker construction of translation planes from flocks is explained. Flocks of ovoids and hyperbolic quadrics, with the classification theorems (without proofs), are the subject of Section 4.3. Flocks of cones, with several important characterization theorems, are handled in Section 4.4; also, the process of derivation of flocks is discussed there. The flocks of quadratic cones related to TGQs are (derived from) the semifield flocks; these are considered in Section 4.5. In the next section, Thas' relation between flocks and generalized quadrangles is explained; this relation, providing a link between generalized quadrangles and translation planes, led to an explosion of the literature on both subjects. The relation between semifield flocks and TGQs is explained in Section 4.7 and some important theorems are mentioned. Then follows a short section on BLT-sets. The direct geometrical constructions of generalized quadrangles from flocks due to Knarr and Thas are given in Section 4.9. A fundamental property for generalized quadrangles of order (s, s^2) is Property (G); this notion was introduced by Payne, who discovered its relation to flock generalized quadrangles. In Section 4.10, we state the results of Thas and Brown characterizing flock generalized quadrangles in terms of Property (G). Finally, there is a short section on the construction of new ovals and hyperovals from flocks in the context of TGQs. Due to Johnson's Theorem, this will not be used later on.

A first addendum to this chapter contains a brief description of Payne's "Fundamental Theorem of q-Clan Geometry" [121], together with some applications.

Given an EGQ with elation point (∞) and elation group G, it is not always the case that the set of all elations about (∞) defines a group. A second, rather extensive, addendum is appended to Chapter 4, containing results on this problem.

Chapter 5

Good eggs are considered in great detail in Chapter 5. The relation between good eggs, Property (G) and subquadrangles is discussed in Section 5.1. An important characterization for q even of the generalized quadrangle $\mathbf{T}_3(\mathcal{O})$ of Tits in terms of good eggs is also given in this section; as a corollary we obtain a pure geometrical proof of Johnson's Theorem alluded to above. Finally, TGQs $\mathbf{T}(\mathcal{O})$, where $\mathcal{O}$ has more than one good element, are considered. In Section 5.2 an interesting classification theorem in terms of Veronese surfaces is stated for odd characteristic. Since the proof of this theorem is quite long and technical we omit it. As an application we characterize eggs in terms of the number of its elements contained in a space $\mathbf{PG}(3n - 1, q)$. In Section 5.3 eggs and their translation duals are coordinatized, and coordinates for the elements of the known eggs are given. Next, some characterizations of eggs in terms of being good at certain elements are stated or proved, and additional properties of known eggs are mentioned.

Chapter 6

Chapter 6 contains a detailed analysis of generalized quadrangles containing regular elements, that is, we characterize these quadrangles in terms of the dual nets corresponding to the regular element(s). In Section 6.1 we assume that (some of) the dual nets satisfy the Axiom of Veblen, and study the implications for the generalized quadrangle. In Section 6.2 we assume in addition that the generalized quadrangles are TGQs, and in Section 6.3 we assume that the generalized quadrangle satisfies Property (G). Flock generalized quadrangles for which the regular base-point defines a dual net satisfying the Axiom of Veblen are handled in Section 6.4; also some interesting corollaries for TGQs are proved. The next section describes the impact of the Axiom of Veblen on the existence of subquadrangles of order s of generalized quadrangles of order (s^2, s). In the final section, nets are used to force a generalized quadrangle to be a TGQ; many useful characterizations of TGQs are then proved.

Chapter 7

Chapter 7 is focussed on ovoids and subquadrangles. We start with a survey of the known ovoids of the classical generalized quadrangle $\mathbf{Q}(4, q)$. In Section 7.2 we show how in a subquadrangle $\mathcal{S}'$ of order (s, t') of a generalized quadrangle $\mathcal{S}$ of order (s, t), with $t' < t$, ovoids of $\mathcal{S}'$ are induced. Next we define the concept of translation ovoid (with respect to some point) of $\mathbf{Q}(4, q)$. Relying on the theory of good eggs we also show how, for q odd, any semifield flock defines a translation ovoid of $\mathbf{Q}(4, q)$ and, conversely, any translation ovoid of $\mathbf{Q}(4, q)$ defines a semifield flock; here we also rely on a paper by Lunardon [100]. This technique led to the discovery of new ovoids of $\mathbf{Q}(4, q)$ and consequently to new translation planes. In Section 7.3 we present a direct construction of the translation ovoid arising from a semifield flock, and conversely, taken from Thas [181], but more details can be found in Lunardon [100]. In the next section we consider a subquadrangle $\mathcal{S}'$ of order q of a TGQ $\mathbf{T}(\mathcal{O})$ of order (q, q^2), where q is even and $\mathcal{O}$ is good. If $\mathcal{S}'$ does not contain the translation point (∞), then it is proved that $\mathbf{T}(\mathcal{O})$ is isomorphic to the classical generalized quadrangle $\mathbf{Q}(5, q)$. If $\mathcal{S}'$ does contain (∞), then we show that $\mathcal{S}' \cong \mathbf{T}(\mathcal{O}')$, with $\mathcal{O}'$ a generalized oval on $\mathcal{O}$. In Section 7.6 the odd case is completely solved. In Section 7.7 subquadrangles of flock generalized quadrangles and of TGQs $\mathbf{T}(\mathcal{O})$, with $\mathcal{O}^*$ good, are handled, and some extra characterization theorems are obtained; here just some proofs are given.

Finally, several characterizations of $\mathbf{Q}(5, q)$ are proved as either a TGQ or an EGQ with a subquadrangle isomorphic to $\mathbf{Q}(4, q)$ that contains either the translation point or elation point. Here q is even, but more general results are obtained in both characteristics.

Chapter 8

We introduce translation generalized ovals and ovoids in Chapter 8, and show in the first section that only for generalized ovals in even characteristic, the defining property makes sense. The next three sections contain several characterizations of translation generalized ovals, and of the $\mathbf{T}_2(\mathcal{O})$ of Tits. In the final section we prove that a 2-transitive generalized oval is of classical type.

Chapter 9

In Chapter 9 we digress to Moufang sets (also known in the literature as rank one groups, or split BN-pairs of rank one, although sometimes an additional assumption on nilpotency of the root groups is required here). These are particular permutation sets which can be viewed as the one-dimensional Moufang buildings. In fact, a Moufang set is basically a permutation group (G, X), where the group G acts on the set X, where each point stabilizer G_x contains a regular normal subgroup (a "root group") acting on $X \setminus \{x\}$. This situation occurs in generalized quadrangles that have at least two translation points, and also in Moufang quadrangles. We define, state and prove the necessary notions and results, and we study the case where two different groups define a Moufang set on the same set. We also review, without proof, the classification of the finite Moufang sets, and provide detailed constructions. We end this chapter with a classification of the sub Moufang sets of the finite proper Moufang sets.

Chapter 10

Chapter 10 contains a sketch of a part of the theory described in the book [206] of K. Thas, especially of Chapter 7 of that work. Since many of the proofs in [206] are long and technical, proofs are not always given. Still, some shorter and more combinatorial proofs are provided. Essentially, Chapter 10 sketches a classification of finite generalized quadrangles based on the possible configurations of axes of symmetry. Since each line incident with a translation point of a TGQ is an axis of symmetry, such a classification is very important for the theory of TGQs. One also notes that the problem that was left open in the approach of Payne and Thas for the classification of finite Moufang quadrangles, is a part of this classification "by symmetry".

Often in the theory of generalized quadrangles, one speaks of "the elation point" and "the translation point". However, in a (rather hidden) part of the paper [116], Payne showed that the Kantor-Knuth TGQs have a line of translation points! So "the translation point" is not always uniquely defined. In the the the series of papers [203, 201, 202, 205, 199, 204], K. Thas started developing a more general theory of quadrangles having noncurrent axes of symmetry. In this theory, it was eventually shown that the class of nonclassical TGQs with more than one translation point coincides with the class of good TGQs in odd characteristic.

The latter theory is tersely sketched in Chapter 10.

Chapter 11

In Chapter 11 we study in detail the Moufang Condition. We start by explaining the relation between generalized quadrangles admitting a certain group of collineations, and groups with a BN-pair (or Tits systems). Then we introduce a lot of Moufang-like conditions, which are all consequences of the Moufang Condition; hence these conditions are a priori weaker ones. Amongst them are the half Moufang Condition, the (half) 2-Moufang and 3-Moufang Conditions, and the Fong-Seitz Condition. We then proceed to show that all these conditions are equivalent. In this chapter, we do not restrict to finite quadrangles, but we treat the general case. One of the crucial tools will be the Moufang sets.

Chapter 12

In Chapter 12 we investigate the implication of a GQ having a translation or elation point on the fact that it contains an ovoid. In general, there is not too much implication at all, but two particular kinds of ovoids inherit elation and translation properties from the ambient EGQ and TGQ, respectively. These two classes of ovoids are the subtended ones and the ovoids arising from a polarity. In this respect we also prove that the elation groups of an EGQ of order s, which is at the same time a dual EGQ in such a way that an elation point is incident with an elation line, are unique (although not necessarily coinciding with the respective sets of elations). Our main result is that in these two particular cases the elation and translation properties of the GQ induce an elation and translation property on the ovoids (which are assumed to contain the elation or translation point).

When such an ovoid contains two elation or translation points, then it can be structured as a Moufang set, and we show that three of the four infinite classes of finite proper Moufang sets arise in this way. Some more attention is given to the case of a polarity — the Suzuki groups and the Suzuki-Tits ovoids.

Chapter 13

In the final Chapter 13 we study (lax) embeddings of translation generalized quadrangles in finite projective spaces. We classify all embedded TGQs which have the property that the translation group is inherited from the collineation group of the ambient projective space. Not only the classical cases turn up here, but also a class of particular projections of $Q(4, q)$, and some sporadic embeddings of small classical generalized quadrangles. We emphasize that the results include all embeddings in a plane — and hence we have here the first classification theorems of certain planar embeddings. Indeed, in the literature, the planar case is always referred to as hopeless — which it is in the general setting without any extra condition. Chapter 13 also contains some general results on the theory of embeddings. For instance we show that full embeddings are automatically polarized, and we show with a counter example that an embedding which is full for one line is not necessarily full for all lines. We also outline the proofs of some other general results obtained by Thas and Van Maldeghem in the fundamental paper [195].

Chapter 13 is followed by an appendix containing a rather substantial list of open problems.

The book ends, respectively, with an extensive bibliography, and a detailed notation and subject index.

Chapter 1

Generalized Quadrangles

In this chapter we recall basic facts on finite generalized quadrangles, including some important results that will often be used in this monograph.

1.1 Finite Generalized Quadrangles

A (finite) *generalized quadrangle* (GQ) is an incidence structure $\mathcal{S} = (\mathcal{P}, \mathcal{B}, \mathrm{I})$ in which $\mathcal{P}$ and $\mathcal{B}$ are disjoint (nonempty) sets of objects called *points* and *lines* respectively, and for which I is a symmetric point-line *incidence relation* satisfying the following axioms.

(i) Each point is incident with $1 + t$ lines ($t \geq 1$) and two distinct points are incident with at most one line.

(ii) Each line is incident with $1 + s$ points ($s \geq 1$) and two distinct lines are incident with at most one point.

(iii) If x is a point and L is a line not incident with x, then there is a unique pair $(y, M) \in \mathcal{P} \times \mathcal{B}$ for which $x\mathrm{I}M\mathrm{I}y\mathrm{I}L$.

Generalized quadrangles were introduced by Tits [217] in the appendix of his celebrated work on triality.

If x and L are as in (iii), then we will denote the unique point y on L collinear with x also by $\mathrm{proj}_L x$, and call it the *projection of x onto L*. Dually we define $\mathrm{proj}_x L := M$. We generalize this to points z on L by putting $z = \mathrm{proj}_L z$, if $z\mathrm{I}L$, and to lines M concurrent with L by denoting the intersection point $\mathrm{proj}_L M$. The dual notation is also used.

The integers s and t are the *parameters* of the GQ and $\mathcal{S}$ is said to have *order* (s, t); if $s = t$, $\mathcal{S}$ is said to have *order* s. There is a *point-line duality* for GQs (of order (s, t)) for which in any definition or theorem the words "point" and "line" are interchanged and the parameters s and t are interchanged. Hence, we assume without further notice that the dual of a given theorem or definition has also been given.

If the point x is not incident with the line L, in a GQ $\mathcal{S} = (\mathcal{P}, \mathcal{B}, \mathrm{I})$, we write $p \mathbin{\cancel{\mathrm{I}}} L$.

Let $\mathcal{S} = (\mathcal{P}, \mathcal{B}, \mathrm{I})$ be a (finite) GQ of order (s, t). Then $\mathcal{S}$ has $v = |\mathcal{P}| = (1 + s)(1 + st)$ points and $b = |\mathcal{B}| = (1 + t)(1 + st)$ lines; see 1.2.1 of Payne and Thas [128]. Also, $s + t$ divides $st(1 + s)(1 + t)$, and, for $s \neq 1 \neq t$, we have $t \leq s^2$ and, dually, $s \leq t^2$ (inequalities of Higman [70]); see 1.2.2 and 1.2.3 of Payne and Thas [128].

Given two (not necessarily distinct) points x, y of $\mathcal{S}$, we write $x \sim y$ and say that x and y are *collinear*, provided that there is some line L for which $x \mathrm{I} L \mathrm{I} y$. And $x \not\sim y$ means that x and y are not collinear. Dually, for $L, M \in \mathcal{B}$, we write $L \sim M$ or $L \not\sim M$ according as L and M are *concurrent* or nonconcurrent, respectively. The line which is incident with distinct collinear points x, y is denoted by xy; the point which is incident with distinct concurrent lines L, M is denoted by either LM or $L \cap M$.

For $x \in \mathcal{P}$, put $x^\perp = \{y \in \mathcal{P} \parallel y \sim x\}$, and note that $x \in x^\perp$. If $A \subseteq \mathcal{P}$, A "perp" is defined by $A^\perp = \cap\{x^\perp \parallel x \in A\}$. Hence, for $x, y \in \mathcal{P}, x \neq y$, we have $\{x, y\}^\perp = x^\perp \cap y^\perp$; we have $|\{x, y\}^\perp| = s + 1$ or $t + 1$ according as $x \sim y$ or $x \not\sim y$. Further, $\{x, y\}^{\perp\perp} = \{u \in \mathcal{P} \parallel u \in z^\perp \; \forall z \in x^\perp \cap y^\perp\}$; we have $|\{x, y\}^{\perp\perp}| = s + 1$ or $|\{x, y\}^{\perp\perp}| \leq t + 1$ according as $x \sim y$ or $x \not\sim y$. The sets $\{x, y\}^\perp$ and $\{x, y\}^{\perp\perp}$ are respectively called the *trace* and the *span* of the pair $\{x, y\}$. If $x \not\sim y$, then $\{x, y\}^{\perp\perp}$ is also called the *hyperbolic line* defined by x and y.

A *triad* of points is a triple of pairwise noncollinear points. Given a triad T, a *center* of T is just an element of $T^\perp$. If a triad has at least one center, then we say that it is *centric*.

A *subquadrangle*, or also *subGQ*, $\mathcal{S}' = (\mathcal{P}', \mathcal{B}', \mathrm{I}')$ of a GQ $\mathcal{S} = (\mathcal{P}, \mathcal{B}, \mathrm{I})$ of order (s, t), is a GQ for which $\mathcal{P}' \subseteq \mathcal{P}, \mathcal{B}' \subseteq \mathcal{B}$, and where I' is the restriction of I to $(\mathcal{P}' \times \mathcal{B}') \cup (\mathcal{B}' \times \mathcal{P}')$. If the GQ $\mathcal{S}'$ of order (s', t') is a subGQ of the GQ $\mathcal{S}$ of order (s, t), with $\mathcal{S} \neq \mathcal{S}'$, then either $s = s'$ or $s \geq s't'$, and, dually, either $t = t'$ or $t \geq s't'$; see 2.2.1 of Payne and Thas [128].

The following result will appear to be very useful; see 2.2.2 of Payne and Thas [128]. When we refer to it, we will often not make a distinction between the statement and its dual.

Lemma 1.1.1 *If the thick* GQ $\mathcal{S}$ *of order* (s, t) *has a proper subquadrangle* $\mathcal{S}'$ *of order* (s, t'), *then* $t \geq st'$; *in particular* $t' \leq s \leq t$. *Hence if* $s = t$, *then* $t' = 1$. *Also, if* $s = t'$, *then* $t = s^2$.

Also, if in addition $\mathcal{S}''$ *is a proper subquadrangle of* $\mathcal{S}'$ *of order* (s, t''), *then* $t'' = 1$, $s = t'$ *and* $t = s^2$. ∎

An *ovoid* of the GQ $\mathcal{S}$ of order (s, t), is a set $\mathcal{O}$ of points of $\mathcal{S}$ such that each line of $\mathcal{S}$ is incident with a unique point of $\mathcal{O}$; a *spread* of $\mathcal{S}$ is a set $\mathcal{T}$ of lines of $\mathcal{S}$ such that each point of $\mathcal{S}$ is incident with a unique line of $\mathcal{T}$. If $\mathcal{O}$ is an ovoid of $\mathcal{S}$, respectively if $\mathcal{T}$ is a spread of $\mathcal{S}$, then $|\mathcal{O}| = |\mathcal{T}| = 1 + st$; see 1.8.1 of Payne and Thas [128].

If the GQ $\mathcal{S}'$ of order (s, t') is a subquadrangle of the GQ $\mathcal{S}$ of order (s, t), with $\mathcal{S} \neq \mathcal{S}'$, then each point of $\mathcal{S}$ not in $\mathcal{S}'$ is collinear with $1 + st'$ points of an ovoid of $\mathcal{S}'$; see 2.2.1 of Payne and Thas [128].

The following result appears to the very useful. Let $\mathcal{S}' = (\mathcal{P}', \mathcal{B}', \mathrm{I}')$ be a substructure of the GQ $\mathcal{S} = (\mathcal{P}, \mathcal{B}, \mathrm{I})$ of order (s, t) for which the following conditions are satisfied.

(i) If $x, y \in \mathcal{P}', x \neq y$, and $x\mathrm{I}L\mathrm{I}y$, then $L \in \mathcal{B}'$.

(ii) Each element of $\mathcal{B}'$ is incident with $1 + s$ elements of $\mathcal{P}'$.

Then by Thas [157], see also 2.3.1 of Payne and Thas [128], there are four possibilities.

(a) $\mathcal{S}'$ is a dual grid (and then $s = 1$); see §1.3 for the definition of a dual grid.

(b) The elements of $\mathcal{B}'$ are lines which are incident with a distinguished point of $\mathcal{P}$, and $\mathcal{P}'$ consists of those points of $\mathcal{P}$ which are incident with these lines.

(c) $\mathcal{B}' = \emptyset$ and $\mathcal{P}'$ is a set of pairwise noncollinear points of $\mathcal{P}$.

(d) $\mathcal{S}'$ is a subquadrangle of order (s, t').

1.2   Automorphisms

Let $\mathcal{S} = (\mathcal{P}, \mathcal{B}, \mathrm{I})$ be a GQ. A *collineation* or *automorphism* of $\mathcal{S}$ is a permutation of $\mathcal{P} \cup \mathcal{B}$ that preserves $\mathcal{P}$, $\mathcal{B}$ and incidence. We will denote the set of all collineations of a given GQ $\mathcal{S}$ by $\mathrm{Aut}(\mathcal{S})$ and call it the *full collineation (automorphism) group of $\mathcal{S}$*, as opposed to an ordinary *collineation (automorphism) group of $\mathcal{S}$*, which is just a subgroup of $\mathrm{Aut}(\mathcal{S})$.

1.3   Grids and Dual Grids

A *grid* is an incidence structure $\mathcal{S} = (\mathcal{P}, \mathcal{B}, \mathrm{I})$ with $\mathcal{P} = \{x_{ij} \,\|\, i = 0, 1, \ldots, s_1$ and $j = 0, 1, \ldots, s_2\}$, $s_1 > 0$ and $s_2 > 0$, $\mathcal{B} = \{L_0, L_1, \ldots, L_{s_1}, M_0, M_1, \ldots, M_{s_2}\}$, $x_{ij} \mathrm{I} L_k$ if and only if $i = k$, and $x_{ij} \mathrm{I} M_k$ if and only if $j = k$. We say that such a grid is an $(s_1 + 1) \times (s_2 + 1)$-*grid*. A grid is a GQ if and only if $s_1 = s_2 = s$; in such a case the GQ has order $(s, 1)$. Any GQ of order $(s, 1)$ is isomorphic to an $(s + 1) \times (s + 1)$-grid. A dual grid has parameters t_1, t_2, and it is a GQ if and only if $t_1 = t_2 = t$, in which case it is a GQ of order $(1, t)$. Any GQ of order $(1, t)$ is isomorphic to a dual $(t + 1) \times (t + 1)$-grid. An ordinary quadrangle is a GQ of order (1,1) and is at the same time a grid and a dual grid. This is the motivation for the term "generalized quadrangle".

A GQ of order (s, t) is called *thin* if either $s = 1$ or $t = 1$; in the other case it is called *thick*.

1.4   The Classical Generalized Quadrangles

We now give a brief description of three families of examples known as the *classical* GQs, all of which are associated with classical groups and were first recognized as GQs by Tits.

(i) Consider a nonsingular quadric $\mathbf{Q}$ of projective index 1, that is, of Witt index 2, of the projective space $\mathbf{PG}(d, q)$, with $d = 3, 4$ or 5. Then the points of $\mathbf{Q}$ together with the lines of $\mathbf{Q}$ (which are the subspaces of maximal dimension on $\mathbf{Q}$) form a GQ $\mathbf{Q}(d, q)$ with parameters

$$s = q, t = 1, \quad v = (q+1)^2, \qquad b = 2(q+1), \qquad \text{when } d = 3,$$
$$s = q, t = q, \quad v = (q+1)(q^2+1), b = (q+1)(q^2+1) \quad \text{when } d = 4,$$
$$s = q, t = q^2, v = (q+1)(q^3+1), b = (q^2+1)(q^3+1), \text{when } d = 5.$$

Notice that $\mathbf{Q}(3, q)$ is a grid.

(ii) Let $\mathbf{H}$ be a nonsingular Hermitian variety of the projective space $\mathbf{PG}(d, q^2)$, $d = 3$ or 4. Then the points of $\mathbf{H}$ together with the lines on $\mathbf{H}$ form a GQ $\mathbf{H}(d, q^2)$ with parameters

$$s = q^2,\, t = q,\ \ v = (q^2 + 1)(q^3 + 1),\, b = (q + 1)(q^3 + 1),\quad \text{when } d = 3,$$
$$s = q^2,\, t = q^3,\, v = (q^2 + 1)(q^5 + 1),\, b = (q^3 + 1)(q^5 + 1),\ \text{when } d = 4.$$

(iii) The points of $\mathbf{PG}(3, q)$, together with the totally isotropic lines with respect to a symplectic polarity, form a GQ $\mathbf{W}(q)$ with parameters

$$s = q,\, t = q,\,,\, v = (q + 1)(q^2 + 1),\, b = (q + 1)(q^2 + 1).$$

Theorem 1.4.1   (i) *The GQ* $\mathbf{Q}(4, q)$ *is isomorphic to the dual of* $\mathbf{W}(q)$. *Also,* $\mathbf{Q}(4, q)$ *(or* $\mathbf{W}(q)$*) is self-dual if and only if* q *is even.*

(ii) *The GQ* $\mathbf{Q}(5, q)$ *is isomorphic to the dual of* $\mathbf{H}(3, q^2)$.

Proof.   See 3.2.1 and 3.2.3 of Payne and Thas [128].   ∎

1.5   The Generalized Quadrangles $\mathbf{T}_2(\mathcal{O})$ and $\mathbf{T}_3(\mathcal{O})$ of Tits

A *k-arc* of $\mathbf{PG}(2, q)$ is a set of k points of $\mathbf{PG}(2, q)$ no three of which are collinear. Then clearly $k \leq q + 2$. By Bose [17], for q odd, $k \leq q + 1$. Further, any nonsingular conic of $\mathbf{PG}(2, q)$ is a $(q + 1)$-arc. It can be shown that each $(q + 1)$-arc $\mathcal{K}$ of $\mathbf{PG}(2, q)$, q even, extends to a $(q + 2)$-arc $\mathcal{K} \cup \{x\}$ (see, e.g., Hirschfeld [73], p. 177); the point x, which is uniquely defined by $\mathcal{K}$, is called the *kernel* or *nucleus* of $\mathcal{K}$. The $(q + 1)$-arcs of $\mathbf{PG}(2, q)$ are called *ovals*; the $(q + 2)$-arcs of $\mathbf{PG}(2, q)$, q even, are called *hyperovals*. By a celebrated theorem of Segre [137], every oval in $\mathbf{PG}(2, q)$, q odd, is a nonsingular conic. For q even, this is valid if and only if $q \in \{2, 4\}$; see, e.g., Thas [178].

We now introduce the notion of "ovoid" as defined by Tits in [221]. An *ovoid* $\mathcal{O}$ of $\mathbf{PG}(3, q)$ is a set of points of $\mathbf{PG}(3, q)$ no three of which are collinear and such that for any point of $\mathcal{O}$ the union of the lines which meet $\mathcal{O}$ only in that point, that is, the *tangent lines* at that point, is a $\mathbf{PG}(2, q)$. If $\mathcal{O}$ is an ovoid, its number of points is $q^2 + 1$. A *k-cap* $\mathcal{K}$ of $\mathbf{PG}(3, q)$ is a set of k points no three of which are collinear. For any k-cap $\mathcal{K}$ of $\mathbf{PG}(3, q)$, with $q \neq 2, k \leq q^2 + 1$; for any k-cap $\mathcal{K}$ of $\mathbf{PG}(3, 2), k \leq 8$ holds and the

8-caps of $\mathbf{PG}(3,2)$ are the complements of planes. For q odd this result is due to Bose [17], for q even to Qvist [132]. A (q^2+1)-cap of $\mathbf{PG}(3,q)$, $q \neq 2$, is precisely an *ovoid* (cf. Barlotti [8] or Hirschfeld [72]); the *ovoids* of $\mathbf{PG}(3,2)$ are the sets of 5 points no 4 of which are coplanar. It is easy to show that each nonsingular elliptic quadric of $\mathbf{PG}(3,q)$ is an ovoid. By a celebrated theorem, due independently to Barlotti [8] and Panella [106], every ovoid in $\mathbf{PG}(3,q)$, q odd or $q = 4$, is an elliptic quadric.

Theorem 1.5.1 (Barlotti [8]; Panella [106]) *Each ovoid of* $\mathbf{PG}(3,q)$, q *odd, is an elliptic quadric.* ∎

To the contrary, in the even case, Tits [221] showed that for any $q = 2^{2e+1}$, with $e \geq 1$, there exists an ovoid which is not an elliptic quadric; these ovoids are called *Tits ovoids*, or also *Suzuki-Tits ovoids*, and are related to the simple Suzuki groups $\mathbf{Sz}(q)$. In fact, for $q = 8$ Segre [138] discovered an ovoid which is not an elliptic quadric, and which was shown to be a Tits ovoid by Fellegara [57]. For even q no other ovoids than the elliptic quadrics and the Tits ovoids are known.

If $\mathcal{O}$ is an ovoid in $\mathbf{PG}(3,q)$, then any plane π of $\mathbf{PG}(3,q)$ intersects $\mathcal{O}$ in either one point or in an oval. If $|\pi \cap \mathcal{O}| = 1$, then we say that π is a *tangent plane* of $\mathcal{O}$. At each of its points $\mathcal{O}$ has exactly one tangent plane. For more details, see Hirschfeld [72]. Finally, a beautiful result due to Brown [23] tells us that any ovoid $\mathcal{O}$ of $\mathbf{PG}(3,q)$ containing at least one conic section, is an elliptic quadric.

Let $d = 2$ (respectively, $d = 3$) and let $\mathcal{O}$ be an oval (respectively, an ovoid) of $\mathbf{PG}(d,q)$. Further, let $\mathbf{PG}(d,q) = H$ be embedded as a hyperplane in $\mathbf{PG}(d+1,q) = P$. Define

- POINTS as

  (i) the points of $P \backslash H$,

  (ii) the hyperplanes X of P for which $|X \cap \mathcal{O}| = 1$, and

  (iii) one new symbol (∞).

- LINES are defined as

  (a) the lines of P which are not contained in H and meet $\mathcal{O}$ (necessarily in a unique point), and

  (b) the points of $\mathcal{O}$.

- INCIDENCE is defined as follows. A point of Type (i) is incident only with lines of Type (a); here the incidence is that of P. A point of Type (ii) is incident with all lines of Type (a) contained in it and with the unique element of $\mathcal{O}$ in it. The point (∞) is incident with no line of Type (a) and all lines of Type (b).

It is an easy exercise to show that the incidence structure $\mathbf{T}_2(\mathcal{O})$ (respectively, $\mathbf{T}_3(\mathcal{O})$) so defined is a GQ. The parameters are

$$s = q, t = q, \quad v = (q+1)(q^2+1), \ b = (q+1)(q^2+1), \quad \text{when } d = 2,$$
$$s = q, t = q^2, \quad v = (q+1)(q^3+1), \ b = (q^2+1)(q^3+1), \quad \text{when } d = 3.$$

Theorem 1.5.2   (i) *The GQ* $\mathbf{T}_2(\mathcal{O})$ *is isomorphic to the GQ* $\mathbf{Q}(4, q)$ *if and only if* $\mathcal{O}$ *is a nonsingular conic. The GQ* $\mathbf{T}_2(\mathcal{O})$ *is isomorphic to the GQ* $\mathbf{W}(q)$ *if and only if* q *is even and* $\mathcal{O}$ *is a conic.*

(ii) *The GQ* $\mathbf{T}_3(\mathcal{O})$ *is isomorphic to the GQ* $\mathbf{Q}(5, q)$ *if and only if* $\mathcal{O}$ *is a nonsingular elliptic quadric.*

Proof.   See 3.2.2 and 3.2.4 of Payne and Thas [128].   ■

Remark 1.5.3   (i) For q odd any oval is a nonsingular conic, hence for q odd we always have $\mathbf{T}_2(\mathcal{O}) \cong \mathbf{Q}(4, q)$.

(ii) For q odd any ovoid is an elliptic quadric, hence for q odd we always have $\mathbf{T}_3(\mathcal{O}) \cong \mathbf{Q}(5, q)$.

1.6   The Generalized Quadrangles $\mathbf{T}_2^*(\mathcal{O})$

Let $\mathcal{O}$ be a hyperoval of $\mathbf{PG}(2, q)$, so q is even. Embed $\mathbf{PG}(2, q) = H$ as a hyperplane in $\mathbf{PG}(3, q) = P$.

- The POINTS of the GQ $\mathbf{T}_2^*(\mathcal{O})$ are the points of $P \backslash H$,

- LINES of the GQ are the lines of P not in H which meet $\mathcal{O}$, and

- the INCIDENCE is inherited from P.

Then $\mathbf{T}_2^*(\mathcal{O})$ is a GQ with parameters

$$s = q - 1, t = q + 1, v = q^3, b = (q+2)q^2.$$

1.7   Orders of the Known Generalized Quadrangles

The orders (s, t) of the known GQs are

$$
\begin{aligned}
&(s, 1),   && s \in \mathbb{N} \backslash \{0\}, \\
&(1, t),   && t \in \mathbb{N} \backslash \{0\}, \\
&(q, q),   && q \text{ any prime power,} \\
&(q, q^2),   && q \text{ any prime power,} \\
&(q^2, q),   && q \text{ any prime power,} \\
&(q^2, q^3),   && q \text{ any prime power,} \\
&(q^3, q^2),   && q \text{ any prime power,} \\
&(q - 1, q + 1),   && q \text{ any prime power,} \\
&(q + 1, q - 1),   && q \text{ any prime power.}
\end{aligned}
$$

1.8   Generalized Quadrangles with Small Parameters

The proofs of all the results in this section are contained in Chapter 6 of Payne and Thas [128].

Let $\mathcal{S} = (\mathcal{P}, \mathcal{B}, \mathrm{I})$ be a finite GQ of order $(s, t), 1 < s \leq t$.

(a) $s = 2$

By Section 1.1, $s + t$ divides $st(s + 1)(t + 1)$ and $t \leq s^2$. Hence $t \in \{2, 4\}$. Up to isomorphism there is only one GQ of order 2 and only one GQ of order (2,4). It follows that the GQs $\mathbf{W}(2)$ and $\mathbf{Q}(4, 2)$ are self-dual and mutually isomorphic. It is easy to show that the GQ of order 2 is unique.

The uniqueness of the GQ of order (2,4) was proved independently at least five times, by Seidel [141], Shult [142], Thas [164], Freudenthal [61] and Dixmier and Zara [53].

(b) $s = 3$

Again by 1.1 we have $t \in \{3, 5, 6, 9\}$. Any GQ of order $(3, 5)$ must be isomorphic to the GQ $\mathbf{T}_2^*(\mathcal{O})$ arising from the unique hyperoval in $\mathbf{PG}(2, 4)$, any GQ of order $(3, 9)$ must be isomorphic to $\mathbf{Q}(5, 3)$, and a GQ of order 3 is isomorphic to either $\mathbf{W}(3)$ or to its dual $\mathbf{Q}(4, 3)$. Finally, there is no GQ of order (3,6).

The uniqueness of the GQ of order (3,5) was proved by Dixmier and Zara [53], the uniqueness of the GQ of order (3,9) was proved independently by Dixmier and Zara [53] and Cameron (see Payne and Thas [126]), the determination of all GQs of order 3 is due independently to Payne [108] and to Dixmier and Zara [53]. Dixmier and Zara [53] proved that there is no GQ of order (3,6).

(c) $s = 4$

Using 1.1 it is easy to check that $t \in \{4, 6, 8, 11, 12, 16\}$. Nothing is known about $t = 11$ or $t = 12$. In the other cases unique examples are known, but the uniqueness question is settled only in the case $t = 4$. The proof of this uniqueness that appears in Payne and Thas [128] is that of Payne [110, 111], with a gap filled in by Tits.

Chapter 2

Regularity, Antiregularity and 3-Regularity

In this chapter the important notions of regularity, antiregularity and 3-regularity are introduced, and the connections with planes, nets, Laguerre planes, inversive planes and subquadrangles are described.

2.1 Regularity

Let $\mathcal{S} = (\mathcal{P}, \mathcal{B}, \mathrm{I})$ be a finite GQ of order (s, t). If $x \sim x'$, $x \neq x'$, or if $x \not\sim x'$ and $|\{x, x'\}^{\perp\perp}| = t + 1$, where $x, x' \in \mathcal{P}$, we say the pair $\{x, x'\}$ is *regular*. The point x is *regular* provided $\{x, x'\}$ is regular for all $x' \in \mathcal{P}, x' \neq x$. Regularity for lines is defined dually. A point x is *coregular* provided each line incident with x is regular.

For the proofs of the following results we refer to 3.3 of Payne and Thas [128].

- All lines of the GQ $\mathbf{Q}(d, q)$ are regular, with $d \in \{3, 4, 5\}$.

- All points of $\mathbf{Q}(3, q)$ are regular, for q even all points of $\mathbf{Q}(4, q)$ are regular, for q odd no point of $\mathbf{Q}(4, q)$ is regular, and for any prime power q no point of $\mathbf{Q}(5, q)$ is regular.

- No point and no line of $\mathbf{H}(4, q^2)$ is regular.

- The point (∞) of the GQ $\mathbf{T}_2(\mathcal{O})$, respectively $\mathbf{T}_3(\mathcal{O})$, is coregular. If $\mathbf{T}_2(\mathcal{O})$ has at least one regular pair of nonconcurrent lines of Type

(a) defining distinct points of $\mathcal{O}$, then $\mathcal{O}$ is a conic and $\mathbf{T}_2(\mathcal{O})$ is isomorphic to $\mathbf{Q}(4, q)$. If $\mathbf{T}_2(\mathcal{O})$ has a regular point of Type (i), then q is even, $\mathcal{O}$ is a conic and $\mathbf{T}_2(\mathcal{O})$ is isomorphic to $\mathbf{Q}(4, q)$. If $\mathbf{T}_3(\mathcal{O})$ has even one regular pair of nonconcurrent lines of Type (a) defining distinct points of $\mathcal{O}$, then $\mathcal{O}$ is an elliptic quadric and $\mathbf{T}_3(\mathcal{O})$ is isomorphic to $\mathbf{Q}(5, q)$ (the proof is similar to the proof of 3.3.3(iii) in Payne and Thas [128], but here we rely on the result of Brown mentioned in 1.5).

- No point and no line of $\mathbf{T}_2^*(\mathcal{O})$ is regular; a pair of nonconcurrent lines of $\mathbf{T}_2^*(\mathcal{O})$ is regular if and only if these lines define the same point of the hyperoval $\mathcal{O}$.

Theorem 2.1.1 *If each point of the* GQ $\mathcal{S}$ *of order* (s, t) *is regular, then* $2(t + 1)$ *divides* $s^2(s^2 - 1)$.

Proof. Count the number of sets $\{x, y\}^{\perp} \cup \{x, y\}^{\perp\perp}$ with $x \not\sim y$. One obtains $(1 + s)(1 + st)s^2 t / (2(t + 1)t)$. Hence $2(t + 1)$ divides $(1 + s)(1 + st)s^2$, hence also $s^2(s^2 - 1)$.   ∎

Remark 2.1.2 The proof of Theorem 2.1.1 is a variation on the proof 1.5.1(iv) of Payne and Thas [128], where it is shown that $t + 1$ divides $s^2(s^2 - 1)$.

Proofs of the following results are also contained in Sections 1.3 and 1.5 of Payne and Thas [128].

Theorem 2.1.3 *Let* $\mathcal{S} = (\mathcal{P}, \mathcal{B}, \mathbf{I})$ *be a finite* GQ *of order* (s, t).

(i) *If* $1 < s < t$, *then no pair of noncollinear points is regular.*

(ii) *Is* $s = t \neq 1$, *then the pair of noncollinear points* $\{x, y\}$ *is regular if and only if any triad* $\{x, y, z\}$ *has either* 1 *or* $s + 1$ *centers.*

Proof. (i). Assume that $s \neq 1$, and that the pair $\{x, y\}$ of noncollinear points is regular. Now we count the number of points incident with lines zu, where $z \in \{x, y\}^{\perp}$ and $u \in \{x, y\}^{\perp\perp}$. We obtain $2(t + 1) + (t + 1)^2(s - 1)$. As this number can be at most $v = |\mathcal{P}| = (s + 1)(st + 1)$, there arises $t(s - t)(s - 1) \geq 0$, and so $s \geq t$.

(ii). Let $s = t \neq 1$, and assume that $\{x, y\}$ is regular, with $x \not\sim y$. Let $z \in \mathcal{P}, z \not\sim x, z \not\sim y$. If $z \in \{x, y\}^{\perp\perp}$, then $|\{x, y, z\}^{\perp}| = s + 1$. If $z \notin \{x, y\}^{\perp}$, then, by the proof of (i), the point z is incident with exactly one line uu', with $u \in \{x, y\}^{\perp}$ and $u' \in \{x, y\}^{\perp\perp}$. It follows that $|\{x, y, z\}^{\perp}| = 1$. Conversely, if $|\{x, y, z\}^{\perp}| \in \{1, s + 1\}$ for any $z \in \mathcal{P}, z \not\sim x, z \not\sim y$, then it is clear that $\{x, y\}^{\perp\perp}$ has size $s + 1$, that is, $\{x, y\}$ is regular. ∎

Theorem 2.1.4 *If x is a coregular point of the thick GQ $\mathcal{S} = (\mathcal{P}, \mathcal{B}, \mathrm{I})$ of order (s, t), then the number of centers of any triad $\{x, y, z\}$ has the same parity as $1 + t$.*

Proof. Let $u_1, u_2, \ldots, u_m$ be all the centers of a triad $\{x, y, z\}$, with $\{x, y\}^{\perp} = \{u_1, u_2, \ldots, u_m, u_{m+1}, \ldots, u_{t+1}\}$. We may suppose that $m < t + 1$. For $i > m$, let L_i be the line xu_i and let M_i be the line yu_i. Let K be the line incident with z and concurrent with L_i, and let N be the line incident with z and concurrent with M_i. Let M be the line incident with y and concurrent with K, and let L be the line incident with x and concurrent with N. Since the line L_i is regular, the pair $\{L_i, N\}$ must be regular, and it follows that M must meet L in some point $u_{i'} \in \{x, y\}^{\perp}$, $m + 1 \leq i' \leq t + 1, i' \neq i$. In this way with each point $u_i \in \{u_{m+1}, u_{m+2}, \ldots, u_{t+1}\}$ corresponds a point $u_{i'} \in \{u_{m+1}, u_{m+2}, \ldots, u_{t+1}\}$, $i' \neq i$, and clearly this correspondence is involutory. Hence the number of points of $\{x, y\}^{\perp}$ that are not centers of $\{x, y, z\}$ is even, proving the theorem. ∎

Theorem 2.1.5 *If x is a coregular point of the thick GQ $\mathcal{S} = (\mathcal{P}, \mathcal{B}, \mathrm{I})$ of order s, then x is regular if and only if s is even.*

Proof. Assume that s is even. Then $|\{x, y, z\}^{\perp}|$ is odd for any triad $\{x, y, z\}$, by Theorem 2.1.4. Let $\{x, y\}^{\perp} = \{u_1, u_2, \ldots, u_{s+1}\}$, with $x \not\sim y$, let $\{x, y, z_i\}$ be any triad containing x, y, and let $|\{x, y, z_i\}^{\perp}| = t_i$. Now count in two ways the number of ordered pairs (u_j, z_i), with $u_j \sim z_i$. We obtain

$$\sum_i t_i = (s+1)(s-1)s.$$

Next we count in two ways the number of ordered triples $(u_j, u_{j'}, z_i)$, with $u_j \sim z_i \sim u_{j'}, j \neq j'$. We obtain

$$\sum_i t_i(t_i - 1) = (s+1)s(s-1).$$

The number of triads $\{x, y, z_i\}$ is equal to

$$(s+1)(s^2+1) - 2(s+1)s + (s+1) - 2 = s^3 - s^2.$$

Hence

$$\sum_i (t_i - 1)(t_i - (s+1)) = 0.$$

As $t_i \leq s + 1$ and t_i is odd, we necessarily have $t_i \in \{1, s+1\}$. Now by Theorem 2.1.3 the pair $\{x, y\}$ is regular. Hence the point x is regular.

Conversely, assume that x is regular. By Theorem 2.1.3 we have $|\{x, y, z\}^\perp| \in \{1, s+1\}$ for any triad $\{x, y, z\}$. Consequently, by Theorem 2.1.4, s is even. ∎

Corollary 2.1.6 *The point (∞) of any GQ $\mathbf{T}_2(\mathcal{O})$ of order q, with q even, is regular.*

Proof. By the beginning of §2.1 the point (∞) of $\mathbf{T}_2(\mathcal{O})$ is coregular, hence by the previous theorem it is regular. ∎

2.2 Regularity and Dual Nets

A (finite) *net* of *order* k (≥ 2) and *degree* r (≥ 2) is an incidence structure $\mathcal{N} = (\mathcal{P}, \mathcal{B}, \mathrm{I})$ of points and lines satisfying the following axioms.

(i) Each point is incident with r lines and two distinct points are incident with at most one line.

(ii) Each line is incident with k points and two distinct lines are incident with at most one point.

(iii) If x is a point and L is a line not incident with x, then there is a unique line M incident with x and not concurrent with L.

For a net of order k and degree r we have $|\mathcal{P}| = k^2$ and $|\mathcal{B}| = kr$. Also, $r \leq k+1$, with $r = k+1$ if and only if the net is an affine plane of order k. For proofs and details we refer to Dembowski [45].

The following theorem gives the relation between regularity in GQs and dual nets; this is 1.3.1 of Payne and Thas [128].

Theorem 2.2.1 *Let x be a regular point of the GQ $\mathcal{S} = (\mathcal{P}, \mathcal{B}, \mathrm{I})$ of order (s, t), $s \neq 1$. Then the incidence structure with*

- POINT SET $x^{\perp} \backslash \{x\}$,

- LINE SET *the set of all spans $\{y, z\}^{\perp\perp}$, where $y, z \in x^{\perp} \backslash \{x\}$, $y \not\sim z$, and*

- INCIDENCE *the natural one,*

is the dual of a net of order s and degree $t + 1$. If in particular $s = t > 1$, then a dual affine plane of order s arises. Also, in the case $s = t > 1$ the incidence structure π_x with

- POINT SET $x^{\perp}$,

- LINE SET *the set of all spans $\{y, z\}^{\perp\perp}$, where $y, z \in x^{\perp}, y \neq z$, and*

- INCIDENCE *the natural one,*

is a projective plane of order s.

Proof. Easy exercise. ∎

Historically, the next result is probably the oldest combinatorial characterization of a class of GQs. A proof is essentially contained in Singleton [144] (although he erroneously thought he had proved a stronger result), but the first satisfactory treatment may have been given by Benson [10]. No doubt it was discovered independently by several authors (e.g. Tallini [149]).

Theorem 2.2.2 *The GQ $\mathcal{S}$ of order s, $s \neq 1$, is isomorphic to the GQ $\mathbf{W}(s)$ if and only if all its points are regular.*

Proof. See 5.2.1 of Payne and Thas [128]. We remark that the proof makes use of Theorems 2.1.3 and 2.2.1. ∎

Now we introduce the *Axiom of Veblen* for dual nets $\mathcal{N}^*$.

Axiom of Veblen. *If $L_1 \mathrm{I} x \mathrm{I} L_2$, with $L_1 \neq L_2, M_1 \, \mathtt{I} \, x \, \mathtt{I} \, M_2$, and if the line L_i is concurrent with the line M_j for all $i, j \in \{1, 2\}$, then M_1 is concurrent with M_2.*

We describe a dual net $\mathcal{N}^*$ which is not a dual affine plane and which satisfies the Axiom of Veblen. This dual net $\mathcal{N}^*$ is denoted by $\mathbf{H}_q^n$, $n > 2$, and constructed as follows:

- the POINTS of $\mathbf{H}_q^n$ are the points of $\mathbf{PG}(n,q)$ which are not contained in a given $\mathbf{PG}(n-2,q) \subset \mathbf{PG}(n,q)$,

- the LINES of $\mathbf{H}_q^n$ are the lines of $\mathbf{PG}(n,q)$ which have no point in common with $\mathbf{PG}(n-2,q)$, and

- the INCIDENCE in $\mathbf{H}_q^n$ is the natural one.

It is easy to check that $\mathbf{H}_q^n$ is the dual of a net of degree $r = q+1$ and order $k = q^{n-1}$; as $n > 2$ we have $r < k + 1$ and so $\mathbf{H}_q^n$ is not an affine plane. Clearly $\mathbf{H}_q^n$ satisfies the Axiom of Veblen.

The following characterization of $\mathbf{H}_q^n$, $n > 2$, is due to Thas and De Clerck [189].

Theorem 2.2.3 *Let $\mathcal{N}^*$ be the dual of a net of order k and degree r, with $r \neq k + 1$. Then $\mathcal{N}^*$ is isomorphic to some dual net $\mathbf{H}_q^n$, with $n > 2$, if and only if it satisfies the Axiom of Veblen.*

Proof.   See Thas and De Clerck [189].   ∎

Corollary 2.2.4 *Let x be a regular point of the GQ $\mathcal{S} = (\mathcal{P}, \mathcal{B}, \mathrm{I})$ of order (s,t), with $s \neq 1$. If the dual net $\mathcal{N}_x^*$ defined by x satisfies the Axiom of Veblen, then either $s = t$ and $\mathcal{N}_x^*$ is a dual affine plane of order s or $s = t^2$ and $\mathcal{N}_x^* \cong \mathbf{H}_q^3$ with $q = t$.*

Proof.   If $s \neq t$, then by Theorem 2.2.3 we have $\mathcal{N}_x^* \cong \mathbf{H}_q^n$, with $n > 2$, $q = t$ and $q^{n-1} = s$. As $s \leq t^2$ we necessarily have $n = 3$.   ∎

2.3   Antiregularity

Let $\mathcal{S} = (\mathcal{P}, \mathcal{B}, \mathrm{I})$ be a GQ of order (s,t). The point pair $\{x,y\}$, with $x \not\sim y$, is *antiregular* provided $|z^{\perp} \cap \{x,y\}^{\perp}| \leq 2$ for all $z \in \mathcal{P} \backslash \{x,y\}$. A point x is *antiregular* provided $\{x,y\}$ is antiregular for all $y \in \mathcal{P} \backslash x^{\perp}$.

For the proofs of the following results we refer to 3.3 of Payne and Thas [128].

- All points of $\mathbf{Q}(3,q)$ are antiregular.

- All points of $\mathbf{Q}(4,q)$ are antiregular if and only if q is odd.

- No point of $\mathbf{Q}(5,q)$ is antiregular.

- No line of $\mathbf{Q}(d,q)$ is antiregular, with $d \in \{3,4,5\}$.

- No point and no line of $\mathbf{H}(4,q^2)$ is antiregular.

- No point of $\mathbf{T}_2^*(\mathcal{O})$ is antiregular. For $q \neq 2$ no line of $\mathbf{T}_2^*(\mathcal{O})$ is antiregular.

Proofs of the following results are also contained in Sections 1.3 and 1.5 of Payne and Thas [128].

Theorem 2.3.1 *Let $\mathcal{S} = (\mathcal{P}, \mathcal{B}, \mathrm{I})$ be a finite GQ of order (s,t).*

(i) *If $s < t$, then no pair of noncollinear points is antiregular.*

(ii) *If $s = t \neq 1$, then the pair of noncollinear points $\{x,y\}$ is antiregular if and only if any triad $\{x,y,z\}$ has either 0 or 2 centers.*

(iii) *If $s = t \neq 1$ and if the pair of noncollinear points $\{x,y\}$ is antiregular, then s is odd.*

Proof. (i). Suppose that the pair $\{x,y\}$ of noncollinear points is antiregular. We may assume that $t \neq 1$. Let L be a line not incident with x or y, but incident with a point z of $\{x,y\}^\perp$. Then every point of $\{x,y\}^\perp \backslash \{z\}$ is collinear with exactly one point u, with $u\mathrm{I}L$ and $u \neq z$, and every point u, with $u\mathrm{I}L$ and $u \neq z$, is collinear with at most one point of $\{x,y\}^\perp \backslash \{z\}$. It follows that $s \geq t$.

(ii). Assume $s = t \neq 1$, and that the pair of noncollinear points $\{x,y\}$ is antiregular. From the proof of (i) easily follows that any triad $\{x,y,z\}$ has either 0 or 2 centers. Conversely, if $|\{x,y,z\}^\perp| \in \{0,2\}$ for any triad $\{x,y,z\}$, then it is clear that the pair $\{x,y\}$ is antiregular.

(iii). Assume $s = t \neq 1$, and that the pair of noncollinear points $\{x,y\}$ is antiregular. Let L be a line not incident with x or y, and not incident with any point of $\{x,y\}^\perp$. The points of L are denoted by $v_0, v_1, \ldots, v_s$ and let $\{x,y\}^\perp = \{u_0, u_1, \ldots, u_s\}$. The notation is chosen in such a way that $xv_0\mathrm{I}u_0$ and $yv_1\mathrm{I}u_1$. By (ii) any triad $\{x,y,v_i\}$, with $i = 2,3,\ldots,s$, has 0 or 2 centers. As each point u_j, with $j = 2,3,\ldots,s$, is collinear with exactly one point of $\{v_2, v_3, \ldots, v_s\}$, it easily follows that $s - 1$ is even. ∎

Theorem 2.3.2 *If x is a coregular point of the thick GQ $\mathcal{S} = (\mathcal{P}, \mathcal{B}, \mathrm{I})$ of order s, then x is antiregular if and only if s is odd.*

Proof. Assume that s is odd. Then $|\{x, y, z\}^{\perp}|$ is even for any triad $\{x, y, z\}$, by Theorem 2.1.4. Let $\{x, y, z_i\}$ be any triad containing the pair of noncollinear points $\{x, y\}$, and let $|\{x, y, z_i\}^{\perp}| = t_i$. By the proof of Theorem 2.1.5 we have

$$\sum_i t_i(t_i - 2) = 0.$$

As t_i is even, we necessarily have $t_i \in \{0, 2\}$. If $z \sim x$ or/and $z \sim y$, with $x \not\sim y$, we clearly have $|\{x, y, z\}^{\perp}| = 1$. Hence the point x is antiregular.

Conversely, assume that x is antiregular. Then by Theorem 2.3.1, s is odd. ∎

Corollary 2.3.3 *The point (∞) of any GQ $\mathbf{T}_2(\mathcal{O})$ of order q, with q odd, is antiregular.*

Proof. By §2.1 the point (∞) of $\mathbf{T}_2(\mathcal{O})$ is coregular, hence by the previous theorem it is antiregular. ∎

2.4   Antiregularity and Laguerre Planes

This section will be about the connection between antiregular points, affine planes and Laguerre planes.

Theorem 2.4.1 *Let x be an antiregular point of the GQ $\mathcal{S} = (\mathcal{P}, \mathcal{B}, \mathrm{I})$ of order s, with $s \neq 1$, and let $y \in x^{\perp} \backslash \{x\}$. Then an affine plane $\pi(x, y)$ of order s may be constructed as follows.*

- POINTS *of $\pi(x, y)$ are just the points of $x^{\perp}$ that are not incident with the line $L = xy$;*

- LINES *of $\pi(x, y)$ are the point sets $\{x, z\}^{\perp} \backslash \{x\}$, with $x \sim z \not\sim y$, and $\{x, u\}^{\perp} \backslash \{y\}$, with $y \sim u \not\sim x$;*

- INCIDENCE *is containment.*

Proof. Easy exercise. ∎

A *Laguerre plane* is an incidence structure $\mathcal{L} = (\mathcal{P}, \mathcal{B}_1 \cup \mathcal{B}_2, \mathrm{I})$ with $\mathcal{P}$ the *point set*, $\mathcal{B}_1$ the *line set*, $\mathcal{B}_2$ the *circle set*, and I the (symmetric) incidence relation, satisfying the following axioms.

(i) Each point is incident with exactly one element of $\mathcal{B}_1$, each line and each circle are incident with exactly one common point.

(ii) Any three distinct points, no two of which are incident with a common line, are incident with exactly one common circle.

(iii) If x and y are distinct noncollinear points of $\mathcal{L}$ and if C is a circle incident with x but not with y, then there is just one circle C' incident with x, y and having only x in common with C (two circles having exactly one point x in common will be called *tangent circles* at x).

(iv) There exist a point x and a circle C with $x \, \mathrm{I} \, C$; each circle is incident with at least three points.

If $\mathcal{L} = (\mathcal{P}, \mathcal{B}, \mathrm{I})$, where $\mathcal{B} = \mathcal{B}_1 \cup \mathcal{B}_2$, is a Laguerre plane and $x \in \mathcal{P}$, then the *derived* or *internal plane* $\mathcal{L}_x = (\mathcal{P}_x, \mathcal{B}_x, \mathrm{I}_x)$ of $\mathcal{L}$ at x is defined as follows:

- $\mathcal{P}_x$ consists of all points of $\mathcal{P}$ not collinear with x,

- $\mathcal{B}_x$ consists of all circles incident with x and all lines not incident with x, and

- I_x is the restriction of I to $(\mathcal{P}_x \times \mathcal{B}_x) \cup (\mathcal{B}_x \times \mathcal{P}_x)$.

Clearly $\mathcal{L}_x$ is an affine plane. If $\mathcal{L}$ is finite, then the order of the affine plane $\mathcal{L}_x$ is independent of the choice of x in $\mathcal{P}$, and this common order n is called the *order* of the Laguerre plane $\mathcal{L}$. For a finite Laguerre plane $\mathcal{L} = (\mathcal{P}, \mathcal{B} = \mathcal{B}_1 \cup \mathcal{B}_2, \mathrm{I})$ of order n we have: $|\mathcal{P}| = n(n+1)$, $|\mathcal{B}_1| = n+1$, $|\mathcal{B}_2| = n^3$, each line is incident with n points, each circle is incident with $n + 1$ points, and each point is incident with one line and n^2 circles.

Let $\mathcal{O}$ be an oval in the plane $\mathbf{PG}(2, q)$ and let $\mathbf{PG}(2, q)$ be embedded in $\mathbf{PG}(3, q)$. Further, let x be a point of $\mathbf{PG}(3, q)$ not in $\mathbf{PG}(2, q)$. If $\mathcal{K}$ is the cone with vertex x and base $\mathcal{O}$, then a Laguerre plane $\mathcal{L}(\mathcal{O})$ of order q is constructed as follows.

- POINTS of $\mathcal{L}(\mathcal{O})$ are the points of $\mathcal{K} \backslash \{x\}$;

- LINES of $\mathcal{L}(\mathcal{O})$ are the generators of $\mathcal{K}$;

- CIRCLES of $\mathcal{L}(\mathcal{O})$ are the intersections $\pi \cap \mathcal{K}$, with π a plane of $\mathbf{PG}(3,q)$ not containing x, and

- INCIDENCE is the natural one.

For any point $y \in \mathcal{K}\backslash\{x\}$ the internal plane $\mathcal{L}(\mathcal{O})_y$ is isomorphic to the Desarguesian affine plane $\mathbf{AG}(2,q)$. If $\mathcal{O}$ is a nonsingular conic, then the Laguerre plane $\mathcal{L}(\mathcal{O})$ is called *classical*.

The following result is due to Payne and Thas [126].

Theorem 2.4.2 *Let x be an antiregular point of the GQ $\mathcal{S} = (\mathcal{P}, \mathcal{B}, \mathrm{I})$ of (odd) order s, with $s \neq 1$. Then the incidence structure $\mathcal{L}(\mathcal{S}) = (\mathcal{P}', \mathcal{B}' = \mathcal{B}'_1 \cup \mathcal{B}'_2, \mathrm{I}')$,*

- *with $\mathcal{P}' = x^{\perp}\backslash\{x\}$,*

- *with $\mathcal{B}'_1$ the set of traces $\{x,y\}^{\perp}$, $x \sim y$ and $x \neq y$,*

- *with $\mathcal{B}'_2$ the set of traces $\{x,y\}^{\perp}$, $x \nsim y$, and*

- *with I' containment,*

is a Laguerre plane of order s. If $y \in \mathcal{P}'$, then the internal plane $\mathcal{L}(\mathcal{S})_y$ is the affine plane $\pi(x,y)$. Conversely, let $\mathcal{L} = (\mathcal{P}', \mathcal{B}' = \mathcal{B}'_1 \cup \mathcal{B}'_2, \mathrm{I}')$ be a finite Laguerre plane of odd order s. Without loss of generality we may assume that I' is containment. Then we define the following incidence structure $\mathcal{S}(\mathcal{L}) = (\mathcal{P}, \mathcal{B}, \mathrm{I})$.

- *The elements of $\mathcal{P}$ (the POINTS of $\mathcal{S}(\mathcal{L})$) are of three types:*

  (i) *the elements of $\mathcal{B}'_2$;*

  (ii) *the elements of $\mathcal{P}'$;*

  (iii) *a symbol (∞).*

- *The elements of $\mathcal{B}$ (the LINES of $\mathcal{S}(\mathcal{L})$) are of two types:*

  (a) *the sets consisting of a point $x \in \mathcal{P}'$ together with a maximal set of mutually tangent circles at x;*

  (b) *the elements of $\mathcal{B}'_1$.*

- INCIDENCE *is defined as follows: a point* C *of Type (i) is incident with all lines of Type (a) containing* C *and with no line of Type (b), a point* x *of Type (ii) is incident with all lines of Type (a) and (b) containing* x, *the point* (∞) *is incident with all lines of Type (b) and no line of Type (a).*

Then $\mathcal{S}(\mathcal{L})$ *is a GQ of (odd) order* s, *for which* (∞) *is an antiregular point.*

Proof. Easy, see Payne and Thas [126]. ∎

Remark 2.4.3 Clearly $\mathcal{S}(\mathcal{L}(\mathcal{S})) \cong \mathcal{S}$ and $\mathcal{L}(\mathcal{S}(\mathcal{L})) \cong \mathcal{L}$. Further, it can be shown that $\mathcal{L}(\mathcal{S})$ (respectively, $\mathcal{L}$) is classical if and only if $\mathcal{S} \cong \mathbf{Q}(4, s)$ (respectively, $\mathcal{S}(\mathcal{L}) \cong \mathbf{Q}(4, s)$); see Payne and Thas [126].

The following characterization is due, independently, to Mazzocca [102], Chen and Kaerlein [39], and Payne and Thas [126].

Theorem 2.4.4 *Let* $\mathcal{S}$ *be a GQ of order* s, $s \neq 1$, *having an antiregular point* x. *Then* $\mathcal{S} \cong \mathbf{Q}(4, s)$ *if and only if there is a point* y, *with* $y \in x^\perp \backslash \{x\}$, *for which the affine plane* $\pi(x, y)$ *is Desarguesian.*

Proof. See 5.2.7 of Payne and Thas [128]. ∎

Remark 2.4.5 Theorem 2.4.4 is equivalent to saying that a Laguerre plane $\mathcal{L}$ of odd order s is classical if and only if at least one of the internal planes $\mathcal{L}_y$ is Desarguesian.

We end this section with a remarkable theorem due to Bagchi, Brouwer and Wilbrink [6], the proof of which is entirely in terms of codes.

Theorem 2.4.6 *If at least one point* x *of the GQ of order* s *is antiregular, then all the points are antiregular.*

Proof. See Bagchi, Brouwer and Wilbrink [6]. ∎

2.5 3-Regularity

Let $\mathcal{S}$ be a GQ order $(s, s^2), s \neq 1$. Then $|\{x, y, z\}^\perp| = s + 1$ for any triad $\{x, y, z\}$ of $\mathcal{S}$; this result is due to Bose and Shrikhande [18], see also 1.2.4 of Payne and Thas [128]. Clearly $|\{x, y, z\}^{\perp\perp}| \leq s + 1$. We say that the triad $\{x, y, z\}$ is *3-regular* provided $|\{x, y, z\}^{\perp\perp}| = s + 1$. The point x is *3-regular* if and only if each triad $\{x, y, z\}$ is 3-regular.

For the proofs of the following results we refer to Payne and Thas [128].

- All points of the GQ $\mathbf{Q}(5, q)$ are 3-regular.

- The point (∞) of the GQ $\mathbf{T}_3(\mathcal{O})$ is 3-regular.

- If $\mathbf{T}_3(\mathcal{O})$ has a 3-regular point other than (∞), then $\mathcal{O}$ is an elliptic quadric and, by Theorem 1.5.2(ii), $\mathbf{T}_3(\mathcal{O})$ is isomorphic to $\mathbf{Q}(5, q)$.

The following lemma was first discovered by Thas [165]; it is also a corollary of Theorem 1.4.1 in Payne and Thas [128]. The proof we give here is a variation on the original proof.

Lemma 2.5.1 *Let $\mathcal{S} = (\mathcal{P}, \mathcal{B}, \mathrm{I})$ be a GQ of order (s, s^2), $s \neq 1$, and let the triad $\{x, y, z\} = T$ be 3-regular. Then any point of $\mathcal{P} \backslash (T^\perp \cup T^{\perp\perp})$ is collinear with exactly two points of $T^\perp \cup T^{\perp\perp}$.*

Proof. Let z_i be any point of $\mathcal{P} \backslash (T^\perp \cup T^{\perp\perp})$ and let t_i be the number of points of $T^\perp \cup T^{\perp\perp}$ collinear with z_i. We count in two ways the number of ordered pairs (z_i, u), with $z_i \in \mathcal{P} \backslash (T^\perp \cup T^{\perp\perp})$, $u \in T^\perp \cup T^{\perp\perp}$, and $z_i \sim u$, and obtain

$$\sum_i t_i = 2(s+1)((s^2-s)s + (s+1)(s-1)) = 2(s^3-1)(s+1).$$

Next we count in two ways the number of ordered triples (z_i, u, u'), with $z_i \in \mathcal{P} \backslash (T^\perp \cup T^{\perp\perp}), u, u' \in T^\perp \cup T^{\perp\perp}, u \sim z_i \sim u', u \neq u'$, and obtain

$$\sum_i t_i(t_i - 1) = 2(s+1)s(s^2-s) + 2(s+1)^2(s-1) = 2(s^3-1)(s+1).$$

Hence

$$\sum_i t_i^2 = 4(s^3 - 1)(s + 1).$$

Let $d = |\mathcal{P}\backslash(T^\perp \cup T^{\perp\perp})|$. Then $d = (s+1)(s^3-1)$. So

$$d\sum_i t_i^2 - \left(\sum_i t_i\right)^2 = 0,$$

that is,

$$\sum_i (t_i - \bar{t})^2 = 0, \text{ with } \bar{t} = \left(\sum_i t_i\right)/d.$$

Hence $t_i = \bar{t} = 2$ for all i. ∎

Remark 2.5.2 In Lemma 2.5.1, if z_i is not incident with a line uu', with $u \in T^\perp$ and $u' \in T^{\perp\perp}$, then the point z_i is either collinear with two points of $T^\perp$ or two points of $T^{\perp\perp}$.

2.6 3-Regularity and Subquadrangles

This is also Section 2.6 of Payne and Thas [128]; the results are due to Thas [171]. As the main result plays an important role in the theory of translation GQs, we will give proofs.

Lemma 2.6.1 *Let $\{x, y, z\}$ be a 3-regular triad of the GQ $\mathcal{S} = (\mathcal{P}, \mathcal{B}, \mathrm{I})$ of order (s, s^2), $s \neq 1$, and let $\mathcal{P}'$ be the set of all points incident with lines of the form uu', $u \in \{x, y, z\}^\perp = U$ and $u' \in \{x, y, z\}^{\perp\perp} = U'$. If L is any line incident with no point of $U \cup U'$ and if k is the number of points in $\mathcal{P}'$ which are incident with L, then $k \in \{0, 2\}$ if s is odd and $k \in \{1, s+1\}$ if s is even.*

Proof. Let L be a line which is incident with no point of $U \cup U'$. If $w \in U$, if $w\mathrm{I}M\mathrm{I}m\mathrm{I}L$, and if M is not a line of the form uu', with $u \in U$ and $u' \in U'$, then there is just one point $w' \in U\backslash\{w\}$ which is collinear with m. Hence the number r of lines uu', $u \in U$ and $u' \in U'$, which are concurrent with L, has the parity of $|U| = s + 1$. Clearly r is also the number of points in $\mathcal{P}'$ which are incident with L.

Let $\{L_1, L_2, \ldots\} = \mathcal{L}$ be the set of all lines which are incident with no point of $U \cup U'$, and let r_i be the number of points in $\mathcal{P}'$ which are incident with L_i. We have $|\mathcal{L}| = s^3(s^2-1)$ and $|\mathcal{P}'\backslash(U \cup U')| = (s+1)^2(s-1)$. We count in two ways the number of ordered pairs (L_i, w), with $L_i \in \mathcal{L}$, $w \in \mathcal{P}'$, $w\mathrm{I}L_i$, and obtain

$$\sum_i r_i = (s+1)^2(s-1)s^2.$$

Next we count in two ways the number of ordered triples (L_i, w, w'), with $L_i \in \mathcal{L}, w, w' \in \mathcal{P}', w \mathrm{I} L_i \mathrm{I} w', w \neq w'$, and obtain

$$\sum_i r_i(r_i - 1) = (s+1)^2 s^2 (s-1).$$

Let s be odd. Then r_i is even, by the first paragraph of this proof. As

$$\sum_i r_i(r_i - 2) = 0,$$

it immediately follows that $r_i \in \{0, 2\}$ for all i.

Let s be even. Then r_i is odd, by the first paragraph of this proof. As

$$\sum_i (r_i - 1)(r_i - (s+1)) = 0,$$

and $r_i \leq s+1$, it immediately follows that $r_i \in \{1, s+1\}$ for all i. ∎

Theorem 2.6.2 *Let $\{x, y, z\}$ be a 3-regular triad of the GQ $\mathcal{S} = (\mathcal{P}, \mathcal{B}, \mathrm{I})$ of order (s, s^2), with s even. If $\mathcal{P}'$ is the set of all points incident with lines of the form uu', $u \in \{x, y, z\}^{\perp}$ and $u' \in \{x, y, z\}^{\perp\perp}$, if $\mathcal{B}'$ is the set of all lines in $\mathcal{B}$ which are incident with at least two points of $\mathcal{P}'$, and if I' is the restriction of I to $(\mathcal{P}' \times \mathcal{B}') \cup (\mathcal{B}' \times \mathcal{P}')$, then $\mathcal{S}' = (\mathcal{P}', \mathcal{B}', \mathrm{I}')$ is a subquadrangle of order s. Also, $\{x, y\}$ is a regular pair of $\mathcal{S}'$, with $\{x, y\}^{\perp'} = \{x, y, z\}^{\perp}$ and $\{x, y\}^{\perp'\perp'} = \{x, y, z\}^{\perp\perp}$.*

Proof. Let $\{x, y, z\}^{\perp} = U$ and $\{x, y, z\}^{\perp\perp} = U'$. We have $|\mathcal{P}'| = (s+1)^2(s-1) + 2(s+1) = (s+1)(s^2+1)$. Let L be a line of $\mathcal{B}'$. If L is incident with some point of $U \cup U'$, then clearly L is of type uu', with $u \in U$ and $u' \in U'$. Then all points incident with L are in $\mathcal{P}'$. If L is incident with no point of $U \cup U'$, then by Lemma 2.6.1, L is again incident with $s+1$ points of $\mathcal{P}'$. Now by Section 1.1, $\mathcal{S}' = (\mathcal{P}', \mathcal{B}', \mathrm{I}')$ is a subquadrangle of order (s, t'). Since $|\mathcal{P}'| = (s+1)(st'+1)$ we have $t' = s$, and so $\mathcal{S}'$ is a subquadrangle of order s. Since $U \cup U' \subseteq \mathcal{P}', |U| = |U'| = s+1$, and each point of U is collinear with each point of U', we have $\{x, y\}^{\perp'} = U$ and $\{x, y\}^{\perp'\perp'} = U'$. ∎

2.7   3-Regularity, Inversive Planes, and Characterizations

An *inversive* or *Möbius plane* is an incidence structure $\mathcal{I} = (\mathcal{P}, \mathcal{B}, \mathrm{I})$ with $\mathcal{P}$ the *point set*, $\mathcal{B}$ the *circle set*, and I the (symmetric) incidence relation, satisfying the following axioms.

(i) Any three distinct points are incident with exactly one common circle.

(ii) If $x, y \in \mathcal{P}, C \in \mathcal{B}, x \text{I} C, y \text{ I } C$, then there is a unique circle incident with x, y and having only x in common with C (two circles having exactly one point x in common will be called *tangent circles* at x).

(iii) There are four points not incident with a common circle, and each circle is incident with at least one point.

If $\mathcal{I} = (\mathcal{P}, \mathcal{B}, \text{I})$ is an inversive plane and $x \in \mathcal{P}$, then the incidence structure $\mathcal{I}_x = (\mathcal{P}_x, \mathcal{B}_x, \text{I}_x)$, with $\mathcal{P}_x = \mathcal{P}\backslash\{x\}, \mathcal{B}_x = \{C \in \mathcal{B} \,\|\, x \text{I} C\}$ and $\text{I}_x = \text{I} \cap ((\mathcal{P}_x \times \mathcal{B}_x) \cup (\mathcal{B}_x \times \mathcal{P}_x))$, is an affine plane. The plane $\mathcal{I}_x$ is also called the *derived* or *internal plane* of $\mathcal{I}$ at x. If $\mathcal{I}$ is finite, then the order of $\mathcal{I}_x$ is independent of the choice of x in $\mathcal{P}$. If this order is n, then we say that the inversive plane $\mathcal{I}$ has *order* n. For a finite inversive plane $\mathcal{I} = (\mathcal{P}, \mathcal{B}, \text{I})$ of order n we have: $|\mathcal{P}| = n^2 + 1, |\mathcal{B}| = n(n^2 + 1)$, each circle is incident with $n + 1$ points, and each point is incident with $n(n + 1)$ circles. In fact an inversive plane of order n is nothing else than a $3 - (n^2 + 1, n + 1, 1)$ design.

If $\mathcal{O}$ is an ovoid of $\mathbf{PG}(3, q)$, then $\mathcal{O}$ together with the intersections $\pi \cap \mathcal{O}$, where π is a plane which is not tangent to $\mathcal{O}$, and with the natural incidence, is an inversive plane of order q. Such an inversive plane is called *egglike* and is denoted by $\mathcal{I}(\mathcal{O})$. Clearly $\mathcal{I}(\mathcal{O})_x$, with $x \in \mathcal{O}$, is the Desarguesian plane $\mathbf{AG}(2, q)$. If $\mathcal{O}$ is an elliptic quadric, then the inversive plane $\mathcal{I}(\mathcal{O})$ is called *classical* or *Miquelian*. By a celebrated theorem of Dembowski [44], each finite inversive plane of even order is egglike. By the theorem of Barlotti [8] and Panella [106], each egglike inversive plane of odd order is Miquelian; no other inversive planes of odd order are known. Thas [175] proved that if for an inversive plane $\mathcal{I}$ of odd order n at least one internal plane is Desarguesian, then $\mathcal{I}$ is Miquelian.

A *flock* $\mathcal{F}$ of an inversive plane $\mathcal{I}$ of order n is a set of $n - 1$ mutually disjoint circles. If $\mathcal{F}$ is a flock of the egglike inversive plane $\mathcal{I}(\mathcal{O})$ for which the $n - 1$ planes containing the circles of $\mathcal{F}$ all pass through a common line, then $\mathcal{F}$ is called *linear*. By theorems of Orr [105] and Thas [159] each flock of $\mathcal{I}(\mathcal{O})$ is linear; see also Fisher and Thas [58].

Let $\mathcal{S} = (\mathcal{P}, \mathcal{B}, \text{I})$ be a GQ of order (s, s^2), $s \neq 1$, and assume the pair $\{x, y\}$ of noncollinear points is such that every triad $\{x, y, z\}$ is 3-regular. Then the incidence structure $\mathcal{I} = (\{x, y\}^\perp, \mathcal{B}', \text{I}')$, with $\mathcal{B}'$ consisting of the sets $\{x, y, z\}^\perp$ where $\{x, y, z\}$ is any triad containing $\{x, y\}$, and with I' the natural incidence, is a $3 - (s^2 + 1, s + 1, 1)$ design, that is, an inversive plane of order s. These inversive planes, and also flocks, were used to prove the following strong characterization theorem.

Theorem 2.7.1   (i) *A GQ of order $(s, s^2), s \neq 1$, is isomorphic to a GQ $\mathbf{T}_3(\mathcal{O})$ of Tits if and only if it has a 3-regular point.*

   (ii) *Let S be a GQ of order (s, s^2), $s > 1$, with s odd. Then S is isomorphic to the classical GQ $\mathbf{Q}(5, s)$ if and only if it has a 3-regular point.*

   (iii) *Let S be a GQ of order (s, s^2), with s even. Then $S \cong \mathbf{Q}(5, s)$ if and only if one of the following holds:*

      (a) *all points are 3-regular;*

      (b) *S has at least one 3-regular point not incident with some regular line.*

Proof.   See Thas [168]; see also Chapter 5 of Payne and Thas [128].   ∎

Remark 2.7.2 Mazzocca [103] proved the following result: a GQ S of order (s, s^2), $s \neq 1$ and s odd, is isomorphic to $\mathbf{Q}(5, s)$ if and only if each point of S is 3-regular.

Chapter 3

Elation and Translation Generalized Quadrangles

In this chapter, we introduce elation and translation generalized quadrangles, and review several basic properties on these types of quadrangles. Also, a key role will be played by pseudo-ovals and pseudo-ovoids. We start with introducing some group theoretical concepts and notation.

3.1 Some Notions from Group Theory

We review some basic notions of group theory. We will generally need only permutation groups, but also some notions from abstract group theory will be needed.

As always in this book, we denote permutation action exponentially and let elements act on the right. We denote the identity element of a group often by id or $\mathbf{1}$, if no special symbol has been introduced for it before. A group G without its identity id is denoted $G^\times$.

Note that we write the action of a group on a set at the right, as an exponent, and as such a *permutation group* (G, X) is a pair consisting of a group G and a set X such that each element g of G defines a permutation $g : X \to X$ of X and the permutation defined by gh, $g, h \in G$, is given by $gh : X \to X : x \mapsto (x^g)^h$.

The *conjugate* of g by h is $g^h = h^{-1}gh$. Let H be a group. The *commutator* of two group elements g, h is equal to $[g, h] = g^{-1}h^{-1}gh$. The *commutator* of two subsets A and B of a group G is the subgroup $[A, B]$ generated by all elements $[a, b]$, with $a \in A$ and $b \in B$. The *commutator subgroup* of G

is $[G, G]$, or sometimes G'. Two subgroups A and B *centralize* each other
if $[A, B] = \{\mathrm{id}\}$. The subgroup A *normalizes* B if $B^a = B$, for all $a \in A$,
which is equivalent with $[A, B] \leq B$.

Inductively, we define the *nth central derivative* $[G, G]_{[n]}$ of a group G
as $[G, [G, G]_{[n-1]}]$, and the *nth normal derivative* $[G, G]_{(n)}$ as $[[G, G]_{(n-1)},$
$[G, G]_{(n-1)}]$. For $n = 0$, the 0th central and normal derivative are by definition equal to G itself. If, for some natural number n, $[G, G]_{(n)} = \{\mathrm{id}\}$,
and $[G, G]_{(n-1)} \neq \{\mathrm{id}\}$, then we say that G is *solvable* (*soluble*) of length n.
If $[G, G]_{[n]} = \{\mathrm{id}\}$ and $[G, G]_{[n-1]} \neq \{\mathrm{id}\}$, then we say that G is *nilpotent*
of class n. The *center* of a group is the set of elements that commute with
every other element, i.e., $Z(G) = \{z \in G \parallel [z, g] = \mathrm{id}, \forall g \in G\}$. Clearly, if
a group G is nilpotent of class n, then the $(n - 1)$th central derivative is a
nontrivial subgroup of $Z(G)$.

A group G is called *perfect* if $G = [G, G] = G'$.

For a prime number p, a *p-group* is a group of order p^n, for some natural
number n. A *Sylow p-subgroup* of a finite group G is a p-subgroup of some
order p^n such that p^{n+1} does not divide $|G|$.

Suppose (G, X) is a permutation group (where G acts on X) which satisfies
the following properties:

(1) G acts transitively but not sharply transitively on X;

(2) there is no nontrivial element of G with more than one fixed point in
     X.

Then (G, X) is a *Frobenius group* (or G is a *Frobenius group* in its action on
X). Define $N \subseteq G$ by:

$$N = \{g \in G \parallel f(g) = 0\} \cup \{\mathbf{1}\},$$

where $f(g)$ is the number of fixed points of g in X. Then N is called the
Frobenius kernel of G (or of (G, X)), and we have the following well-known
result.

Theorem 3.1.1 (Theorem of Frobenius) *N is a normal regular subgroup
of G.*   ■

3.2 Elation Generalized Quadrangles

3.2.1 Elation Generalized Quadrangles: Definition and Two Fundamental Structure Theorems

Let $S = (\mathcal{P}, \mathcal{B}, \mathrm{I})$ be a finite generalized quadrangle. A *whorl about a point* p of the GQ S is a collineation fixing p linewise. A *whorl about a line* L is a collineation fixing L pointwise. An *elation about the point* $p \in \mathcal{P}$ is a whorl about p that fixes no point of $\mathcal{P} \setminus p^{\perp}$. Dually, one defines *elations about lines*. If θ is an elation about p, then we will often say that p is the *center* of θ. By definition, the identical permutation is an elation (about every point). If p is a point of the GQ S, for which there exists a group of elations G about p which acts sharply transitively on the points of $\mathcal{P} \setminus p^{\perp}$, then S is said to be an *elation generalized quadrangle* (EGQ) with *base-point* or *center* or *elation point* p and *elation group* (or *base-group*) G, and we sometimes write $(S^{(p)}, G)$ or $S^{(p)}$ for S. Dually, we define the *base-line* of an EGQ.

We only work with EGQs that are thick, and therefore we will not bother each time to mention this.

The next theorem proves itself to be quite useful when working with whorls.

Theorem 3.2.1 ([128], 8.1.1) *Let θ be a nontrivial whorl about p of the GQ $S = (\mathcal{P}, \mathcal{B}, \mathrm{I})$ of order (s, t), $s \ne 1 \ne t$. Then one of the following must hold for the fixed elements structure $S_{\theta} = (\mathcal{P}_{\theta}, \mathcal{B}_{\theta}, \mathrm{I}_{\theta})$.*

(1) *$y^{\theta} \ne y$ for each $y \in \mathcal{P} \setminus p^{\perp}$.*

(2) *There is a point y, $y \not\sim p$, for which $y^{\theta} = y$. Put $V = \{p, y\}^{\perp}$ and $U = V^{\perp}$. Then $V \cup \{p, y\} \subseteq \mathcal{P}_{\theta} \subseteq V \cup U$, and $L \in \mathcal{B}_{\theta}$ if and only if L joins a point of V with a point of $U \cap \mathcal{P}_{\theta}$.*

(3) *S_{θ} is a subGQ of order (s', t), where $2 \le s' \le s/t \le t$, and hence $t < s$.*

Proof. If we are not in Cases (1) and (2), the fixed elements structure of θ is a subGQ of order (s', t), $s' > 1$ (just apply the dual of the result at the end of Section 1.1). Now 2.2.1 of Payne and Thas [128], see also Section 1.1, ends it. ∎

Theorem 3.2.2 ([128], 8.2.4) *Let $S = (\mathcal{P}, \mathcal{B}, \mathrm{I})$ be a GQ of order (s, t) with $s \le t$ and $s > 1$, and let p be a point for which $\{p, x\}^{\perp\perp} = \{p, x\}$ for all $x \in \mathcal{P} \setminus p^{\perp}$. Let G be a group of whorls about p.*

(1) *If $y \sim p$, $y \neq p$, and if θ is a nonidentity whorl about p and y, then all points fixed by θ lie on py and all lines fixed by θ meet py.*

(2) *If θ is a nonidentity whorl about p, then θ fixes at most one point of $\mathcal{P} \setminus p^{\perp}$.*

(3) *If G is generated by elations about p, then G is a group of elations, and so the set of elations about p is a group.*

(4) *If G acts transitively on $\mathcal{P} \setminus p^{\perp}$ and $|G| > s^2 t$, then G is a Frobenius group on $\mathcal{P} \setminus p^{\perp}$, so that the set of all elations about p is a normal subgroup of G of order $s^2 t$ acting sharply transitively on $\mathcal{P} \setminus p^{\perp}$, i.e. $\mathcal{S}^{(p)}$ is an EGQ with some normal subgroup of G as elation group.*

(5) *If G is transitive on $\mathcal{P} \setminus p^{\perp}$ and G is generated by elations about p, then $(\mathcal{S}^{(p)}, G)$ is an EGQ.*

Proof. (1) If there is a fixed point not on py, then Theorem 3.2.1 leads us to a contradiction. If $L \not\sim py$ is fixed by θ, $\mathrm{proj}_L y$ is a fixed point not on py, a contradiction.

(2) This again follows directly from Theorem 3.2.1.

(3) Suppose G is as in (3) of the statement. Suppose $x \in \mathcal{P} \setminus p^{\perp}$ is a point for which $|G_x| \neq 1$. Then G is a Frobenius group in its action on the G-orbit X of x. So the Frobenius kernel N of G acts sharply transitively on X. As G is generated by elations about p, G itself acts sharply transitively on X. It follows that G is a group of elations about p.

(4) This follows from (3).

(5) This follows immediately from (3). ∎

3.2.2   4-Gonal Families and EGQs

In this section it is explained how EGQs can always be represented as a group coset geometry, and vice versa.

Suppose $(\mathcal{S}^{(p)}, G)$ is an EGQ of order (s, t), $s \neq 1 \neq t$, with elation point p and elation group G, and let q be a point of $\mathcal{P} \setminus p^{\perp}$. Let $L_0, L_1, \ldots, L_t$ be the lines incident with p, and define r_i and M_i by $L_i \mathrm{I} r_i \mathrm{I} M_i \mathrm{I} q$, $0 \le i \le t$. Put $H_i = \{\theta \in G \parallel M_i^\theta = M_i\}$, $H_i^* = \{\theta \in G \parallel r_i^\theta = r_i\}$, and $\mathcal{J} = \{H_i \parallel 0 \le i \le t\}$. Also, put $\mathcal{J}^* = \{H_i^* \parallel 0 \le i \le t\}$. Then $|G| = s^2 t$ and $\mathcal{J}$ is a set of $t+1$ subgroups of G, each of order s. Also, for each i, H_i^* is a subgroup of G of order st containing H_i as a subgroup. Moreover, the following two conditions are satisfied:

(K1) $H_i H_j \cap H_k = \{\mathbf{1}\}$ for distinct i, j and k;

(K2) $H_i^* \cap H_j = \{\mathbf{1}\}$ for distinct i and j.

Conversely, if G is a group of order $s^2 t$ and $\mathcal{J}$ (respectively $\mathcal{J}^*$) is a set of $t+1$ subgroups H_i (respectively H_i^*) of G of order s (respectively of order st), and if the Conditions (K1) and (K2) are satisfied, then the H_i^*s are uniquely defined by the H_is as

$$H_i^* = H_i \cup \{g \in G \parallel H_i g \cap H_j = \emptyset \text{ for } 0 \le i \le t\}$$

(H_i^* is sometimes called the *tangent space* at H_i), and $(\mathcal{J}, \mathcal{J}^*)$ is said to be a 4-*gonal family of type* (s, t) *in* G. Sometimes we will also say that $\mathcal{J}$ is a 4-*gonal family of type* (s, t) *in* G if this seems convenient.

Let $(\mathcal{J}, \mathcal{J}^*)$ be a 4-gonal family of type (s, t) in the group G of order $s^2 t$, $s \ne 1 \ne t$. Define an incidence structure $\mathcal{S}(G, \mathcal{J})$ as follows.

- POINTS of $\mathcal{S}(G, \mathcal{J})$ are of three kinds:

  (i) elements of G;

  (ii) right cosets $H_i^* g$, $g \in G$, $i \in \{0, 1, \ldots, t\}$;

  (iii) a symbol (∞).

- LINES are of two kinds:

  (a) right cosets $H_i g$, $g \in G$, $i \in \{0, 1, \ldots, t\}$;

  (b) symbols $[H_i]$, $i \in \{0, 1, \ldots, t\}$.

- INCIDENCE. A point g of Type (i) is incident with each line $H_i g$, $0 \le i \le t$. A point $H_i^* g$ of Type (ii) is incident with $[H_i]$ and with each line $H_i h$ contained in $H_i^* g$. The point (∞) is incident with each line $[H_i]$ of Type (b). There are no further incidences.

It is straightforward to check that the incidence structure $\mathcal{S}(G, \mathcal{J})$ is a GQ of order (s, t).

For any $h \in G$, define θ_h by $g^{\theta_h} = gh$, $(H_i g)^{\theta_h} = H_i gh$, $(H_i^* g)^{\theta_h} = H_i^* gh$, $[H_i]^{\theta_h} = [H_i]$, $(\infty)^{\theta_h} = (\infty)$, with $g \in G$, $H_i \in \mathcal{J}$, $H_i^* \in \mathcal{J}^*$. Then θ_h is an automorphism of $\mathcal{S}(G, \mathcal{J})$ which fixes the point (∞) and all lines of Type (b). If

$$G' = \{\theta_h \parallel h \in G\},$$

then clearly $G' \cong G$ and G' acts sharply transitively on the points of Type (i). Hence $\mathcal{S}(G, \mathcal{J})$ is an EGQ with elation point (∞) and elation group isomorphic to G.

Moreover, if we start with an EGQ $(\mathcal{S}^{(p)}, G)$ to obtain the family $\mathcal{J}$ as above, then we have that

$$(\mathcal{S}^{(p)}, G) \cong \mathcal{S}(G, \mathcal{J}).$$

(Of course q^g corresponds to g, r_i^g corresponds to $H_i^* g$, p corresponds to (∞), M_i^g corresponds to $H_i g$, and L_i corresponds to $[H_i]$.)

Theorem 3.2.3 *A group of order $s^2 t$ admitting a 4-gonal family is an elation group for a suitable elation generalized quadrangle.* ∎

These results were first noted by Kantor [81].

Finally we notice that many of the known GQs can be shown to be of Type $\mathcal{S}(G, \mathcal{J})$; see Payne and Thas [128], Johnson and Payne [79] and Payne [122].

3.3   Translation Generalized Quadrangles

Before introducing the concept of symmetry, we recall a result of Benson, taken from [10]. The proof makes use of matrix techniques.

Theorem 3.3.1 (C. T. Benson [10]) *Let S be a thick finite GQ of order (s, t), and let θ be a nonidentity automorphism of S. Let f be the number of fixed points of θ, and g the number of points which are mapped onto a collinear but different point. Then*

$$(t + 1)f + g \equiv st + 1 \mod s + t.$$

Proof.   See, e.g., 1.9.1 of [128]. ∎

3.3.1 Symmetry

Let $S = (\mathcal{P}, \mathcal{B}, \mathbf{I})$ be a GQ of order (s, t) with $s \neq 1 \neq t$, and let L be a line of S. A *symmetry about* L is an automorphism of S that fixes each element of $L^\perp$. The line L is an *axis of symmetry* if the group of symmetries about L has size s. By Theorem 3.2.1 it is easy to see that this is the maximal size for such a group. The dual notions are *symmetry about a point* and *center of symmetry*. A GQ has a *translation point* p if each line incident with p is an axis of symmetry. Dually, we speak of *translation lines*.

Let us first remark that the only fixed points of a nontrivial symmetry about a line clearly are on that line (again by Theorem 3.2.1).

Theorem 3.3.2 *If a thick GQ S of order (s, t) has a nonidentical symmetry θ about some line, then $st(s + 1) \equiv 0 \mod s + t$.*

Proof. Using the notation of Theorem 3.3.1, we have $f = s + 1$ and $g = (s + 1)ts$. Now apply that theorem. ∎

Theorem 3.3.3 *Let S be a GQ of order (s, t), $s \neq 1 \neq t$, and suppose that L and M are distinct concurrent lines. If α is a symmetry about L and β a symmetry about M, then $\alpha\beta = \beta\alpha$.*

Proof. The commutator $[\alpha, \beta]$ fixes $L^\perp$ and $M^\perp$ linewise, so S is fixed pointwise. Hence α and β commute. ∎

Theorem 3.3.4 *Any center of symmetry is regular. Dually, any axis of symmetry is regular.*

Proof. Let x be a center of symmetry of S. Further, let $x' \nsim x$, and $r, r' \in \{x, x'\}^\perp$, with $r \neq r'$, $x'' \in \{r, r'\}^\perp$, with $x'' \notin \{x, x'\}$, and $r'' \in \{x, x'\}^\perp$, with $r'' \notin \{r, r'\}$. We must prove that $x'' \sim r''$.
Let θ be the symmetry about x with $r'x'^\theta = r'x''$. Then $rx'^\theta = rx''$, and so $x'^\theta = x''$. Hence $(r''x')^\theta = r''x''$ and so $r'' \sim x''$. ∎

Corollary 3.3.5 *If the GQ S admits a center a symmetry, then $s \geq t$. Dually, if S has an axis of symmetry, $s \leq t$.*

Proof. Follows from Theorem 3.3.4 and Theorem 2.1.3. ∎

3.3.2   Translation Generalized Quadrangles: Definition

If a GQ $(\mathcal{S}^{(p)}, G)$ is an EGQ with elation point p, and if G is abelian, then we say that $\mathcal{S}$ is a *translation generalized quadrangle* (TGQ) with *base-point* p and *translation group* (or *base-group*) G. The elements of a translation group are called *translations*.

3.3.3   Symmetries and Translations

Assume that the GQ $\mathcal{S} = (\mathcal{P}, \mathcal{B}, \mathrm{I})$ is a TGQ $(\mathcal{S}^{(x)}, G)$ of order (s, t), with $s \neq 1 \neq t$.

Theorem 3.3.6 *If $\theta \in G$ and $y^\theta = y$, $y \sim x$, $x \neq y$, then xy is pointwise fixed.*

Proof. As $\mathcal{S}$ is an EGQ, G acts transitively on $xy \setminus \{x\}$. Take any $\phi \in G$. Then $y^{\theta\phi} = y^\phi = y^{\phi\theta}$, so θ fixes y^ϕ. The theorem follows.   ■

Theorem 3.3.7 *Each line incident with x is an axis of symmetry.*

Proof. Let L be any line incident with x. Let $N \sim L$, with $N \,\mathrm{I}\!\!\!/\, x$, and let $r\mathrm{I}N\mathrm{I}r'$, with $r \not\sim x \not\sim r'$. Let $\theta \in G$ with $r^\theta = r'$. Then $N^\theta = N$. Choose any line W with $W \sim L$ and $x \,\mathrm{I}\!\!\!/\, W$, and let ξ be any element of G mapping W onto N. Then $\xi\theta\xi^{-1} = \theta$ fixes W. Hence θ fixes each line concurrent with L. It immediately follows that θ is a symmetry about L and that L is an axis of symmetry.   ■

Corollary 3.3.8 *Each line incident with x is regular and so $t \geq s$.*

Proof. Follows from Theorem 3.3.7, Theorem 3.3.4 and Corollary 3.3.5.

■

Corollary 3.3.9 *If $s = t$, then for s even the point x is regular and for s odd the point x is antiregular.*

Proof. From Corollary 3.3.8, Theorem 2.1.5 and Theorem 2.3.2.   ■

Theorem 3.3.10 *The translation group G is generated by all symmetries about all lines incident with x, hence is uniquely defined by the translation point x.*

Proof. In the proof of Theorem 3.3.7 it was shown that all symmetries about all lines incident with x are elements of the group G. Let u, u' be distinct points of $\mathcal{P}\backslash x^{\perp}$, and let θ be the element of G which maps u onto u'. If $u \sim u'$, then θ is a symmetry about some line incident with x. So let $u \not\sim u'$. It is easy to show that there are points $w, w' \notin x^{\perp}$, not necessarily distinct, with $u \sim w \sim w' \sim u'$. As there are symmetries $\theta_1, \theta_2, \theta_3$ about lines incident with x for which $u^{\theta_1} = w, w^{\theta_2} = w', w'^{\theta_3} = u'$, we have $\theta_1\theta_2\theta_3 = \theta' \in G$ and $u^{\theta'} = u'$. Hence $\theta = \theta'$, and so θ is a product of symmetries about lines incident with x.   ∎

Remark 3.3.11 If a GQ $\mathcal{S}$ is an EGQ about a point x, then the elation group is not always uniquely defined by x; see the addendum to Chapter 4.

The next theorem is taken from K. Thas [198], see also the monograph [206].

Theorem 3.3.12 *A thick EGQ $(\mathcal{S}^{(x)}, G)$ is a TGQ if and only if the point x is coregular.*

Proof. If $(\mathcal{S}^{(x)}, G)$ is a TGQ, then, by Corollary 3.3.8, the point x is coregular.

Conversely, assume that the point x is coregular. Let M be a line not incident with x, let $L \sim M$ with $x\mathrm{I}L$, and let S be the subgroup (of size s) of G fixing M. Let H be a Sylow subgroup of S, with $|H| = p^l$ and p prime. As $(s - 1, p^l) = 1$, there is a point $z\mathrm{I}L$ with $z \not{\mathrm{I}} M, z \neq x$, which is fixed by all elements of H. Further, let N be a line incident with z, with $N \neq L$. Choose an element $\theta \in H$. Let U be a line concurrent with M, N, with $U \neq L$, and let L' be the line incident with x and concurrent with U. Let $g = M \cap U$ and let $g^\theta = h$; then $h\mathrm{I}M$. By the regularity of L there is a line W incident with h and concurrent with N and L'. As $z^\theta = z$, we have $N^\theta = N$. Now assume that N' is incident with z', $z'\mathrm{I}L$, $z' \neq z, z' \neq x, N' \neq L$. Then consider two distinct $(s + 1) \times (s + 1)$-grids containing N' and L. As each of these grids contains distinct nonconcurrent fixed lines for θ, these grids are also fixed and so N' is fixed. It follows that θ is a symmetry about L. As S is generated by its Sylow subgroups, any element of S is a symmetry about L, and consequently L is an axis of symmetry.

Let σ and τ be symmetries about distinct axes, L and L' respectively, incident with x. Then by Theorem 3.3.3, we have $\tau\sigma = \sigma\tau$. Now let σ, σ' be symmetries with axis L. For $s = 2$ the GQ is either $\mathbf{Q}(4, 2)$ or

$\mathbf{Q}(5, 2)$, see Section 1.8. Hence in such a case $\sigma\sigma' = \sigma'\sigma$. For $s \neq 2$ it is easy to show that σ' can be written as $\sigma_1\sigma_2\sigma_3$, with $\sigma_1, \sigma_2, \sigma_3$ symmetries having as axes lines incident with x but different from L. Hence $\sigma\sigma' = \sigma\sigma_1\sigma_2\sigma_3 = \sigma_1\sigma_2\sigma_3\sigma = \sigma'\sigma$. Consequently symmetries σ, τ, each with an axis incident with x, commute. As any element of G is a product of at most three symmetries about lines incident with x (see the proof of Theorem 3.3.10), it follows that G is abelian. We conclude that $(\mathcal{S}^{(x)}, G)$ is a TGQ. ∎

Note that from the proof of the previous result, one can actually derive the next proposition (first noted in K. Thas [198], see also [206]).

Theorem 3.3.13 *Let $\mathcal{S}$ be a thick GQ with regular line L, let $x\mathrm{I}L$ and $x\mathrm{I}M \sim L \neq M$. If H is a group of whorls about x that acts transitively on $M \setminus \{M \cap L\}$, then L is an axis of symmetry, and H contains all symmetries about L.*

Proof. Let (s, t) be the order of $\mathcal{S}$. Take an arbitrary element θ of H of order p^h, where p is a prime and p^h divides s. Then by the proof of Theorem 3.3.12, θ is a symmetry about L. It follows easily that L is an axis of symmetry (the group generated by all such θs has size s), and all symmetries about L are in H. ∎

Let $(\mathcal{S}^{(p)}, G)$ be an EGQ, and define H_i, with $i = 0, 1, \ldots, t$, and $\mathcal{J}$ as in Section 3.2.2.

Theorem 3.3.14 *H_i is a group of symmetries for all i about the line L_i if and only if $H_i \trianglelefteq G$ (and hence $\mathcal{S}^{(p)}$ is a TGQ if and only if $H_i \trianglelefteq G$ for each i), only if L_i is a regular line. The line L_i is regular if and only if $H_iH_j = H_jH_i$ for all $H_j \in \mathcal{J}$.*

Proof. Straightforward exercise (see 8.2.2 of [128] and Theorem 3.3.12). ∎

K. Thas completed Theorem 3.3.14 by proving the following result.

Theorem 3.3.15 *H_i is a group of symmetries if and only if L_i is a regular line.*

Proof. See K. Thas [198], or also the monograph [206]. In fact, we proved a more general version by Theorem 3.3.13. ∎

Now recall the following result of Chen and Frohardt.

Theorem 3.3.16 *Let $(\mathcal{J}, \mathcal{J}^*)$ be a 4-gonal family of Type (s, t) in the group G. If there are two distinct members of $\mathcal{J}$ that are normal in G, then G is abelian, that is, the corresponding EGQ is a TGQ.*

Proof. See Chen and Frohardt [38]. ∎

Together with Theorem 3.3.14 and Theorem 3.3.15, this gives

Corollary 3.3.17 *Let $(\mathcal{S}^{(p)}, G)$ be an EGQ of order (s, t). Then $\mathcal{S}^{(p)}$ is a TGQ if and only if there are at least two distinct regular lines on p.* ∎

Hachenberger could strengthen Theorem 3.3.16 when $s^2 t$ is even.

Theorem 3.3.18 *Let $(\mathcal{J}, \mathcal{J}^*)$ be a 4-gonal family of Type (s, t) in the group G, where $s^2 t$ is even or $s = t$. If there is at least one element of $\mathcal{J}$ normal in G, then G is abelian, that is, the corresponding EGQ is a TGQ.*

Proof. See Hachenberger [65]. ∎

Together with Theorem 3.3.14 and Theorem 3.3.15, this gives

Corollary 3.3.19 *Let $(\mathcal{S}^{(p)}, G)$ be an EGQ of order (s, t) with $s^2 t$ even. Then $\mathcal{S}^{(p)}$ is a TGQ if and only if there is at least one regular line on p.* ∎

3.4 The Kernel of a Translation Generalized Quadrangle

Each TGQ $\mathcal{S}$ of order (s, t) with translation point (∞) has a *kernel* $\mathbb{K}$, which is a field with multiplicative group isomorphic to the group of all collineations of $\mathcal{S}$ fixing the point (∞) and any given point not collinear with (∞) linewise. We will introduce the kernel in detail in this section.

Let $(\mathcal{S}^{(p)}, G)$ be a TGQ with translation group G and with $H_i, H_i^*, \mathcal{J}$, etc., as before. The *kernel* $\mathbb{K}$ of $\mathcal{S}^{(p)}$ (or of $(\mathcal{S}^{(p)}, G)$ or of $\mathcal{J}$) is the set of all endomorphisms α of G for which $H_i^\alpha \subseteq H_i$, $0 \leq i \leq t$. With the usual addition and multiplication of endomorphisms $\mathbb{K}$ is a ring.

Theorem 3.4.1 ([128], 8.5.1) *The ring $\mathbb{K}$ is a field, so that $H_i^\alpha = H_i$, $(H_i^*)^\alpha = H_i^*$ for all $i = 0, 1, \ldots, t$ and all $\alpha \in \mathbb{K}^0 = \mathbb{K}\backslash\{0\}$.*

Proof. The only GQs with $s = 2$ and $t > 1$ are $\mathbf{W}(2)$ and $\mathbf{Q}(5,2)$, in which cases we can check the theorem (in these cases $\mathbb{K} = \{0, 1\}$). So from now on we may assume that $s > 2$.

If each $\alpha \in \mathbb{K}^0$ is an automorphism of G, then clearly $\mathbb{K}$ is a field. So suppose some $\alpha \in \mathbb{K}^0$ is not an automorphism. Then $\langle H_0, H_1, \ldots, H_t \rangle = G \supseteq G^\alpha = \langle H_0^\alpha, H_1^\alpha, \ldots, H_t^\alpha \rangle$, with $G \neq G^\alpha$, implying $H_i^\alpha \neq H_i$ for some i. Let $g^\alpha = \mathbf{1}$, $G \in H_i\backslash\{\mathbf{1}\}$. If i, j, k are mutually distinct and $g' \in H_j$ with $\{g'\} \neq H_j \cap H_k^* g^{-1}$, then we have $gg' = hh'$ with $h \in H_k, h' \in H_l$, for a uniquely defined l, with $l \neq k, j$. Hence $h^\alpha h'^\alpha = g'^\alpha$, implying that $h^\alpha = h'^\alpha = g'^\alpha = \mathbf{1}$ (by (K1)). Since g' was any one of $s - 1$ elements of H_j, $|\ker(\alpha) \cap H_j| \geq s - 1 > s/2$, implying $H_j \subseteq \ker(\alpha)$. This implies $H_j \subseteq \ker(\alpha)$ for each j, with $j \neq i$, so that $G_i \subseteq \ker(\alpha)$, where $G_i = \langle H_j \parallel j \in \{0, 1, \ldots, t\} \setminus \{i\}\rangle$. As each $\sigma \in H_i$ can be written as $\sigma_1\sigma_2\sigma_3$, with $\sigma_1, \sigma_2, \sigma_3$ elements of respectively $H_j, H_{j'}, H_{j''}$ for some $j, j', j'' \in \{0, 1, \ldots, t\} \setminus \{i\}$, we have $G = G_i$. This says $\alpha = 0$, a contradiction. Hence we have shown that $\mathbb{K}$ is a field and $H_i^* = H_i$ for $i = 0, 1, \ldots, t$ and $\alpha \in \mathbb{K}^0$. Since H_i^* is the set theoretic union of H_i together with all those cosets of H_i disjoint from $\bigcup\{H_i \parallel 0 \leq i \leq t\}$, we also have $(H_i^*)^\alpha = H_i^*$. ∎

For each subfield $\mathbb{F}$ of $\mathbb{K}$ there is a vector space $(G, \mathbb{F})$ whose vectors are the elements of G, and whose scalars are the elements of $\mathbb{F}$. Vector addition is the group operation in G, and scalar multiplication is defined by $g\alpha = g^\alpha, g \in G, \alpha \in \mathbb{F}$. It is easy to verify that $(G, \mathbb{F})$ is indeed a vector space. As H_i is a subspace of $(G, \mathbb{F})$, we have $|H_i| \geq |\mathbb{F}|$. It follows that $s \geq |\mathbb{K}|$. There is an interesting corollary of $(G, \mathbb{F})$ being a vector space.

Theorem 3.4.2 ([128], 8.5.2) *The group G is elementary abelian, so s and t must be powers of the same prime.*

Proof. Let $|\mathbb{F}| = q$, so q is a prime power. As H_i and H_i^* may be viewed as subspaces of the vector space $(G, \mathbb{F})$, we have $s = |H_i| = q^n$ and $st = |H_i^*| = q^{n+m}$, hence $t = q^m$. ∎

Let $\mathcal{S}^{(\infty)}$ be a TGQ with translation group G. We keep using the notation of above. Since every element of $\mathbb{K}^0$ is an automorphism of G fixing the 4-gonal family $\mathcal{J}$ elementwise, it is straightforward to see that such an element induces a whorl of the TGQ about (∞). Moreover, each such element must fix the identity of G, so it induces an element stabilizing some fixed point y not collinear with the translation point. Let κ be any whorl about (∞), and σ any symmetry about a line through (∞). Then clearly σ^κ is also a symmetry about that line. As G is generated by the symmetries about the lines incident with (∞), it follows that G is a normal subgroup of the group of all whorls about (∞). Now consider any whorl φ about (∞) and y. Then $G^\varphi = G$, and so φ induces an automorphism of G that fixes $\mathcal{J}$ elementwise. Hence $\varphi \in \mathbb{K}$. We have proved the following theorem.

Theorem 3.4.3 *The multiplicative group $\mathbb{K}^0$ induces the group of all whorls about (∞) and y.* ∎

3.5 $\mathbf{T}(n, m, q)s$ and Translation Generalized Quadrangles

In this section, we introduce the notion of $\mathbf{T}(n, m, q)$, which is a natural generalization of the $\mathbf{T}_d(\mathcal{O})$ constructions of Tits, $d \in \{2, 3\}$.

Suppose $H = \mathbf{PG}(2n + m - 1, q)$ is the finite projective $(2n + m - 1)$-space over $\mathbf{GF}(q)$. Now define a set $\mathcal{O} = \mathcal{O}(n, m, q)$ of subspaces as follows: $\mathcal{O}$ is a set of $q^m + 1$ $(n-1)$-dimensional subspaces of H, denoted $\mathbf{PG}^{(i)}(n-1, q)$, and often also by π_i, so that

  (i) every three generate a $\mathbf{PG}(3n - 1, q)$;

  (ii) for every $i = 0, 1, \ldots, q^m$, there is a subspace $\mathbf{PG}^{(i)}(n + m - 1, q)$, also denoted by τ_i, of H of dimension $n + m - 1$, which contains $\mathbf{PG}^{(i)}(n-1, q)$ and which is disjoint from any $\mathbf{PG}^{(j)}(n-1, q)$ if $j \neq i$.

If $\mathcal{O}$ satisfies these conditions for $n = m$, then $\mathcal{O}$ is called a *pseudo-oval* or a *generalized oval* or an $[n - 1]$-*oval* of $\mathbf{PG}(3n - 1, q)$. A $[0]$-oval of $\mathbf{PG}(2, q)$ is just an oval of $\mathbf{PG}(2, q)$. For $n \neq m$, $\mathcal{O}(n, m, q)$ is called a *pseudo-ovoid* or a *generalized ovoid* or an $[n - 1]$-*ovoid* or an *egg* of $\mathbf{PG}(2n + m - 1, q)$. A $[0]$-ovoid of $\mathbf{PG}(3, q)$ is just an ovoid of $\mathbf{PG}(3, q)$.

The space $\mathbf{PG}^{(i)}(n + m - 1, q)$ is the *tangent space* of $\mathcal{O}(n, m, q)$ at $\mathbf{PG}^{(i)}(n - 1, q)$; it is uniquely determined by $\mathcal{O}(n, m, q)$ and $\mathbf{PG}^{(i)}(n - 1, q)$. Sometimes we will call an $\mathcal{O}(n, n, q)$ also an "egg" or a "generalized ovoid" for the sake of convenience.

From any egg $\mathcal{O}(n, m, q)$ arises a GQ $\mathbf{T}(n, m, q) = \mathbf{T}(\mathcal{O})$ which is a TGQ of order (q^n, q^m) for some base-point (∞). This goes as follows. Let H be embedded in a $\mathbf{PG}(2n + m, q) = H'$.

- The POINTS are of three types.

  (i) The points of $H' \setminus H$.
  (ii) The subspaces $\mathbf{PG}(n + m, q)$ of H' which intersect H in a $\mathbf{PG}^{(i)}(n + m - 1, q)$.
  (iii) A symbol (∞).

- The LINES are of two types.

  (a) The subspaces $\mathbf{PG}(n, q)$ of $\mathbf{PG}(2n + m, q)$ which intersect H in an element of the egg.
  (b) The elements of the egg $\mathcal{O}(n, m, q)$.

- INCIDENCE is defined as follows. The point (∞) is incident with all the lines of Type (b) and with no other lines. A point of Type (ii) is incident with the unique line of Type (b) contained in it and with all the lines of Type (a) contained in it. Finally, a point of Type (i) is incident with the lines of Type (a) containing it.

Conversely, any TGQ can be seen in this way, that is, as a $\mathbf{T}(n, m, q)$ associated to an $\mathcal{O}(n, m, q)$ in $\mathbf{PG}(2n + m - 1, q)$.

Theorem 3.5.1 ([128], 8.7.1) *The geometry* $\mathbf{T}(n, m, q)$ *is a TGQ of order* (q^n, q^m) *with translation point* (∞) *and for which* $\mathbf{GF}(q)$ *is a subfield of the kernel. Moreover, the translations of* $\mathbf{T}(n, m, q)$ *induce translations of the affine space* $\mathbf{AG}(2n+m, q) = \mathbf{PG}(2n+m, q) \backslash \mathbf{PG}(2n+m-1, q)$. *Conversely, every TGQ for which* $\mathbf{GF}(q)$ *is a subfield of the kernel is isomorphic to a* $\mathbf{T}(n, m, q)$. *It follows that the theory of TGQs is "equivalent" to the theory of the sets* $\mathcal{O}(n, m, q)$.

Proof. It is routine to show that $\mathbf{T}(n, m, q)$ is a GQ of order (q^n, q^m). A translation of $\mathbf{AG}(2n + m, q)$ defines in a natural way an elation about (∞) of $\mathbf{T}(n, m, q)$. It follows that $\mathbf{T}(n, m, q)$ is an EGQ with abelian elation group G, where G is isomorphic to the translation group of $\mathbf{AG}(2n+m, q)$, and hence $\mathbf{T}(n, m, q)$ is a TGQ with translation group G. (With the q^n translations of $\mathbf{AG}(2n + m, q)$ having center in $\mathbf{PG}^{(i)}(n - 1, q)$ correspond the q^n symmetries of $\mathbf{T}(n, m, q)$ about the line $\mathbf{PG}^{(i)}(n - 1, q)$ of Type (b).) It also follows that $\mathbf{GF}(q)$ is a subfield of the kernel of $\mathbf{T}(n, m, q)$: with the group of all homologies of $\mathbf{PG}(2n + m, q)$ having a center y not in $\mathbf{PG}(2n + m - 1, q)$ and axis $\mathbf{PG}(2n + m - 1, q)$ corresponds in a natural way the multiplicative group of a subfield of the kernel (recall Theorem 3.4.3).

Conversely, consider a TGQ $\mathcal{S}^{(p)}$ with translation group G for which $\mathbf{GF}(q) = \mathbb{F}$ is a subfield of the kernel. If $s = q^n$ and $t = q^m$, then $[(G, \mathbb{F}) : \mathbb{F}] = 2n + m$. Hence with $\mathcal{S}^{(p)}$ corresponds an affine space $\mathbf{AG}(2n + m, q)$. The cosets $H_i g$ of a fixed H_i are the elements of a parallel class of n-dimensional subspaces of $\mathbf{AG}(2n + m, q)$, and the cosets $H_i^* g$ of a fixed H_i^* are the elements of a parallel class of $(n + m)$-dimensional subspaces of $\mathbf{AG}(2n + m, q)$. The interpretation in $\mathbf{PG}(2n + m, q)$ together with (K1) and (K2) prove the last part of the theorem. ∎

Remark 3.5.2 We emphasize that with the group of all homologies of $\mathbf{PG}(2n + m, q)$ having a center y not in $\mathbf{PG}(2n + m - 1, q)$ and axis $\mathbf{PG}(2n + m - 1, q)$ corresponds in a natural way the multiplicative group of a subfield of the kernel.

Corollary 3.5.3 *If $\mathcal{S} \cong \mathbf{T}(\mathcal{O})$, with $\mathcal{O} = \mathcal{O}(n, m, q)$, then for fixed n, m, q, we have that $\mathcal{O}$ is uniquely defined by $\mathcal{S}$.*

Proof. Immediate from the interpretation of $\mathcal{O}$ in terms of $\mathcal{S}$. ∎

Remark 3.5.4 It is clear that $\mathbf{T}(1, 1, q)$ is a $\mathbf{T}_2(\mathcal{O})$ of Tits, and that $\mathbf{T}(1, 2, q)$ is a $\mathbf{T}_3(\mathcal{O})$ of Tits.

Corollary 3.5.5 *For any $\mathcal{O}(n, m, q)$ we have $n \leq m \leq 2n$.*

Proof. The GQ $\mathbf{T}(n, m, q)$ is a TGQ, and so, by Corollary 3.3.8, $t \geq s$, hence $m \geq n$. By the inequality of Higman we have that $t \leq s^2$, and so $m \leq 2n$. ∎

Corollary 3.5.6 *Let $\mathcal{S}^{(x)}$ be a TGQ of order (s, t), s a prime. Then either $\mathcal{S}^{(x)} \cong \mathbf{Q}(4, s)$ or $\mathcal{S}^{(x)} \cong \mathbf{Q}(5, s)$.*

Proof. Represent $\mathcal{S}^{(x)}$ as $\mathbf{T}(\mathcal{O})$, with $\mathcal{O} = \mathcal{O}(n, m, q)$ for some n, m, q. Then $n = 1$ and $q = p$, so that $\mathcal{O} = \mathcal{O}(1, m, p)$, and by Corollary 3.5.5, $m \in \{1, 2\}$. As each oval in $\mathbf{PG}(2, p)$, respectively ovoid in $\mathbf{PG}(3, p)$, is a conic, respectively an elliptic quadric (see, e.g., Section 1.1), the corollary follows. ∎

Theorem 3.5.7 *Let $(\mathcal{S}^{(x)}, G)$ be the* TGQ *arising from the generalized ovoid $\mathcal{O} = \mathcal{O}(n, m, q)$ with $m \in \{n, 2n\}$. Then $\mathcal{S}^{(x)} \cong \mathbf{T}_d(\mathcal{O})$ for some oval, respectively ovoid, $\mathcal{O}$ of $\mathbf{PG}(d, q^n)$, if and only if for a fixed point y, $y \not\sim x$, the group of all whorls about x fixing y has order $q^n - 1$, that is, the kernel has size q^n.*

Proof. By Theorem 3.4.3, the group of all whorls about x fixing y has order $q^n - 1$ if and only if the kernel has size q^n. If $\mathcal{S}^{(x)} \cong \mathbf{T}_d(\mathcal{O})$ for some oval, respectively ovoid, $\mathcal{O}$ of $\mathbf{PG}(d, q^n)$, then clearly the kernel has size q^n. Conversely, if the kernel has size q^n, then $\mathcal{S}^{(x)} \cong \mathbf{T}(1, m, q^n)$ and so $\mathcal{S}^{(x)}$ is isomorphic to a GQ of Tits. ∎

3.6 Regular Pseudo-Ovals and Regular Pseudo-Ovoids

In the extension $\mathbf{PG}(2n+m-1, q^n)$ of $\mathbf{PG}(2n+m-1, q)$, with $m \in \{n, 2n\}$, we consider n ($\frac{m}{n} + 1$)-dimensional spaces $\mathbf{PG}^{(i)}(\frac{m}{n} + 1, q^n) = \xi_i$, with $i = 1, 2, \ldots, n$, which are conjugate with respect to the extension $\mathbf{GF}(q^n)$ of $\mathbf{GF}(q)$, that is, which form an orbit of the Galois group corresponding to this extension, and which span $\mathbf{PG}(2n + m - 1, q^n)$. In ξ_1 we consider an oval $\mathcal{O}_1$ for $m = n$ and an ovoid $\mathcal{O}_1$ for $m = 2n$. Let $\mathcal{O}_1 = \{x_0^{(1)}, x_1^{(1)}, \ldots, x_{q^m}^{(1)}\}$. Further, let $x_i^{(1)}, x_i^{(2)}, \ldots, x_i^{(n)}$, with $i = 0, 1, \ldots, q^m$, be conjugate with respect to the extension $\mathbf{GF}(q^n)$ of $\mathbf{GF}(q)$. The points $x_i^{(1)}, x_i^{(2)}, \ldots, x_i^{(n)}$ define an $(n-1)$-dimensional space $\mathbf{PG}^{(i)}(n-1, q) = \pi_i$ over $\mathbf{GF}(q)$, with $i = 0, 1, \ldots, q^m$. Then $\mathcal{O} = \{\pi_0, \pi_1, \ldots, \pi_{q^m}\}$ is a generalized oval of $\mathbf{PG}(3n - 1, q)$ for $m = n$, and a generalized ovoid of $\mathbf{PG}(4n - 1, q)$ for $m = 2n$. Here, we speak of a *regular* or *elementary pseudo-oval*, respectively a *regular* or *elementary pseudo-ovoid*.

If $m = n$, then we will show that in the regular case $\mathbf{T}(n, n, q) \cong \mathbf{T}_2(\mathcal{O}_1)$; similarly, for $m = 2n$, we have $\mathbf{T}(n, 2n, q) \cong \mathbf{T}_3(\mathcal{O}_1)$. As this is not explicit in the monograph by Payne and Thas [128], and the reasonings are useful in other situations, we will sketch this in two ways.

With Coordinates

Firstly, let $\mathbf{GF}(q^n) = \{a_0 + a_1\alpha + a_2\alpha^2 + \cdots + a_{n-1}\alpha^{n-1} \| a_i \in \mathbf{GF}(q)$, with $i = 0, 1, \ldots, n - 1\}$. Consider the GQ $\mathbf{T}_2(\mathcal{O}_1)$ of Tits. Then $\mathbf{T}_2(\mathcal{O}_1)$ is constructed in a $\mathbf{PG}(3, q^n)$. Let (z_1, z_2, z_3) be any point of $\mathbf{AG}(3, q^n) =$

$\mathbf{PG}(3, q^n)\backslash\xi_1$, with $\mathcal{O}_1 \subseteq \xi_1$. Further, let

$$z_j = z_{j0} + z_{j1}\alpha + \cdots + z_{j,n-1}\alpha^{n-1},$$

with

$$z_{jl} \in \mathbf{GF}(q), j = 1, 2, 3, \text{ and } l = 0, 1, \ldots, n - 1.$$

Now we consider the bijection

$$\zeta : \mathbf{AG}(3, q^n) \to \mathbf{AG}(3n, q),$$
$$(z_1, z_2, z_3) \mapsto (z_{10}, \ldots, z_{1,n-1}, z_{20}, \ldots, z_{2,n-1}, z_{30}, \ldots, z_{3,n-1}).$$

Then ζ maps a line of $\mathbf{AG}(3, q^n)$ onto a subspace $\mathbf{AG}(n, q)$ of $\mathbf{AG}(3n, q)$, etc. Applying this ζ to $\mathbf{T}_2(\mathcal{O}_1)$ one sees that $\mathbf{T}_2(\mathcal{O}_1)$ is isomorphic to the regular $\mathbf{T}(n, n, q)$ arising from $\mathcal{O}_1$.

Geometrically

Secondly, consider the extension $\mathbf{PG}(4n - 1, q^n)$ of $\mathbf{PG}(4n - 1, q)$ and choose a set of n 3-dimensional spaces $\mathbf{PG}^{(i)}(3, q^n) = \xi'_i$, with $i = 1, 2, \ldots, n$, which are conjugate with respect to the extension $\mathbf{GF}(q^n)$ of $\mathbf{GF}(q)$, and which generate $\mathbf{PG}(4n - 1, q^n)$. If x_1 is any point of ξ'_1, then the n conjugate points $x_1, x_2, \ldots, x_n$, with $x_i \in \xi'_i$, define an $(n - 1)$-dimensional space over $\mathbf{GF}(q)$. In this way a bijection ζ' arises from ξ'_1 onto an $(n - 1)$-spread of $\mathbf{PG}(4n - 1, q)$. Choose a plane ξ_1 of ξ'_1, and let $\mathcal{O}_1$ be an oval in ξ_1. Then $\mathcal{O}_1^{\zeta'}$ is a regular $\mathcal{O}(n, n, q)$ which is contained in the $\mathbf{PG}(3n - 1, q)$ defined by $\xi_1, \xi_2, \ldots, \xi_n$, where $\xi_1, \xi_2, \ldots, \xi_n$ are conjugate. Now we consider a $\mathbf{PG}(3n, q)$, with $\mathbf{PG}(3n - 1, q) \subseteq \mathbf{PG}(3n, q) \subseteq \mathbf{PG}(4n - 1, q)$. Each $x_1^{\zeta'}$, with $x_1 \in \xi'_1 \backslash \xi_1$, intersects $\mathbf{PG}(3n, q)$ in a point. In such a way a bijection $\bar{\zeta}'$ of $\xi'_1\backslash\xi_1$ onto $\mathbf{PG}(3n, q)\backslash\mathbf{PG}(3n - 1, q)$ arises. With a line of $\mathbf{AG}(3, q^n) = \xi'_1\backslash\xi_1$ defining the point x_1 of $\mathcal{O}_1$, corresponds under $\bar{\zeta}'$ an affine n-space over $\mathbf{GF}(q)$ defining the element $x_1^{\zeta'}$ of $\mathcal{O}(n, n, q)$. Now it becomes clear that the GQ $\mathbf{T}_2(\mathcal{O}_1)$ constructed in ξ'_1 is isomorphic to the GQ $\mathbf{T}(n, n, q)$ constructed in $\mathbf{PG}(3n, q)$.

For $m = n$, any known $[n - 1]$-oval is regular, for $m = 2n$ and q even any known $[n - 1]$-ovoid is regular, but for $m = 2n$ and q odd there are $[n - 1]$-ovoids which are not regular, which we will encounter later. A regular pseudo-oval for which $\mathcal{O}_1$ is a conic is called either a *pseudo-conic* or a *classical pseudo-oval*, and a regular pseudo-ovoid for which $\mathcal{O}_1$ is an elliptic quadric is called a *pseudo-quadric* or a *classical pseudo-ovoid* or a *classical egg.* If $\mathcal{O}$ is a pseudo-conic, then, by Theorem 1.5.2(i) we have $\mathbf{T}(\mathcal{O}) \cong \mathbf{T}_2(\mathcal{O}_1) \cong \mathbf{Q}(4, q^n)$; if $\mathcal{O}$ is a pseudo-quadric, then, by Theorem 1.5.2(ii), we have $\mathbf{T}(\mathcal{O}) \cong \mathbf{T}_3(\mathcal{O}_1) \cong \mathbf{Q}(5, q^n)$. For q odd, by the theorems of Segre [137], Barlotti [8] and Panella [106], we have $\mathbf{T}(\mathcal{O}) \cong \mathbf{Q}(4, q^n)$ for any regular pseudo-oval $\mathcal{O}$ and $\mathbf{T}(\mathcal{O}) \cong \mathbf{Q}(5, q^n)$ for any regular pseudo-ovoid $\mathcal{O}$.

3.7 Automorphisms of Translation Generalized Quadrangles

As any translation generalized quadrangle $\mathcal{S}$ can be represented in a projective space $\mathbf{PG}(n, q)$ for suitable n and q, a natural question is whether each automorphism of $\mathcal{S}$ which fixes a translation point is induced by an element of $\mathbf{P\Gamma L}_{n+1}(q)$. This problem was solved by Bader, Lunardon and Pinneri [4] with a general argument holding for coset geometries. In [191] the authors gave an alternative proof which led in its turn to an interesting generalization.

Lemma 3.7.1 *If G is the translation group of the TGQ $\mathcal{S}^{(x)}$, then G is a normal subgroup of the group of all automorphisms of $\mathcal{S}$ fixing x.*

Proof.   This follows immediately from Theorem 3.3.10 and Theorem 3.3.14. ∎

Theorem 3.7.2 *Suppose $\mathcal{S} = \mathbf{T}(\mathcal{O})$ is a TGQ of order (q^n, q^m) with translation point (∞), and let $\mathbf{GF}(q)$ be a subfield of the kernel $\mathbf{GF}(q')$ of $\mathbf{T}(\mathcal{O})$, where $\mathcal{O}$ either is a generalized ovoid $(n \neq m)$ or a generalized oval $(n = m)$ in $\mathbf{PG}(2n + m - 1, q) \subseteq \mathbf{PG}(2n + m, q)$. Then every automorphism of $\mathcal{S}$ which fixes (∞) is induced by an automorphism of $\mathbf{PG}(2n+m, q)$ which fixes $\mathcal{O}$, and conversely.*

Proof.   Consider a point x in $\mathbf{PG}(2n + m, q) \setminus \mathbf{PG}(2n + m - 1, q)$, and suppose that ϕ is an automorphism of $\mathbf{T}(\mathcal{O})$ which fixes the points (∞) and x. Denote by $\mathcal{C} = \mathcal{C}((\infty), x)$ the group of automorphisms of $\mathbf{T}(\mathcal{O})$ which fix (∞) and x linewise. Then by Theorem 3.4.3 $\mathcal{C}$ is isomorphic to the multiplicative group of $\mathbf{GF}(q')$, where $\mathbf{GF}(q')$ is the kernel of $\mathbf{T}(\mathcal{O})$. Let $\mathcal{C}'$ be the subgroup of $\mathcal{C}$ induced by the perspectivities of $\mathbf{PG}(2n + m, q)$ with axis $\mathbf{PG}(2n + m - 1, q)$ and center x. Take a nontrivial $\sigma \in \mathcal{C}'$. Then

$\phi\sigma\phi^{-1}$ is an element of $\mathrm{Aut}(\mathcal{S})$ which fixes (∞) and x linewise, so belongs to $\mathcal{C}$. As $\mathcal{C}'$ is a subgroup of the cyclic group $\mathcal{C}$ and $\sigma \in \mathcal{C}'$, we have that $\sigma' = \phi\sigma\phi^{-1} \in \mathcal{C}'$.

First suppose $q \neq 2$. Consider a line L through x in $\mathbf{PG}(2n + m, q)$, and suppose y and z are points of L, both not in $\mathbf{PG}(2n + m - 1, q)$, where x, y and z are distinct points, and so that $z = y^{\sigma'} = y^{\phi\sigma\phi^{-1}}$. As $z^{\phi} = (y^{\phi})^{\sigma}$ and σ fixes affine lines of $\mathbf{AG}(2n+m, q) = \mathbf{PG}(2n+m, q) \setminus \mathbf{PG}(2n+m-1, q)$ through x, the points x, y^{ϕ} and z^{ϕ} are on the same line of $\mathbf{AG}(2n+m, q)$, and hence ϕ maps affine lines of $\mathbf{AG}(2n + m, q)$ through x onto affine lines (through x). Now suppose θ' is an arbitrary nontrivial element of the translation group G of $\mathbf{T}(\mathcal{O})$, and note that by Theorem 3.5.1, every element of G is induced by an elation of $\mathbf{PG}(2n+m, q)$ with axis $\mathbf{PG}(2n+m-1, q)$. Then $\phi^{-1}\theta'\phi = \theta$ is also an element of the translation group; see Lemma 3.7.1. Consider an arbitrary affine line M of $\mathbf{AG}(2n+m, q)$ not through x, and suppose that $\theta' \in G$ is so that M is mapped onto some affine line of $\mathbf{AG}(2n + m, q)$ through x. Then $M^{\phi} = M^{\theta'\phi\theta^{-1}}$, and hence M^{ϕ} is also an affine line of $\mathbf{AG}(2n+m, q)$. Thus, any element ϕ of $\mathrm{Aut}(\mathcal{S})$ which fixes (∞) and some point x of $\mathbf{PG}(2n+m, q) \setminus \mathbf{PG}(2n+m-1, q)$ induces an element of the stabilizer of $\mathbf{PG}(2n + m - 1, q)$ in $\mathbf{P\Gamma L}_{2n+m+1}(q)$ which fixes $\mathcal{O}$. Since G is a normal subgroup of $\mathrm{Aut}(\mathcal{S})_{(\infty)}$, we have $\mathrm{Aut}(\mathcal{S})_{(\infty)} = GH = HG$, where H is the stabilizer in $\mathrm{Aut}(\mathcal{S})_{(\infty)}$ of an arbitrary point in $\mathcal{P}\setminus(\infty)^{\perp}$ ($\mathcal{P}$ being the point set of $\mathcal{S}$). Since each element of G, respectively of H, maps affine lines of $\mathbf{AG}(2n+m, q)$ onto affine lines of $\mathbf{AG}(2n+m, q)$, the theorem follows.

Now assume $q = 2$. Suppose that ϕ is an automorphism of $\mathbf{T}(\mathcal{O})$ which fixes (∞). Let L and M be lines of $\mathbf{PG}(2n + m, 2)$ not contained in $\mathbf{PG}(2n + m - 1, 2)$, but containing a common point $u \in \mathbf{PG}(2n + m - 1, 2)$. Let θ be the translation of $\mathbf{T}(\mathcal{O})$ defined by $l_1^{\theta} = l_2$, with l_1, l_2 the points of L not in $\mathbf{PG}(2n + m - 1, 2)$. Then θ is induced by a translation of $\mathbf{AG}(2n + m, 2) = \mathbf{PG}(2n + m, 2) \setminus \mathbf{PG}(2n + m - 1, 2)$. If m_1, m_2 are the points of M not in $\mathbf{PG}(2n + m - 1, 2)$, then clearly $m_1^{\theta} = m_2$. By Lemma 3.7.1, the automorphism $\phi^{-1}\theta\phi = \theta'$ is a translation of $\mathbf{T}(\mathcal{O})$. We have $(l_1^{\phi})^{\theta'} = l_2^{\phi}$ and $(m_1^{\phi})^{\theta'} = m_2^{\phi}$. Hence the lines $l_1^{\phi}l_2^{\phi}$ and $m_1^{\phi}m_2^{\phi}$ of $\mathbf{PG}(2n + m, 2)$ contain a common point u' of $\mathbf{PG}(2n+m-1, 2)$. If we put $u' = u^{\phi}$, then ϕ defines a bijection of $\mathbf{PG}(2n + m, 2)$ onto itself, such that lines of $\mathbf{PG}(2n + m, 2)$ not contained in $\mathbf{PG}(2n + m - 1, 2)$ are mapped onto lines. It easily follows that also the lines of $\mathbf{PG}(2n + m - 1, 2)$ are mapped onto lines. So ϕ is induced by an automorphism of $\mathbf{PG}(2n + m, 2)$ which fixes $\mathcal{O}$. ∎

Theorem 3.7.2 is also contained in O'Keefe and Penttila [92], for the special case of the $\mathbf{T}_d(\mathcal{O})$ of Tits, $d \in \{2,3\}$.

In a similar way as Theorem 3.7.2, one proves the following theorem.

Theorem 3.7.3 *Suppose $\mathcal{S}_i = \mathbf{T}(\mathcal{O}_i)$ is a TGQ of order (q^n, q^m) with translation point $(\infty)_i$ and $\mathbf{GF}(q)$ a subfield of the kernel, where $\mathcal{O}_i$ either is a generalized ovoid $(n \neq m)$ or a generalized oval $(n = m)$ in $\mathbf{PG}^{(i)}(2n+m-1, q) \subseteq \mathbf{PG}^{(i)}(2n+m, q)$, with $i = 1, 2$. Then every isomorphism of $\mathcal{S}_1$ onto $\mathcal{S}_2$ which maps $(\infty)_1$ onto $(\infty)_2$ is induced by an isomorphism of $\mathbf{PG}^{(1)}(2n + m, q)$ onto $\mathbf{PG}^{(2)}(2n + m, q)$ which maps $\mathcal{O}_1$ onto $\mathcal{O}_2$, and conversely.* ∎

The following theorem is a direct corollary.

Theorem 3.7.4 *If $\mathbf{T}(\mathcal{O})$ is a TGQ for which $\mathbf{T}(\mathcal{O}) \cong \mathbf{T}_d(\mathcal{O}')$, where either $d = 2$ and $\mathcal{O}'$ is an oval of $\mathbf{PG}(2, q^n)$, or $d = 3$ and $\mathcal{O}'$ is an ovoid of $\mathbf{PG}(3, q^n)$, then $\mathcal{O}$ is regular.*

Proof.   Easily derived from Theorem 3.7.3 and Section 3.6.   ∎

Let $\Gamma = (\mathcal{P}, \mathcal{B}, \mathbf{I})$ be a point-line incidence geometry. A *generalized linear representation* of Γ in $\mathbf{AG}(n, q)$ is defined similarly as the usual linear representation of Γ in $\mathbf{AG}(n, q)$, but where we allow subspaces of $\mathbf{AG}(n, q)$ instead of lines as in a usual linear representation. Thus, it is a monomorphism θ of Γ into the geometry of points and subspaces of the affine space $\mathbf{AG}(n, q)$, in such a way that $\mathcal{P}^\theta$ is the set of all points of $\mathbf{AG}(n, q)$, that $\mathcal{B}^\theta$ is a union of parallel classes of subspaces (not necessarily of the same dimension) of $\mathbf{AG}(n, q)$, and that each point of L^θ is the image of some point of L for any line L in $\mathcal{B}$. We usually identify Γ with its image Γ^θ.

The following general theorem has a proof completely similar to the proof of Theorem 3.7.2.

Theorem 3.7.5 *Let Γ be a point-line incidence geometry having a generalized linear representation θ in $\mathbf{AG}(n, q)$, $n > 1$. Assume that G is a subgroup of $\mathrm{Aut}(\Gamma)$ so that the group of translations of $\mathbf{AG}(n, q)$ is a normal subgroup of G, and so that the group of homologies of the projective completion $\mathbf{PG}(n, q)$ of $\mathbf{AG}(n, q)$ with axis $\mathbf{PG}(n - 1, q) = \mathbf{PG}(n, q) \setminus \mathbf{AG}(n, q)$ and center $x \in$*

$\mathbf{AG}(n,q)$ *is a normal subgroup of* G_x. *Then* G *is induced by a subgroup of* $\mathbf{A\Gamma L}(n+1,q)$ *fixing the set of all spaces at infinity of the elements of* Γ^θ, *and vice versa.*   ■

In particular, Theorem 3.7.5 applies to translation planes, see Section 1 of Kallaher [80] (Chapter 5 of [31]).

3.8   Important Properties of $\mathcal{O}(n,m,q)$

In this section we will deduce some important properties of sets of type $\mathcal{O}(n,m,q)$. The following theorem is due to Payne and Thas [128], but now we give the elementary proof contained in Payne and Thas [127], in particular avoiding 1.9.1 in their monograph, see Theorem 3.3.1. Also the corollary is due to Payne and Thas [128].

Theorem 3.8.1 *The following hold for any* $\mathcal{O}(n,m,q)$.

(i) *Each hyperplane of* $\mathbf{PG}(2n+m-1,q)$ *which does not contain a tangent space of* $\mathcal{O}(n,m,q)$, *contains either* 0 *or* $1+q^{m-n}$ *elements of* $\mathcal{O}(n,m,q)$. *If* $m=2n$, *then each hyperplane of* $\mathbf{PG}(4n-1,q)$ *which does not contain a tangent space of* $\mathcal{O}(n,2n,q)$ *contains exactly* $1+q^n$ *elements of* $\mathcal{O}(n,2n,q)$. *If* $m \neq 2n$, *then there are hyperplanes which contain no element of* $\mathcal{O}(n,m,q)$.

(ii) *Either* $n=m$ *or* $n(a+1)=ma$ *with* $a \in \mathbb{N}_0$ *and* a *odd.*

Proof. (i) Let π be a hyperplane of $\mathbf{PG}(2n+m-1,q)$ which does not contain a tangent space of $\mathcal{O}(n,m,q)$, but which contains at least one element of $\mathcal{O}(n,m,q) = \{\pi_0, \pi_1, \ldots, \pi_{q^m}\}$, say π_0. If τ_i is the tangent space of $\mathcal{O}(n,m,q)$ at π_i, with $i = 0, 1, \ldots, q^m$, then $\tau_0 \cap \pi = \mathbf{PG}(n+m-2,q)$. Let β be the number of elements of $\mathcal{O}(n,m,q)$ in π. Count the number of n-dimensional spaces in π which contain π_0 and a point of $\widetilde{\mathcal{O}} \backslash \pi_0$, with $\widetilde{\mathcal{O}} = \pi_0 \cup \pi_1 \cup \cdots \cup \pi_{q^m}$. One obtains

$$(\beta - 1)\frac{q^n - 1}{q - 1} + (q^m + 1 - \beta)\frac{q^{n-1} - 1}{q - 1}.$$

These n-dimensional spaces are exactly the n-dimensional spaces in π which contain π_0 and are not contained in $\mathbf{PG}(n+m-2,q)$. So this number also equals

$$\frac{q^{m+n-1}-1}{q-1} - \frac{q^{m-1}-1}{q-1}.$$

It follows that $\beta = 1 + q^{m-n}$.

Now let $m = 2n$. Let γ be the number of hyperplanes of $\mathbf{PG}(4n-1,q)$ containing just one element of $\mathcal{O}(n,2n,q)$, and let δ be the number of hyperplanes of $\mathbf{PG}(4n-1,q)$ containing at least two (and so exactly q^n+1) elements of $\mathcal{O}(n,2n,q)$. Then we have

$$\gamma = (q^{2n}+1)\frac{q^n-1}{q-1} \text{ and } \delta = \frac{(q^{2n}+1)q^{2n}}{(q^n+1)q^n}\cdot\frac{q^{2n}-1}{q-1} = (q^{2n}+1)q^n\frac{q^n-1}{q-1}.$$

Hence $\gamma + \delta = (q^{4n}-1)/(q-1)$, that is, the number of hyperplanes in $\mathbf{PG}(4n-1,q)$. So there is no hyperplane in $\mathbf{PG}(4n-1,q)$ containing no element of $\mathcal{O}(n,2n,q)$.

Assume that $m \neq 2n$. Counting the number of hyperplanes of $\mathbf{PG}(2n+m-1,q)$ which contain at least one element of $\mathcal{O}(n,m,q)$, one obtains a number less than $(q^{2n+m}-1)/(q-1)$, that is, the total number of hyperplanes in $\mathbf{PG}(2n+m-1,q)$.

(ii) Let $n \neq m$. For $n = 1$ we have $m = 2$, by Corollary 3.5.5. Assume $n \neq 1$. Let π' be a hyperplane of $\pi_0 \in \mathcal{O}(n,m,q)$. Now we count the number of hyperplanes of $\mathbf{PG}(2n+m-1,q)$ containing π', not containing π_0, and containing $q^{m-n}+1$ elements of $\mathcal{O}(n,m,q)$. This number is equal to

$$q^m(q^m-1)/(q^{m-n}+1).$$

So $q^{m-n}+1$ divides q^m-1. Hence $m = 2l(m-n)$ with $l \in \mathbb{N}_0$, so $ma = n(a+1)$ with $a \in \mathbb{N}_0$ and a odd. ∎

Corollary 3.8.2 *Let $\widetilde{\mathcal{O}}$ be the union of all elements of any $\mathcal{O}(n,2n,q)$ and let π be any hyperplane of $\mathbf{PG}(4n-1,q)$. Then $|\widetilde{\mathcal{O}} \cap \pi| \in \{\gamma_1, \gamma_2\}$, with $\gamma_1 = (q^n-1)(q^{2n-1}+1)/(q-1)$ and $\gamma_1 - \gamma_2 = q^{2n-1}$. Hence $\widetilde{\mathcal{O}}$ defines a linear projective two-weight code and a strongly regular graph.*

Proof. It follows directly from Theorem 3.8.1(i) that $\widetilde{\mathcal{O}}$ has two intersection numbers with respect to hyperplanes, and so $\widetilde{\mathcal{O}}$ defines a linear projective two-weight code and a strongly regular graph; see Calderbank and Kantor [34]. ∎

Corollary 3.8.3 *Let S be a* TGQ *of order* (s,t). *If all lines are regular, then either* $S \cong \mathbf{Q}(4,s)$ *or* $t = s^2$.

Proof. By Theorems 3.8.1, 2.2.2 and 1.4.1, we have either $S \cong \mathbf{Q}(4,s)$ or $s = q^n, t = q^m, ma = n(a+1)$ with $a \in \mathbb{N}_0$ and a odd. By Theorem 2.1.1 we have that $2(q^n + 1)$ divides $q^{2m}(q^{2m} - 1)$. So $q^n + 1$ divides $q^{2m} - 1$, hence divides $q^{2(n+(n/a))} - 1$. Consequently $q^n + 1$ divides $q^{2n/a} - 1$, implying $n < 2n/a$, from which $a = 1$, that is, $t = s^2$. $\blacksquare$

3.9 Pseudo-Ovals

In this section we mention important results known about pseudo-ovals. The next theorem is taken from Payne and Thas [128]; Part (i) is already contained in Thas [156] and Part (ii) in Payne and Thas [126].

Theorem 3.9.1 *Let $\mathcal{O} = \mathcal{O}(n,n,q)$ be a pseudo-oval in* $\mathbf{PG}(3n - 1, q)$.

(i) *If q is even, then all tangent spaces of $\mathcal{O}$ contain a common $(n - 1)$-dimensional space, called the* kernel *or* nucleus *of $\mathcal{O}$.*

(ii) *If q is odd, then each point of $\mathbf{PG}(3n - 1, q)$ not in an element of $\mathcal{O}$ is contained in either 0 or 2 tangent spaces of $\mathcal{O}$, and each hyperplane of $\mathbf{PG}(3n - 1, q)$ not containing a tangent space of $\mathcal{O}$ contains either 0 or 2 elements of $\mathcal{O}$.*

Proof. (i) Let q be even. By Corollary 3.3.8 the point (∞) of the TGQ $\mathbf{T}(\mathcal{O})$ is coregular, and by Corollary 3.3.9 the point (∞) is regular. So if x and y are points of the GQ with $(\infty) \not\sim x \not\sim y \not\sim (\infty)$, then $|\{x, y, (\infty)\}^\perp| \in \{1, q^n + 1\}$. It follows that each point of $\mathbf{PG}(3n - 1, q)$ not in an element of $\mathcal{O}$ is contained in either 1 or $q^n + 1$ tangent spaces of $\mathcal{O}$. So the $(n - 1)$-dimensional space common to two tangent spaces, is common to all tangent spaces.

(ii) Let q be odd. By Corollary 3.3.8 the point (∞) of the TGQ $\mathbf{T}(\mathcal{O})$ is coregular, and by Corollary 3.3.9 the point (∞) is antiregular. So for any triad $\{(\infty), x, y\}$ we have $|\{(\infty), x, y\}^\perp| \in \{0, 2\}$. It follows that any point of $\mathbf{PG}(3n-1, q)$ not in an element of $\mathcal{O}$ is contained in either 0 or 2 tangent spaces of $\mathcal{O}$. It easily follows that in the dual space of $\mathbf{PG}(3n - 1, q)$ the $q^n + 1$ tangent spaces of $\mathcal{O}$ form a pseudo-oval $\mathcal{O}^* = \mathcal{O}^*(n, n, q)$. The elements of $\mathcal{O}$ are the tangent spaces of $\mathcal{O}^*$. By the foregoing, any point of the dual space not in an element of $\mathcal{O}^*$ is contained in either 0 or 2 tangent spaces of $\mathcal{O}^*$, that is, each hyperplane of $\mathbf{PG}(3n - 1, q)$ not containing a tangent space of $\mathcal{O}$ contains either 0 or 2 elements of $\mathcal{O}$. ∎

Remark 3.9.2 The second part in (ii) of Theorem 3.9.1 is also included in Theorem 3.8.1(i).

Let q be odd. In the proof of Theorem 3.9.1(ii) we showed that the tangent spaces of a pseudo-oval $\mathcal{O} = \mathcal{O}(n, n, q)$ are the elements of a pseudo-oval $\mathcal{O}^* = \mathcal{O}^*(n, n, q)$ in the dual space of $\mathbf{PG}(3n - 1, q)$.

Definition 3.9.3 *The pseudo-oval $\mathcal{O}^*$ is called the* translation dual *of the pseudo-oval $\mathcal{O}$. The TGQ $\mathbf{T}(\mathcal{O}^*) = \mathbf{T}^*(\mathcal{O})$ is also called the* translation dual *of the TGQ $\mathbf{T}(\mathcal{O})$.*

Remark 3.9.4 The translation dual $\mathcal{S}^*$ of the TGQ $\mathcal{S}$ is (up to isomorphism) uniquely defined by $\mathcal{S}$, that is, is independent of the chosen subfield of the kernel to construct $\mathcal{O}(n, m, q)$.

It is not known whether or not we have always $\mathcal{O} \cong \mathcal{O}^*$, respectively $\mathbf{T}(\mathcal{O}) \cong \mathbf{T}^*(\mathcal{O})$. In any case, for each known $\mathbf{T}(\mathcal{O})$, with $\mathcal{O} = \mathcal{O}(n, n, q)$ and q odd, we have $\mathbf{T}(\mathcal{O}) \cong \mathbf{Q}(4, q^n)$. The following result is taken from Thas and K. Thas [191].

Theorem 3.9.5 *The TGQ $\mathcal{S}$ of order s, with s odd, and its translation dual $\mathcal{S}^*$ have isomorphic kernels.*

Proof. Let $\mathcal{S} \cong \mathbf{T}(\mathcal{O})$, with $\mathcal{O} = \mathcal{O}(n, n, q)$ be a pseudo-oval in $\mathbf{PG}(3n - 1, q)$, and asssume that $\mathbf{GF}(q)$ is the kernel of $\mathcal{S}$. Then $\mathcal{O}^* = \mathcal{O}^*(n, n, q)$ is also a pseudo-oval in a $\mathbf{PG}(3n - 1, q)$, and so, by Theorem 3.5.1, $\mathbf{GF}(q)$ is a subfield of the kernel of $\mathcal{S}^*$. Interchanging roles of $\mathcal{S}$ and $\mathcal{S}^*$ yields the result. ∎

Let $\mathcal{O}(n, n, q) = \{\pi_0, \pi_1, \ldots, \pi_{q^n}\}$ be a pseudo-oval in $\mathbf{PG}(3n - 1, q)$. The tangent space of $\mathcal{O}(n, n, q)$ at π_i will be denoted by τ_i, with $i = 0, 1, \ldots, q^n$. Choose $\pi_i, i \in \{0, 1, \ldots, q^n\}$, and let $\mathbf{PG}(2n - 1, q) \subseteq \mathbf{PG}(3n - 1, q)$ be skew to π_i. Further, let $\tau_i \cap \mathbf{PG}(2n - 1, q) = \xi_i$ and $\langle \pi_i, \pi_j \rangle \cap \mathbf{PG}(2n - 1, q) = \xi_j$, with $j \neq i$. Then $\{\xi_0, \xi_1, \ldots, \xi_{q^n}\} = \mathcal{T}_i$ is an $(n - 1)$-spread of $\mathbf{PG}(2n - 1, q)$.

Now let q be even and let η be the nucleus of $\mathcal{O}(n, n, q)$. Let $\mathbf{PG}(2n - 1, q) \subseteq \mathbf{PG}(3n - 1, q)$ be skew to η. If $\zeta_j = \mathbf{PG}(2n - 1, q) \cap \langle \eta, \pi_j \rangle$, then $\{\zeta_0, \zeta_1, \ldots, \zeta_{q^n}\} = \mathcal{T}$ is an $(n - 1)$-spread of $\mathbf{PG}(2n - 1, q)$.

Next, let q be odd. Choose $\tau_i, i \in \{0, 1, \ldots, q^n\}$. If $\tau_i \cap \tau_j = \delta_j$, with $j \neq i$, then, by Theorem 3.9.1(ii), $\{\delta_0, \delta_1, \ldots, \delta_{i-1}, \pi_i, \delta_{i+1}, \ldots, \delta_{q^n}\} = \mathcal{T}_i^*$ is an $(n - 1)$-spread of τ_i.

The following characterization theorem, for q odd, is due to Casse, Thas and Wild [37].

Theorem 3.9.6 *Consider a pseudo-oval $\mathcal{O}(n, n, q)$, with q odd. Then at least one of the $(n - 1)$-spreads $\mathcal{T}_0, \mathcal{T}_1, \ldots, \mathcal{T}_{q^n}, \mathcal{T}_0^*, \mathcal{T}_1^*, \ldots, \mathcal{T}_{q^n}^*$ is regular if and only if they all are regular if and only if the corresponding GQ $\mathbf{T}(n, n, q)$ is isomorphic to the classical GQ $\mathbf{Q}(4, q^n)$.*

Proof. If $\mathbf{T}(n, n, q) \cong \mathbf{Q}(4, q^n)$, then all planes defined by the regular and antiregular elements of $\mathbf{Q}(4, q^n)$ are Desarguesian. It immediately follows that all $(n - 1)$-spreads $\mathcal{T}_i$ and $\mathcal{T}_i^*$, with $i = 0, 1, \ldots, q^n$, are regular.

Now assume that $\mathcal{T}_i^*$, with $i \in \{0, 1, \ldots, q^n\}$, is regular. By Corolllary 3.3.9 the point (∞) of $\mathbf{T}(n, n, q)$ is antiregular. Let φ be a $2n$-dimensional subspace of the projective space $\mathbf{PG}(3n, q)$ containing $\mathbf{T}(n, n, q)$, where φ contains τ_i but is not contained in the hyperplane $\mathbf{PG}(3n - 1, q)$ containing $\mathcal{O}(n, n, q)$. Then the projective completion of the affine plane $\pi((\infty), \varphi)$, defined in Theorem 2.4.1, is the dual of the projective plane of order q^n defined by the $(n - 1)$-spread $\mathcal{T}_i^*$. As $\mathcal{T}_i^*$ is regular, these planes are Desarguesian. Now, by Theorem 2.4.4, $\mathbf{T}(n, n, q) \cong \mathbf{Q}(4, q^n)$ and so all $(n - 1)$-spreads $\mathcal{T}_j$ and $\mathcal{T}_j^*$, with $j = 0, 1, \ldots, q^n$, are regular.

Next, assume that $\mathcal{T}_i$ is regular, with $i \in \{0, 1, \ldots, q^n\}$. Let $\mathcal{O}^*(n, n, q)$ be the translation dual of $\mathcal{O}(n, n, q)$. By applying the reasoning of the foregoing section to $\mathcal{O}^*(n, n, q)$ and $\mathcal{T}_i$, it follows that all $(n - 1)$-spreads $\mathcal{T}_j$ and $\mathcal{T}_j^*$, with $j = 0, 1, \ldots, q^n$, are regular, and so $\mathbf{T}(n, n, q) \cong \mathbf{Q}(4, q^n)$. ■

We end this section on pseudo-ovals with a characterization by Thas and K. Thas [192] of the GQ $\mathbf{T}_2(\mathcal{O})$ of Tits, for q even.

Theorem 3.9.7 *Let $\mathcal{S} = \mathbf{T}(n, n, q) = \mathbf{T}(\mathcal{O})$, with $\mathcal{O} = \mathcal{O}(n, n, q)$ and q even. Let η be the nucleus of $\mathcal{O} = \{\pi_0, \pi_1, \ldots, \pi_{q^n}\}$ and let $\xi_{ijk} = \tau_i \cap \langle \pi_j, \pi_k \rangle$, with $\tau_i = \langle \eta, \pi_i \rangle$ the tangent space of $\mathcal{O}$ at π_i, with $j \neq k$ and $i, j, k = 0, 1, \ldots, q^n$. Further, let $\widetilde{\mathcal{T}}_{ij} = \{\xi_{ijk} \| k = 0, 1, \ldots, j-1, j+1, \ldots, q^n\} \cup \{\eta\}$ with $i \neq j$ and $i, j = 0, 1, \ldots, q^n$ (so $\widetilde{\mathcal{T}}_{ij}$ is the $(n-1)$-spread $\mathcal{T}_j$ of the previous section, where*

$\mathbf{PG}(2n-1,q) = \tau_i$). *If for some i and at least two distinct j and j' we have* $\widetilde{T}_{ij} = \widetilde{T}_{ij'}$ *with* $\widetilde{T}_{ij}$ *regular, then* $\mathcal{S} \cong \mathbf{T}_2(\mathcal{O}')$ *for some oval $\mathcal{O}'$ of $\mathbf{PG}(2,s)$ with $s = q^n$. Conversely, if $\mathcal{S} \cong \mathbf{T}_2(\mathcal{O}')$ for some oval $\mathcal{O}'$ of $\mathbf{PG}(2,s)$, then* $\widetilde{T}_{ij} = \widetilde{T}_{ij'}$ *for all i,j,j', with $i \neq j \neq j' \neq i$ and all $\widetilde{T}_{ij}$ are regular.*

Proof. If $\mathcal{S} \cong \mathbf{T}_2(\mathcal{O}')$ for some oval $\mathcal{O}'$ of $\mathbf{PG}(2,s)$, then by §3.6 and Corollary 3.5.3, $\mathcal{O}(n,n,q)$ is regular and so $\widetilde{T}_{ij} = \widetilde{T}_{ij'}$ for all i,j,j', with $i \neq j \neq j' \neq i$, and all $\widetilde{T}_{ij}$ are regular.

Conversely, suppose that for some i and at least two distinct j and j' we have $\widetilde{T}_{ij} = \widetilde{T}_{ij'}$ with $\widetilde{T}_{ij}$ regular. First of all, as $\widetilde{T}_{ij}$ is regular, there are precisely n lines over $\mathbf{GF}(q^n)$ which intersect the extensions of the spread elements of $\widetilde{T}_{ij}$ in precisely one point. Fix such a line L^*. Let M^* be the unique line over $\mathbf{GF}(q^n)$ through the point of L^* in the extension of $\xi_{ijj'} = \langle \pi_j, \pi_{j'} \rangle \cap \tau_i$ to $\mathbf{GF}(q^n)$, which intersects the extensions of π_j and $\pi_{j'}$. Let Γ^* be the Desarguesian projective plane over $\mathbf{GF}(q^n)$ which is generated by L^* and M^*. Now let π_k be an element of $\mathcal{O}$, with $i \neq k \neq j$ and $k \neq j'$. Then ξ_{ijk} and $\xi_{ij'k}$ are elements of $\widetilde{T}_{ij} = \widetilde{T}_{ij'}$, distinct from π_i, η and $\xi_{ijj'}$. Let U^* be the line of Γ^* which is incident with the point of L^* in the extension of ξ_{ijk} to $\mathbf{GF}(q^n)$, and with the point of M^* in the extension of π_j to $\mathbf{GF}(q^n)$. Let V^* be the line of Γ^* which is incident with the point of L^* in the extension of $\xi_{ij'k}$ to $\mathbf{GF}(q^n)$, and with the point of M^* in the extension of $\pi_{j'}$ to $\mathbf{GF}(q^n)$. Then over $\mathbf{GF}(q^n)$, $U^* \cap V^*$ is a point of the extension of $\langle \pi_j, \pi_k \rangle \cap \langle \pi_{j'}, \pi_k \rangle = \pi_k$ to $\mathbf{GF}(q^n)$. We have shown that over $\mathbf{GF}(q^n)$, each element of $\mathcal{O}$ meets Γ^*. Hence over $\mathbf{GF}(q^n)$, there are n planes which all meet each element of $\mathcal{O}$, implying that $\mathcal{O}$ is regular. So by Section 3.6 we have $\mathcal{S} \cong \mathbf{T}_2(\mathcal{O}')$ for some oval of $\mathbf{PG}(2,q^n)$. ∎

3.10   Eggs

In this section we prove fundamental results on eggs. Also, some interesting and useful characterizations of regular eggs will be given.

Theorem 3.10.1 *Let $\mathcal{O}(n,m,q)$ be an egg, so $n \neq m$.*

(i) *Each point of $\mathbf{PG}(2n+m-1,q)$ which is not contained in an element of $\mathcal{O}(n,m,q)$ belongs to either 0 or $q^{m-n} + 1$ tangent spaces of the egg. If $m = 2n$, then each point of $\mathbf{PG}(4n-1,q)$ not contained in an element of $\mathcal{O}(n,2n,q)$ belongs to exactly $q^n + 1$ tangent spaces of the egg. If $m \neq 2n$, then there are points contained in no tangent space of $\mathcal{O}(n,m,q)$.*

(ii) *For q even we necessarily have $m = 2n$. Hence for any $\mathcal{O}(n, m, q)$ with q even, we have $m \in \{n, 2n\}$, that is, for any TGQ $\mathcal{S}$ of order (s, t), with s and/or t even, we have $t \in \{s, s^2\}$.*

Proof. (i) Let $\mathcal{O}(n, m, q)$ be an egg, so $n \neq m$. Take a point x not in an element of $\mathcal{O}(n, m, q)$. Firstly we will show that the number of tangent spaces containing x is either 0 or at least $q^{m-n} + 1$.

The number of hyperplanes on x containing just one element of the egg is denoted by α; the number of hyperplanes on x containing $q^{m-n} + 1$ elements of the egg is denoted by β. Now we count in two ways the number of pairs (γ, π), with γ an hyperplane on x and π an element of $\mathcal{O}(n, m, q)$ contained in γ. We obtain

$$\alpha + \beta(q^{m-n} + 1) = (q^m + 1)\frac{q^{m+n-1} - 1}{q - 1}. \tag{3.1}$$

Let ρ be the number of tangent spaces of $\mathcal{O}(n, m, q)$ through x. Now we count in two ways the number of pairs (γ', π), with γ' a hyperplane on x containing just one element of the egg and π the element of the egg in γ'. We obtain

$$\alpha = \rho\frac{q^n - 1}{q - 1} + (q^m + 1 - \rho)\frac{q^{n-1} - 1}{q - 1}.$$

So

$$\alpha = \rho q^{n-1} + (q^m + 1)\frac{q^{n-1} - 1}{q - 1}. \tag{3.2}$$

From (3.1) and (3.2) it follows that

$$\rho q^{n-1} + \beta(q^{m-n} + 1) = \frac{q^{n-1}(q^{2m} - 1)}{q - 1}. \tag{3.3}$$

As $ma = n(a + 1)$, with $a \in \mathbb{N}_0$ and a odd, we have that $q^{m-n} + 1$ divides $q^m - 1$, and so, by (3.3), $q^{m-n} + 1$ divides ρ. It follows that either ρ is zero or at least $q^{m-n} + 1$. Next, we will prove that $\rho \in \{0, q^{m-n} + 1\}$.

Let z_i be any point of $\mathbf{PG}(2n + m - 1, q)$ not in an element of the egg, and let ρ_i be the number of tangent spaces containing z_i. We count in two ways the number of pairs (z_i, τ), with τ a tangent space containing z_i. We obtain

$$\sum_i \rho_i = (q^m + 1)q^n(q^m - 1)/(q - 1). \tag{3.4}$$

Next, we count in two ways the number of ordered triples (z_i, τ, τ'), with τ, τ' distinct tangent spaces containing z_i. We obtain

$$\sum_i \rho_i(\rho_i - 1) = (q^m + 1)q^m(q^m - 1)/(q - 1). \tag{3.5}$$

From (3.4) and (3.5) it follows that

$$\sum_i \rho_i(\rho_i - (q^{m-n} + 1)) = 0.$$

As ρ_i is either zero or at least $q^{m-n} + 1$, we necessarily have $\rho_i \in \{0, q^{m-n} + 1\}$. That proves the first part of (i).

Now assume that $m = 2n$. Counting the number of points contained in $q^n + 1$ tangent spaces, we obtain

$$\frac{(q^{2n} + 1)(q^{3n} - q^n)}{(q - 1)(q^n + 1)} = \frac{q^{4n} - 1}{q - 1} - (q^{2n} + 1)\frac{q^n - 1}{q - 1},$$

and so each point not contained in an element of $\mathcal{O}(n, 2n, q)$, is contained in exactly $q^n + 1$ tangent spaces. This proves the second part of (i).

Finally, assume $m \neq 2n$. Counting the number of points contained in at least one tangent space, we see that there are points contained in no tangent space.

(ii)   Let q be even, and assume by way of contradiction that $m \neq 2n$. By (i) there is a point z of $\mathbf{PG}(2n + m - 1, q)$ contained in no tangent space of $\mathcal{O}(n, m, q)$. Let $\mathcal{O}(n, m, q) = \{\pi_0, \pi_1, \ldots, \pi_{q^m}\}$. Then the space $\langle z, \pi_i \rangle$ intersects some π_j; also $\langle z, \pi_j \rangle$ intersects π_i. Hence we obtain a pairing in $\mathcal{O}(n, m, q)$, implying that $q^m + 1$ is even, a contradiction. We conclude that $m = 2n$. ∎

Remark 3.10.2 Theorem 3.10.1 is due to Payne and Thas [128], but here we present a simpler proof. Also Theorem 3.10.3 below is due to Payne and Thas [128].

Theorem 3.10.3 *Every three distinct tangent spaces of $\mathcal{O}(n, m, q)$, $m \neq n$, have as intersection a space of dimension $m - n - 1$.*

Proof. By Theorem 3.10.1 the number of points contained in exactly $q^{m-n} + 1$ tangent spaces is equal to

$$\frac{(q^m + 1)q^n(q^m - 1)}{(q^{m-n} + 1)(q - 1)}.$$

Now we count in two ways the number of ordered 4-tuples $(z, \tau_i, \tau_j, \tau_k)$, with τ_i, τ_j, τ_k distinct tangent spaces of $\mathcal{O}(n, m, q)$ and $u \in \tau_i \cap \tau_j \cap \tau_k$. If $t_{ijk} = |\tau_i \cap \tau_j \cap \tau_k|$, with i, j, k distinct, we obtain

$$\sum_{i,j,k} t_{ijk} = \frac{(q^m + 1)q^n(q^m - 1)}{(q^{m-n} + 1)(q - 1)} \cdot (q^{m-n} + 1)q^{m-n}(q^{m-n} - 1).$$

As $t_{ijk} \geq \frac{q^{m-n}-1}{q-1}$, and the number of ordered triples (τ_i, τ_j, τ_k), with i, j, k distinct, is equal to $(q^m + 1)q^m(q^m - 1)$, we necessarily have $t_{ijk} = (q^{m-n} - 1)/(q - 1)$ for all distinct i, j, k. ∎

By Theorem 3.10.3 the tangent spaces of $\mathcal{O}(n, m, q)$, with $m \neq n$, are the elements of an egg $\mathcal{O}^*(n, m, q)$ in the dual space of $\mathbf{PG}(2n + m - 1, q)$.

Definition 3.10.4 *The tangent spaces of an egg $\mathcal{O}(n, m, q)$ in $\mathbf{PG}(2n + m - 1, q)$ form an egg $\mathcal{O}^*(n, m, q)$ in the dual space of $\mathbf{PG}(2n + m - 1, q)$. So in addition to the TGQ $\mathbf{T}(n, m, q)$, a TGQ $\mathbf{T}^*(n, m, q)$ arises. The egg $\mathcal{O}^*(n, m, q) = \mathcal{O}^*$ will be called the* translation dual *of $\mathcal{O}(n, m, q) = \mathcal{O}$, and $\mathbf{T}^*(n, m, q) = \mathbf{T}^*(\mathcal{O}) = \mathbf{T}(\mathcal{O}^*)$ will be called the* translation dual *of the GQ $\mathbf{T}(n, m, q) = \mathbf{T}(\mathcal{O})$.*

Remark 3.10.5 The translation dual $\mathcal{S}^*$ of the TGQ $\mathcal{S}$ is (up to isomorphism) uniquely defined by $\mathcal{S}$, that is, is independent of the chosen subfield of the kernel to construct $\mathcal{O}(n, m, q)$.

For regular eggs $\mathcal{O}$, we clearly have $\mathcal{O} \cong \mathcal{O}^*$, respectively $\mathbf{T}(\mathcal{O}) \cong \mathbf{T}^*(\mathcal{O})$. For q odd there are examples with $\mathcal{O} \not\cong \mathcal{O}^*$, respectively $\mathbf{T}(\mathcal{O}) \not\cong \mathbf{T}^*(\mathcal{O})$; see Chapter 10. For q even we have $\mathcal{O} \cong \mathcal{O}^*$ for all known examples (all known examples are regular). The following result is taken from Thas and K. Thas [191].

Theorem 3.10.6 *The TGQ $\mathcal{S}$ of order (s, t), with $s \neq t$, and its translation dual $\mathcal{S}^*$ have isomorphic kernels.*

Proof. Let $\mathcal{S} \cong \mathbf{T}(\mathcal{O})$, with $\mathcal{O} = \mathcal{O}(n, m, q)$ an egg in $\mathbf{PG}(2n + m - 1, q)$, and assume that $\mathbf{GF}(q)$ is the kernel of $\mathcal{S}$. Then $\mathcal{O}^* = \mathcal{O}^*(n, m, q)$ is also an egg in a $\mathbf{PG}(2n + m - 1, q)$, and so, by Theorem 3.5.1, $\mathbf{GF}(q)$ is a subfield of the kernel of $\mathcal{S}^*$. Interchanging roles of $\mathcal{S}$ and $\mathcal{S}^*$ yields the result. ∎

To end this section on eggs we give some interesting and elegant characterizations of regular and classical eggs. Characterization Theorem 3.10.7 is contained in Payne and Thas [128], but as the proof is quite short we repeat it here.

Theorem 3.10.7 *The egg $\mathcal{O}(n, 2n, q)$ is regular if and only if one of the following holds.*

(i) *For each point z not contained in an element of $\mathcal{O}(n, 2n, q)$, the $q^n + 1$ tangent spaces containing z have exactly $(q^n - 1)/(q - 1)$ points in common.*

(ii) *Each $\mathbf{PG}(3n - 1, q)$ containing at least three elements of $\mathcal{O}(n, 2n, q)$, contains exactly $q^n + 1$ elements of $\mathcal{O}(n, 2n, q)$.*

Proof. (i) If $\mathcal{O}(n, 2n, q)$ is regular, then it is clear that for each point z not contained in an element of $\mathcal{O}(n, 2n, q)$, the $q^n + 1$ tangent spaces containing z have exactly $(q^n - 1)/(q - 1)$ points in common.

Conversely, assume that for any point z not contained in an element of $\mathcal{O}(n, 2n, q)$, the $q^n + 1$ tangent spaces containing z have exactly $(q^n - 1)/(q - 1)$ points in common. This means that for any triad $\{(\infty), x, y\}$ of the corresponding $\mathbf{T}(n, 2n, q)$ we have $|\{(\infty), x, y\}^{\perp\perp}| = q^n + 1$. Hence the point (∞) of the GQ $\mathbf{T}(n, 2n, q)$ is 3-regular. So by Theorem 2.7.1(i) it follows that $\mathbf{T}(n, 2n, q) \cong \mathbf{T}_3(\mathcal{O}')$ for some ovoid $\mathcal{O}'$ of $\mathbf{PG}(3, q^n)$. Consequently, by §3.6 and Corollary 3.5.3, $\mathcal{O}(n, 2n, q)$ is regular.

(ii) If $\mathcal{O}(n, 2n, q)$ is regular, then clearly each $\mathbf{PG}(3n - 1, q)$ containing at least three elements of $\mathcal{O}(n, 2n, q)$, contains exactly $q^n + 1$ elements of $\mathcal{O}(n, 2n, q)$.

Conversely, assume that any $\mathbf{PG}(3n - 1, q)$ containing at least three elements of $\mathcal{O}(n, 2n, q)$, contains exactly $q^n + 1$ elements of $\mathcal{O}(n, 2n, q)$. This means that the translation dual $\mathcal{O}^*(n, 2n, q)$ of $\mathcal{O}(n, 2n, q)$ satisfies the condition in (i). So the GQ $\mathbf{T}^*(n, 2n, q)$ is isomorphic to a $\mathbf{T}_3(\mathcal{O}')$ of Tits, with $\mathcal{O}'$ some ovoid in $\mathbf{PG}(3, q^n)$. It follows that $\mathbf{T}(n, 2n, q)$ is isomorphic to $\mathbf{T}_3(\mathcal{O}'^*)$, with $\mathcal{O}'^*$ the translation dual of $\mathcal{O}'$. Clearly $\mathcal{O}'^*$ is also an ovoid of some $\mathbf{PG}(3, q^n)$; in fact $\mathcal{O}'$ and $\mathcal{O}'^*$ are isomorphic; see Hirschfeld [72]. Consequently, by §3.6 and Corollary 3.5.3, $\mathcal{O}(n, 2n, q)$ is regular. We can also argue as follows. As $\mathbf{T}^*(n, 2n, q) \cong \mathbf{T}_3(\mathcal{O}')$, the kernel has size q^n. So by Theorem 3.10.6 the kernel of $\mathbf{T}(n, 2n, q)$ has size q^n, that is, $\mathbf{T}(n, 2n, q)$ is isomorphic to a $\mathbf{T}_3(\mathcal{O}'')$ of Tits, with $\mathcal{O}''$ an ovoid of some $\mathbf{PG}(3, q^n)$. Hence $\mathcal{O}(n, 2n, q)$ is regular. ∎

3.11   The Stabilizer of the Base-Point of a Translation Generalized Quadrangle

From each translation generalized quadrangle of suitable order $\mathcal{S}$ one can construct another translation generalized quadrangle $\mathcal{S}^*$ with the same parameters, namely the translation dual of $\mathcal{S}$. Essential for the classification of translation generalized quadrangles is the isomorphism problem between $\mathcal{S}$ and $\mathcal{S}^*$. Only two (infinite) classes of translation generalized quadrangles $\mathcal{S}$ are known for which indeed $\mathcal{S} \cong \mathcal{S}^*$, namely the $\mathbf{T}_d(\mathcal{O})$ of Tits, where $d \in \{2, 3\}$ and the characteristic is not two if $d = 2$, and the dual Kantor-Knuth generalized quadrangles, see Theorems 4.7.3 and 5.1.13. Using Theorem 3.7.2, we will start to investigate this problem by completely explaining the connection between the full automorphism groups of $\mathcal{S}$ and $\mathcal{S}^*$, see Theorem 3.11.1. In fact, the only classes of translation generalized quadrangles where the sizes of these groups are different are *precisely* the translation generalized quadrangles which are the point-line dual of a flock generalized quadrangle $\mathcal{S}(\mathcal{F})$, $\mathcal{F}$ not a Kantor-Knuth semifield flock, a result which will be described in Chapter 10. (This observation was first done in work of K. Thas — see [201], the extensive paper [202], and [205], or [206] as a general reference, where it was first shown that these examples indeed have different automorphism groups than their translation dual.)

We now have the following result which is an important step towards the determination of all translation generalized quadrangles.

Theorem 3.11.1 *Suppose $\mathcal{S} = \mathbf{T}(\mathcal{O})$ is a TGQ of order (q^n, q^m) with base-point (∞), where q is odd if $n = m$, and suppose that $G_{(\infty)}$ is the stabilizer of (∞) in the automorphism group G of $\mathcal{S}$. Furthermore, suppose $(\infty)'$ is the base-point of $\mathbf{T}(\mathcal{O}^*) = \mathcal{S}^*$, and let $G'_{(\infty)'}$ be the stabilizer of $(\infty)'$ in the automorphism group G' of $\mathcal{S}^*$. Suppose u and v are arbitrary points of $\mathbf{T}(\mathcal{O})$ and $\mathbf{T}(\mathcal{O}^*)$, not collinear with (∞) and $(\infty)'$, respectively. Then $[G_{(\infty)}]_u \cong [G'_{(\infty)'}]_v$. We also have that*

$$|G_{(\infty)}| = |G'_{(\infty)'}|.$$

Proof.   Suppose $\mathcal{O} = \{\pi_0, \pi_1, \ldots, \pi_{q^m}\}$ is contained in $\Pi = \mathbf{PG}(2n + m - 1, q)$, and embed Π in the $(2n + m)$-space $\Pi' = \mathbf{PG}(2n + m, q)$. Let $\tau_0, \tau_1, \ldots, \tau_{q^m}$ be the tangent spaces at respectively $\pi_0, \pi_1, \ldots, \pi_{q^m}$. Now consider the dual space Π'^* of Π'; then the set of tangent spaces interpreted in Π'^*, say $\mathcal{U}^* = \{\tau_0^*, \tau_1^*, \ldots, \tau_{q^m}^*\}$, forms a set of $q^m + 1$ spaces of dimension n which satisfy the following properties:

(1) they intersect two by two in the same fixed point z (which corresponds to Π);

(2) for distinct i, j and k, $\tau_i^* \tau_j^* \tau_k^*$ has dimension $3n$;

(3) for each i, there is an $(n+m)$-space π_i^* which contains τ_i^*, and so that $\pi_i^* \cap \tau_j^* = \{z\}$ if $i \neq j$.

Now consider an arbitrary point $x \in \Pi' \setminus \Pi$. Then the corresponding hyperplane Π_x in the dual space Π'^* of Π' is so that $\Pi_x \cap \mathcal{U}^*$ is an egg $\mathcal{O}^*$ in Π_x, which is clearly isomorphic to the dual egg of $\mathcal{O}$. Now consider an arbitrary collineation θ of $\mathbf{T}(\mathcal{O})$ which fixes (∞) and the point x. Then by Theorem 3.7.2, θ induces a collineation of $\Pi' = \mathbf{PG}(2n+m, q)$ which fixes $\mathcal{O}$ and x. Now interpret θ naturally as a collineation of Π'^*, that is, if η is an r-dimensional space in Π'^* and if η^* is the corresponding $(2n+m-r-1)$-dimensional space in the dual space Π, then η^θ is the r-dimensional space of Π'^* which corresponds to $\eta^{*\theta}$. Then θ fixes z, Π_x, and hence $\mathcal{O}^*$, and thus θ induces a collineation of $\mathbf{T}(\mathcal{O}^*)$ which fixes $(\infty)'$ and z, where $(\infty)'$ is the translation point of $\mathbf{T}(\mathcal{O}^*)$. It follows immediately that

$$[G_{(\infty)}]_u \cong [G'_{(\infty)'}]_v,$$

where u, respectively v, is an arbitrary point of $\mathbf{T}(\mathcal{O})$, respectively $\mathbf{T}(\mathcal{O}^*)$, not collinear with (∞), respectively $(\infty)'$. For any TGQ $\mathcal{S}^{(y)} = (\mathcal{P}, \mathcal{B}, \mathrm{I})$ with translation point y, translation group H and full automorphism group G, it holds, by Lemma 3.7.1, that $G_y = (G_y)_z H = H(G_y)_z$, where z is arbitrary in $\mathcal{P} \setminus y^\perp$. As the translation groups of $\mathbf{T}(\mathcal{O})$ and $\mathbf{T}(\mathcal{O}^*)$ are isomorphic, the theorem now readily follows. ∎

It may be useful for other purposes to define the following object $\mathcal{U}$ in $\mathbf{PG}(2n+m, q)$: $\mathcal{U} = \{\zeta^0, \zeta^1, \ldots, \zeta^{q^m}\}$ is a set of $q^m + 1$ spaces of dimension n which satisfy the following properties:

(1) the elements of $\mathcal{U}$ intersect two by two in the same fixed point x;

(2) for distinct i, j and k, $\zeta^i \zeta^j \zeta^k$ has dimension $3n$;

(3) for each i, there is an $(n+m)$-space π^i which contains ζ^i, and so that $\pi^i \cap \zeta^j = \{x\}$ if $i \neq j$.

We call $\mathcal{U}$ a *generalized ovoid cone*, respectively *generalized oval cone*, with vertex x if $n \neq m$, respectively $n = m$.

3.12 Structure of the Automorphism Group of a Translation Quadrangle

Let $\mathcal{S} = \mathbf{T}(n, m, q) = \mathbf{T}(\mathcal{O})$ be a TGQ of order (q^n, q^m) for the generalized ovoid $\mathcal{O}$ $(n \neq m)$, respectively generalized oval $\mathcal{O}$ $(n = m)$, with full group of automorphisms $\mathrm{Aut}(\mathcal{S})$ and base-point (∞). By Theorem 3.7.2, we know that $\mathrm{Aut}(\mathcal{S})_{(\infty)} \cong \mathbf{P\Gamma L}_{2n+m+1}(q)_{\mathcal{O}}$, where $\langle \mathcal{O} \rangle = \mathbf{PG}(2n + m - 1, q) \subseteq \mathbf{PG}(2n + m, q)$. The following theorem analyses more closely the structure of $\mathrm{Aut}(\mathcal{S})_{(\infty)}$.

Theorem 3.12.1 *Let $\mathcal{S} = \mathbf{T}(n, m, q) = \mathbf{T}(\mathcal{O})$ be a TGQ of order (q^n, q^m) for the generalized ovoid $\mathcal{O}$ $(n \neq m)$, respectively generalized oval $\mathcal{O}$ $(n = m)$, with full group of automorphisms $\mathrm{Aut}(\mathcal{S})$ and base-point (∞). Let T be the translation group of $\mathcal{S}$, and suppose x is an arbitrary point of $\mathcal{P} \setminus (\infty)^{\perp}$, where $\mathcal{P}$ is the point set of $\mathcal{S}$. Let $H = \mathrm{Aut}(\mathcal{S})_{(\infty)}$ be the stabilizer of (∞) in $\mathrm{Aut}(\mathcal{S})$. Then we have that*

$$H \cong T \rtimes H_x,$$

where $\rtimes$ denotes the natural semidirect product.

Proof. The theorem follows from the fact that $H_x \cap T = \{\mathbf{1}\}$, that T is a normal subgroup of $H = \mathrm{Aut}(\mathcal{S})_{(\infty)}$ by Lemma 3.7.1, and that $H = TH_x = H_xT$. ∎

Later, in Chapter 10, we will apply Theorem 3.7.2 to obtain a classification result on TGQs, and to explain precisely the connection between the size of the automorphism group of a TGQ, and the size of the automorphism group of its translation dual, if it is defined.

Chapter 4

Generalized Quadrangles and Flocks

Dembowski [45] used the word flock to refer to the partition of all but two points of an inversive plane into disjoint circles. The concept was naturally extended to other circle geometries, and significant applications have been discovered in each of its settings. For surveys of the existence problems and their solutions and of the different applications we refer to Fisher and Thas [58], Thas [172, 174, 178, 187], and the webpage of Cherowitzo (http://www-math.cudenver.edu/~wcherowi).

4.1 Flocks

Let $\mathcal{O}$ be an ovoid of $\mathbf{PG}(3, q)$. A partition of all but two points of $\mathcal{O}$ into $q - 1$ disjoint ovals is called a *flock* of $\mathcal{O}$. The two remaining points x, y are called the *carriers* of the flock. If L is a line of $\mathbf{PG}(3, q)$ having no points in common with $\mathcal{O}$, then the $q - 1$ planes through L which are not tangent to $\mathcal{O}$ intersect $\mathcal{O}$ in the elements of a flock $\mathcal{F}$. Such a flock is called *linear*.

Next, let $\mathcal{H}$ be a hyperbolic quadric of $\mathbf{PG}(3, q)$. A partition of $\mathcal{H}$ into $q + 1$ disjoint nonsingular conics is called a *flock* of $\mathcal{H}$. If L is a line of $\mathbf{PG}(3, q)$ having no points in common with $\mathcal{H}$, then the $q+1$ planes through L intersect $\mathcal{H}$ in the elements of a flock $\mathcal{F}$. Such a flock is called *linear*.

Finally, let $\mathcal{O}$ be either an oval in $\mathbf{PG}(2, q)$ for any q or a hyperoval in $\mathbf{PG}(2, q)$ for q even, and let $\mathcal{K}$ be the cone with vertex a point x of $\mathbf{PG}(3, q) \supset \mathbf{PG}(2, q)$, with $x \notin \mathbf{PG}(2, q)$, and base $\mathcal{O}$. A partition of $\mathcal{K} \backslash \{x\}$ into q disjoint ovals or hyperovals in the respective cases is called a *flock* of $\mathcal{K}$. If L is a line of $\mathbf{PG}(3, q)$ having no points in common with $\mathcal{K}$, then the

q planes through L but not through x intersect $\mathcal{K}$ in the elements of a flock $\mathcal{F}$. Such a flock is called *linear*.

4.2   Flocks and Translation Planes

We next present a construction, discovered independently by Walker [237] and Thas [167], of a class of translation planes.

Let $\mathcal{F} = \{C_1, C_2, \ldots, C_\alpha\}$ be a flock of a quadric $\mathbf{Q}$ of $\mathbf{PG}(3, q)$, with $\alpha = q - 1, q + 1$, or q according as $\mathbf{Q}$ is elliptic, hyperbolic or a cone. We now embed $\mathbf{Q}$ in the nonsingular hyperbolic (or Klein) quadric $\mathcal{H}$ of $\mathbf{PG}(5, q)$. Denote the plane of C_i by π_i, the polar plane of π_i with respect to $\mathcal{H}$ by π_i^*, and the conic $\pi_i^* \cap \mathcal{H}$ by C_i^*, with $i = 1, 2, \ldots, \alpha$. Note that since, for $i \neq j$, $\pi_i \cap \pi_j$ is a line exterior to $\mathcal{H}$, $\pi_i^* \cup \pi_j^*$ lies in a 3-space that intersects $\mathcal{H}$ in an elliptic quadric; in particular, no two points of $C_i^* \cup C_j^*$ are on a common line of $\mathcal{H}$.

(a)   When $\mathbf{Q}$ is elliptic and $\mathcal{F}$ has carriers x and y, the set $C_1^* \cup C_2^* \cup \cdots \cup C_{q-1}^* \cup \{x, y\} = \mathcal{O}$ has size $q^2 + 1$ and no two points of this set are on a common line of $\mathcal{H}$;

(b)   When $\mathbf{Q}$ is hyperbolic, the set $C_1^* \cup C_2^* \cup \cdots \cup C_{q+1}^* = \mathcal{O}$ has size $q^2 + 1$ and no two points of this set are on a common line of $\mathcal{H}$;

(c)   When $\mathbf{Q}$ is a quadratic cone with vertex x, the set $C_1^* \cup C_2^* \cup \cdots \cup C_q^* = \mathcal{O}$ contains x, has size $q^2 + 1$ and no two points of this set are on a common line of $\mathcal{H}$.

A subset $\mathcal{O}$ of size $q^2 + 1$ of $\mathcal{H}$ such that no two of its points are on a common line of $\mathcal{H}$ is called an *ovoid* of $\mathcal{H}$; see e.g. Thas [178]. Equivalently, an *ovoid* of $\mathcal{H}$ is a subset of $\mathcal{H}$ intersecting each plane of $\mathcal{H}$ in exactly one point.

Let β be the Klein mapping from the points of $\mathcal{H}$ to the lines of $\mathbf{PG}(3, q)$; see e.g. Chapter 5 of Hirschfeld [72]. Then in each of the three cases $\mathcal{O}^\beta = \mathcal{T}$ is a set of $q^2 + 1$ mutually skew lines of $\mathbf{PG}(3, q)$. Hence, $\mathcal{T}$ is a spread of $\mathbf{PG}(3, q)$. According to the standard argument, see e.g. Section 3.1 of Dembowski [45], to $\mathcal{T}$ there is an associated translation plane π of order q^2 and dimension at most two over its kernel. The plane π is Desarguesian if and only if the spread $\mathcal{T}$ is regular, if and only if the ovoid $\mathcal{O}$ is an elliptic quadric, if and only if the planes $\pi_1, \pi_2, \ldots, \pi_\alpha$ have a common line, if and only if the flock $\mathcal{F}$ is linear.

4.3 Flocks of Ovoids and Hyperbolic Quadrics

In the context of generalized quadrangles flocks of quadratic cones are much more important than flocks of ovoids or hyperbolic quadrics; it will appear that flocks of cones play a key role. However, we emphasize that e.g. the fundamental theorem on flocks of ovoids appeared to be very useful in the proof of strong characterizations of certain classes of GQs, see e.g. Theorem 5.3.1 in Payne and Thas [128].

The following theorem was proved in the odd case by Orr [105], in the even case by Thas [159]. A short proof of Orr's result can be found in Fisher and Thas [58].

Theorem 4.3.1 *Any flock of an ovoid $\mathcal{O}$ of $\mathbf{PG}(3, q)$ is linear.*   ∎

The next theorem is due to Thas [166].

Theorem 4.3.2 *For q even any flock of a hyperbolic quadric of $\mathbf{PG}(3, q)$ is linear.*   ∎

From now on assume that q is odd. Let $\mathbf{Q}$ be a hyperbolic quadric of $\mathbf{PG}(3, q)$, with q odd. In the set of all nonsingular conics of $\mathbf{Q}$ we define the following equivalence relation, see Thas [166, 169]: two conics $\mathcal{C}_1$ and $\mathcal{C}_2$ are *equivalent* if and only if there is a nonsingular conic $\mathcal{C}$ on $\mathbf{Q}$ which is tangent to both $\mathcal{C}_1$ and $\mathcal{C}_2$. There are two equivalence classes, denoted by I and II. Let L be a line having no point in common with $\mathbf{Q}$, and let L' be the polar line of L with respect to $\mathbf{Q}$. The set of all conics of Class I, respectively II, whose plane contains L is denoted by V, respectively V'. For $q \equiv 1 \mod 4$ the set of all conics of Class II, respectively I, whose plane contains L' is denoted by W, respectively W'; for $q \equiv -1 \mod 4$ the set of all conics of Class I, respectively II, containing L' is denoted by W, respectively W'. Then it was shown by Thas [166] that $V \cup W$ and $V' \cup W'$ are nonlinear flocks. By most authors these flocks are called *Thas flocks*.

Further, Bader [1] showed that for $q = 11, 23, 59$ the hyperbolic quadric $\mathbf{Q}$ of $\mathbf{PG}(3, q)$ has a flock which is neither linear nor a Thas flock. These flocks were independently discovered by Johnson [77], and for $q = 11, 23$ also by Baker and Ebert [7]. Since these flocks were derived by Bader from exceptional nearfield planes, exploiting the relationship between flocks and translation planes mentioned in Section 4.2, Bader calls these flocks the *exceptional flocks*.

The following theorem classifying all flocks of the hyperbolic quadric $\mathbf{Q}$ of $\mathbf{PG}(3, q)$, with q odd, is a combination of results by Bader and Lunardon [2] and Thas [173].

Theorem 4.3.3 *Any flock of the nonsingular hyperbolic quadric $\mathbf{Q}$ of $\mathbf{PG}(3, q)$, with q odd, either is a linear flock, a Thas flock, or an exceptional flock.*   ∎

4.4   Flocks of Cones

Let $\mathcal{O}^*$ be a hyperoval of $\mathbf{PG}(2, q)$, with $q = 2^h$, and let $\mathcal{K}^*$ be the cone of $\mathbf{PG}(3, q) \supset \mathbf{PG}(2, q)$ which projects $\mathcal{O}^*$ from a point x of $\mathbf{PG}(3, q) \backslash \mathbf{PG}(2, q)$. Further, assume that $\mathcal{F}^* = \{\mathcal{C}_1^*, \mathcal{C}_2^*, \ldots, \mathcal{C}_q^*\}$ is a flock of $\mathcal{K}^*$. Now, let $y \in \mathcal{O}^*$, $\mathcal{O} = \mathcal{O}^* \backslash \{y\}$, and let $\mathcal{K}$ be the cone which projects the oval $\mathcal{O}$ from x. Then it is clear that the planes $\pi_1, \pi_2, \ldots, \pi_q$, with $\mathcal{C}_i^* \subset \pi_i$, define a flock $\mathcal{F} = \{\mathcal{C}_1, \mathcal{C}_2, \ldots, \mathcal{C}_q\}$ with $\mathcal{C}_i \subset \mathcal{C}_i^*$, of $\mathcal{K}$. In the next theorem we prove the converse.

Theorem 4.4.1 *Let $\mathcal{O}$ be an oval of $\mathbf{PG}(2, q)$, with $q = 2^h$, and let $\mathcal{K}$ be the cone which projects $\mathcal{O}$ from a point x of $\mathbf{PG}(3, q) \setminus \mathbf{PG}(2, q)$. Further, assume that $\mathcal{F} = \{\mathcal{C}_1, \mathcal{C}_2, \ldots, \mathcal{C}_q\}$ is a flock of $\mathcal{K}$. Let n be the nucleus (kernel) of $\mathcal{O}$, and let $\mathcal{O}^*$ be the hyperoval $\mathcal{O} \cup \{n\}$. If $\mathcal{K}^*$ is the cone which projects $\mathcal{O}^*$ from x, then the planes $\pi_1, \pi_2, \ldots, \pi_q$, with $\mathcal{C}_i \subset \pi_i$, define a flock $\mathcal{F}^* = \{\mathcal{C}_1^*, \mathcal{C}_2^*, \ldots, \mathcal{C}_q^*\}$, with $\mathcal{C}_i \subset \mathcal{C}_i^*$, of $\mathcal{K}^*$.*

Proof.   Suppose that $z \in \mathcal{C}_i^* \cap \mathcal{C}_j^*$, with $i \neq j$. Since $\mathcal{O}^*$ is a hyperoval, the line $\pi_i \cap \pi_j$ has exactly two distinct points y, z in common with $\mathcal{K}^*$. Clearly $\{y, z\} = \mathcal{C}_i^* \cap \mathcal{C}_j^*$. Hence $|\mathcal{C}_i \cap \mathcal{C}_j| \geq 1$, a contradiction. So $\mathcal{C}_i^* \cap \mathcal{C}_j^* = \emptyset$, for all $i \neq j$, that is, $\{\mathcal{C}_1^*, \mathcal{C}_2^*, \ldots, \mathcal{C}_q^*\}$ is a flock of $\mathcal{K}^*$.   ∎

Let $\mathcal{F} = \{\mathcal{C}_1, \mathcal{C}_2, \ldots, \mathcal{C}_q\}$ be a flock of the quadratic cone $\mathcal{K}$ in $\mathbf{PG}(3, q)$. Let x be the vertex of $\mathcal{K}$ and let π_i be the plane of $\mathcal{C}_i$, with $i = 1, 2, \ldots, q$. Dualizing in $\mathbf{PG}(3, q)$, the point x becomes a plane $\mathbf{PG}(2, q)$, the $q+1$ lines on $\mathcal{K}$ become the $q + 1$ elements of a nonsingular dual conic Γ in $\mathbf{PG}(2, q)$, and the plane π_i becomes a point x_i not in $\mathbf{PG}(2, q)$, with $= 1, 2, \ldots, q$. Then $\mathcal{C}_i \cap \mathcal{C}_j = \emptyset$, with $i \neq j$, is equivalent with $x_i x_j \cap \mathbf{PG}(2, q)$ not lying on an element of Γ. We say that $\overline{\mathcal{F}} = \{x_1, x_2, \ldots, x_q\}$ is a *flock* of the dual conic Γ; $\overline{\mathcal{F}}$ is called *linear* if and only if $\mathcal{F}$ is linear. If q is odd, then let $y_1, y_2, \ldots, y_{q+1}$ be the tangent points of the elements of Γ, that is, the points of $\mathbf{PG}(2, q)$ lying on exactly one element of Γ. Then $\{y_1, y_2, \ldots, y_{q+1}\} = C$ is a nonsingular conic in $\mathbf{PG}(2, q)$. In such a case we also say that $\overline{\mathcal{F}}$ is a *flock* of the conic C.

In the remaining part of this section we will only consider flocks of quadratic cones. We shall use the following notation: the set of all elements m of $\mathbf{GF}(q)$, q even, for which the polynomial $X^2 + X + m$ is reducible, respectively irreducible, over $\mathbf{GF}(q)$ is denoted by C_0, respectively C_1.

Theorem 4.4.2 *With each flock $\mathcal{F}$ of the quadratic cone $\mathcal{K}$ of $\mathbf{PG}(3, q)$ corresponds a set of q ordered triples $(a_i, b_i, c_i), a_i, b_i, c_i \in \mathbf{GF}(q)$, such that for q odd*

$$(c_i - c_j)^2 - 4(a_i - a_j)(b_i - b_j) \text{ is a nonsquare}$$

whenever $i \neq j$, and for q even

$$c_i \neq c_j \text{ and } (a_i + a_j)(b_i + b_j)(c_i + c_j)^{-2} \in C_1$$

whenever $i \neq j$. Conversely, with each such set of q ordered triples corresponds a flock $\mathcal{F}$ of the quadratic cone $\mathcal{K}$.

Proof. Let $X_0 X_1 = X_2^2$ be the equation of the cone $\mathcal{K}$. Now we consider q planes π_i, with $i = 1, 2, \ldots, q$, which do not contain the vertex $x(0, 0, 0, 1)$ of $\mathcal{K}$. Such a plane π_i has equation $a_i X_0 + b_i X_1 + c_i X_2 + X_3 = 0$. The plane π_{ij} projecting $\pi_i \cap \pi_j$, with $i \neq j$, from the vertex x has equation $(a_i - a_j)X_0 + (b_i - b_j)X_1 + (c_i - c_j)X_2 = 0$.

Let q be odd. Then the plane π_{ij} has no line in common with $\mathcal{K}$ if and only if $(c_i - c_j)^2 - 4(a_i - a_j)(b_i - b_j)$ is a nonsquare. Hence the conics $\pi_i \cap \mathcal{K}$ form a flock of $\mathcal{K}$ if and only if $(c_i - c_j)^2 - 4(a_i - a_j)(b_i - b_j)$ is a nonsquare whenever $i \neq j$.

Let q be even. Then the plane π_{ij} has no line in common with $\mathcal{K}$ if and only if the homogeneous quadratic polynomial $X^2(a_i + a_j)^2 + XY(c_i + c_j)^2 + Y^2(b_i + b_j)^2$ is irreducible, that is, if and only if $c_i \neq c_j$ and $(a_i + a_j)^2(b_i + b_j)^2(c_i + c_j)^{-4} \in C_1$, if and only if $c_i \neq c_j$ and $(a_i + a_j)(b_i + b_j)(c_i + c_j)^{-2} \in C_1$. Hence the conics $\pi_i \cap \mathcal{K}$ form a flock if and only if $c_i \neq c_j$ and $(a_i + a_j)(b_i + b_j)(c_i + c_j)^{-2} \in C_1$ whenever $i \neq j$. ∎

Remark 4.4.3 Clearly coordinates can be chosen in such a way that π_1 has equation $X_3 = 0$, that is, $(a_1, b_1, c_1) = (0, 0, 0)$.

Now we will classifiy all flocks $\mathcal{F} = \{C_1, C_2, \ldots, C_q\}$ for which the planes $\pi_1, \pi_2, \ldots, \pi_q$ of the respective conics $C_1, C_2, \ldots, C_q$ all contain a common point.

Theorem 4.4.4 *Let $\mathcal{F} = \{C_1, C_2, \ldots, C_q\}$ be a flock of the quadratic cone $\mathcal{K}$ of $\mathbf{PG}(3, q)$, with q even. If the planes $\pi_1, \pi_2, \ldots, \pi_q$ of the respective conics $C_1, C_2, \ldots, C_q$ all contain a common point, then the flock $\mathcal{F}$ is linear.*

Proof. For $q = 2$ the statement is trivial, so assume that $q > 2$. Let the cone $\mathcal{K}$ be embedded in the Klein quadric $\mathcal{H}$ of $\mathbf{PG}(5, q)$. If π_i^* is the polar plane of π_i with respect to $\mathcal{H}$ and if $\pi_i^* \cap \mathcal{H} = C_i^*$, with $i = 1, 2, \ldots, q$, then by Section 4.2 $C_1^* \cup C_2^* \cup \cdots \cup C_q^* = \mathcal{O}$ is an ovoid of $\mathcal{H}$. The planes $\pi_1^*, \pi_2^*, \ldots, \pi_q^*$ contain a common tangent line L of $\mathcal{H}$, where L is the polar line of $\mathbf{PG}(3, q)$ with respect to $\mathcal{H}$. Since the planes $\pi_1, \pi_2, \ldots, \pi_q$ contain a common point y, the planes $\pi_1^*, \pi_2^*, \ldots, \pi_q^*$ are contained in the polar space $\mathbf{PG}(4, q)$ of y with respect to $\mathcal{H}$. Also, y is the nucleus of the quadric $\mathbf{Q} = \mathbf{PG}(4, q) \cap \mathcal{H}$. Notice that each line of $\mathbf{Q}$ has exactly one point in common with $\mathcal{O}$. Now consider distinct points z_1, z_2 of $\mathcal{O}$, and let C^* be the intersection of $\mathbf{Q}$ and the plane yz_1z_2. Let $C^* \cap \mathcal{O} = V$, with $|V| = \alpha$. Now we count in different ways the number of ordered pairs (u, z), with $u \in \mathcal{O} \backslash V, z \in C^* \backslash V$, and uz a line of $\mathbf{Q}$. If $u \in \mathcal{O} \backslash V$, then the tangent 3-space of $\mathbf{Q}$ at u contains exactly one of the lines joining y to a point of C^*. Hence we obtain $q^2 + 1 - \alpha = (q + 1 - \alpha)(q + 1)$, and so $\alpha = 2$. Consequently the plane yz_1z_2 only contains the points z_1, z_2 of $\mathcal{O}$. Now we project $\mathcal{O}$ from y onto a $\mathbf{PG}(3, q) \subset \mathbf{PG}(4, q)$ not containing y. By the foregoing no three points of this projection $\overline{\mathcal{O}}$ of $\mathcal{O}$ are collinear, and so (since $q > 2$) $\overline{\mathcal{O}}$ is an ovoid of $\mathbf{PG}(3, q)$; see also Section 1.5. The projections of $C_1^*, C_2^*, \ldots, C_q^*$ are conics $\overline{C}_1, \overline{C}_2, \ldots, \overline{C}_q$ having a common tangent line at their common point. By a theorem of Glynn [63], a short proof of which can be found in Storme and Thas [148], $\overline{\mathcal{O}}$ is an elliptic quadric. Let $\overline{C}$ be a conic on $\overline{\mathcal{O}}$ containing the common point of $\overline{C}_1, \overline{C}_2, \ldots, \overline{C}_q$, but distinct from any of these conics. The conic $\overline{C}$ is the projection from y of a conic C^* on $\mathcal{O}$ through the common point of $C_1^*, C_2^*, \ldots, C_q^*$, which is the vertex x of $\mathcal{K}$. As $|C_i^* \cap C^*| = \{x, u_i\}$, with $x \neq u_i$, the conic C_i^* belongs to the 3-space $\langle C^*, L \rangle$, with $i = 1, 2, \ldots, q$. So $\mathcal{O}$ is contained in $\langle C^*, L \rangle$. Hence $\mathcal{O}$ is an elliptic quadric, and so the planes $\pi_1, \pi_2, \ldots, \pi_q$ have a common line. We conclude that the flock is linear. ∎

In the next two theorems we will handle the odd case.

Theorem 4.4.5 *Let $\mathcal{F} = \{C_1, C_2, \ldots, C_q\}$ be a flock of the quadratic cone $\mathcal{K}$ of $\mathbf{PG}(3, q)$, with q odd. If the planes $\pi_1, \pi_2, \ldots, \pi_q$ of the respective conics $C_1, C_2, \ldots, C_q$ all contain a common interior point x of $\mathcal{K}$, then $\mathcal{F}$ is linear.*

Proof. Let $X_0 X_1 = X_2^2$ be the equation of the cone $\mathcal{K}$, and let $a_i X_0 + b_i X_1 + c_i X_2 + X_3 = 0$, with $i = 1, 2 \ldots, q$, be the equation of the plane π_i. Since the automorphism group of $\mathbf{PG}(3, q)$ leaving $\mathcal{K}$ invariant is transitive on the interior points of $\mathcal{K}$, we may assume that $x = (1, -m, 0, 0)$, with m a nonsquare of $\mathbf{GF}(q)$. As x is contained in π_i, we have $a_i - b_i m = 0$, with $i = 1, 2, \ldots, q$.

Let $\mathbf{GF}(q^2)$ be the field obtained by adjoining a root ω of $X^2 = m$ to $\mathbf{GF}(q)$. If $\alpha \in \mathbf{GF}(q^2)$ with $\alpha = a + b\omega$, and $a, b \in \mathbf{GF}(q)$, then let $\overline{\alpha} = a - b\omega$. With the q planes π_i we now let correspond the q elements $c_i + 2b_i\omega = \alpha_i$ of $\mathbf{GF}(q^2)$. Assume that $\alpha_i - \alpha_j$, with $i \neq j$, is a square in $\mathbf{GF}(q^2)$, so let $\alpha_i - \alpha_j = (r + s\omega)^2$ with $r, s \in \mathbf{GF}(q)$. Then we have

$$(c_i - c_j)^2 - 4(a_i - a_j)(b_i - b_j) = (c_i - c_j)^2 - 4(b_i - b_j)^2 m$$
$$= (c_i - c_j)^2 - 4(b_i - b_j)^2 \omega^2 = (c_i - c_j + 2(b_i - b_j)\omega)(c_i - c_j - 2(b_i - b_j)\omega)$$
$$= (\alpha_i - \alpha_j)\overline{(\alpha_i - \alpha_j)} = (r + s\omega)^2 (r - s\omega)^2 = (r^2 - s^2 m)^2.$$

Hence $(c_i - c_j)^2 - 4(a_i - a_j)(b_i - b_j)$ is a square in $\mathbf{GF}(q)$, contradicting Theorem 4.4.2. Consequently $\alpha_i - \alpha_j$ is a nonsquare in $\mathbf{GF}(q^2)$ whenever $i \neq j$.

Let ν be a given nonsquare of $\mathbf{GF}(q^2)$. Then the set V consisting of the q elements $\nu(\alpha_i - \alpha_1)$ contains 0 and has the property that $\delta - \gamma$ is a square whenever $\delta, \gamma \in V$. Now by a theorem of Blokhuis [15] we have $V = \{d\alpha \,\|\, d \in \mathbf{GF}(q)\}$ with α some nonzero square in $\mathbf{GF}(q^2)$. Hence $\nu(\alpha_i - \alpha_1)/\alpha \in \mathbf{GF}(q)$, for $i = 1, 2, \ldots, q$. If $\nu/\alpha = u + u'\omega$, with $u, u' \in \mathbf{GF}(q)$ and $(u, u') \neq (0, 0)$, then $u'c_i + 2b_i u = u'c_1 + 2b_1 u, i = 1, 2, \ldots, q$. Consequently the q planes π_i all contain the point $(0, 2u, u', -u'c_1 - 2b_1 u)$. So the q planes π_i all contain a common line, that is, the flock $\mathcal{F}$ is linear.   ∎

Theorem 4.4.6 *Let $\mathcal{F} = \{\mathcal{C}_1, \mathcal{C}_2, \ldots, \mathcal{C}_q\}$, with q odd, be a flock of the quadratic cone $\mathcal{K}$ and suppose that the planes $\pi_1, \pi_2, \ldots, \pi_q$, with $\mathcal{C}_i \subset \pi_i$, contain a common exterior point of $\mathcal{K}$. If m is a given nonsquare of $\mathbf{GF}(q)$, then coordinates can be chosen in such a way that $\mathcal{K}$ has equation $X_0 X_1 = X_2^2$ and the planes π_i have equation $a_i X_0 - m a_i^\sigma X_1 + X_3 = 0$, with $\{a_1, a_2, \ldots, a_q\} = \mathbf{GF}(q)$ and σ an automorphism of $\mathbf{GF}(q)$. Conversely, given a nonsquare m of $\mathbf{GF}(q)$, the planes π_i with equation $a_i X_0 - m a_i^\sigma X_1 + X_3 = 0$, with $\{a_1, a_2, \ldots, a_q\} = \mathbf{GF}(q)$ and σ any automorphism of $\mathbf{GF}(q)$, define a flock $\mathcal{F}$ of the cone $\mathcal{K}$ with equation $X_0 X_1 = X_2^2$. Also, the planes π_i all contain the exterior point $(0, 0, 1, 0)$ of $\mathcal{K}$. Finally, the flock is linear if and only if $\sigma = 1$.*

Proof. Consider the cone $\mathcal{K}$ with equation $X_0X_1 = X_2^2$ and the q planes π_i with equation $a_iX_0 - ma_i^\sigma X_1 + X_3 = 0$, with $\{a_1, a_2, \ldots, a_q\} = \mathbf{GF}(q)$, m a given nonsquare of $\mathbf{GF}(q)$, and σ any automorphism of $\mathbf{GF}(q)$. All these planes contain the exterior point $(0, 0, 1, 0)$ of $\mathcal{K}$. Since $-4(a_i - a_j)(-ma_i^\sigma + ma_j^\sigma) = 4m(a_i - a_j)^{\sigma+1}$ is a nonsquare whenever $i \neq j$, the planes $\pi_1, \pi_2, \ldots, \pi_q$ define a flock of $\mathcal{K}$ by Theorem 4.4.2. It is easy to show that the planes $\pi_1, \pi_2, \ldots, \pi_q$ have a line in common if and only if $\sigma = 1$. This proves the second part of the theorem.

Conversely, consider any flock $\mathcal{F} = \{\mathcal{C}_1, \mathcal{C}_2, \ldots, \mathcal{C}_q\}$ of a quadratic cone $\mathcal{K}$ where the planes $\pi_1, \pi_2, \ldots, \pi_q$, with $\mathcal{C}_i \subset \pi_i$, contain a common exterior point x of $\mathcal{K}$. Coordinates can be chosen in such a way that $x = (0, 0, 1, 0)$, that π_1 is the plane $X_3 = 0$, and that $\mathcal{K}$ has equation $X_0X_1 = X_2^2$. Then the plane π_i has an equation of the form $a_iX_0 + b_iX_1 + X_3 = 0$, with $a_1 = b_1 = 0$. Assume, by way of contradiction, that $a_i = a_j$ for $i \neq j$. Then $\pi_i \cap \pi_j$ lies in the plane $X_1 = 0$. Since $X_1 = 0$ and $\mathcal{K}$ have a line in common, the line $\pi_i \cap \pi_j$ and $\mathcal{K}$ are not disjoint, a contradiction. Hence $a_i \neq a_j$ whenever $i \neq j$; similarly $b_i \neq b_j$ whenever $i \neq j$. Consequently $\theta : a_i \mapsto b_i$, with $i = 1, 2, \ldots, q$, is a permutation of $\mathbf{GF}(q)$, with $0^\theta = 0$. By Theorem 4.4.2 $-(a_i - a_j)(a_i^\theta - a_j^\theta)$ is a nonsquare whenever $i \neq j$. Putting $a_j = 0$ and $a_i = 1$, we see that -1^θ is a nonsquare. Hence $((a_i^\theta/1^\theta) - (a_j^\theta/1^\theta))/(a_i - a_j)$ is a nonzero square whenever $i \neq j$. Let $\sigma : a_i \mapsto (a_i^\theta/1^\theta)$, with $i = 1, 2, \ldots, q$. Then σ is a permutation of $\mathbf{GF}(q)$ for which $0^\sigma = 0$ and $1^\sigma = 1$. Moreover $(a_i^\sigma - a_j^\sigma)/(a_i - a_j)$ is a nonzero square whenever $i \neq j$. Then by a theorem of Carlitz [36] σ is an automorphism of the field $\mathbf{GF}(q)$. The plane π_i has equation $a_iX_0 + 1^\theta a_i^\sigma X_1 + X_3 = 0$. Now let m be any given nonsquare of $\mathbf{GF}(q)$, and consider the following change of coordinates: $x_0 = x_0', x_1 = (-m/1^\theta)x_1', x_2 = sx_2', x_3 = x_3'$ with $-m/1^\theta = s^2$. Then $\mathcal{K}$ has still equation $X_0X_1 = X_2^2$ and π_i has equation $a_iX_0 - ma_i^\sigma X_1 + X_3 = 0$, with $i = 1, 2, \ldots, q$. Now the theorem is completely proved. ∎

Let $\mathcal{F} = \{\mathcal{C}_1, \mathcal{C}_2, \ldots, \mathcal{C}_q\}$ be a flock of the quadratic cone $\mathcal{K}$ of $\mathbf{PG}(3, q)$, q odd, and suppose that the planes $\pi_1, \pi_2, \ldots, \pi_q$, with $\mathcal{C}_i \subset \pi_i$, contain a common exterior point of $\mathcal{K}$. By Section 4.2 the flock $\mathcal{F}$ defines an ovoid $\mathcal{O}$ of the Klein quadric $\mathcal{H}$. Since all planes $\pi_1, \pi_2, \ldots, \pi_q$ contain a common point, the ovoid $\mathcal{O}$ is contained in a hyperplane $\mathbf{PG}(4, q)$ of the space $\mathbf{PG}(5, q)$ containing $\mathcal{H}$. Hence $\mathcal{O}$ is an ovoid of the GQ $\mathbf{Q}(4, q)$ with point set $\mathbf{PG}(4, q) \cap \mathcal{H}$. Consequently, if β is the Klein mapping, then $\mathcal{O}^\beta = \mathcal{T}$ is a spread of the GQ $\mathbf{Q}(4, q)^\beta = \mathbf{W}(q)$. It is easy to check that the ovoids $\mathcal{O}$ corresponding to the flocks described in Theorem 4.4.6 are ovoids described by Kantor [82], for which the translation planes defined by the corresponding spreads T are Knuth semifield planes. For this reason these flocks will be called *Kantor-Knuth flocks*.

We conclude this section on flocks of cones with a short description of the process of *derivation* introduced by Bader, Lunardon and Thas [5]. Let $\mathcal{F} = \{\mathcal{C}_1, \mathcal{C}_2, \ldots, \mathcal{C}_q\}$ be a flock of the quadratic cone $\mathcal{K}$ with vertex x of $\mathbf{PG}(3, q)$, with q odd. The plane of $\mathcal{C}_i$ is denoted by π_i, $i = 1, 2, \ldots, q$. Let $\mathcal{K}$ be embedded in the nonsingular quadric $\mathbf{Q}$ of $\mathbf{PG}(4, q)$. Let the polar line of π_i with respect to $\mathbf{Q}$ be denoted by L_i and let $L_i \cap \mathbf{Q} = \{x, x_i\}, i = 1, 2, \ldots, q$. If H_i is the tangent hyperplane of $\mathbf{Q}$ at x_i, then put

$$H_i \cap \mathbf{Q} = \mathcal{K}_i, H_i \cap H_j \cap \mathbf{Q} = \mathcal{K}_i \cap \mathcal{K}_j = \mathcal{K}_j \cap H_i = \mathcal{C}_{ij} = \mathcal{C}_{ji}$$

and

$$\mathcal{C}_{ii} = \mathcal{C}_i, \quad i, j = 1, 2, \ldots, q \text{ and } i \neq j.$$

Then Bader, Lunardon and Thas [5] prove that $\mathcal{F}_i = \{\mathcal{C}_{i1}, \mathcal{C}_{i2}, \ldots, \mathcal{C}_{iq}\}$ is a flock of $\mathcal{K}_i$, with $i = 1, 2, \ldots, q$. We say that the flocks $\mathcal{F}_1, \mathcal{F}_2, \ldots, \mathcal{F}_q$ are *derived* from the flock $\mathcal{F}$. It is clear that $\mathcal{F}, \mathcal{F}_1, \ldots, \mathcal{F}_{i-1}, \mathcal{F}_{i+1}, \ldots, \mathcal{F}_q$ are derived from the flock $\mathcal{F}_i$. There is an extensive literature on the subject, see e.g. Thas [178], Johnson and Payne [79], Payne [122], and the recent paper by Johnson [78].

Remark 4.4.7 Derivation for any q can be described algebraically; see Bader, Lunardon and Payne [3], Payne [121] and Section 2.9 of Cardinali and Payne [35]

4.5   Semifield Flocks

Casse, Thas and Wild [37] give the following construction of flocks.

Let $\mathbf{GF}(q^n)$ be an extension of $\mathbf{GF}(q)$ and let $\mathbf{PG}(3n - 1, q^n)$ be the corresponding extension of $\mathbf{PG}(3n - 1, q)$. Let π be a plane of $\mathbf{PG}(3n - 1, q^n)$ with the property that π and its $n - 1$ conjugates with respect to the extension $\mathbf{GF}(q^n)$ of $\mathbf{GF}(q)$, generate $\mathbf{PG}(3n-1, q^n)$. Then each point x of π, together with the $n-1$ conjugates of x, define an $(n-1)$-dimensional subspace of $\mathbf{PG}(3n - 1, q)$. These $(n - 1)$-dimensional subspaces of $\mathbf{PG}(3n - 1, q)$ form an $(n - 1)$-spread T of $\mathbf{PG}(3n - 1, q)$. Now embed $\mathbf{PG}(3n - 1, q)$ in $\mathbf{PG}(3n, q)$. Then the incidence structure with as point set $\mathbf{PG}(3n, q) \setminus$

$\mathbf{PG}(3n - 1, q)$, with as lines the n-dimensional subspaces of $\mathbf{PG}(3n, q)$ which contain an element of $\mathcal{T}$ but are not contained in $\mathbf{PG}(3n - 1, q)$, and with as incidence the natural one, is isomorphic to the design of points and lines of $\mathbf{AG}(3, q^n)$. Now consider a line L of π. Then L and its $n - 1$ conjugates define a $(2n - 1)$-dimensional subspace of $\mathbf{PG}(3n - 1, q)$. The set of these $(2n - 1)$-dimensional subspaces is denoted by $\mathcal{T}'$. The planes of $\mathbf{AG}(3, q^n)$ correspond to the $2n$-dimensional subspaces of $\mathbf{PG}(3n, q)$ which contain an element of $\mathcal{T}'$ but are not contained in $\mathbf{PG}(3n - 1, q)$. See also Section 5.1 of Dembowski [45]; for an algebraic description of this representation of $\mathbf{AG}(3, q^n)$ we refer to our Section 3.6.

Next, let Γ be a nonsingular dual conic of π and, for q odd, let $\mathcal{C}$ be the nonsingular conic consisting of the tangent points of Γ. With the $q^n + 1$ lines of Γ correspond $q^n + 1$ elements of $\mathcal{T}'$; this subset of $\mathcal{T}'$ will be denoted by $\mathcal{C}_{2n}$. For q odd, with the $q^n + 1$ points of $\mathcal{C}$ correspond $q^n + 1$ elements of $\mathcal{T}$; this subset of $\mathcal{T}$ will be denoted by $\mathcal{C}_n$. For q odd, the set $\mathcal{C}_n$ is a pseudo-conic $\mathcal{O}(n, n, q)$ and $\mathcal{C}_{2n}$ is the set of all tangent spaces of $\mathcal{O}(n, n, q)$. We remark that $\mathcal{C}_{2n}$ is a pseudo-conic in the dual space of $\mathbf{PG}(3n - 1, q)$. Let $\mathbf{PG}(n - 1, q)$ be an $(n - 1)$-dimensional subspace of $\mathbf{PG}(3n-1, q)$ intersecting no element of $\mathcal{C}_{2n}$. Further, let $\mathbf{PG}(n, q)$ be an n-dimensional subspace of $\mathbf{PG}(3n, q)$ which contains $\mathbf{PG}(n - 1, q)$ but which is not contained in $\mathbf{PG}(3n-1, q)$. If $\mathbf{PG}(3, q^n)$ is the projective completion of $\mathbf{AG}(3, q^n)$, with $\overline{\pi} = \mathbf{PG}(3, q^n) \backslash \mathbf{AG}(3, q^n)$, then Γ defines a nonsingular dual conic $\overline{\Gamma}$ of $\overline{\pi}$ and, for q odd, $\mathcal{C}$ defines a nonsingular conic $\overline{\mathcal{C}}$ of $\overline{\pi}$. If $\overline{\mathcal{F}}$ is the point set of $\mathbf{AG}(3, q^n)$ which corresponds to $\mathbf{PG}(n, q) \backslash \mathbf{PG}(n-1, q)$, then the line of $\mathbf{PG}(3, q^n)$ joining any two distinct points of $\overline{\mathcal{F}}$ has no point in common with the union of the lines of $\overline{\Gamma}$; for q odd this line intersects $\overline{\pi}$ in an interior point of the conic $\overline{\mathcal{C}}$. Hence $\overline{\mathcal{F}}$ is a flock of the dual conic $\overline{\Gamma}$. This flock $\overline{\mathcal{F}}$ is not linear if and only if $\mathbf{PG}(n - 1, q) \notin \mathcal{T}$.

Assume that $\mathbf{PG}(n - 1, q)$ is contained in an element η of $\mathcal{T}'$. Then the set $\overline{\mathcal{F}}$ is contained in a plane. Dualizing, we obtain a flock $\mathcal{F}$ of the quadratic cone for which the planes of the conics of $\mathcal{F}$ all contain a common point. Hence, by Theorems 4.4.4 up to 4.4.6, either $\mathbf{PG}(n - 1, q) \in \mathcal{T}$ and $\overline{\mathcal{F}}$ is linear, or q is odd and $\mathcal{F}$ is a Kantor-Knuth flock; if $\overline{\mathcal{F}}$ is not linear, then η contains two distinct elements of $\mathcal{C}_n$.

Conversely, let q be odd and let $\mathcal{F}$ be a flock of the quadratic cone $\mathcal{K}$ for which the planes π_i all contain a common point. Then, using the equations of the planes π_i given in the statement of Theorem 4.4.6, dualizing, and applying the mapping ζ of Section 3.6 from $\mathbf{AG}(3, q)$ onto $\mathbf{AG}(3n, q')$, with $q'^n = q$ and $\mathbf{GF}(q')$ a subfield of the field of fixed elements of the automorphism σ, one easily sees that we have a flock of the type described above with $\mathbf{PG}(n-1,q)$ in an element of $\mathcal{T}'$.

Let $X_0 X_1 = X_2^2$ be the equation of the quadratic cone $\mathcal{K}$ in $\mathbf{PG}(3,q)$, and consider a flock $\mathcal{F} = \{\mathcal{C}_1, \mathcal{C}_2, \ldots, \mathcal{C}_q\}$ of $\mathcal{K}$. Let the plane π_i of $\mathcal{C}_i$ have equation $a_i X_0 + b_i X_1 + c_i X_2 + X_3 = 0$, with $i = 1, 2, \ldots, q$. Clearly we may put $a_i = t, b_i = z_t, c_i = y_t$, with $t \in \mathbf{GF}(q)$ and $y_0 = z_0 = 0$. As $\mathcal{F}$ is a flock, by Theorem 4.4.2, $t \mapsto z_t$ defines a permutation of $\mathbf{GF}(q)$, and for q even also $t \mapsto y_t$ is a permutation of $\mathbf{GF}(q)$. The construction of Casse, Thas and Wild yields a flock $\mathcal{F}$ if and only if the points (t, z_t, y_t) of $\mathbf{AG}(3,q)$ define an $\mathbf{AG}(n, q')$ for some subfield $\mathbf{GF}(q')$ of $\mathbf{GF}(q)$, that is, if and only if y_t and z_t are additive. On the other hand the translation plane defined by the flock $\mathcal{F}$ (see Section 4.2) is a semifield plane if and only if y_t and z_t are additive; see Kallaher [80]. For this reason these flocks are called *semifield flocks*. For more details we refer to Lundardon [100].

The following fundamental result is due to Johnson [76].

Theorem 4.5.1 *A semifield flock of a quadratic cone in* $\mathbf{PG}(3,q)$ *for* q *even is linear.* ∎

Remark on the Proof. The original proof of Theorem 4.5.1 relies on the connection between the flock and the semifield which defines the corresponding semifield plane. A geometrical proof will be given in Section 5.1 (Theorem 5.1.11).

Known Examples of Semifield Flocks

With the notation introduced above, we will give for each flock the set of triples (t, z_t, y_t), with $t \in \mathbf{GF}(q)$ and q odd.

(a) **The Kantor-Knuth flocks**

$$(t, -mt^\sigma, 0), t \in \mathbf{GF}(q), q \text{ odd},$$

m a given nonsquare of $\mathbf{GF}(q)$, σ any automorphism of $\mathbf{GF}(q)$.

All planes of the conics of the flock contain a common exterior point of the cone. The flock is linear if and only if $\sigma = 1$. Derivation yields no new flock [122].

(b) **The Ganley flocks**

$$(t, t^3, -mt - m^{-1}t^9), t \in \mathbf{GF}(q), q = 3^h,$$

m a given nonsquare of $\mathbf{GF}(q)$.

This flock is linear if and only if $q = 3$; it is isomorphic to a Kantor-Knuth flock if and only if $q \in \{3, 9\}$. For $q > 9$ derivation yields one extra (non-semifield) flock [122].

(c) **The Penttila-Williams-Bader-Lunardon-Pinneri flock**

$$(t, -t^9, t^{27}), t \in \mathbf{GF}(3^5).$$

This flock is not of the previous types, and derivation yields one extra (non-semifield) flock [122].

This example was found by an interesting mix of theory and computer assistance. Penttila and Williams [131], with a slight assist of a computer, discovered a translation ovoid of the GQ $\mathbf{Q}(4, 3^5)$; see also Chapter 7. In 1994 (see the proceedings of the Oberwolfach meeting on Designs and Codes in 1994) Thas [181] had shown how to interpret a translation ovoid of a GQ $\mathbf{Q}(4, q)$ as a semifield flock; a detailed and more complete treatment of this correspondence was given by Lunardon [100]. Using this technique Bader, Lunardon and Pinneri [4] succeeded to determine the exact form of this semifield flock over $\mathbf{GF}(3^5)$.

4.6 Generalized Quadrangles and Flocks

In this section we will consider a particular class of GQs $\mathcal{S}(G, \mathcal{J})$. If $u = (u_1, u_2), v = (v_1, v_2) \in \mathbf{GF}(q) \times \mathbf{GF}(q)$, then put $uv = u_1v_1 + u_2v_2$. Let $G = \{(\alpha, c, \beta) \| \alpha, \beta \in \mathbf{GF}(q) \times \mathbf{GF}(q), c \in \mathbf{GF}(q)\}$, with the group operation defined by

$$(\alpha, c, \beta) \circ (\alpha', c', \beta') = (\alpha + \alpha', c + c' + \beta\alpha', \beta + \beta').$$

This operation makes G into a group of order q^5.

Put $A(\infty) = \{(0, 0, \beta) \in G \| \beta \in \mathbf{GF}(q) \times \mathbf{GF}(q)\}$.

For each $t \in \mathbf{GF}(q)$, let

$$A_t = \begin{pmatrix} x_t & y_t \\ 0 & z_t \end{pmatrix}$$

be an upper triangular 2×2-matrix over $\mathbf{GF}(q)$, with the convention that A_0 be the zero matrix. For each $t \in \mathbf{GF}(q)$, put $K_t = A_t + A_t^T$, and let

$$A(t) = \{(\alpha, \alpha A_t \alpha^T, \alpha K_t) \,\|\, \alpha \in \mathbf{GF}(q) \times \mathbf{GF}(q)\}.$$

It follows that $A(t)$ is a commutative subgroup of G having order q^2, for each $t \in \mathbf{GF}(q) \cup \{\infty\}$.

Let $C = \{(0, c, 0) \in G \,\|\, c \in \mathbf{GF}(q)\}$, and put $A^*(t) = A(t)C$, $t \in \mathbf{GF}(q) \cup \{\infty\}$. Then $A^*(t)$ is a commutative subgroup of G having order q^3, $t \in \mathbf{GF}(q) \cup \{\infty\}$. Moreover $A^*(t) \supset A(t)$. Further, let $\mathcal{J} = \{A(t) \,\|\, t \in \mathbf{GF}(q) \cup \{\infty\}\}$ and $\mathcal{J}^* = \{A^*(t) \,\|\, t \in \mathbf{GF}(q) \cup \{\infty\}\}$.

With the foregoing notation we have the following two important theorems.

Theorem 4.6.1 (Kantor [83]) *If q is odd, then $G, \mathcal{J}, \mathcal{J}^*$ satisfy Conditions (K1) and (K2) of Section 3.2.2, so that $\mathcal{S}(G, \mathcal{J})$ is a GQ of order (q^2, q), if and only if*

$$-det(K_t - K_u) = (y_t - y_u)^2 - 4(x_t - x_u)(z_t - z_u)$$

is a nonsquare of $\mathbf{GF}(q)$ whenever $t, u \in \mathbf{GF}(q)$, $t \neq u$. ∎

Theorem 4.6.2 (Payne [117]) *If q is even, then $G, \mathcal{J}, \mathcal{J}^*$ satisfy Conditions (K1) and (K2) of Section 3.2.2, so that $\mathcal{S}(G, \mathcal{J})$ is a GQ of order (q^2, q), if and only if*

$$y_t \neq y_u \text{ and } (x_t + x_u)(z_t + z_u)(y_t + y_u)^{-2} \in \mathcal{C}_1$$

whenever $t, u \in \mathbf{GF}(q), t \neq u$. ∎

A set of q upper triangular 2×2-matrices over $\mathbf{GF}(q)$ satisfying the conditions of Theorems 4.6.1 and 4.6.2 is called a *q-clan*.

Comparing Theorems 4.6.1 and 4.6.2 with Theorem 4.4.2 yields the following fundamental result.

Theorem 4.6.3 (Thas [172]) *The group G and the sets $\mathcal{J}, \mathcal{J}^*$ satisfy Conditions (K1) and (K2) of Section 3.2.2, that is, $\mathcal{S}(G, \mathcal{J})$ is a GQ of order (q^2, q), if and only if the planes $x_t X_0 + z_t X_1 + y_t X_2 + X_3 = 0$, $t \in \mathbf{GF}(q)$, define a flock $\mathcal{F}$ of the quadratic cone with equation $X_0 X_1 = X_2^2$ in $\mathbf{GF}(q)$. Each flock $\mathcal{F}$ of the quadratic cone of $\mathbf{PG}(3, q)$ gives us a GQ, and each GQ $\mathcal{S}(G, \mathcal{J})$ of the type described above gives us a flock of the quadratic cone; the GQ $\mathcal{S}(G, \mathcal{J})$ is also denoted $\mathcal{S}(\mathcal{F})$ and is called a* flock GQ. ∎

Payne and Thas [128] show that if the flock $\mathcal{F}$ is linear then $\mathcal{S}(G, \mathcal{J})$ is isomorphic to the classical GQ $\mathbf{H}(3, q^2)$; conversely, it follows from Payne [120, 121] that if $\mathcal{S}(\mathcal{F}) \cong \mathbf{H}(3, q^2)$, then $\mathcal{F}$ is linear. In [129] Payne and Thas show that if the GQ $\mathcal{S}(\mathcal{F})$ is not classical, then the elation point (∞) of $\mathcal{S}(\mathcal{F})$ is fixed by the full automorphism group of $\mathcal{S}(\mathcal{F})$. Finally, as C is a normal subgroup of G it follows from 8.2.2(v) of Payne and Thas [128] that (∞) is a center of symmetry of $\mathcal{S}(\mathcal{F})$.

Finally, by recoordinatization, every line of $\mathcal{S}(\mathcal{F})$ incident with the elation point (∞) can be taken as line $[A(\infty)]$; see Payne and Rogers [125] and Payne [120, 121].

4.7 Semifield Flocks and Translation Generalized Quadrangles

4.7.1 Position of a Translation Line in a Flock Quadrangle

Let L be a translation line of the flock GQ $\mathcal{S}(\mathcal{F})$ of order (q^2, q), q odd, and suppose that L is not incident with (∞). Then L is incident with a point $x \notin (\infty)^{\perp}$. As x is a center of symmetry, there is a symmetry θ with center x mapping (∞) onto $(\infty)' \neq (\infty)$. So by Payne and Thas [129], $\mathcal{S}(\mathcal{F}) \cong \mathbf{H}(3, q^2)$ and consequently $\mathcal{F}$ is linear. Hence when $\mathcal{F}$ is nonlinear, any translation line is incident with (∞). When $\mathcal{F}$ is linear, each line is a translation line and so also in that case a given translation line may be taken to be incident with (∞).

4.7.2 Additive q-Clans and Translation Generalized Quadrangles

Suppose q is a prime power. A q-clan $\{A_t = \begin{pmatrix} t & y_t \\ 0 & z_t \end{pmatrix} \parallel t \in \mathbf{GF}(q)\}$ is *additive* if for each $t, t' \in \mathbf{GF}(q)$ the following property holds:

$$A_t + A_{t'} = A_{t+t'}.$$

Let $\mathcal{C} = \{A_t \parallel t \in \mathbf{GF}(q)\}$ be a q-clan, put $K_t = A_t + A_t^T$, and define $g_t(\lambda) = \lambda A_t \lambda^T$ and $\lambda^{\delta_t} = \lambda K_t$ for $\lambda \in \mathbf{GF}(q)^2$. Put

$$G = \{(\alpha, c, \beta) \parallel \alpha, \beta \in \mathbf{GF}(q)^2, c \in \mathbf{GF}(q)\},$$

and define a binary operation on G by

$$(\alpha, c, \beta) \circ (\alpha', c', \beta') = (\alpha + \alpha', c + c' + \beta\alpha'^T, \beta + \beta').$$

This makes G into a group. Define a family of subgroups $\mathcal{J}$ of G as follows

$$A(t) = \{(\alpha, g_t(\alpha), \alpha^{\delta_t}) \mid\mid \alpha \in \mathbf{GF}(q)^2\}, \; t \in \mathbf{GF}(q),$$

and

$$A(\infty) = \{(0, 0, \beta) \mid\mid \beta \in \mathbf{GF}(q)^2\}.$$

Define a family of subgroups $\mathcal{J}^*$ of G by

$$A^*(t) = \{(\alpha, c, \alpha^{\delta_t}) \mid\mid \alpha \in \mathbf{GF}(q)^2, c \in \mathbf{GF}(q)\}, \; t \in \mathbf{GF}(q),$$

and

$$A^*(\infty) = \{(0, c, \beta) \mid\mid c \in \mathbf{GF}(q), \beta \in \mathbf{GF}(q)^2\}.$$

Then $\mathcal{J}$ is a 4-gonal family which defines a flock GQ of order (q^2, q); see Section 4.6.

The next theorem is due to Payne [118].

Theorem 4.7.1 *A q-clan* $\mathcal{C} = \{A_t = \begin{pmatrix} t & y_t \\ 0 & z_t \end{pmatrix} \mid\mid t \in \mathbf{GF}(q)\}$ *is additive if and only if the point-line dual of the corresponding flock* GQ *is a* TGQ *with translation point corresponding to* $[A(\infty)]$.

Proof. Let $\mathcal{S}$ be the GQ of order (q^2, q) defined by $\mathcal{C}$, and let $\mathcal{S}^D$ be its point-line dual. Suppose $\mathcal{S}^D$ is a TGQ with translation point $[A(\infty)]$ and translation group G. As automorphism group of $\mathcal{S}$, G induces a regular permutation group on $\{[A(t)] \mid\mid t \in \mathbf{GF}(q)\}$, which is an elementary abelian p-group, where $q = p^h$ for the prime p, of size q. Denote this group by $G^\#$. Let $\mathcal{F}$ be the flock of the quadratic cone $\mathcal{K}$ (with vertex v) in $\mathbf{PG}(3, q)$ defined by the A_ts. Then $G^\#$ can be interpreted as a subgroup of $\mathbf{P\Gamma L}_4(q)$ that acts regularly on the flock planes (cf. Addendum A (§4.12) to this chapter). In the dual space of $\mathbf{PG}(3, q)$, the flock planes become q points lying in the affine 3-space obtained by deleting the plane corresponding to v, on which an elementary abelian p-subgroup of $\mathbf{P\Gamma L}_4(q)$ acts regularly. So this point set can be considered as the point set of an affine space over some subfield of $\mathbf{GF}(q)$. Whence $\mathcal{F}$ is a semifield flock by Section 4.5, and therefore

$$\Theta : t \mapsto A_t$$

is additive.

Conversely, suppose Θ is additive. Then $\delta_{t+r} = \delta_t + \delta_r$. Now define, for $r \in \mathbf{GF}(q)$,

$$\phi_r : (\alpha, c, \beta) \mapsto (\alpha, c + g_r(\alpha), \beta + \alpha^{\delta_r}).$$

It is routine to check that ϕ_r is a symmetry of the flock GQ about the point $A^*(\infty)$ sending $A(t)$ to $A(t + r)$. So $A^*(\infty)$ is a center of symmetry, and then each point of $[A(\infty)]$ is as such, in view of the fact that (∞) is an elation point and that (∞) is a center of symmetry (see Section 4.6). The theorem follows. ∎

Theorem 4.7.2 *The flock $\mathcal{F}$ defined by the q-clan $\mathcal{C} = \{A_t = \begin{pmatrix} t & y_t \\ 0 & z_t \end{pmatrix} \| t \in \mathbf{GF}(q)\}$ is a semifield flock if and only if the line $A[(\infty)]$ of the flock GQ $\mathcal{S}(\mathcal{F})$ is a translation line.*

Proof. Follows from Theorem 4.7.1 and Section 4.5. ∎

4.7.3   Known Cases

With each semifield flock corresponds a TGQ, and hence an egg. We give an overview of some terminology on these eggs and their translation duals.

KANTOR-KNUTH EGGS.   Let $\mathcal{F}$ be a Kantor-Knuth flock. The corresponding GQ $\mathcal{S}(\mathcal{F})$, which is a TGQ for some base-line, was first discovered by Kantor [83], and is called a *Kantor-Knuth generalized quadrangle*. The kernel $\mathbb{K}$ of the TGQ is the fixed field of σ, see [133]. The following was shown by Payne in [118].

Theorem 4.7.3 (Payne [118]) *Suppose a TGQ $\mathcal{S} = \mathbf{T}(\mathcal{O})$ is the point-line dual of a flock GQ $\mathcal{S}(\mathcal{F})$, $\mathcal{F}$ a Kantor-Knuth semifield flock. Then $\mathbf{T}(\mathcal{O})$ is isomorphic to its translation dual $\mathbf{T}(\mathcal{O}^*)$.* ∎

GANLEY EGGS AND ROMAN (GANLEY-PAYNE) EGGS.   Let $\mathcal{F}$ be a Ganley flock. The point-line dual of the corresponding GQ $\mathcal{S}(\mathcal{F})$ is a TGQ for some base-line, and so the dual $\mathcal{S}(\mathcal{F})^D$ of $\mathcal{S}(\mathcal{F})$ is isomorphic to some $\mathbf{T}(\mathcal{O})$. The GQ $\mathbf{T}(\mathcal{O})$ is called a *Ganley generalized quadrangle*. By [133], the kernel $\mathbb{K}$ is isomorphic to $\mathbf{GF}(3)$. Payne [118] shows the following.

Theorem 4.7.4 (Payne [118]) *For $q > 9$, $\mathbf{T}(\mathcal{O})$ is not isomorphic to its translation dual $\mathbf{T}(\mathcal{O}^*)$.* ∎

Further, Payne proves in [118] that $\mathbf{T}(\mathcal{O}^*)$ is a TGQ which does not arise from a flock. In [118], the GQs $\mathbf{T}(\mathcal{O}^*)$ were called the *Roman generalized quadrangles*.

THE PENTTILA-WILLIAMS-BADER-LUNARDON-PINNERI EGG AND ITS TRANS-LATION DUAL. Let $\mathcal{F}$ be the Penttila-Williams-Bader-Lunardon-Pinneri flock. The translation dual $\mathbf{T}(\mathcal{O})$ of $\mathcal{S}(\mathcal{F})^D$ is referred to as the *Penttila-Williams-Bader-Lunardon-Pinneri generalized quadrangle*. The kernel of this GQ is isomorphic to $\mathbf{GF}(3)$ [122]. For a detailed study we refer to Bader, Lunardon and Pinneri [4].

Theorem 4.7.5 (Bader, Lunardon and Pinneri [4]) $\mathbf{T}(\mathcal{O})$ *is not isomorphic to its translation dual* $\mathbf{T}(\mathcal{O}^*)$. ∎

4.8 Derivation and BLT-Sets

Let $\mathcal{F} = \{\mathcal{C}_1, \mathcal{C}_2, \ldots, \mathcal{C}_q\}$ be a flock of the quadratic cone $\mathcal{K}$ with vertex x of $\mathbf{PG}(3, q)$, with q odd. The plane of $\mathcal{C}_i$ is denoted by $\pi_i, i = 1, 2, \ldots, q$. Let $\mathcal{K}$ be embedded in the nonsingular quadric $\mathbf{Q}$ of $\mathbf{PG}(4, q)$. Let the polar line of π_i with respect to $\mathbf{Q}$ be denoted by L_i and let $L_i \cap \mathbf{Q} = \{x, x_i\}$, $i = 1, 2, \ldots, q$. If H_i is the tangent hyperplane of $\mathbf{Q}$ at x_i, then the process of derivation, see Section 4.4, yields a flock $\mathcal{F}_i$ of the cone $H_i \cap \mathbf{Q} = \mathcal{K}_i$, with $i = 1, 2, \ldots, q$. The following important theorem on derivation of flocks is due to Payne and Rogers [125].

Theorem 4.8.1 *The process of derivation produces new flocks and new planes, but never new GQs.* ∎

The set of points $\{x, x_1, x_2, \ldots, x_q\} = \mathcal{B}$ is called a *dual* BLT-*set* of the quadric $\mathbf{Q}$, following a suggestion of Kantor [84]. So a dual BLT-set of a nonsingular quadric $\mathbf{Q}$ in $\mathbf{PG}(4, q)$, with q odd, is a set $\mathcal{B}$ of size $q + 1$, so that for any three distinct points y_i, y_j, y_k of $\mathcal{B}$ we have $|\{y_i, y_j, y_k\}^\perp| = 0$ on $\mathbf{Q}$. By Bader, Lunardon and Thas [5] it is sufficient that $|\mathcal{B}| = q + 1$ and $|\{y_i, y_j, y_k\}^\perp| = 0$ for $y_i \neq y_j \neq y_k \neq y_i$ and y_i fixed. Since the GQ $\mathbf{Q}(4, q)$ arising from $\mathbf{Q}$ is isomorphic to the dual of the GQ $\mathbf{W}(q)$, to a dual BLT-set corresponds a set $\mathcal{V}$ of $q + 1$ lines of $\mathbf{W}(q)$ with the property that no line of $\mathbf{W}(q)$ is concurrent with three distinct lines of $\mathcal{V}$; such a set $\mathcal{V}$ is called a BLT-*set*.

Remark 4.8.2 Derivation for any q can be described algebraically. In the flock GQ it amounts to choose any given line incident with (∞) as the line $[A(\infty)]$. From this approach it is clear that the process of derivation does not produce new GQs, but, as in the odd case, also in the even case new flocks and planes arise. For details we refer to Bader, Lunardon and Payne [3], Payne [121] and Section 2.9 of Cardinali and Payne [35].

4.9   Constructions

(a) The Construction of Knarr

Start with a symplectic polarity ζ of $\mathbf{PG}(5,q)$. Let $p \in \mathbf{PG}(5,q)$ and let $\mathbf{PG}(3,q)$ be a 3-dimensional subspace of $\mathbf{PG}(5,q)$ for which $p \notin \mathbf{PG}(3,q) \subset p^\zeta$. In $\mathbf{PG}(3,q)$ the polarity ζ induces a symplectic polarity ζ', and hence a GQ $\mathbf{W}(q)$. Let $\mathcal{V}$ be a BLT-set of the GQ $\mathbf{W}(q)$ and construct a geometry $\mathcal{S} = (\mathcal{P}, \mathcal{B}, \mathrm{I})$ as follows.

- POINTS are of three types:

    (i)   the points of $\mathbf{PG}(5,q)$ not in p^ζ;

    (ii) the lines of $\mathbf{PG}(5,q)$ not containing p but contained in one of the planes $\pi_t = pL_t$, with $L_t \in \mathcal{V}$;

    (iii) p.

- LINES are of two types:

    (a)   the totally isotropic planes of ζ not contained in p^ζ and meeting some $\pi_t = pL_t$, with $L_t \in \mathcal{V}$, in a line (not through p);

    (b)   the planes $\pi_t = pL_t$, with $L_t \in \mathcal{V}$.

- The INCIDENCE RELATION I is just the natural incidence inherited from $\mathbf{PG}(5,q)$.

Then Knarr [89] proves that $\mathcal{S}$ is a GQ of order (q^2, q) isomorphic to one of the GQs arising from any of the $q+1$ flocks defined by the BLT-set $\mathcal{V}$. If $\mathcal{F}$ is some flock corresponding to $\mathcal{V}$, then the elation point (∞) of the GQ $\mathcal{S}(\mathcal{F}) \cong \mathcal{S}$ corresponds to the point p of $\mathcal{S}$.

The problem of finding a geometric construction of a flock GQ was open for quite some time. As the years passed after Knarr found such a construction for q odd, many researchers came to believe that no such construction could exist for q even. However, as a byproduct of the proof of a fundamental characterization of flock GQs, see Section 4.10, Thas [183] presented such a construction, with refinement of related material in Thas [182, 186].

(b) The Construction of Thas

Let $\mathcal{K}$ be a quadratic cone with vertex x of $\mathbf{PG}(3, q)$. Further, let y be a point of $\mathcal{K}\backslash\{x\}$ and let π be a plane of $\mathbf{PG}(3, q)$ not containing y nor x. Now we project $\mathcal{K}\backslash\{y\}$ from y onto π. Let τ be the tangent plane of $\mathcal{K}$ at the line xy and let $\tau \cap \pi = T$. Then with the q^2 points of $\mathcal{K}\backslash xy$ correspond the q^2 points of the affine plane $\pi\backslash T = \pi'$, with any point of $xy\backslash\{y\}$ corresponds the intersection ∞ of xy and π, with the generators of $\mathcal{K}$ distinct from xy correspond the lines of π distinct from T containing ∞, with the (nonsingular) conics on $\mathcal{K}$ passing through y correspond the affine parts of the q^2 lines of π not passing through ∞, and with the (nonsingular) conics on $\mathcal{K}$ not passing through y correspond the $q^2(q-1)$ (nonsingular) conics of π which are tangent to T at ∞.

Let $\mathcal{F} = \{\mathcal{C}_1^*, \mathcal{C}_2^*, \ldots, \mathcal{C}_q^*\}$ be a flock of the cone $\mathcal{K}$. Now consider the set $\widetilde{\mathcal{F}} = \{\mathcal{C}_1, \mathcal{C}_2, \ldots, \mathcal{C}_{q-1}, N\}$ consisting of the $q-1$ nonsingular conics $\mathcal{C}_1, \mathcal{C}_2, \ldots, \mathcal{C}_{q-1}$ and the line N of π, which is obtained by projecting the elements of $\mathcal{F}$ from y onto π. So $\mathcal{C}_1, \mathcal{C}_2, \ldots, \mathcal{C}_{q-1}$ are conics which are mutually tangent at ∞ (with common tangent line T) and N is a line of π not containing ∞.

Now we consider planes $\pi_\infty \neq \pi$ and $\mu \neq \pi$ of $\mathbf{PG}(3, q)$, respectively containing T and N; in μ we consider a point r, with $r \notin \pi \cup \pi_\infty$. Next, let $\mathcal{O}_i$ be the nonsingular quadric which contains $\mathcal{C}_i$, which is tangent to π_∞ at ∞ and which is tangent to μ at r, with $i = 1, 2, \ldots, q-1$. As $\mathcal{C}_i \cap N = \emptyset$, the quadric $\mathcal{O}_i$ is elliptic, $i = 1, 2, \ldots, q-1$.

Next let $\mathcal{S}$ be the following incidence structure.

- POINTS are of five types.

  (a) The $q^3(q-1)$ nonsingular elliptic quadrics $\mathcal{O}$ containing $\mathcal{O}_i \cap \pi_\infty = L_\infty^{(i)} \cup M_\infty^{(i)}$ (over $\mathbf{GF}(q^2)$) such that the intersection multiplicity of $\mathcal{O}_i$ and $\mathcal{O}$ at ∞ is at least three (these are $\mathcal{O}_i$, the nonsingular elliptic quadrics $\mathcal{O} \neq \mathcal{O}_i$ containing $L_\infty^{(i)} \cup M_\infty^{(i)}$ (over $\mathbf{GF}(q^2)$) and intersecting $\mathcal{O}_i$ over $\mathbf{GF}(q)$ in a nonsingular conic containing ∞, and the nonsingular elliptic quadrics $\mathcal{O} \neq \mathcal{O}_i$ for which $\mathcal{O} \cap \mathcal{O}_i$ over $\mathbf{GF}(q^2)$ is $L_\infty^{(i)} \cup M_\infty^{(i)}$ counted twice), with $i = 1, 2, \ldots, q-1$.

(b) The q^3 points of $\mathbf{PG}(3,q) \backslash \pi_\infty$.

(c) The q^3 planes of $\mathbf{PG}(3,q)$ not containing ∞.

(d) The $q-1$ sets $\mathbf{O}_i$, where $\mathbf{O}_i$ consists of the q^3 quadrics $\mathcal{O}$ of Type (a) corresponding to $\mathcal{O}_i$, with $i = 1, 2 \ldots, q-1$.

- LINES are of five types.

  (i) Let (w, γ) be a point-plane flag of $\mathbf{PG}(3,q)$, with $w \notin \pi_\infty$ and $\infty \notin \gamma$. Then all quadrics $\mathcal{O}$ of Type (a) which are tangent to γ at w, together with w and γ, form a line of Type (i). Any two distinct quadrics of such a line have exactly two points (∞ and w) in common. The total number of lines of Type (i) is q^5.

  (ii) Let $\mathcal{O}$ be a point of Type (a) which corresponds to the quadric $\mathcal{O}_i$, $i \in \{1, 2, \ldots, q-1\}$. If $\mathcal{O} \cap \pi_\infty = \mathcal{O}_i \cap \pi_\infty = L_\infty^{(i)} \cup M_\infty^{(i)}$ (over $\mathbf{GF}(q^2)$), then all points $\mathcal{O}'$ of Type (a) for which $\mathcal{O}' \cap \mathcal{O}$ over $\mathbf{GF}(q^2)$ is $L_\infty^{(i)} \cup M_\infty^{(i)}$ counted twice, together with $\mathcal{O}$ and $\mathcal{O}_i$, form a line of Type (ii). There are $q^2(q-1)$ lines of Type (ii).

  (iii) A set of q parallel planes of $\mathbf{AG}(3,q) = \mathbf{PG}(3,q) \setminus \pi_\infty$, where the line at infinity does not contain ∞, together with the plane π_∞, is a line of Type (iii).

  (iv) Lines of Type (iv) are the lines of $\mathbf{PG}(3,q)$, not in π_∞, containing ∞.

  (v) $\{\infty, \pi_\infty, \mathbf{O}_1, \mathbf{O}_2, \ldots, \mathbf{O}_{q-1}\}$ is the unique line of Type (v).

- INCIDENCE is containment.

Then it is proved in Thas [183] that $\mathcal{S}$ is a GQ isomorphic to the point-line dual of the flock GQ $\mathcal{S}(\mathcal{F})$. We emphasize that this construction works for any prime power q.

4.10 Property (G) for Generalized Quadrangles of Order (s, s^2)

In Section 2.5 we introduced 3-regularity in GQs of order (s, s^2). Here we will weaken this definition so as to obtain a strong characterizing property for flock GQs.

Let $\mathcal{S} = (\mathcal{P}, \mathcal{B}, \mathrm{I})$ be a GQ of order (s, s^2), $s \neq 1$. Let x_1, y_1 be distinct collinear points. We say that the pair $\{x_1, y_1\}$ has *Property* (G) or that $\mathcal{S}$ has *Property* (G) at $\{x_1, y_1\}$, if every triad $\{x_1, x_2, x_3\}$, with $y_1 \in \{x_1, x_2, x_3\}^{\perp}$, is 3-regular. Then also every triad $\{y_1, y_2, y_3\}$, with $x_1 \in \{y_1, y_2, y_3\}^{\perp}$, is 3-regular. The GQ $\mathcal{S}$ has *Property* (G) *at the line* L, or the line L has Property (G), if each pair of points $\{x, y\}$, $x \neq y$, and $x\mathrm{I}L\mathrm{I}y$, has Property (G). If (x, L) is a flag, that is, if $x\mathrm{I}L$, then we say that $\mathcal{S}$ has *Property* (G) at (x, L), or that (x, L) has Property (G), if every pair $\{x, y\}$, $x \neq y$, and $y\mathrm{I}L$, has Property (G).

The point x of the GQ $\mathcal{S}$ of order (s, s^2), with $s \neq 1$, is 3-regular if and only if each flag (x, L) has Property (G).

If s is even and if $\mathcal{S}$ contains a pair of points having Property (G), then by Theorem 2.6.2, $\mathcal{S}$ contains subquadrangles of order s.

In Payne [118] Property (G) is defined for any GQ of order (s, t), with $t > s > 1$.

Motivation for introducing Property (G) is the following result of Payne [118].

Theorem 4.10.1 *If $\mathcal{S}(\mathcal{F})$ is the GQ of order (q^2, q) arising from the flock $\mathcal{F}$ of the quadratic cone $\mathcal{K}$ of* $\mathbf{PG}(3, q)$*, then $\mathcal{S}(\mathcal{F})$ has Property* (G) *at its elation point* (∞). ∎

Now we prove that if $\mathcal{S}$ is a GQ of order (q, q^2), $q \neq 1$, for which at least one pair of points $\{x, y\}$, $x \sim y$, has Property (G), then $\mathcal{S}$ defines a projective space $\mathbf{PG}(3, q)$. This is due to Payne and Thas; see Thas [176].

Let $\mathcal{S} = (\mathcal{P}, \mathcal{B}, \mathrm{I})$ be a GQ of order $(q, q^2), q \neq 1$, for which the pair of distinct points $\{x, y\}, x \sim y$, has Property (G). We introduce the following incidence structure $\mathcal{S}_{xy} = (\mathcal{P}_{xy}, \mathcal{B}_{xy}, \mathrm{I}_{xy})$;

(i) $\mathcal{P}_{xy} = x^{\perp} \backslash \{x, y\}^{\perp}$.

(ii) Elements of $\mathcal{B}_{xy}$ are of two types:

(a) the sets $\{y, z, u\}^{\perp\perp}\backslash\{y\}$, with $\{y, z, u\}$ a triad having x as center, and

(b) the sets $\{x, w\}^{\perp}\backslash\{x\}$, with $x \sim w \not\sim y$.

(iii) $\mathbf{I}_{xy}$ is containment.

Then $|\mathcal{P}_{xy}| = q^3$, $\mathcal{B}_{xy}$ contains $q^3(q^3 - q)/(q(q-1)) = q^4 + q^3$ elements of Type (a) and q^2 elements of Type (b).

Lemma 4.10.2 *Let M_1, M_2 be different elements of Type (a) of $\mathcal{B}_{xy}$, let L_1, L_2 be different elements of Type (b) of $\mathcal{B}_{xy}$, and assume that $L_i \cap M_j \neq \emptyset$ for all $i, j \in \{1, 2\}$. Then each element of Type (b) of $\mathcal{B}_{xy}$ which contains a point of M_1 also contains a point of M_2.*

Proof. Let $L_i \cap M_j = \{x_{ij}\}$. Since $M_1 \neq M_2$, we have $\{x_{11}, x_{21}\} \neq \{x_{12}, x_{22}\}$. Suppose, e.g., that $x_{21} \neq x_{22}$. Let $u \in M_1, u \notin L_1, u \notin L_2, N = \{x, u\}^{\perp}\backslash\{x\}$. If N does not contain a point of $\{y, x_{11}, x_{22}\}^{\perp\perp}$ then by Lemma 2.5.1, and since the point $x \in \{y, x_{11}, x_{22}\}^{\perp}$ is collinear with u, the point u is collinear with a second point x' of $\{y, x_{11}, x_{22}\}^{\perp}$. So x' is collinear with y, x_{11}, x_{22}, u. Hence $x' \in \{y, x_{11}, u\}^{\perp}$ and so $x' \sim x_{21}$, giving a triangle $\{x', x_{21}, x_{22}\}$, a contradiction. Consequently N contains a point of $\{y, x_{11}, x_{22}\}^{\perp\perp}$. If $x_{11} \neq x_{12}$, then a similar argument shows that N contains a point of M_2; if $x_{11} = x_{12}$, then, as $\{y, x_{11}, x_{22}\}^{\perp\perp} = M_2 \cup \{y\}$, it is clear that N contains a point of M_2. $\blacksquare$

Now we introduce the set $\mathcal{H}_{xy}$. The elements of $\mathcal{H}_{xy}$ are of two types.

(a) The sets $\{x, z\}^{\perp}\backslash\{y\}$, with $x \not\sim z$ and $y \in \{x, z\}^{\perp}$.

(b) Each set which is the union of all elements of Type (b) of $\mathcal{B}_{xy}$ containing a point of some line M of Type (a) of $\mathcal{B}_{xy}$.

The set $\mathcal{H}_{xy}$ contains q^3 elements of Type (a), and by Lemma 4.10.2 it contains $q^2 + q$ elements of Type (b).

Lemma 4.10.3 *The elements of $\mathcal{P}_{xy}$ and $\mathcal{B}_{xy}$ in an element of $\mathcal{H}_{xy}$ are the points and lines of a 2-$(q^2, q, 1)$ design, that is, an affine plane of order q.*

Proof. Easy, using Property (G) and Lemma 4.10.2. $\blacksquare$

Theorem 4.10.4 *The elements of* $\mathcal{P}_{xy}, \mathcal{B}_{xy}$ *and* $\mathcal{H}_{xy}$, *respectively, are the points, lines and planes of an affine space* $\mathbf{AG}(3, q)$.

Proof. We have $|\mathcal{P}_{xy}| = q^3$, each element of $\mathcal{B}_{xy}$ contains q elements of $\mathcal{P}_{xy}$, and any two distinct elements of $\mathcal{P}_{xy}$ are contained in a unique element of $\mathcal{B}_{xy}$. Hence $\mathcal{S}_{xy}$ is a 2-$(q^3, q, 1)$-design.

If $q = 3$, then by Section 1.8 we have $\mathcal{S} \cong \mathbf{Q}(5, 3)$. By Theorem 1.5.2(ii) $\mathbf{Q}(5, 3) \cong \mathbf{T}_3(\mathcal{O})$, with $\mathcal{O}$ an elliptic quadric of $\mathbf{PG}(3, 3)$. Now the statement of the theorem is easily checked by taking for y the point (∞) of $\mathbf{T}_3(\mathcal{O})$, and for x any point distinct from (∞) but collinear with (∞).

Now assume $q > 3$. By a theorem of Buekenhout [30] it is sufficient to show that any three points of $\mathcal{P}_{xy}$, not contained in a common element of $\mathcal{B}_{xy}$, are contained in a common element of $\mathcal{H}_{xy}$. Let z, u, w be three distinct points of $\mathcal{P}_{xy}$, not in a common element of $\mathcal{B}_{xy}$, and not in a common element of Type (b) of $\mathcal{H}_{xy}$. Then no two points of $\{z, u, w\}$ are collinear in $\mathcal{S}$ (that is, no two points of $\{z, u, w\}$ are incident with a common line of $\mathcal{S}$ through x), and $\{y, z, u\}^{\perp\perp}$ has no point in common with $\{x, w\}^\perp$. Since $w \sim x$ and $x \in \{y, z, u\}^\perp$, by Lemma 2.5.1 w is collinear with a second point $x' \in \{y, z, u\}^\perp$. Clearly $\{x, x'\}$ contains y, z, u, w, and hence z, u, w are contained in an element of Type (a) of $\mathcal{H}_{xy}$. ∎

Corollary 4.10.5 *If* $\mathcal{S} = (\mathcal{P}, \mathcal{B}, \mathbf{I})$ *is a GQ of order* (q, q^2), $q \neq 1$, *for which at least one pair of points* $\{x, y\}$, $x \sim y$, *has Property (G), then* q *is a prime power.*

Proof. Immediate from Theorem 4.10.4. ∎

Let $\mathcal{S}(\mathcal{F})$ be the GQ of order (q^2, q) arising from a flock $\mathcal{F}$ of the quadratic cone $\mathcal{K}$ of $\mathbf{PG}(3, q)$. If $\mathcal{S}(\mathcal{F})^D$ is the point-line dual of $\mathcal{S}(\mathcal{F})$, then by Theorem 4.10.1 the GQ $\mathcal{S}(\mathcal{F})^D$ has Property (G) at its elation line L. Let x, y be distinct points of L and consider $\mathcal{P}_{xy}, \mathcal{B}_{xy}, \mathcal{H}_{xy}$ and the corresponding affine space $\mathbf{AG}(3, q)$. Then by interpreting the points and lines of $\mathcal{S}(\mathcal{F})^D$ in $\mathbf{AG}(3, q)$, Construction (b) of Section 4.9 is obtained.

Let $\mathcal{S} = (\mathcal{P}, \mathcal{B}, \mathbf{I})$ be a GQ of order (q, q^2), $q \neq 1$, which satisfies Property (G) at the flag (x, L). Let $y \mathbf{I} L, x \neq y$, and as above we consider the affine space $\mathbf{AG}(3, q)$. The lines of Type (b) of $\mathbf{AG}(3, q)$ are parallel, so define a common point at infinity ∞. Let $\mathbf{PG}(3, q)$ be the projective completion of $\mathbf{AG}(3, q)$, with $\mathbf{PG}(3, q) \setminus \mathbf{AG}(3, q) = \pi_\infty$. If $x \not\sim z \not\sim y$, then $(\{x, z\}^\perp \setminus \{u\}) \cup \{\infty\}$, with $z \sim u \mathbf{I} L$, is an ovoid $\mathcal{O}_z$ of $\mathbf{PG}(3, q)$ which is tangent to π_∞ at the point ∞; see Thas [183]. Then in Thas [183] the following fundamental characterization theorem of flock GQs is obtained.

Theorem 4.10.6 Let $\mathcal{S} = (\mathcal{P}, \mathcal{B}, \mathrm{I})$ be a GQ of order (q, q^2), $q > 1$, and assume that $\mathcal{S}$ satisfies Property (G) at the flag (x, L). If q is odd, then $\mathcal{S}$ is the point-line dual of a flock GQ. If q is even and all ovoids $\mathcal{O}_z$ are elliptic quadrics, then we have the same conclusion. ∎

If $\mathcal{S}$ is the point-line dual of a flock GQ and we construct the affine space $\mathbf{AG}(3, q)$, then all ovoids $\mathcal{O}_z$ are elliptic quadrics; see Thas [183]. Now consider the GQ $\mathbf{T}_3(\mathcal{O})$, with $\mathcal{O}$ a Tits ovoid in $\mathbf{PG}(3, q)$, with $q = 2^{2e+1}$ and $e \geq 1$. As by Section 2.5 the point (∞) of $\mathbf{T}_3(\mathcal{O})$ is 3-regular, the GQ $\mathbf{T}_3(\mathcal{O})$ satisfies Property (G) at any flag $((\infty), L)$. We leave it as an exercise to check that here the ovoids $\mathcal{O}_z$ are Tits ovoids isomorphic to $\mathcal{O}$. Hence $\mathbf{T}_3(\mathcal{O})$ is not the point-line dual of a flock GQ. This also follows from Theorem 4.5.1 and Section 4.7. So in Thas [183] it was conjectured that a GQ $\mathcal{S} = (\mathcal{P}, \mathcal{B}, \mathrm{I})$ of order (q, q^2), q even, is the point-line dual of a flock GQ if and only if it satisfies Property (G) at some line L.

Relying on Thas [183], Barwick, Brown and Penttila [9] obtain the following generalization of Theorem 4.10.6.

Theorem 4.10.7 Let $\mathcal{S} = (\mathcal{P}, \mathcal{B}, \mathrm{I})$ be a GQ of order (q, q^2), $q > 1$, and assume that $\mathcal{S}$ satisfies Property (G) at the pair $\{x, y\}$, with x and y distinct collinear points. If q is odd, then $\mathcal{S}$ is the point-line dual of a flock GQ. If q is even and all ovoids $\mathcal{O}_z$ are elliptic quadrics, then we have the same conclusion. ∎

Finally, Brown [25] proved the conjecture of Thas in the even case.

Theorem 4.10.8 Let $\mathcal{S} = (\mathcal{P}, \mathcal{B}, \mathrm{I})$ be a GQ of order (q, q^2), q even, and assume that $\mathcal{S}$ satisfies Property (G) at the line L. Then $\mathcal{S}$ is the point-line dual of a flock GQ. ∎

Now we show that Theorem 4.10.7 can be deduced easily from Theorem 4.10.6.

Proposition 4.10.9 Let $\mathcal{S} = (\mathcal{P}, \mathcal{B}, \mathrm{I})$ be a GQ of order (q, q^2), $q > 1$, and assume that $\mathcal{S}$ satisfies Property (G) at the pair $\{x, y\}$, with x and y distinct collinear points. If q is odd, then $\mathcal{S}$ satisfies Property (G) at the line $L = xy$. If q is even and all ovoids $\mathcal{O}_z$ are elliptic quadrics, then we have the same conclusion.

Proof. We use the notation $\mathcal{P}_{xy}$, $\mathcal{B}_{xy}$, $\mathcal{H}_{xy}$, π_∞, ∞ introduced in this section. Let $\pi_{xy} \in \mathcal{H}_{xy}$ be a plane of Type (b) of the affine space $\mathbf{AG}(3, q)$. Then the point ∞ belongs to the projective completion $\overline{\pi_{xy}}$ of π_{xy}. All elliptic quadrics $\mathcal{O}_z$ contain ∞ and are tangent to the plane π_∞ at ∞. Now we consider the conics $\mathcal{O}_z \cap \overline{\pi_{xy}} = \mathcal{C}_{xy}$. Then $\mathcal{C}_{xy}$ contains ∞ and is tangent to $L_\infty = \overline{\pi_{xy}} \cap \pi_\infty$ at ∞. The number of quadrics $\mathcal{O}_z$ is $q^4 - q^3$, and there are exactly $q^3 - q^2$ nonsingular conics in $\overline{\pi_{xy}}$ which are tangent to L_∞ at ∞. In a GQ of order (q, q^2), for each set X of size $q + 1$ consisting of mutually noncollinear points, we have $|X^\perp| \leq q+1$. Hence each of the $q^3 - q^2$ conics arises from at most q quadrics $\mathcal{O}_z$. Consequently each nonsingular conic of $\overline{\pi_{xy}}$ which is tangent to L_∞ at ∞, arises from exactly q quadrics $\mathcal{O}_z$. It easily follows that $\mathcal{S}$ satisfies Property (G) at the line $L = xy$. ∎

Remark 4.10.10 (a) Lavrauw and Penttila [97] give an alternative proof of Theorem 4.10.6 in the particular case that $\mathcal{S}$ is a TGQ.

(b) From the proof of Proposition 4.10.9 follows that over $\mathbf{GF}(q^2)$ all elliptic quadrics $\mathcal{O}_z$, z being collinear with a fixed point $u \mathrm{I} L$, intersect the plane π_∞ in the same two lines M_∞ and N_∞. This provides a short proof of Lemma 7.1.1 in Thas [183].

4.11 Flocks, Subquadrangles and Ovals

If $\mathcal{S}$ is a GQ of order (q^2, q), with q even, having Property (G) at a pair of concurrent lines, then, by Theorem 2.6.2, it contains subquadrangles of order q. So by Theorem 4.10.1 any flock GQ $\mathcal{S}(\mathcal{F})$ contains subquadrangles of order q containing the elation point (∞). Also each such subquadrangle is a GQ $\mathbf{T}_2(\mathcal{O})$ of Tits, with $\mathcal{O}$ some oval of $\mathbf{PG}(2, q)$; see Payne and Maneri [124] and Payne [113, 117]. In this way Payne [117] discovered a new infinite class of ovals. By Cherowitzo, Penttila, Pinneri and Royle [40], for any flock $\mathcal{F}$ of the quadratic cone $\mathcal{K}$ of $\mathbf{PG}(3, q)$, with q even, the hyperovals defined by the ovals arising from $\mathcal{S}(\mathcal{F})$, can also be described as follows.

Theorem 4.11.1 *The q planes $x_t X_0 + z_t X_1 + t X_2 + X_3 = 0$, with $t \in \mathbf{GF}(q)$ and $x_0 = z_0 = 0$, define a flock $\mathcal{F}$ of the cone $\mathcal{K}$ with equation $X_0 X_1 = X_2^2$ of $\mathbf{PG}(3, q)$, with q even, if and only if for all $(a_1, a_2) \in \mathbf{GF}(q)^2 \backslash \{(0, 0)\}$, the set*

$$\mathcal{K}_{(a_1, a_2)} = \{(1, \sqrt{a_1^2 x_t + a_1 a_2 t + a_2^2 z_t}, t) \, \| \, t \in \mathbf{GF}(q)\} \cup \{(0, 1, 0), (0, 0, 1)\}$$

is a hyperoval of $\mathbf{PG}(2, q)$. *Such a set of hyperovals is called a* herd *of hyperovals.*   ∎

In Storme and Thas [148] Theorem 4.11.1 is generalized to *partial flocks*, that is, sets of k disjoint conics on the quadratic cone of $\mathbf{PG}(3, q)$, and *herds* of $(k + 2)$-arcs in $\mathbf{PG}(2, q)$, with q even. In that paper also a short proof of Johnson's Theorem 4.5.1 is given, avoiding the classification of certain classes of semifields but relying on the connection between flocks and herds of hyperovals. In Thas [186] it is shown how in a purely geometric way the herd can be constructed directly from the flock $\mathcal{F}$. For q even, the connection between flocks, spreads, q-clans, GQs of order (q^2, q), subquadrangles of order q and herds of hyperovals is given in detail in the survey by Johnson and Payne [79].

In this monograph, herds of hyperovals will not be studied in detail as by Theorem 5.1.11 any flock TGQ of order (q^2, q), q even, is classical, so that the corresponding flock is linear, and consequently all hyperovals of the herd are conics together with their nucleus.

Addendum A: Isomorphisms of Flock Quadrangles and Associated Geometries

In this section, we describe the so-called "Fundamental Theorem of q-Clan Geometry". Starting with a natural definition for *equivalent q-clans* (defined below), it interprets the equivalence of q-clans C_1 and C_2 as an isomorphism between $\mathcal{G}(C_1)$ and $\mathcal{G}(C_2)$, where $\mathcal{G}(C_i)$ is either the flock $\mathcal{F}(C_i)$ associated to C_i, or the flock quadrangle $\mathcal{S}(\mathcal{F}(C_i))$, $i = 1, 2$.

4.12   The Fundamental Theorem of q-Clan Geometry, and Applications

Suppose q is a prime power, and suppose $A = \begin{pmatrix} x & y \\ z & w \end{pmatrix}$ and $B = \begin{pmatrix} r & s \\ t & u \end{pmatrix}$ be 2×2-matrices over $\mathbf{GF}(q)$. We write $A \equiv B$ to mean that $x = r$, $w = u$ and $y + z = s + t$. It follows that $A \equiv B$ if and only if $\alpha A \alpha^T = \alpha B \alpha^T$ for all $\alpha \in \mathbf{GF}(q)^2$. If $C = \{A_t \parallel t \in \mathbf{GF}(q)\}$ is a q-clan, the matrices $A_t \in C$ are used to construct quadratic forms $\alpha A_t \alpha^T$, with $\alpha \in \mathbf{GF}(q)^2$, so an $A_t \in C$ may be replaced by an A_t' whenever $A_t \equiv A_t'$ without "effectively changing" C.

If q is odd, one could adjust each $A_t \in C$ to be symmetric, say, for instance,

$$A_t = \begin{pmatrix} x_t & y_t/2 \\ y_t/2 & z_t \end{pmatrix}, \quad t \in \mathbf{GF}(q).$$

For any q we can adjust each $A_t \in C$ to be upper triangular, so, for instance,

$$A_t = \begin{pmatrix} x_t & y_t \\ 0 & z_t \end{pmatrix}, \quad t \in \mathbf{GF}(q).$$

The q-clan is *normalized* if A_0 is the zero matrix.

For any prime power q, let

$$C = \{A_t \equiv \begin{pmatrix} x_t & y_t \\ 0 & z_t \end{pmatrix} \parallel t \in \mathbf{GF}(q)\}$$

and

$$C' = \{A_t' \equiv \begin{pmatrix} x_t' & y_t' \\ 0 & z_t' \end{pmatrix} \parallel t \in \mathbf{GF}(q)\}$$

be two not necessarily distinct q-clans. We say that C and C' are *equivalent* and write $C \sim C'$ provided there exist $0 \neq \mu \in \mathbf{GF}(q)$, $B \in \mathbf{GL}_2(q)$, $\sigma \in \mathrm{Aut}(\mathbf{GF}(q))$, M a 2×2-matrix over $\mathbf{GF}(q)$ and a permutation $\pi : t \mapsto \bar{t}$ on $\mathbf{GF}(q)$ such that the following holds:

$$A'_{\bar{t}} \equiv \mu B A^\sigma_t B^T + M \text{ for all } t \in \mathbf{GF}(q).$$

If for a q-clan C each A_t is replaced by $A^*_t = A_t - A_0$, then there arises a normalized q-clan equivalent to C.

Theorem 4.12.1 (Fundamental Theorem of q-Clan Geometry)   *Assume that $C = \{A_t \parallel t \in \mathbf{GF}(q)\}$ and $C' = \{A'_t \parallel t \in \mathbf{GF}(q)\}$ are normalized q-clans.*
The following are equivalent:

(i) $C \sim C'$.

(ii) *The flocks $\mathcal{F}(C)$ and $\mathcal{F}(C')$ are projectively equivalent.*

(iii) *The GQs $\mathcal{S}(\mathcal{F}(C))$ and $\mathcal{S}(\mathcal{F}(C'))$ are isomorphic by an isomorphism mapping (∞) to (∞), $[A(\infty)]$ to $[A'(\infty)]$, and $(\bar{0}, 0, \bar{0})$ to $(\bar{0}, 0, \bar{0})$.*

These three equivalent conditions hold if and only if the following exist:

(i) *a permutation $\pi : t \mapsto \bar{t}$ of the elements of $\mathbf{GF}(q)$;*

(ii) $\tau \in \mathrm{Aut}(\mathbf{GF}(q))$;

(iii) $\lambda \in \mathbf{GF}(q)$, $\lambda \neq 0$;

(iv) $D \in \mathbf{GL}_2(q)$ *for which* $A_{\bar{t}} - \lambda D^T A^\tau_t D - A_{\bar{0}}$ *is skew-symmetric (with zero diagonal) for all $t \in \mathbf{GF}(q)$.*

Given τ, D, λ and the permutation $\pi : x \mapsto \bar{x}$ satisfying Condition (iv), an isomorphism $\theta = \theta(\tau, D, \lambda, \pi)$ of the GQ $\mathcal{S}(\mathcal{F}(C))$ onto $\mathcal{S}(\mathcal{F}(C'))$ arises, as follows:

$$\begin{aligned}
\theta = \theta(\tau, D, \lambda, \pi) &: (\alpha, c, \beta) \\
&\mapsto (\lambda^{-1}\alpha^\tau D^{-T}, \lambda^{-1}c^\tau + \lambda^{-2}\alpha^\tau(D^{-T}A_{\bar{0}}D^{-1})(\alpha^\tau)^T, \beta^\tau D \\
&\quad + \lambda^{-1}\alpha^\tau D^{-T}K_{\bar{0}}).
\end{aligned}$$

(We write D^{-T} for the matrix $(D^{-1})^T = (D^T)^{-1}$.)

For $\theta = \theta(\tau, D, \lambda, \pi)$, write $D = \begin{pmatrix} a & b \\ c & d \end{pmatrix}$. Now let $\mathcal{F}(\mathcal{C}) = \mathcal{F}(\mathcal{C}') = \mathcal{F}$. Define a projective semilinear collineation T_θ of $\mathbf{PG}(3, q)$ as follows (defined on the planes of $\mathbf{PG}(3, q)$, and with a general plane having equation $xX_0 + zX_1 + yX_2 + X_3 = 0$):

$$T_\theta : \begin{bmatrix} x \\ y \\ z \\ 1 \end{bmatrix} \mapsto \begin{pmatrix} \lambda a^2 & \lambda ab & \lambda b^2 & x_{\overline{0}} \\ 2\lambda ac & \lambda(ad + bc) & 2\lambda bd & y_{\overline{0}} \\ \lambda c^2 & \lambda cd & \lambda d^2 & z_{\overline{0}} \\ 0 & 0 & 0 & 1 \end{pmatrix} \begin{bmatrix} x^\tau \\ y^\tau \\ z^\tau \\ 1 \end{bmatrix}.$$

Then T_θ fixes the cone $\mathcal{K}$ with equation $X_0 X_1 = X_2^2$, and leaves invariant $\mathcal{F} = \mathcal{F}(\mathcal{C})$ precisely when θ defines a collineation of $\mathcal{S}(\mathcal{F})$ fixing (∞), $[A(\infty)]$ and $(\overline{0}, 0, \overline{0})$. ∎

The map

$$T : \theta \mapsto T_\theta$$

is a homomorphism from the subgroup of $\mathrm{Aut}(\mathcal{S}(\mathcal{F}))$ leaving (∞), $[A(\infty)]$ and $(\overline{0}, 0, \overline{0})$ invariant onto the subgroup of $\mathbf{P\Gamma O}_4(q)$ leaving the flock $\mathcal{F}$ invariant. The kernel $N(T)$ of T is

$$N(T) = \{\theta_a \parallel (\alpha, c, \beta) \mapsto (a\alpha, a^2 c, a\beta), 0 \neq a \in \mathbf{GF}(q)\}.$$

Note that $N(T)$ is a group of whorls about (∞) and $(\overline{0}, 0, \overline{0})$.

The following statement is a direct corollary of the Fundamental Theorem.

Corollary 4.12.2 *Let $\mathcal{F}$ and $\mathcal{F}'$ be projectively equivalent flocks of the quadratic cone in $\mathbf{PG}(3, q)$. Then $\mathcal{S}(\mathcal{F}) \cong \mathcal{S}(\mathcal{F}')$.*

Proof. Let $\mathcal{C}$ be the q-clan associated to $\mathcal{F}$ and $\mathcal{C}'$ the q-clan associated to $\mathcal{F}'$. Then $\mathcal{C} \sim \mathcal{C}'$, and $\mathcal{S}(\mathcal{F}) = \mathcal{S}(\mathcal{F}(\mathcal{C})) \cong \mathcal{S}(\mathcal{F}(\mathcal{C}')) = \mathcal{S}(\mathcal{F}')$. ∎

Let $\mathcal{F}$ be a flock of the quadratic cone in $\mathbf{PG}(3, q)$, q odd. Then $\mathcal{F}$ together with its q derived flocks correspond to the $q + 1$ lines incident with (∞), cf. Remark 4.8.2. Let $\mathcal{F}_0 = \mathcal{F}, \mathcal{F}_1, \ldots, \mathcal{F}_q$ be these $q + 1$ flocks. Then in this section we denote by $L(i)$ the line corresponding to $\mathcal{F}_i$.

The following result can be found in Payne and Rogers [125].

Theorem 4.12.3 *Let $\mathcal{F}$ and $\mathcal{S}(\mathcal{F})$, etc. be as above. Then there is an automorphism of $\mathcal{S}(\mathcal{F})$ fixing (∞) and mapping $L(i)$ to $L(j)$, with $i, j \in \{0, 1, \ldots, q\}$, if and only if $\mathcal{F}_i$ is projectively equivalent to $\mathcal{F}_j$.* ∎

The converse of Corollary 4.12.2 is not necessarily true; this is because nonequivalent q-clans can be associated to the same flock generalized quadrangle. For, let $\mathcal{F} = \mathcal{F}_0$ be a flock of the quadratic cone in $\mathbf{PG}(3,q)$, q odd, and let $\mathcal{F}_1, \mathcal{F}_2, \ldots, \mathcal{F}_q$ be its derived flocks. Then $\mathcal{S}(\mathcal{F}) \cong \mathcal{S}(\mathcal{F}_i)$, $i \in \{0, 1, \ldots, q\}$, according to Section 4.8. On the other hand, $\mathcal{F}_i$ is projectively equivalent to $\mathcal{F}_j$, with $i, j \in \{0, 1, \ldots, q\}$, if and only if there is an automorphism of $\mathcal{S}(\mathcal{F})$ that fixes (∞) and maps $L(i)$ to $L(j)$. But many examples of flock GQs are known for which $\mathrm{Aut}(\mathcal{S})_{(\infty)}$ does not act transitively on the lines incident with (∞).

Theorem 4.12.5 determines such examples for flock GQs that are dual TGQs. Note that, in any case, if $\mathcal{S}(\mathcal{F})$ is a dual TGQ of order (q^2, q), q odd, then there are at most two isomorphism classes in the set $\{\mathcal{F}_0, \mathcal{F}_1, \ldots, \mathcal{F}_q\}$, as the translation group of $\mathcal{S}(\mathcal{F})^D$ induces an automorphism group of $\mathcal{S}(\mathcal{F})$ that fixes the translation line and acts transitively on the other lines incident with (∞).

We first mention the following result, which can be found in [206], and which we will encounter later in Chapter 10, cf. Theorem 10.7.2 in combination with Theorem 6.4.4.

Theorem 4.12.4 *Let S be a TGQ of order (s, s^2) which contains a line of translation points, and which is the point-line dual of a flock GQ $\mathcal{S}(\mathcal{F})$. Then either s is even and $\mathcal{F}$ is linear, or s is odd and $\mathcal{F}$ is a Kantor-Knuth flock.* ∎

Theorem 4.12.5 *Let $\mathcal{F}$ be a nonlinear flock of the quadratic cone in $\mathbf{PG}(3,q)$, q odd, with $\mathcal{S}(\mathcal{F})$ the point-line dual of a TGQ. Then $\mathrm{Aut}(\mathcal{S})_{(\infty)}$ acts transitively on the lines incident with (∞) if and only if $\mathcal{F}$ is of Kantor-Knuth type.*

Proof. By the assumption, each line incident with (∞) is a translation line. So $\mathcal{S}(\mathcal{F})^D$ contains a line of translation points. By Theorem 4.12.4, it now follows that $\mathcal{F}$ is of Kantor-Knuth type. ∎

Hence a flock GQ in odd characteristic of which the point-line dual is a TGQ, but which does not arise from a Kantor-Knuth flock, admits precisely two isomorphism classes of flocks.

We end this section with the following useful observation by Payne [121]:

Theorem 4.12.6 (Payne [121]) *Suppose $\mathcal{S}(\mathcal{F})$ is a nonclassical flock GQ of order (q^2, q), $q > 1$, and let H be the full group of automorphisms of $\mathcal{S}(\mathcal{F})$ fixing (∞) and $(\overline{0}, 0, \overline{0})$ linewise.*

(i) *If $q = 2^h$, then $H = N(T)$.*

(ii) *If q is odd, and $H^* = \{\theta(\tau, D, \lambda, \pi) \in H \parallel \tau = \mathbf{1}\}$, then $H^* = N(T)$, except if $\mathcal{F}$ is a Kantor-Knuth semifield flock. In that case, $H = H^*$ if $\sigma^2 \neq \mathbf{1}$, and then $[H : N(T)] = 2$, where σ is the automorphism of $\mathbf{GF}(q)$ which is used to define $\mathcal{F}$. If $\sigma^2 = \mathbf{1} \neq \sigma$, then $[H : H^*] = [H^* : N(T)] = 2$.* ∎

Addendum B: Basic Questions on Elation Groups

In this addendum, we pose four fundamental and old questions on (elation) generalized quadrangles, and survey tersely answers on these questions coming from work of Payne and K. Thas [130], K. Thas and Payne [214], Rostermundt [134] and K. Thas [207]. The importance of collineations that fix proper (but not necessarily thick) subGQs will be underlined, and in particular this will be clear when the Kantor-Knuth GQs will serve as counter examples for one of the conjectures.

Not all proofs of [130, 214, 134] and [207] will be given. But we will give the ones which reveal the important nature of those collineations fixing subGQs pointwise.

For more details on the theory sketched in this addendum, we refer the reader to the monograph "Lectures on Elation Quadrangles" [212] of K. Thas.

4.13   The Standard Conjectures and Questions on Elation Generalized Quadrangles

In Chapter 8 of [128], the following is quoted:

> "In general it seems to be an open question as to whether or not the set of elations about a point must be a group."

In the same chapter of *loc. cit.*, the authors study TGQs, and show that all elations about the elation point are in the translation group (cf. 8.6.4 of [128]). We will call the aforementioned question "QUESTION (1)".

Most of the known GQs are, up to duality, EGQs with at least one elation point. (In fact, each known GQ is as such, or is constructed from an EGQ — see [206], and in particular Chapter 3 of that book, for details.) We therefore formulate the following specialization of Question (1):

QUESTION (2).   *Given an EGQ $\mathcal{S}^{(x)}$, is the set of elations about x a group?*

The following question makes sense if the answer is "not always".

QUESTION (2′). *Given an EGQ $S^{(x)}$, when is the set of elations about x a group?*

Let $\mathcal{F}$ be a Kantor-Knuth semifield flock of the quadratic cone $\mathcal{K}$ in $\mathbf{PG}(3, q)$, $q = p^h$, with p an odd prime. Let $S(\mathcal{F})$ be the corresponding flock GQ of order (q^2, q). Recall that each flock GQ has a "special" point (∞) for which there is a group of elations K making it into in an EGQ with elation point (∞). In [214], Payne and K. Thas constructed an elation θ about (∞) which generates a group not only consisting of elations. So θ is not an element of K.

The construction is as follows. We use the standard notation of §4.6.

Let $Q = \begin{pmatrix} 1 & 0 \\ 0 & -1 \end{pmatrix}$, so $Q = Q^T = Q^{-1}$. It is noted in [119] that the map

$$\tau : (\alpha, c, \beta) \mapsto (\alpha Q, c, \beta Q)$$

is an automorphism of K which induces a collineation of S that fixes $(\bar{0}, 0, \bar{0})$ and (∞) linewise. Let $g = (\alpha', c', \beta')$ be a fixed element of K to be determined later. Put

$$\theta = \tau \circ \pi(g) = \tau \circ \pi(\alpha', c', \beta').$$

For the appropriate choice of g we will establish the following:

(1) θ is an elation about (∞) that is not in K;

(2) θ^p is an involution whose fixed element structure is a subquadrangle of order q, so that in particular θ^p is not an elation about (∞);

(3) $\theta^2 \in K$.

Lemma 4.13.1 *If j is a positive integer, let j_2 denote the element of $\{0, 1\}$ to which j is congruent modulo 2. Similarly, let $-j_2$ denote -1 if j is odd and 0 if j is even. Also, I is the 2×2 identity matrix.*

$$I + Q + Q^2 + \cdots + Q^j = \begin{pmatrix} j + 1 & 0 \\ 0 & (j+1)_2 \end{pmatrix}. \qquad (4.1)$$

$$Q + Q^2 + \cdots + Q^j = \begin{pmatrix} j & 0 \\ 0 & -j_2 \end{pmatrix}. \qquad (4.2)$$

$$Q^{j-1} + 2Q^{j-2} + 3Q^{j-3} + \cdots + (j-1)Q^1 = \begin{pmatrix} \frac{(j-1)j}{2} & 0 \\ 0 & -\lfloor \frac{j}{2} \rfloor \end{pmatrix}. \quad (4.3)$$

In Equations (4.1) and (4.2) we take $j \geq 1$. In Equation (4.3) we must take $j \geq 2$.

Proof. All three equations are established by routine induction arguments (note that $\lfloor \frac{j}{2} \rfloor = \frac{j - j_2}{2}$). ∎

By definition we have

$$\theta = \tau \circ \pi(\alpha', c', \beta') : (\alpha, c, \beta) \mapsto (\alpha Q + \alpha', c + c' + \beta Q \circ \alpha', \beta Q + \beta'). \quad (4.4)$$

Then by an easy calculation we have

$$\theta^2 : (\alpha, c, \beta) \mapsto (\alpha Q^2 + \alpha'(Q + I), c + 2c' + \beta(Q + Q^2) \circ \alpha'$$
$$+ \beta' Q \circ \alpha', \beta Q^2 + \beta'(Q + I)). \quad (4.5)$$

It now follows by a routine induction that

$$\theta^i : (\alpha, c, \beta) \mapsto (\alpha Q^i + \alpha'(Q^{i-1} + Q^{i-2} + \cdots + Q + I), \quad (4.6)$$

$$c + ic' + \beta(Q + Q^2 + Q^3 + \cdots + Q^i) \circ \alpha'$$
$$+ \beta'(Q^{i-1} + 2Q^{i-2} + 3Q^{i-3} + \cdots + (i-1)Q^1) \circ \alpha',$$

$$\beta Q^i + \beta'(Q^{i-1} + Q^{i-2} + \cdots + Q + I))$$
$$= (\alpha \begin{pmatrix} 1 & 0 \\ 0 & (-1)^i \end{pmatrix} + \alpha' \begin{pmatrix} i & 0 \\ 0 & i_2 \end{pmatrix},$$

$$c + ic' + \beta \begin{pmatrix} i & 0 \\ 0 & -i_2 \end{pmatrix} \circ \alpha' + \beta' \begin{pmatrix} \frac{(i-1)i}{2} & 0 \\ 0 & -\lfloor \frac{i}{2} \rfloor \end{pmatrix} \circ \alpha',$$

$$\beta \begin{pmatrix} 1 & 0 \\ 0 & (-1)^i \end{pmatrix} + \beta' \begin{pmatrix} i & 0 \\ 0 & i_2 \end{pmatrix}).$$

At this stage it is convenient to have written out the image of θ^i separately for odd and even i.

$$\theta^{2k} : (\alpha, c, \beta) \mapsto (\alpha + \alpha' \begin{pmatrix} 2k & 0 \\ 0 & 0 \end{pmatrix}, \quad (4.7)$$

$$c + 2kc' + \beta \begin{pmatrix} 2k & 0 \\ 0 & 0 \end{pmatrix} \circ \alpha' + \beta' \begin{pmatrix} (2k-1)k & 0 \\ 0 & -k \end{pmatrix} \circ \alpha', \beta + \beta' \begin{pmatrix} 2k & 0 \\ 0 & 0 \end{pmatrix}).$$

$$\theta^{2k+1} : (\alpha, c, \beta) \mapsto (\alpha \begin{pmatrix} 1 & 0 \\ 0 & -1 \end{pmatrix} + \alpha' \begin{pmatrix} 2k+1 & 0 \\ 0 & 1 \end{pmatrix}, \tag{4.8}$$

$$c + (2k+1)c' + \beta \begin{pmatrix} 2k+1 & 0 \\ 0 & -1 \end{pmatrix} \circ \alpha' + \beta' \begin{pmatrix} k(2k+1) & 0 \\ 0 & -k \end{pmatrix} \circ \alpha',$$

$$\beta \begin{pmatrix} 1 & 0 \\ 0 & -1 \end{pmatrix} + \beta' \begin{pmatrix} 2k+1 & 0 \\ 0 & 1 \end{pmatrix}).$$

We now determine whether or not θ^i fixes some point (α, c, β). First, θ^{2k} fixes (α, c, β) if and only if

(i) $\alpha' \begin{pmatrix} 2k & 0 \\ 0 & 0 \end{pmatrix} = (0, 0)$;

(ii) $2kc' + \beta \begin{pmatrix} 2k & 0 \\ 0 & 0 \end{pmatrix} \circ \alpha' + \beta' \begin{pmatrix} (2k-1)k & 0 \\ 0 & -k \end{pmatrix} \circ \alpha' = 0$;

and

(iii) $\beta' \begin{pmatrix} 2k & 0 \\ 0 & 0 \end{pmatrix} = (0, 0)$.

It follows readily that if $\alpha' = (a_1, a_2)$ with $a_1 \neq 0$, then we have proved the following lemma.

Lemma 4.13.2 *Assume that $a_1 \neq 0$. Then θ^{2k} fixes some (α, c, β) if and only if $k \equiv 0 \mod p$, in which case $\theta^{2k} = \text{id}$. In particular, θ^2 is an elation for which $\langle \theta^2 \rangle$ is a group of elations.* ∎

Note that it is also easy to check that $\theta^2 \in K$.

Now we consider whether or not θ^{2k+1} fixes some (α, c, β). Since we are assuming that $\alpha' = (a_1, a_2)$ with $a_1 \neq 0$, it follows easily that if θ^{2k+1} fixes some (α, c, β) it must be the case that $\alpha \begin{pmatrix} 0 & 0 \\ 0 & 2 \end{pmatrix} = \alpha' \begin{pmatrix} 2k+1 & 0 \\ 0 & 1 \end{pmatrix}$, which forces $2k+1 \equiv 0 \mod p$. So consider the fixed points of θ^p. It is routine to check that θ^p fixes the point (α, c, β) if and only if $(\alpha, c, \beta) = ((a, -ka_2), c, (b, -kb_2))$, $a, c, b \in \mathbf{GF}(q)$, where $\alpha' = (a_1, a_2)$, $\beta' = (b_1, b_2)$

and $p = 2k + 1$.

It now follows that θ^p is an involution with a subquadrangle of order q as its fixed element structure.

This suggests the definition of a "standard elation" about a point x: this is an elation ϕ for which $\langle\phi\rangle$ is a group of elations (ϕ acts freely on the points not collinear with x). The following natural question arises:

QUESTION (3).   *Given an EGQ $S^{(x)}$, is the set of standard elations about x a group?*

In the same way as for Question (2), one could now also formulate Question (3′).

QUESTION (3′).   *Given an EGQ $S^{(x)}$, when is the set of standard elations about x a group?*

4.14   Some Results by Payne and K. Thas

For nonclassical flock GQs there is a complete answer to Question (3):

Theorem 4.14.1 *Let $S(\mathcal{F})$ be a nonclassical flock GQ of order (q^2, q). Then the set of standard elations about (∞) is a group. This group is the usual elation group K. When q is odd, the same conclusion holds in the classical case.*

Proof.   See the proof of Theorem 6.1 of [130].   ∎

Remark 4.14.2 In [130], the condition that $S(\mathcal{F})$ is nonclassical for q even was forgotten in the statement of the theorem (cf. Theorem 2.4 and Theorem 6.1).

Generalizing Theorem 4.14.1, the following was also obtained in [130]. First define a *skew translation generalized quadrangle* (STGQ) to be an EGQ $(S^{(p)}, G)$ for which the elation point p is a center of symmetry about which the group of symmetries is contained in G.

Theorem 4.14.3 *Let $\mathcal{S}^{(p)}$ be an STGQ of order (s,t), $s,t > 1$. Then we have two possibilities:*

(a) *the set of standard elations about p is a group;*

(b) *$s = t^2$, s is a power of 2, and there is a subGQ isomorphic to $\mathbf{W}(t)$ containing p which is fixed pointwise by an involution of $\mathcal{S}^{(p)}$.*

Proof.   See the proof of Theorem 7.2 of [130].   ∎

Remark 4.14.4 Examples of (b) yield counter examples to Question (3). Such STGQs will play a central role in §4.15.

From the next theorem of [214], it will follow that "most of the time", the answer to Question (2) is that the set of elations about an elation point is *not* a group, and it also explains *precisely* why.

SETTING.   *In this section, $\mathcal{S} = (\mathcal{P}, \mathcal{B}, \mathbf{I})$ is a GQ of order (s,t), $s,t > 1$, and x is an elation point for the elation group G. We let W be the group of all whorls about x.*

Now let $H \le W$ be any subgroup of W that acts transitively on $\mathcal{P} \setminus x^{\perp}$. We apply Burnside's Lemma on the permutation group $(H, \mathcal{P} \setminus x^{\perp})$ to obtain

$$|H| = s^2 t + \sum_{i=1}^{s^2 t - 1} i\delta(i),$$

where $\delta(j)$ denotes the number of elements of H that fix *precisely* j points of $\mathcal{P} \setminus x^{\perp}$.

Let E be the number of elations in H (not including $\mathbf{1}$) about x; then, as

$$1 + E + \sum_{i=1}^{s^2 t - 1} \delta(i) = s^2 t + \sum_{i=1}^{s^2 t - 1} i\delta(i),$$

it follows that the number of elations (now including $\mathbf{1} = \text{id}!$) is given by the following:

$$1 + E = s^2 t + \sum_{i=2}^{s^2 t - 1} (i - 1)\delta(i),$$

which is clearly at least $s^2 t$.

The following theorem now easily follows.

Theorem 4.14.5 *Let* $(\mathcal{S}^{(p)}, G)$ *be an EGQ, and let* W *be the group of all whorls about* p. *Then the set of elations about* p *is a group if and only if there is no nontrivial element in* W *fixing more than one point not collinear with* p *if and only if* W *is a Frobenius group.* ∎

The Known Examples

It is convenient to mention the known (classes) of examples of EGQs with elation point p for which the set of elations about p is *not* a group. These classes are treated in detail in [214].

- *The classical and dual classical examples.* $\mathbf{W}(q)$ with q odd; $\mathbf{H}(3, q^2)$; $\mathbf{H}(4, q^2)$; $\mathbf{H}(4, q^2)^D$.

- *Flock GQs.* The flock GQs with an even number of points on a line.

- *Dual flock GQs which are EGQs.* As the order is (q, q^2) for some q, it can be shown that not more than one point not collinear with the elation point can be fixed by a nonidentity collineation. So there are no examples possible.

- *TGQs.* There are no examples possible.

- *Dual TGQs which are EGQs.* GQs $\mathcal{S}^D$, where $\mathcal{S}$ is a TGQ of order (q, q^2), q odd, that is the translation dual of the point-line dual of a flock GQ.

4.15 Elation Generalized Quadrangles with Nonisomorphic Elation Groups

The following fundamental question (especially in construction theory for GQs) was posed by Payne [123] during the 2004 Pingree Park Conference "Finite Geometries, Groups and Computation":

QUESTION (4). *Let* $\mathcal{S}^{(x)}$ *be an EGQ. Can* x *be an elation point for nonisomorphic elation groups?*

The following theorem considers a class of GQs which do admit nonisomorphic elation groups, thus answering Payne's question affirmatively. The only known examples of this class are the GQs $\mathbf{H}(3, q^2)$ with q even.

Theorem 4.15.1 *Let* $S = (S^{(p)}, H)$ *be an EGQ of order* (q^2, q)*, where* q *is even, which contains a subGQ* S' *of order* (s, q)*,* $s > 1$*, fixed pointwise by a nontrivial automorphism* θ *of* S*. If* $H^2 \leq Z(H)$*, and if for all* $L \mathrm{I} p$ *and* $x \mathrm{I} L \mathrm{I} p$*, with* $p \neq x$*, we have that* $|H(x, L, p)| = q^2$*, where* $H(x, L, p)$ *is the group of whorls about* x*,* L *and* p*, then there is an automorphism group* H' *of* S *such that* $H' \not\cong H$ *and* $(S^{(p)}, H')$ *is an EGQ.*

Proof. See K. Thas [207]. ∎

Corollary 4.15.2 *Let* $S = (S^{(p)}, H)$ *be an EGQ of order* (q^2, q)*, where* q *is even, which contains a subGQ* S' *of order* (s, q)*,* $s > 1$*, which is fixed pointwise by a nontrivial automorphism* θ *of* S*. Let* $z \not\sim p$ *and suppose* $z \sim z_i \sim p$ *for* $i = 0, 1, \ldots, q$*. If all groups* $H(p, pz_i, z_i) \cap H$ *are elementary abelian and have size* q^2*, then there is an automorphism group* H' *of* S *such that* $H' \not\cong H$ *and* $(S^{(p)}, H')$ *is an EGQ.*

Proof. The groups $H(p, pz_i, z_i)$ are elementary abelian if and only if $H^2 \subseteq Z(H)$ (exercise). ∎

In the next section, we will show that $\mathbf{H}(3, q^2)$ with q even satisfies the assumptions of Theorem 4.15.1, therefore providing a "concrete" answer to the question of Payne.

An Example of Theorem 4.15.1: $\mathbf{H}(3, q^2)$, q **even**

Consider $S \cong \mathbf{H}(3, q^2)$, q even, and suppose p is a point of S. We will show that all the assumptions of Theorem 4.15.1 are satisfied.

Suppose $L \mathrm{I} p$, and let $x \mathrm{I} L \mathrm{I} p$, with $p \neq x$; then the group $H(x, L, p)$ has size q^2, and is isomorphic to the additive group of $\mathbf{GF}(q^2)$. By putting H equal to the group generated by all such elations (so that $(S^{(p)}, H)$ is an EGQ), the assumptions of Theorem 4.15.1 are satisfied ($H^2 = Z(H)$ for this H).

Remark 4.15.3 The previous result was independently obtained by Rostermundt [134] in an entirely different fashion. He represents $\mathbf{H}(3, q^2)$ (q even) as a group coset geometry in the extra-special group $K = \{(\alpha, c, \beta) \parallel \alpha, \beta \in \mathbf{GF}(q^2), c \in \mathbf{GF}(q)\}$, where the group operation is given by

$$(\alpha, c, \beta) \circ (\alpha', c', \beta') = (\alpha + \alpha', c + c' + \beta \alpha'^T, \beta + \beta').$$

He then constructs $q^2 - 1$ distinct elation groups $K_i = 1, 2, \ldots, q^2 - 1$ of size q^5 with base-point (∞), and shows that all K_i are mutually isomorphic. The K_i's have nilpotency class 3, while K has nilpotency class 2, so that $K \ncong K_i$ for all i. The proofs are long and technical. For details and several other results, see Rostermundt [134].

Chapter 5

Good Eggs

In this chapter we will define good eggs, explain the relationship with Property (G) and flock GQs, and give some interesting characterization theorems. There is also a section on the coordinatization of good eggs and their translation duals, and coordinates of all known examples are given.

5.1 Good Eggs and Good Translation Generalized Quadrangles

Definition 5.1.1 The egg $\mathcal{O}(n, 2n, q) = \mathcal{O}$ is *good at its element* π if any $\mathbf{PG}(3n-1, q)$ containing π and at least two other elements of $\mathcal{O}(n, 2n, q)$ contains exactly $q^n + 1$ elements of $\mathcal{O}(n, 2n, q)$; the space π is called a *good element* of $\mathcal{O}(n, 2n, q)$. In such a case we also say that $\mathcal{O}(n, 2n, q)$ is *good*. If $\mathcal{O}$ is good at π, then we say that the TGQ $\mathbf{T}(\mathcal{O})$ is *good at its line* π, and π is a *good line* of $\mathbf{T}(\mathcal{O})$; shortly we say that the TGQ $\mathbf{T}(\mathcal{O})$ is *good*.

We emphasize that each known egg $\mathcal{O}(n, 2n, q)$ or its translation dual is good.

Now we give the relation between good and Property (G); see Thas [176].

Theorem 5.1.2 The egg $\mathcal{O}(n, 2n, q) = \mathcal{O}$ satisfies Property (G) at the flag $((\infty), \pi)$ if and only if the translation dual $\mathbf{T}(\mathcal{O}^*)$ of the TGQ $\mathbf{T}(\mathcal{O})$ is good at its element τ with τ the tangent space of $\mathcal{O}$ at π.

Proof. Suppose that $\mathbf{T}(\mathcal{O})$ satisfies Property (G) at the flag $((\infty), \pi)$, with $\pi \in \mathcal{O}$. The tangent space of $\mathcal{O}$ at π is denoted by τ. Let π_j, π_k be distinct elements of $\mathcal{O}\backslash\{\pi\}$ and put $\mathbf{PG}(n-1,q) = \tau \cap \tau_j \cap \tau_k$, with τ_l the tangent space of $\mathcal{O}$ at π_l, $l \in \{j,k\}$. By Theorem 3.10.1(i) each point of $\mathbf{PG}(n-1,q)$ is contained in exactly $q^n + 1$ tangent spaces of $\mathcal{O}$. Let $\mathbf{PG}(4n,q)$ be the space in which $\mathbf{T}(\mathcal{O})$ is defined, and let $\mathbf{PG}(4n-1,q)$ be the space of $\mathcal{O}$. Further, let $u \in \mathbf{PG}(n-1,q)$, let N be a line of $\mathbf{PG}(4n,q)$ through u not contained in $\mathbf{PG}(4n-1,q)$, and let u_1, u_2 be distinct points of N not contained in $\mathbf{PG}(4n-1,q)$. Then each of the points $\langle N,\tau\rangle, \langle N,\tau_j\rangle, \langle N,\tau_k\rangle$ of Type (ii) of $\mathbf{T}(\mathcal{O})$ is collinear with each of the points (∞), u_1, u_2 of $\mathbf{T}(\mathcal{O})$. Clearly $\{(\infty), u_1, u_2\}^\perp$ is the set $\{\langle N,\tau\rangle, \langle N,\tau_j\rangle, \langle N,\tau_k\rangle, \cdots\}$ with $\tau, \tau_j, \tau_k, \ldots$ the $q^n + 1$ tangent spaces of $\mathcal{O}$ through u. As $\mathbf{T}(\mathcal{O})$ satisfies Property (G) at the flag $((\infty), \pi)$, we have $|\{(\infty), u_1, u_2\}^{\perp\perp}| = q^n + 1$. Hence $\langle N,\tau\rangle \cap \langle N,\tau_j\rangle \cap \langle N,\tau_k\rangle \cap \cdots = \eta$ contains q^n points of $\mathbf{PG}(4n,q) \setminus \mathbf{PG}(4n-1,q)$. Consequently η is an n-dimensional space, that is, $\tau \cap \tau_j \cap \tau_k \cap \cdots$ is $(n-1)$-dimensional. So the intersection of these $q^n + 1$ tangent spaces is $\mathbf{PG}(n-1,q)$. We conclude that $\mathbf{PG}(n-1,q)$ is contained in exactly $q^n + 1$ tangent spaces of $\mathcal{O}$. Hence the translation dual $\mathcal{O}^*$ of $\mathcal{O}$ is good at its element τ.

Conversely, let $\mathcal{O}^*$ be good at its element τ. So for any two distinct elements $\tau_j, \tau_k \in \mathcal{O}^*\backslash\{\tau\}$ the space $\mathbf{PG}(n-1,q) = \tau \cap \tau_j \cap \tau_k$ is contained in exactly $q^n + 1$ tangent spaces of $\mathcal{O}$. Let $\bar{\tau}, \bar{\tau}_j, \bar{\tau}_k$ be points of Type (ii) of $\mathbf{T}(\mathcal{O})$ collinear with (∞), where $\tau \subset \bar{\tau}$, $\tau_j \subset \bar{\tau}_j$, $\tau_k \subset \bar{\tau}_k$. Then $\bar{\tau} \cap \bar{\tau}_j \cap \bar{\tau}_k$ is an n-dimensional space $\mathbf{PG}(n,q)$. Clearly $\mathbf{PG}(n-1,q) = \mathbf{PG}(n,q) \cap \mathbf{PG}(4n-1,q)$. If $\tau, \tau_j, \tau_k, \ldots$ are the $q^n + 1$ tangent spaces of $\mathcal{O}$ containing $\mathbf{PG}(n-1,q)$, then the $q^n + 1$ spaces $\langle \mathbf{PG}(n,q), \tau \rangle = \bar{\tau}$, $\langle \mathbf{PG}(n,q), \tau_j \rangle = \bar{\tau}_j$, $\langle \mathbf{PG}(n,q), \tau_k \rangle = \bar{\tau}_k, \ldots$ are points of Type (ii) of $\mathbf{T}(\mathcal{O})$, each of which is collinear in $\mathbf{T}(\mathcal{O})$ with (∞) and the q^n points of Type (i) in $\mathbf{PG}(n,q)$. Hence the triad $\{\bar{\tau}, \bar{\tau}_j, \bar{\tau}_k\}$ is 3-regular. So we conclude that $\mathbf{T}(\mathcal{O})$ satisfies Property (G) at the flag $((\infty), \pi)$.   ∎

Remark 5.1.3 As the translation group of $\mathbf{T}(\mathcal{O})$ is transitive on the points of Type (ii) of $\mathbf{T}(\mathcal{O})$ incident with π, in the statement of Theorem 5.1.2 the flag $((\infty), \pi)$ may be replaced by a point pair $((\infty), \tilde{\pi})$, with $\tilde{\pi}$ a point of Type (ii) incident with π.

Corollary 5.1.4 *If the egg $\mathcal{O}(n, 2n, q) = \mathcal{O}$, with q odd, is good at its element π, then $\mathbf{T}(\mathcal{O}^*)$ is the point-line dual of a flock GQ with elation point τ, where τ is the tangent space of $\mathcal{O}$ at π. Conversely, if the flock GQ S has a translation line, then the point-line dual of S is isomorphic to a $\mathbf{T}(\mathcal{O})$ for which $\mathcal{O}^*$ is good at its element which corresponds to the elation point of S.*

Proof. From Theorems 4.10.1, 4.10.6 and 5.1.2. ■

Theorem 5.1.5 *If q is even and if the* TGQ $\mathbf{T}(\mathcal{O})$ *satisfies Property* (G) *at the flag* $((\infty), \pi)$, *with $\pi \in \mathcal{O}$, then the translation dual $\mathbf{T}(\mathcal{O}^*)$ satisfies Property* (G) *at the flag* $((\infty), \tau)$, *with τ the tangent space of $\mathcal{O}$ at π.*

Proof. Suppose that $\mathbf{T}(\mathcal{O})$ satisfies Property (G) at the flag $((\infty), \pi)$, with q even. Let $\bar{\tau}, \bar{\tau}_i, \bar{\tau}_j$ be points of Type (ii) of $\mathbf{T}(\mathcal{O})$ collinear with (∞), where $\tau \subset \bar{\tau}, \tau_i \subset \bar{\tau}_i, \tau_j \subset \bar{\tau}_j$, with τ, τ_i, τ_j the tangent spaces of $\mathcal{O}$ at the respective (distinct) elements π, π_i, π_j of $\mathcal{O}$. Then, by Theorem 2.6.2, $\{\bar{\tau}, \bar{\tau}_i, \bar{\tau}_j\}^{\perp}$ and $\{\bar{\tau}, \bar{\tau}_i, \bar{\tau}_j\}^{\perp\perp}$ are contained in a subquadrangle $\mathcal{S}'$ of order q^n of $\mathbf{T}(\mathcal{O})$. The space $\bar{\tau} \cap \bar{\tau}_i \cap \bar{\tau}_j$ is an n-dimensional space $\mathbf{PG}(n, q)$, and the space $\mathbf{PG}(n-1, q) = \mathbf{PG}(n, q) \cap \mathbf{PG}(4n-1, q) = \tau \cap \tau_i \cap \tau_j$ is contained in exactly $q^n + 1$ tangent spaces $\tau, \tau_i, \tau_j, \ldots$ of $\mathcal{O}$. Then the points of $\mathbf{PG}(4n, q) \setminus \mathbf{PG}(4n-1, q)$ in $\langle \mathbf{PG}(n, q), \pi \rangle, \langle \mathbf{PG}(n, q), \pi_i \rangle, \langle \mathbf{PG}(n, q), \pi_j \rangle, \cdots$ are the points of Type (i) in $\mathcal{S}'$. Now we consider a line of Type (a) concurrent with π and incident with a point of Type (i) in the space $\langle \mathbf{PG}(n, q), \pi_i \rangle$. As this line is also a line of $\mathcal{S}'$, it is incident with a point of Type (i) in each of the $q^n - 1$ spaces $\langle \mathbf{PG}(n, q), \pi_j \rangle, \cdots$. Hence the spaces $\langle \mathbf{PG}(n, q), \pi_j \rangle, \cdots$ are contained in the $3n$-dimensional space $\langle \mathbf{PG}(n, q), \pi, \pi_i \rangle = \mathbf{PG}(3n, q)$. Consequently the spaces $\mathbf{PG}(n-1, q), \pi, \pi_i, \pi_j, \ldots$ are contained in a common $\mathbf{PG}(3n-1, q) = \mathbf{PG}(3n, q) \cap \mathbf{PG}(4n-1, q)$. So, by Theorem 3.8.1(i), the $(3n-1)$-dimensional space $\langle \pi, \pi_i, \pi_j \rangle$ contains exactly $q^n + 1$ elements of $\mathcal{O}$. Now, by Theorem 5.1.2, the GQ $\mathbf{T}(\mathcal{O}^*)$ has Property (G) at the flag $((\infty), \tau)$. ■

Corollary 5.1.6 *If q is even and if the egg $\mathcal{O}(n, 2n, q)$ is good at its element π, then the translation dual $\mathcal{O}^*(n, 2n, q)$ is good at the tangent space τ of $\mathcal{O}(n, 2n, q)$ at π.*

Proof. Directly from Theorems 5.1.2 and 5.1.5. ■

Remark 5.1.7 An alternative proof of Corollary 5.1.6, without using GQs, goes as follows. Assume that q is even and that the egg $\mathcal{O}(n, 2n, q) = \mathcal{O}$ is good at its element π. Let π_i, π_j be distinct elements of $\mathcal{O} \setminus \{\pi\}$, and let τ, τ_i, τ_j be the respective tangent spaces of $\mathcal{O}$ at π, π_i, π_j. Further, let $\pi, \pi_i, \pi_j, \ldots$ be the $q^n + 1$ elements of $\mathcal{O}$ in $\langle \pi, \pi_i, \pi_j \rangle = \mathbf{PG}(3n-1, q)$ and let $\mathcal{O}'$ be the pseudo-oval of $\mathbf{PG}(3n-1, q)$ consisting of these elements. By Theorem 3.9.1 the tangent spaces of $\mathcal{O}'$ contain a common $(n-1)$-dimensional space η, which was there called the kernel of $\mathcal{O}'$. Clearly the tangent spaces of $\mathcal{O}$ at the elements of $\mathcal{O}'$ all contain η. It follows that the translation dual $\mathcal{O}^*$ of $\mathcal{O}$ is good at its element τ.

Assume that the egg $\mathcal{O}(n, 2n, q) = \mathcal{O}$, with $\mathcal{O} = \{\pi, \pi_1, \ldots, \pi_{q^{2n}}\}$, is good at its element π, let τ be the tangent space of $\mathcal{O}$ at π, and let τ_i be the tangent space of $\mathcal{O}$ at π_i, with $i = 1, 2, \ldots, q^{2n}$. Further, let ζ be a $(3n - 1)$-dimensional subspace of the $\mathbf{PG}(4n - 1, q)$ containing $\mathcal{O}$, with $\pi \cap \zeta = \emptyset$. Now we project the elements of $\mathcal{O}\backslash\{\pi\}$ from π onto ζ. The projection of π_i from π onto ζ is denoted by π_i', that is, $\pi_i' = \langle \pi, \pi_i\rangle \cap \zeta$, with $i = 1, 2, \ldots, q^{2n}$. Then $\pi_1' \cup \pi_2' \cup \cdots \cup \pi_{q^{2n}}'$ is $\zeta\backslash\tau'$, where τ' is the $(2n - 1)$-dimensional space $\tau \cap \zeta$. The set of all intersections $\langle \pi, \pi_i, \pi_j \rangle \cap \zeta$, $i \neq j$, is denoted by $\mathcal{B}$. Then, with respect to inclusion, the elements of $\mathcal{P} = \{\pi_1', \pi_2', \ldots, \pi_{q^{2n}}'\}$ and $\mathcal{B}$, are the points and lines, respectively, of a $2 - (q^{2n}, q^n, 1)$ design $\mathcal{A}$, that is, of an affine plane of order $q^n = s$. The affine plane $\mathcal{A}$ will be called the *derived* or *internal plane* of the egg $\mathcal{O}$ at the good element π. If ρ' and ρ'' are distinct parallel lines of $\mathcal{A}$, then $\rho' \cap \rho''$ is necessarily an $(n-1)$-dimensional subspace of τ'. Hence all lines of $\mathcal{A}$ parallel to ρ' contain a common $(n-1)$-dimensional subspace of τ'. These $q^n + 1$ (distinct) subspaces of τ' are denoted by $\nu_0, \nu_1, \ldots, \nu_{q^n}$. If for some $i \neq j$ we have $\nu_i \cap \nu_j \neq \emptyset$, say $x \in \nu_i \cap \nu_j$, then the lines ρ' and ρ'' of $\mathcal{A}$ containing respectively ν_i and ν_j, do not intersect in a point of $\mathcal{A}$, so are parallel, that is, $\nu_i = \nu_j$, a contradiction. Hence $\{\nu_0, \nu_1, \ldots, \nu_{q^n}\}$ is a partition (an $(n - 1)$-spread) of τ'. The elements $\nu_0, \nu_1, \ldots, \nu_{q^n}$ and τ' extend $\mathcal{A}$ to a projective plane Π of order q^n. Now by a theorem of Segre [140] such a plane Π is Desarguesian, so in ζ there are n planes $\xi_1, \xi_2, \ldots, \xi_n$ over $\mathbf{GF}(q^n)$, which are conjugate with respect to the n-th extension $\mathbf{GF}(q^n)$ of $\mathbf{GF}(q)$ and which generate ζ, such that each plane ξ_i has a point in common with the extension to $\mathbf{GF}(q^n)$ of each element of the point set of Π, with $i = 1, 2, \ldots, n$; the extension to $\mathbf{GF}(q^n)$ of any element of the point set of Π is generated by a point of π_1 and its conjugates.

The following lemma on subquadrangles appears to be very useful.

Lemma 5.1.8 *Let $\mathcal{S} = (\mathcal{P}, \mathcal{B}, \mathbf{I})$ be a GQ of order $(s, s^2), s \neq 1$. If L is any line of $\mathcal{S}$, then $\mathcal{S}$ contains at most $s^3 + s^2$ subquadrangles of order s containing the line L.*

Proof. Any two lines M_1, M_2, with $M_1 \sim L \sim M_2$, $M_1 \nsim M_2$, belong to at most $s + 1$ subquadrangles of order s; see 2.2 of Payne and Thas [128]. It easily follows that L is contained in at most $s^3 + s^2$ subquadrangles of order s. ∎

Assume again that the egg $\mathcal{O} = \mathcal{O}(n, 2n, q)$ is good at its element π. Let τ be the tangent space of $\mathcal{O}$ at π, and let τ_i be the tangent space of $\mathcal{O}$ at π_i, with $\pi_i \in \mathcal{O}\backslash\{\pi\}$ and $i = 1, 2, \ldots, q^{2n}$. Further, let $\mathcal{O}$ be contained in $\mathbf{PG}(4n-1, q)$, let $\mathbf{PG}(4n-1, q)$ be a hyperplane of $\mathbf{PG}(4n, q)$, and let $\mathbf{T}(\mathcal{O})$ be the corresponding TGQ of order (q^n, q^{2n}). If π_k, π_j are distinct elements of $\mathcal{O}\backslash\{\pi\}$ then $\mathbf{PG}(3n - 1, q) = \langle \pi, \pi_j, \pi_k \rangle$ contains exactly $q^n + 1$ elements of $\mathcal{O}$. Let $\mathbf{PG}(3n - 1, q) \subset \mathbf{PG}(3n, q) \subset \mathbf{PG}(4n, q)$, with $\mathbf{PG}(3n, q) \not\subset \mathbf{PG}(4n - 1, q)$. Then the point (∞) of $\mathbf{T}(\mathcal{O})$, the points of Type (i) of $\mathbf{T}(\mathcal{O})$ in $\mathbf{PG}(3n, q)$, the lines of Type (a) and (b) of $\mathbf{T}(\mathcal{O})$ in $\mathbf{PG}(3n, q)$, and the points of Type (ii) of $\mathbf{T}(\mathcal{O})$ incident with an element of $\mathcal{O}$ in $\mathbf{PG}(3n - 1, q)$, form a subGQ $\mathcal{S}'$ of $\mathbf{T}(\mathcal{O})$ of order q^n which is isomorphic to $\mathbf{T}(\mathcal{O}')$ with $\mathcal{O}'$ the pseudo-oval consisting of the $q^n + 1$ elements of $\mathcal{O}$ in $\mathbf{PG}(3n - 1, q)$.

The following theorem of Thas [176] has important applications; see Chapter 6.

Theorem 5.1.9 *Assume that the egg $\mathcal{O} = \mathcal{O}(n, 2n, q)$ is good at its element π. Then the TGQ $\mathbf{T}(\mathcal{O})$ contains exactly $q^{3n} + q^{2n}$ subquadrangles of order q^n containing the line π of $\mathbf{T}(\mathcal{O})$. These subquadrangles are TGQs and for q odd each of these subquadrangles is isomorphic to the classical GQ $\mathbf{Q}(4, q^n)$.*

Proof. The space π is contained in $q^{2n} + q^n$ pseudo-ovals on $\mathcal{O}$ and each such pseudo-oval defines q^n subquadrangles of order q^n of $\mathbf{T}(\mathcal{O})$, each of which contains the line π of $\mathbf{T}(\mathcal{O})$. So there arise $q^{3n} + q^{2n}$ subquadrangles of order q^n, containing π, of $\mathbf{T}(\mathcal{O})$. By Lemma 5.1.8 $\mathbf{T}(\mathcal{O})$ contains exactly $q^{3n} + q^{2n}$ subquadrangles of order q^n containing the line π of $\mathbf{T}(\mathcal{O})$. Any of these subquadrangles is isomorphic to a TGQ $\mathbf{T}(\mathcal{O}')$, with $\mathcal{O}'$ a pseudo-oval on $\mathcal{O}$ containing π.

Let $\mathcal{O}'$ be one of the pseudo-ovals on $\mathcal{O}$ containing π, let $\mathcal{O}'$ be contained in $\mathbf{PG}(3n-1, q)$, and let γ be a $(2n-1)$-dimensional subspace of $\mathbf{PG}(3n-1, q)$ skew to π. Assume that τ' is the tangent space of $\mathcal{O}'$ at π, and that e.g. $\pi, \pi_1, \pi_2, \ldots, \pi_{q^n}$ are the elements of $\mathcal{O}$ in $\mathcal{O}'$. Put $\tau' \cap \gamma = \pi'$ and $\langle \pi, \pi_i \rangle \cap \gamma = \pi'_i$, with $i = 1, 2, \ldots, q^n$. Then $\{\pi', \pi'_1, \pi'_2, \ldots, \pi'_{q^n}\} = T'$ is an $(n - 1)$-spread of γ. From the section preceding Lemma 5.1.8 follows that the $(n - 1)$-spread T' is regular. So, by Theorem 3.9.6, for q odd $\mathbf{T}(\mathcal{O}') \cong \mathbf{Q}(4, q^n)$. ∎

In the next theorem we will consider good eggs $\mathcal{O}(n, 2n, q)$ for which all pseudo-ovals containing the good element are regular. This theorem is taken from Thas and K. Thas [192].

Theorem 5.1.10 *Let $\mathcal{O}(n, 2n, q) = \mathcal{O}$ be an egg in $\mathbf{PG}(4n - 1, q)$, q even, which is good at $\pi \in \mathcal{O}$. If the $q^{2n} + q^n$ pseudo-ovals on $\mathcal{O}$ containing π are regular, then $\mathcal{O}$ is regular, so the corresponding TGQ $\mathbf{T}(\mathcal{O})$ is isomorphic to a $\mathbf{T}_3(\mathcal{O}')$ of Tits.*

Proof. Let $\mathcal{O} = \{\pi, \pi_1, \pi_2, \ldots, \pi_{q^{2n}}\}$ be an egg in $\mathbf{PG}(4n - 1, q)$, q even, which is good at π. Suppose that the $q^{2n} + q^n$ pseudo-ovals on $\mathcal{O}$ containing π are regular. Clearly we may assume that $n > 1$. Next, assume that $q^n = 4$, so $q = n = 2$. As $q^n = 4$, the $q^{2n} + q^n$ regular pseudo-ovals on $\mathcal{O}$ containing π are pseudo-conics. Now, by Brown and Lavrauw [26], cf. Corollary 7.8.4 below, the egg is classical, hence regular. From now on we assume that $q^n > 4$.

Let $\mathbf{PG}(3n - 1, q)$ be a subspace of $\mathbf{PG}(4n - 1, q)$ which is skew to π and let $\langle \pi, \pi_i \rangle \cap \mathbf{PG}(3n - 1, q) = \pi'_i$, with $i = 1, 2, \ldots, q^{2n}$. Let τ be the tangent space of $\mathcal{O}$ at π, and let τ_i be the tangent space of $\mathcal{O}$ at π_i, with $i = 1, 2, \ldots, q^{2n}$. If $\tau' = \tau \cap \mathbf{PG}(3n - 1, q)$, then the intersections of τ' with the tangent spaces at π of the pseudo-ovals on $\mathcal{O}$ containing π, form an $(n-1)$-spread $\mathcal{T}^*$ of τ'. Then $\mathcal{T}^* \cup \{\pi'_1, \pi'_2, \ldots, \pi'_{q^{2n}}\}$ is a regular $(n-1)$-spread $\mathcal{T}$ of $\mathbf{PG}(3n - 1, q)$. This spread $\mathcal{T}$ is the point set of the projective completion Π of the derived plane $\mathcal{A}$ of $\mathcal{O}$ at π. If we extend the elements of $\mathcal{T}$ to $\mathbf{GF}(q^n)$, then these extensions have a point in common with each of n planes $\xi_1, \xi_2, \ldots, \xi_n$ over $\mathbf{GF}(q^n)$.

Let $\overline{\pi}'_i \cap \xi_j = \{p_{ij}\}$, with $\overline{\pi}'_i$ the extension of π'_i to $\mathbf{GF}(q^n)$, $i = 1, 2, \ldots, q^{2n}$ and $j = 1, 2, \ldots, n$. Then $\xi_j = \{p_{1j}, p_{2j}, \ldots, p_{q^{2n}j}\} \cup L_j$, with L_j the intersection of ξ_j with the extension of τ' to $\mathbf{GF}(q^n)$, with $j = 1, 2, \ldots, n$. Let M_j be a line of ξ_j, with $L_j \neq M_j$. Then $\langle M_j, \overline{\pi} \rangle$, with $\overline{\pi}$ the extension of π to $\mathbf{GF}(q^n)$, intersects the extensions to $\mathbf{GF}(q^n)$ of all elements of a pseudo-oval $\mathcal{O}'$ on $\mathcal{O}$ containing π. As $\mathcal{O}'$ is a regular pseudo-oval, there are planes γ_k over $\mathbf{GF}(q^n)$ intersecting the extensions of all elements of $\mathcal{O}'$, with $k = 1, 2, \ldots, n$. Then $\{\langle \overline{\pi}, \gamma_k \rangle \cap \mathbf{PG}(3n - 1, q^n) \| k = 1, 2, \ldots, n\} = \{\langle \mathcal{O}' \rangle \cap \xi_j \| j = 1, 2, \ldots, n\}$; here we rely on the fact that a regular $(n-1)$-spread in a $\mathbf{PG}(2n - 1, q)$ uniquely defines n transversals over $\mathbf{GF}(q^n)$. So we may assume that $\langle \overline{\pi}, \gamma_j \rangle \cap \mathbf{PG}(3n - 1, q^n) = \langle \mathcal{O}' \rangle \cap \xi_j = M_j$, with $j = 1, 2, \ldots, n$. Clearly the extensions of the elements of $\mathcal{O}'$ intersect γ_j in the points of an oval $\mathcal{O}'_j$ over $\mathbf{GF}(q^n)$, with $\mathcal{O}'_j \cap \overline{\pi} = \{r_j\}, j = 1, 2, \ldots, n$. This oval $\mathcal{O}'_j$ is contained in $\langle \overline{\pi}, \xi_j \rangle$. Now let N_j be the tangent of $\mathcal{O}'_j$ at r_j. Then $\langle N_j, \overline{\pi} \rangle$ intersects ξ_j in a point of L_j. If $\mathcal{O}'_j$ and $\mathcal{O}''_j$ are two such ovals, with $\mathcal{O}'_j \neq \mathcal{O}''_j$, then we have the following cases.

- If the corresponding lines of ξ_j intersect in a point not on L_j, then $(\mathcal{O}'_j \cap \mathcal{O}''_j)\backslash\overline{\pi}$ is a point.

- If the corresponding lines of ξ_j intersect in a point of L_j, then $(\mathcal{O}'_j \cap \mathcal{O}''_j)\backslash\overline{\pi}$ is empty.

Let the distinct points $p_{ij}, p_{i'j}, p_{i''j}$ define a triangle of ξ_j. The ovals defined by the lines $p_{ij}p_{i'j}, p_{i'j}p_{i''j}, p_{i''j}p_{ij}$ as above, are denoted by $\mathcal{O}_j^{(3)}$, $\mathcal{O}_j^{(1)}$, $\mathcal{O}_j^{(2)}$, respectively. Then one easily sees that all $q^{2n} + q^n$ ovals $\mathcal{O}'_j$ are contained in the space $\langle \mathcal{O}_j^{(1)}, \mathcal{O}_j^{(2)}, \mathcal{O}_j^{(3)} \rangle$. So all these ovals generate a $\mathbf{PG}^{(j)}(m, q^n)$, with $3 \le m \le 5$.

(a) The Case $m = 3$

Then $\langle \overline{\pi}, \mathbf{PG}^{(j)}(3, q^n) \rangle = \langle \overline{\pi}, \xi_j \rangle$ is $(n+2)$-dimensional, so $\overline{\pi} \cap \mathbf{PG}^{(j)}(3, q^n)$ is a point u. This point u belongs to the $q^{2n}+q^n$ ovals $\mathcal{O}'_j$. The union of these ovals is a set W of size $q^{2n}+1$. As no three points of W are collinear, this set W is an ovoid $\overline{\mathcal{O}}$ of $\mathbf{PG}^{(j)}(3, q^n)$. Hence the extensions of the elements of $\mathcal{O}$ meet the n spaces $\mathbf{PG}^{(1)}(3, q^n), \mathbf{PG}^{(2)}(3, q^n), \dots, \mathbf{PG}^{(n)}(3, q^n)$ and the egg $\mathcal{O}$ is regular (the TGQ $\mathbf{T}(\mathcal{O})$ is isomorphic to the TGQ $\mathbf{T}_3(\overline{\mathcal{O}})$ of Tits).

(b) The Case $m = 5$

Then $\langle \overline{\pi}, \mathbf{PG}^{(j)}(5, q^n) \rangle = \langle \overline{\pi}, \xi_j \rangle$ is $(n+2)$-dimensional, so $\overline{\pi} \cap \mathbf{PG}^{(j)}(5, q^n)$ is a plane δ_j. Let $\overline{\pi}_i$ be the extension of π_i to $\mathbf{GF}(q^n)$ and let $\overline{\pi}_i \cap \mathbf{PG}^{(j)}(5, q^n) = \{s_i\}$, with $i = 1, 2, \dots, q^{2n}$. Now we project from $s_1 s_2$ onto a subspace $\mathbf{PG}(3, q^n)$ of $\mathbf{PG}^{(j)}(5, q^n)$ which is skew to $s_1 s_2$. Let the q^n ovals containing s_1 but not s_2 be $\mathcal{O}_{11}, \mathcal{O}_{12}, \dots, \mathcal{O}_{1q^n}$, and let the q^n ovals containing s_2 but not s_1 be $\mathcal{O}_{21}, \mathcal{O}_{22}, \dots, \mathcal{O}_{2q^n}$. Further, let L_{li} be the tangent line of $\mathcal{O}_{li}$ at s_l, with $l = 1, 2$ and $i = 1, 2, \dots, q^n$. By projection of $\mathcal{O}_{li}\backslash\{s_l\}$ from $s_1 s_2$ onto $\mathbf{PG}(3, q^n)$, there arise q^n points of a line T_{li}. Let $\langle s_1, s_2, L_{li} \rangle \cap \mathbf{PG}(3, q^n) = \{t_{li}\}$; then $t_{li} \in T_{li}$. Assume, by way of contradiction, that $T_{li} \cap T_{lk} \ne \emptyset$ for $i \ne k$. Then all ovals in $\mathbf{PG}^{(j)}(5, q^n)$ are contained in $\langle T_{li}, T_{lk}, s_1, s_2 \rangle$ which is at most 4-dimensional, a contradiction. So $T_{l1}, T_{l2}, \dots, T_{lq^n}$ are mutually disjoint, with $l = 1, 2$. If $\widetilde{\mathcal{O}}$ is the oval containing s_1 and s_2, then $\widetilde{\mathcal{O}}\backslash\{s_1, s_2\}$ is projected onto a point t of $\mathbf{PG}(3, q^n)$. The notation is chosen in such a way that $\mathcal{O}_{2i}$ is the oval on s_2 which contains no point of $\mathcal{O}_{1i}\backslash\overline{\pi}$, with $i = 1, 2, \dots, q^n$.

So $T_{1i} \cap T_{2k} \neq \emptyset$ for $k \neq i$. As $q^n > 4$, it follows that there is a hyperbolic quadric $\mathcal{H}$ in $\mathbf{PG}(3, q^n)$ such that $\{T_{l1}, T_{l2}, \ldots, T_{lq^n}\}$ is a subset of a regulus $\mathcal{R}_l$ of $\mathcal{H}$, with $l = 1, 2$. Let $\mathcal{O}_{li} \cap \bar{\pi} = \mathcal{O}_{li} \cap \delta_j = \{r_{li}\}$, and $\langle s_1, s_2, r_{li} \rangle \cap \mathbf{PG}(3, q^n) = \{s_{li}\}$. If $\bar{\tau}_i$ is the extension of τ_i to $\mathbf{GF}(q^n)$, then $\bar{\tau}_i \cap \mathbf{PG}^{(j)}(5, q^n)$ is a plane ρ_{ji}, with $j = 1, 2, \ldots, n$ and $i = 1, 2, \ldots, q^{2n}$. So the lines $L_{l1}, L_{l2}, \ldots, L_{lq^n}$ are contained in the plane ρ_{jl}. It follows that $t_{l1}, t_{l2}, \ldots, t_{lq^n}$ are points of some line T_l', with $l = 1, 2$. Clearly $\{t\} = T_1' \cap T_2'$ and $s_{1i} = s_{2i} = s_i'$, with $i = 1, 2, \ldots, q^n$. Hence ovals which define a common point of L_j intersect δ_j in a common point. The points $s_1', s_2', \ldots, s_{q^n}'$ belong to $\langle \delta_j, s_1, s_2 \rangle \cap \mathbf{PG}(3, q^n)$, so are contained in a plane ψ. Hence $s_1', s_2', \ldots, s_{q^n}'$ belong to the nonsingular conic $\mathcal{C} = \psi \cap \mathcal{H}$. Also, we have $t \in \mathcal{C}$.

Now we consider an oval $\mathcal{O}_j'$ not containing s_1 nor s_2. By projection of $\mathcal{O}_j'$ from $s_1 s_2$ we obtain the intersection of $\mathcal{H}$ with a plane of $\mathbf{PG}(3, q^n)$ through t, but not containing T_1' nor T_2'. Hence this intersection is a nonsingular conic, and so $\mathcal{O}_j'$ is a conic. It follows that the corresponding pseudo-oval $\mathcal{O}$ is a pseudo-conic. Now, by Brown and Lavrauw [26], see also Corollary 7.8.4, the egg is classical. But in such a case m is necessarily three, a contradiction.

(c) The Case $m = 4$

Then $\langle \bar{\pi}, \mathbf{PG}^{(j)}(4, q^n) \rangle = \langle \bar{\pi}, \xi_j \rangle$ is $(n+2)$-dimensional, so $\bar{\pi} \cap \mathbf{PG}^{(j)}(4, q^n)$ is a line W_j. Let $\bar{\pi}_i$ be the extension of π_i to $\mathbf{GF}(q^n)$ and let $\bar{\pi}_i \cap \mathbf{PG}^{(j)}(4, q^n) = \{s_i\}$, with $i = 1, 2, \ldots, q^{2n}$.

Let $w \in L_j$. With the q^n lines of ξ_j through w, but distinct from L_j, correspond q^n ovals. The tangent lines of these ovals at their intersection point with W_j are contained in the plane $\mathbf{PG}^{(j)}(4, q^n) \cap \langle \bar{\pi}, w \rangle$. Assume, by way of contradiction, that the point $u \in U = W_j$ is on at least $q^n + 1$ ovals $\mathcal{O}_j'$. So at least two of these ovals, say $\mathcal{O}_j^{(1)}$ and $\mathcal{O}_j^{(2)}$, have a point $s_i \notin U$ in common. Let $\mathcal{O}_j^{(3)}$ be a third oval containing u. As all ovals generate $\mathbf{PG}^{(j)}(4, q^n)$, the oval $\mathcal{O}_j^{(3)}$ cannot contain a point of both $\mathcal{O}_j^{(1)} \setminus \{u, s_i\}$ and $\mathcal{O}_j^{(2)} \setminus \{u, s_i\}$.

First, assume that $\mathcal{O}_j^{(3)}$ contains s_i. If $\bar{\tau}_i$ is the extension of τ_i to $\mathbf{GF}(q^n)$, then $\bar{\tau}_i \cap \mathbf{PG}^{(j)}(4, q^n)$ is a plane ρ_{ji}, with $j = 1, 2 \ldots, n$ and $i = 1, 2, \ldots, q^{2n}$. So the tangent lines at s_i of the ovals $\mathcal{O}_j^{(1)}, \mathcal{O}_j^{(2)}, \mathcal{O}_j^{(3)}$ are coplanar. It follows that these three ovals are contained in the threedimensional space $\langle \rho_{ji}, u \rangle$, and so all $q^{2n} + q^n$ ovals $\mathcal{O}'_j$ are contained in $\langle \rho_{ji}, u \rangle$, clearly a contradiction. So any oval $\mathcal{O}^{(k)}_j$ containing u, with $\mathcal{O}^{(1)}_j \neq \mathcal{O}^{(k)}_j \neq \mathcal{O}^{(2)}_j$, does not contain s_i. Let $\mathcal{O}^{(k)}_j$ and $\mathcal{O}^{(k')}_j$ be distinct ovals containing u, with $|\mathcal{O}^{(k)}_j \cap \mathcal{O}^{(l)}_j| = |\mathcal{O}^{(k')}_j \cap \mathcal{O}^{(l)}_j| = 2$ and $l \in \{1, 2\}$. Then $\mathcal{O}^{(k)}_j \cap \mathcal{O}^{(k')}_j = \{u\}$, as otherwise $\langle \mathcal{O}^{(l)}_j, \mathcal{O}^{(k)}_j, \mathcal{O}^{(k')}_j \rangle$ is 3-dimensional. That means that all ovals containing u are elements of two classes W_1 and W_2, where $\mathcal{O}^{(l)}_j \in W_l$ and any two distinct ovals of W_l just have u in common, with $i = 1, 2$. Now we project from u onto a $\mathbf{PG}(3, q^n)$ in $\mathbf{PG}^{(j)}(4, q^n)$, with $u \notin \mathbf{PG}(3, q^n)$. The ovals of W_l are denoted by $\mathcal{O}^{(l)}_{j1} = \mathcal{O}^{(l)}_j, \mathcal{O}^{(l)}_{j2}, \ldots$, with $l = 1, 2$. Let $T^{(l)}_k$ be the tangent line of $\mathcal{O}^{(l)}_{jk}$ at u, and let $T^{(l)}_k \cap \mathbf{PG}(3, q^n) = \{t^{(l)}_k\}$. Then the points $t^{(l)}_1, t^{(l)}_2, \ldots$ are distinct, as otherwise all ovals are contained in a 3-dimensional space, with $i = 1, 2$; as ovals of different classes W_1 and W_2 always have a point of $\mathbf{PG}^{(j)}(4, q^n) \setminus U$ in common, we clearly have $t^{(1)}_k \neq t^{(2)}_{k'}$. Let the projection from u onto $\mathbf{PG}(3, q^n)$ of $\mathcal{O}^{(l)}_{jk} \setminus \{u\}$ belong to the line $V^{(l)}_k$. Then $t^{(l)}_k \in V^{(l)}_k$. Now we shall prove that the lines $V^{(l)}_1, V^{(l)}_2, \ldots$, with $l = 1, 2$, all belong to a common hyperbolic quadric $\mathcal{H}$ of $\mathbf{PG}(3, q^n)$. Suppose, by way of contradiction, that these lines do not belong to a common hyperbolic quadric. As $T^{(l)}_1, T^{(l)}_2, \ldots$ define a common point on L_j, the points $t^{(l)}_1, t^{(l)}_2, \ldots$ are collinear, with $l = 1, 2$. Assume that $|W_1| \leq |W_2|$. Then the only possibility is that $|W_1| = 1$ and $|W_2| = q^n$, as otherwise all lines $V^{(l)}_k$ belong to a common hyperbolic quadric. The points $t^{(2)}_1, t^{(2)}_2, \ldots$ are on a line R; let u' be the $(q^n + 1)$-th point of that line. Then $\{u'\} = U \cap \mathbf{PG}(3, q^n)$. With the $q^{2n} - 1$ ovals not containing u correspond, by projection from u, $q^{2n} - 1$ ovals through u'. Any such projection is a subset of $V^{(2)}_1 \cup V^{(2)}_2 \cup \cdots \cup V^{(2)}_{q^n} \cup \{u'\}$. The planes of these projections, together with the $q^n + 1$ planes on R in $\mathbf{PG}(3, q^n)$ and the plane $\langle u', V^{(1)}_1 \rangle$ are the $q^{2n} + q^n + 1$ planes of $\mathbf{PG}(3, q^n)$ through u'. Now consider a line V, with $V^{(1)}_1 \neq V \neq R$, which intersects $V^{(2)}_1, V^{(2)}_2, V^{(2)}_3$. Then $\langle V, u' \rangle$ intersects $V^{(2)}_1 \cup V^{(2)}_2 \cup \cdots \cup V^{(2)}_{q^n} \cup \{u'\}$ in $q^n + 1$ points which form an oval, a contradiction as this intersection contains three distinct collinear points. Consequently the lines $V^{(l)}_k$ all belong to a common hyperbolic quadric $\mathcal{H}$ of $\mathbf{PG}(3, q^n)$.

Assume $|W_1| \leq |W_2|$ and let $U \cap \mathbf{PG}(3, q^n) = \{u'\}$; from the foregoing it follows that $|W_1| \geq 2$. Then the points $u', t^{(l)}_1, t^{(l)}_2, \ldots$ are on a common line R_l, with $l = 1, 2$. Assume, by way of contradiction, that we are not in the case $|W_1| = |W_2| = q^n$. Then there exists an oval $\mathcal{O}'_j$, with $\mathcal{O}'_j \cap \mathcal{O}^{(1)}_j =$

$\emptyset$. The projection $\widetilde{\mathcal{O}}'_j$ of $\mathcal{O}'_j$ from u onto $\mathbf{PG}(3, q^n)$ contains u', and the projection of the tangent line of $\mathcal{O}'_j$ at $\mathcal{O}'_j \cap U$ is the line R_1. It follows that $\langle \widetilde{\mathcal{O}}'_j \rangle$ contains R_1. As $R_1 \subset \mathcal{H}$, the oval $\widetilde{\mathcal{O}}'_j$ contains $|W_2| \geq q^n - 1$ points of a line of $\mathcal{H}$, clearly a contradiction. Hence $|W_1| = |W_2| = q^n$. Any oval $\mathcal{O}'_j$ not in $W_1 \cup W_2$ is then projected onto a conic of $\mathcal{H}$, and so $\mathcal{O}'_j$ is a conic. It follows that the corresponding pseudo-oval $\mathcal{O}$ is a pseudo-conic. Now, by Brown and Lavrauw [26], see Corollary 7.8.4, the egg is classical. But in such a case m is necessarily three, a contradiction.

So we have shown that each point u of U is contained in exactly q^n of the $q^{2n} + q^n$ ovals.

Consider again distinct ovals $\mathcal{O}^{(1)}_j, \mathcal{O}^{(2)}_j$ containing the point $u \in U$ and assume that $\mathcal{O}^{(1)}_j \cap \mathcal{O}^{(2)}_j = \{u, s_i\}$, with $s_i \notin U$. Then as before all ovals containing u are elements of two classes W_1 and W_2, where $\mathcal{O}^{(l)}_j \in W_l$ and any two distinct ovals of W_l just have u in common, with $l = 1, 2$. Assume that $|W_1| \leq |W_2|$. We proceed as before, using similar notation. Then, if the lines $V^{(l)}_k$ do not belong to a common hyperbolic quadric $\mathcal{H}$, we have $|W_1| = 1$ and $|W_2| = q^n - 1$. In such a case the planes of the q^{2n} ovals not on u together with the $q^n + 1$ planes through R in $\mathbf{PG}(3, q^n)$ and the plane $\langle u', V^{(1)}_1 \rangle$, are the $q^{2n} + q^n + 1$ distinct planes in $\mathbf{PG}(3, q^n)$ containing u'; the unique oval $\mathcal{O}^*$ having no point in common with the union of the elements of W_2 defines a plane of $\mathbf{PG}(3, q^n)$ which contains R (the projection $\widetilde{\mathcal{O}}^*$ onto $\mathbf{PG}(3, q^n)$ of this oval is tangent to R at u'). Now consider a line V, with $V^{(1)}_1 \neq V \neq R$, which intersects $V^{(2)}_1, V^{(2)}_2, V^{(2)}_3$. Then $\langle V, u' \rangle$ intersects $V^{(2)}_1 \cup V^{(2)}_2 \cup \cdots \cup V^{(2)}_{q^n-1} \cup \widetilde{\mathcal{O}}^* \cup \{u'\}$ in $q^n + 1$ points which form an oval, a contradiction as this intersection contains three distinct collinear points. So the lines $V^{(l)}_k$ belong to a common hyperbolic quadric $\mathcal{H}$ of $\mathbf{PG}(3, q^n)$. Also, from the foregoing reasoning it follows that $|W_1| \geq 2$. Consider an oval $\mathcal{O}'_j$ with $\mathcal{O}^{(1)}_j \cap \mathcal{O}'_j = \emptyset$. The projection $\widetilde{\mathcal{O}}'_j$ of $\mathcal{O}'_j$ from u onto $\mathbf{PG}(3, q^n)$ contains u', and the projection of the tangent line of $\mathcal{O}'_j$ at $\mathcal{O}'_j \cap U$ is the line R_1 containing $u', t^{(1)}_1, t^{(1)}_2, \ldots$. It follows that $\langle \widetilde{\mathcal{O}}'_j \rangle$ contains R_1. As R_1 contains at least three points of $\mathcal{H}$, it belongs to $\mathcal{H}$. It follows that $\widetilde{\mathcal{O}}'_j$ contains $|W_2| \geq q^n - 2$ points of a line of $\mathcal{H}$, clearly a contradiction.

So if $\mathcal{O}^{(1)}_j, \mathcal{O}^{(2)}_j$ are any two distinct ovals containing a point u of U, then $\mathcal{O}^{(1)}_j \cap \mathcal{O}^{(2)}_j = \{u\}$. Let $u \in U$ and project from u onto a $\mathbf{PG}(3, q^n) \subset \mathbf{PG}^{(j)}(4, q^n)$, with $u \notin \mathbf{PG}(3, q^n)$. Let $\mathcal{O}^{(1)}_j, \mathcal{O}^{(2)}_j, \ldots, \mathcal{O}^{(q^n)}_j$ be the ovals containing u, and let T_i be the tangent line of $\mathcal{O}^{(i)}_j$ at u. Further, let

$T_i \cap \mathbf{PG}(3, q^n) = \{t_i\}$ and let the projection of $\mathcal{O}_j^{(i)} \backslash \{u\}$ belong to the line V_i, with $i = 1, 2, \ldots, q^n$. Then $t_i \in V_i$. The points $t_1, t_2, \ldots, t_{q^n}$ are not necessarily distinct. Also, as $T_1, T_2, \ldots, T_{q^n}$ define a common point on L_j, the points $t_1, t_2, \ldots, t_{q^n}$ are collinear. If $U \cap \mathbf{PG}(3, q^n) = \{u'\}$, then the points $u', t_1, t_2, \ldots, t_{q^n}$ are on a common line U' of $\mathbf{PG}(3, q^n)$. Let $u_l \in U \backslash \{u\}$, with $l = 1, 2, \ldots, q^n$. The tangent lines at u_l of the q^n ovals on u_l are coplanar, so intersect $\mathbf{PG}(3, q^n)$ in points on a line U'_l containing u'. In this way arise q^n lines $U'_1, U'_2, \ldots, U'_{q^n}$, which, together with U', are the $q^n + 1$ lines through u' in a plane β. The projections of the q^{2n} ovals not through u are projected onto ovals in the q^{2n} planes of $\mathbf{PG}(3, q^n)$ which contain the point u' but not the line U'. Each of these ovals in $\mathbf{PG}(3, q^n)$ contains u' and a point of $V_i \backslash U'$, with $i = 1, 2, \ldots, q^n$. If $t_i = t_k$, with $i \neq k$, then $\langle \mathcal{O}_j^{(i)}, \mathcal{O}_j^{(k)} \rangle$ does not contain U, as otherwise all ovals would be contained in a 3-dimensional space. If $t_i \neq t_k$, with $i \neq k$, then we can find three collinear points on $(V_1 \cup V_2 \cup \cdots \cup V_{q^n}) \backslash U'$ such that the line W' joining them has no point in common with U' (if e.g. $t_i \in V_i, t_i = t_l \in V_l, t_k \in V_k$, with i, j, k distinct, then in the plane $\langle V_i, V_l \rangle$ there are three collinear points on $V_i \cup V_l \cup V_k$ satisfying the requirement). Then $\langle W', u' \rangle$ contains the projection $\tilde{\mathcal{O}}'_j$ of an oval not containing u, and $\tilde{\mathcal{O}}'_j$ has three collinear points, a contradiction. Consequently $t_1 = t_2 = \cdots = t_{q^n}$. Hence $T_1 = T_2 = \cdots = T_{q^n}$. A similar conclusion holds for each point $u_l \in U$, with $l = 1, 2, \ldots, q^n$. So in $\mathbf{PG}^{(j)}(4, q^n)$ the nuclei of the $q^{2n} + q^n$ ovals are on $q^n + 1$ lines $B_1, B_2, \ldots, B_{q^n+1}$. As $B_1, B_2, \ldots, B_{q^n+1}$ define all $q^n + 1$ points of L_j, these lines are mutually skew. In the tangent space τ of $\mathcal{O}$ at π, the space π together with the nuclei of the pseudo-ovals on $\mathcal{O}$ containing π form a regular $(n-1)$-spread $\mathcal{T}^*$ in τ; see Remark 5.1.7 and the text preceding Lemma 5.1.8. As the nuclei of the $q^{2n} + q^n$ pseudo-ovals are distinct, also the nuclei of the $q^{2n} + q^n$ ovals are distinct. The extensions to $\mathbf{GF}(q^n)$ of the elements of $\mathcal{T}^*$ meet n planes $\xi_1^*, \xi_2^*, \ldots, \xi_n^*$ over $\mathbf{GF}(q^n)$. Hence $B_1, B_2, \ldots, B_{q^n+1}$ are contained in $q^n + 1$ distinct planes ξ_i^*, a contradiction as $q^n + 1 > n$. Now the theorem is completely proved. ∎

As a corollary of the previous theorem we will prove in a geometric way a result which is equivalent to Johnson's Theorem 4.5.1.

Theorem 5.1.11 *If the TGQ* $\mathbf{T}(\mathcal{O})$ *of order* (s, s^2), *with* s *even, is the dual of a flock GQ, then* $\mathbf{T}(\mathcal{O})$ *is classical.*

Proof. Let $\mathbf{T}(\mathcal{O})$ be a TGQ of order (s, s^2), s even, which is the dual of a flock GQ. By Theorems 4.10.1, 5.1.2 and 5.1.5 the egg $\mathcal{O}$ is good at some element $\pi = \mathbf{PG}(n-1, q)$, where $s = q^n$. Let $\mathcal{O}'$ be any of the pseudo-ovals on $\mathcal{O}$ through π. By Payne [117] the point-line dual of the TGQ $\mathbf{T}(\mathcal{O}')$ of order q^n is isomorphic to a $\mathbf{T}_2(\widetilde{\mathcal{O}}')$ of Tits, with $\widetilde{\mathcal{O}}'$ some oval of $\mathbf{PG}(2, q^n)$; the translation point of the point-line dual of $\mathbf{T}_2(\widetilde{\mathcal{O}}')$ is the line π of $\mathbf{T}(\mathcal{O}')$. As the elements of $\mathcal{O}'$ are regular points of the point-line dual of $\mathbf{T}_2(\widetilde{\mathcal{O}}')$, the GQ $\mathbf{T}(\mathcal{O}')$ is a $\mathbf{T}_2(\mathcal{O}'')$ of Tits, where $\mathcal{O}''$ is a translation oval of $\mathbf{PG}(2, s)$; see Payne [109] and Section 12.5 of Payne and Thas [128]. By Corollary 3.5.3 and Section 3.6 the pseudo-oval $\mathcal{O}'$ is regular. Now by Theorem 5.1.10 the egg $\mathcal{O}$ is regular, that is, $\mathbf{T}(\mathcal{O})$ is isomorphic to a $\mathbf{T}_3(\widetilde{\mathcal{O}})$ of Tits, with $\widetilde{\mathcal{O}}$ some ovoid of $\mathbf{PG}(3, q^n)$. With the notation of Section 4.10, the ovoids $\mathcal{O}_z$ are isomorphic to the ovoid $\widetilde{\mathcal{O}}$. As $\mathbf{T}(\mathcal{O})$ is the dual of a flock GQ, all ovoids $\mathcal{O}_z$ are elliptic quadrics; see also the paragraph preceding Theorem 4.10.7. Hence $\widetilde{\mathcal{O}}$ is an elliptic quadric, that is, $\mathbf{T}_3(\widetilde{\mathcal{O}}) \cong \mathbf{T}(\mathcal{O})$ is classical. ∎

If $\mathcal{F}$ is a semifield flock of the quadratic cone $\mathcal{K}$ in $\mathbf{PG}(3, q)$ with q even, then, by Section 4.7, the corresponding flock GQ $\mathcal{S}(\mathcal{F})$ has a translation line. So by Theorem 5.1.11 $\mathcal{S}(\mathcal{F})$ is classical, so $\mathcal{F}$ is linear (see e.g. Payne [120, 121]). This is Johnson's Theorem 4.5.1.

Theorem 5.1.11 also follows from the theorem of Johnson. Consider a flock GQ $\mathcal{S}(\mathcal{F})$ having a translation line L, with $\mathcal{F}$ a flock of the quadratic cone $\mathcal{K}$ of $\mathbf{PG}(3, q)$, with q even. If $\mathcal{S}(\mathcal{F})$ is not classical, then, by Payne and Thas [129], the elation point (∞) of $\mathcal{S}(\mathcal{F})$ is fixed by the full automorphism group of $\mathcal{S}(\mathcal{F})$. So we may assume that L contains the point (∞). By Payne [121] we have $\mathcal{S}(\mathcal{F}) = \mathcal{S}(\mathcal{F}')$, where in $\mathcal{S}(\mathcal{F}')$ the line L is the line $[\infty]$ (we "recoordinatize" the flock GQ). Then, by Section 4.7, $\mathcal{F}'$ is a semifield flock. Hence, by Theorem 4.5.1, $\mathcal{F}'$ is linear, and consequently, by Payne and Thas [128], the GQ $\mathcal{S}(\mathcal{F}) = \mathcal{S}(\mathcal{F}')$ is classical. ∎

We now consider TGQs $\mathbf{T}(\mathcal{O})$ of order (s, s^2) for which the egg $\mathcal{O}$ has more than one good element.

Theorem 5.1.12 *If q is odd and the egg $\mathcal{O}(n, 2n, q) = \mathcal{O}$ has at least two good elements, then $\mathcal{O}$ is classical and so the TGQ $\mathbf{T}(\mathcal{O})$ is classical. If q is even and the egg $\mathcal{O}(n, 2n, q) = \mathcal{O}$ has at least four good elements, not contained in a common pseudo-oval on $\mathcal{O}$, then $\mathcal{O}$ is regular and so $\mathbf{T}(\mathcal{O})$ is a $\mathbf{T}_3(\mathcal{O}')$ of Tits.*

Proof. Assume that q is odd and that the egg $\mathcal{O}(n, 2n, q) = \mathcal{O}$ has at least two good elements. Let π be a good element of $\mathcal{O}$. By Corollary 5.1.4 the translation dual $\mathbf{T}(\mathcal{O}^*)$ of $\mathbf{T}(\mathcal{O})$ is the point-line dual of a flock GQ with elation point τ, where τ is the tangent space of $\mathcal{O}$ at π. Let $\pi' \in \mathcal{O} \backslash \{\pi\}$ be good, and let $\pi'' \in \mathcal{O} \backslash \{\pi, \pi'\}$. There is an elation of $\mathbf{T}(\mathcal{O}^*)$ with center τ, mapping the tangent space τ' of $\mathcal{O}$ at π' onto the tangent space τ'' of $\mathcal{O}$ at π''. By Theorem 3.7.2 there is an automorphism θ of $\mathbf{PG}(4n-1, q) \supset \mathcal{O}$, fixing $\mathcal{O}^*$, fixing τ and mapping τ' onto τ''. Clearly θ fixes $\mathcal{O}$, fixes π and maps π' onto π''. Hence all elements of $\mathcal{O}$ are good. By Theorem 3.10.7(ii) the egg is regular, and so by Section 3.6 the egg $\mathcal{O}$ is classical.

Next, assume that q is even and that the egg $\mathcal{O}(n, 2n, q) = \mathcal{O}$ has at least four good elements, not contained in a common pseudo-oval on $\mathcal{O}$. Let π and π' be distinct good elements of $\mathcal{O}$. Now we consider a pseudo-oval $\mathcal{O}_1$ on $\mathcal{O}$ containing π, but not containing π'. Let $\mathbf{PG}(3n-1, q)$ be the subspace of $\mathbf{PG}(4n-1, q) \supset \mathcal{O}$ generated by $\mathcal{O}_1$. By projection of $\mathcal{O} \backslash \{\pi'\}$ from π' onto $\mathbf{PG}(3n-1, q)$ there arises the derived plane $\mathcal{A}$ of $\mathcal{O}$ at π'; see Section 5.1. Clearly $\mathcal{O}_1$ is an oval of $\mathcal{A}$, and so $\mathcal{O}_1$ is regular. As there are four good elements, not contained in a common pseudo-oval on $\mathcal{O}$, it easily follows that all pseudo-ovals on $\mathcal{O}$ containing π are regular. Now by Theorem 5.1.10 the egg $\mathcal{O}$ is regular. ∎

From Theorem 4.7.3 and Theorem 5.1.12, one can now deduce the following result, which is essentially contained in Bader, Lunardon and Pinneri [4].

Theorem 5.1.13 *Suppose $\mathcal{F}$ is a Kantor-Knuth semifield flock of the quadratic cone in $\mathbf{PG}(3, q)$, q odd, and let $\mathcal{S}(\mathcal{F})^D = \mathbf{T}(\mathcal{O})$ be the associated dual flock GQ. Let π be a good element of $\mathcal{O}$, and let π^* be a good element of $\mathcal{O}^*$. Then there exists an isomorphism between $\mathbf{T}(\mathcal{O})$ and $\mathbf{T}(\mathcal{O}^*)$ that maps π onto π^*. If $\mathcal{F}$ is nonlinear, then any isomorphism between $\mathbf{T}(\mathcal{O})$ and $\mathbf{T}(\mathcal{O}^*)$ has this property. Moreover, in the latter situation, the good element of $\mathcal{O}^*$ is the tangent space of $\mathcal{O}$ at π.*

Proof. If $\mathcal{F}$ is linear, then $\mathbf{T}(\mathcal{O}) \cong \mathbf{T}(\mathcal{O}^*) \cong \mathbf{Q}(5, q)$, and the theorem easily follows.

Let $\mathcal{F}$ be nonlinear. Then $\mathcal{O}$ and $\mathcal{O}^*$ have precisely one good element by Theorem 5.1.12, say, respectively, π and π^*. So any isomorphism α between $\mathbf{T}(\mathcal{O})$ and $\mathbf{T}(\mathcal{O}^*)$ maps π onto π^*. Let δ be an anti-automorphism of the space $\mathbf{PG}(4n-1, q)$ containing $\mathcal{O}$, for which $\mathcal{O}^\delta = \mathcal{O}^*$. Suppose

$\pi^\delta = \pi^* \neq \tau$, with τ the tangent space of $\mathcal{O}$ at π; if π^* is tangent to $\mathcal{O}$ at π', then, by Corollary 5.1.4, π' corresponds to the point (∞) of $\mathcal{S}(\mathcal{F})$. Let $LI(\infty)$ correspond to the translation point of $\mathbf{T}(\mathcal{O})$. Then since $\mathcal{S}(\mathcal{F})$ is an EGQ, the stabilizer of the flag $((\infty), L)$ in the automorphism group of $\mathcal{S}(\mathcal{F})$ acts transitively on the points incident with L and different from (∞). It readily follows that $\mathcal{O}^*$ contains more than one good element, contradicting the hypothesis that $\mathcal{F}$ is nonlinear. So $\pi^* = \tau$, and the theorem is proved. ∎

The next theorem provides an interesting condition for a good TGQ to be a Kantor-Knuth GQ.

Theorem 5.1.14 (Blokhuis, Lavrauw and Ball [16]) *Assume that* $\mathbf{T}(\mathcal{O})$ *is a good TGQ of order* (q^n, q^{2n}), *q odd, where* $\mathbf{GF}(q)$ *is the kernel of the TGQ, with the additional condition that*

$$q \geq 4n^2 - 8n + 2.$$

Then $\mathbf{T}(\mathcal{O})$ *is isomorphic to a Kantor-Knuth GQ.*

Sketch of Proof. Since $\mathbf{T}(\mathcal{O})$ is good, by Corollary 5.1.4 $\mathbf{T}(\mathcal{O}^*)^D$ is a flock GQ. By Theorem 4.7.2, $\mathbf{T}(\mathcal{O}^*)^D \cong \mathcal{S}(\mathcal{F})$, where $\mathcal{F}$ is a semifield flock of the quadratic cone $\mathcal{K}$ with equation $X_2^2 - X_0X_1 = 0$ in $\mathbf{PG}(3, q^n)$, and where the kernel of the corresponding TGQ is $\mathbf{GF}(q)$. In the dual space of $\mathbf{PG}(3, q)$, the generators of $\mathcal{K}$ become a set of $q^n + 1$ lines in the plane π_v which is the dual of the vertex v of the cone. These lines are the tangents to a conic $\mathcal{C}'$. If the dual is taken w.r.t. the standard inner product, then $\mathcal{C}'$ is given by the equations

$$X_3 = X_2^2 - 4X_0X_1 = 0.$$

In the dual space, let $\overline{\mathcal{F}}$ be the set of q^n points which corresponds to $\mathcal{F}$. Since every intersection line of distinct flock planes is skew to every generator of $\mathcal{K}$, it follows that in the dual space the line joining two points of $\overline{\mathcal{F}}$ meets π_v in an internal point of $\mathcal{C}'$ (the external points and the points of $\mathcal{C}'$ are incident with a tangent); see also Section 4.4. Let $\mathcal{I}$ be the set of these internal points. With the notation of Section 4.6, and taking into account that $f(t) = -z_t$ and $g(t) = y_t$, $t \in \mathbf{GF}(q^n)$, are additive, we have

$$\mathcal{I} = \{(t, -f(t), g(t), 0) \parallel t \in \mathbf{GF}(q^n)\}.$$

A nondegenerate quadratic form is either always a square on the external points of the conic it defines, and a nonsquare on the internal points, or vice versa. The quadratic form

$$\mathbf{Q}(X_0, X_1, X_2) = X_2^2 - 4X_0X_1$$

is a square on all external points of the conic $\mathcal{C}'$; $(0, 0, 1)$ is incident with a tangent line and $Q(0, 0, 1) = 1$. So it is a nonsquare on the set of internal points $\mathcal{I}$, implying that

$$g(t)^2 + 4tf(t)$$

is a nonsquare for all $t \in \mathbf{GF}(q^n)$.

The argument just given can be reversed, so that any pair of functions f and g which are $\mathbf{GF}(q)$-linear with the property above, gives rise to a semifield flock and TGQ with kernel $\mathbf{GF}(q)$.

There are two cases to consider now.

CASE (1): $\mathcal{I}$ IS CONTAINED IN A LINE.   Then the flock planes share a common point. Theorems 4.4.5 and 4.4.6 imply that the flock is a Kantor-Knuth flock.

CASE (2): WE ARE NOT IN CASE 1.   Then $n \geq 3$ and $\mathcal{I}$ generates the plane π_v (over $\mathbf{GF}(q^n)$). The set $\mathcal{I}$ is a $\mathbf{PG}(n-1, q)$ when interpreted over $\mathbf{GF}(q)$, so $\mathcal{I}$ contains a subplane of order q of π_v which generates π_v and which is contained in the set of internal points of the conic $\mathcal{C}'$.

The following lemma of Weil can be found in [135].

Lemma 5.1.15 (Weil) *The number N of solutions in $\mathbf{GF}(q)$ of the hyperelliptic equation*

$$Y^2 = g(X),$$

where $g \in \mathbf{GF}(q)[X]$ is not a square and has degree $2m > 2$, satisfies

$$|N - q + 1| < (2m - 2)\sqrt{q}.$$

∎

Blokhuis, Lavrauw and Ball [16] use Lemma 5.1.15 to obtain

Lemma 5.1.16 (Blokhuis, Lavrauw and Ball [16]) *Let* $f(X) = X^2 + uX + v \in \mathbf{GF}(q^n)[X]$ *be a nonzero square in* $\mathbf{GF}(q^n)$ *for all* $X = x \in \mathbf{GF}(q)$, *where* q *is odd and* $q \geq 4n^2 - 8n + 2$. *At least one of the following holds:*

(i) $f(X)$ *is the square of a linear polynomial;*

(ii) n *is even and* f *has two distinct roots in* $\mathbf{GF}(q^{n/2})$;

(iii) *the roots of* f *are* α *and* α^σ *for some* $\sigma \in \mathrm{Gal}(\mathbf{GF}(q^n)/\mathbf{GF}(q))$, *where* $\alpha \in \mathbf{GF}(q^n)$. ∎

Using the latter lemma, it is then shown that when $q \geq 4n^2 - 8n + 2$, no subplanes of order q can be contained in the internal points of a conic in $\mathbf{PG}(2, q^n)$. ∎

5.2 Good Eggs and Veronese Surfaces

There is a strong connection between good eggs and the Veronese surface $\mathcal{V}_2^4$ of all conics of $\mathbf{PG}(2,q)$. So we include a short section on Veronese varieties; a good reference is Chapter 25 of Hirschfeld and Thas [74].

The *Veronese variety* of all quadrics of $\mathbf{PG}(n, \mathbb{K})$, $n \geq 1$ and $\mathbb{K}$ any commutative field, is the variety

$$\mathcal{V} = \{(x_0^2, x_1^2, \ldots, x_n^2, x_0x_1, x_0x_2, \ldots, x_0x_n, x_1x_2, \ldots, x_1x_n, \ldots, x_{n-1}x_n) \,\| \\ (x_0, x_1, \ldots, x_n) \text{ is a point of } \mathbf{PG}(n, \mathbb{K})\}$$

of $\mathbf{PG}(N, \mathbb{K})$ with $N = n(n+3)/2$. The variety $\mathcal{V}$ has dimension n and order 2^n [74]; for $\mathcal{V}$ we also write $\mathcal{V}_n$ or $\mathcal{V}_n^{2^n}$. It is also called the *Veronesean of quadrics* of $\mathbf{PG}(n, \mathbb{K})$, or simply the *quadric Veronesean* of $\mathbf{PG}(n, \mathbb{K})$. It can be shown that the quadric Veronesean is absolutely irreducible and nonsingular [74].

Let $\mathbf{PG}(N, \mathbb{K})$, with $N = n(n+3)/2$, consist of all points

$$(y_{00}, y_{11}, \ldots, y_{nn}, y_{01}, y_{02}, \ldots, y_{0n}, y_{12}, \ldots, y_{1n}, \ldots, y_{n-1,n});$$

for y_{ij} we also write y_{ji}. Let $\zeta : \mathbf{PG}(n, \mathbb{K}) \mapsto \mathbf{PG}(N, \mathbb{K})$, with $n \geq 1$, be defined by

$$(x_0, x_1, \ldots, x_n) \mapsto (y_{00}, y_{11}, \ldots, y_{n-1,n}),$$

with $y_{ij} = x_i x_j$. Then ζ is a bijection of $\mathbf{PG}(n, \mathbb{K})$ onto the quadric Veronesean $\mathcal{V}$ of $\mathbf{PG}(n, \mathbb{K})$. It then follows that the variety $\mathcal{V}$ is rational.

The quadrics of $\mathbf{PG}(n, \mathbb{K})$ are mapped by ζ onto all hyperplane sections of $\mathcal{V}$. It follows that no hyperplane of $\mathbf{PG}(N, \mathbb{K})$ contains the quadric Veronesean $\mathcal{V}$. The lines of $\mathbf{PG}(n, \mathbb{K})$ are mapped onto irreducible conics of $\mathcal{V}$; for $\mathbb{K} \neq \mathbf{GF}(2)$ each irreducible conic of $\mathcal{V}$ is the image of a line of $\mathbf{PG}(n, \mathbb{K})$. Hence for $|\mathbb{K}| > 2$ any two distinct points of $\mathcal{V}$ are contained in a unique irreducible conic of $\mathcal{V}$.

If $\mathbb{K} = \mathbf{GF}(q)$, then it is clear that $\mathcal{V}_n$ contains $\theta(n) = q^n + q^{n-1} + \cdots + q + 1$ points. As no three points of $\mathcal{V}_n$ are collinear, the quadric Veronesean $\mathcal{V}_n$ is a $\theta(n)$-cap of $\mathbf{PG}(N, q)$, with $N = n(n + 3)/2$.

Let $n = 2$ and assume from now on that $\mathbb{K} = \mathbf{GF}(q)$. Then $\mathcal{V}$ is a surface of order 4 in $\mathbf{PG}(5, \mathbb{K})$. Apart from $\mathcal{V}_1^2$, which is a conic, the variety $\mathcal{V}_2^4$ is the quadric Veronesean which is most studied and characterized. With the conics (irreducible or not) of $\mathbf{PG}(2, q)$ correspond all hyperplanes sections of $\mathcal{V}_2^4$. The hyperplane is uniquely determined by the conic if and only if the latter is not a single point. If the conic of $\mathbf{PG}(2, q)$ is one line, then the corresponding hyperplane of $\mathbf{PG}(5, q)$ meets $\mathcal{V}_2^4$ in an irreducible conic; the surface $\mathcal{V}_2^4$ contains no other irreducible conics. It follows that $\mathcal{V}_2^4$ contains exactly $q^2 + q + 1$ irreducible conics, that any two distinct points of $\mathcal{V}_2^4$ are contained in a unique irreducible conic of $\mathcal{V}_2^4$, and that any two distinct irreducible conics on $\mathcal{V}_2^4$ meet in a unique point. If the conic $\mathcal{C}$ of $\mathbf{PG}(2, q)$ is two distinct lines, then the corresponding hyperplane $\mathbf{PG}(4, q)$ meets $\mathcal{V}_2^4$ in two irreducible conics with exactly one point in common; if $\mathcal{C}$ is irreducible, then $\mathbf{PG}(4, q)$ meets $\mathcal{V}_2^4$ in a rational quartic curve. The planes of $\mathbf{PG}(5, q)$ which meet $\mathcal{V}_2^4$ in an irreducible conic are called the *conic planes* of $\mathcal{V}_2^4$.

Any two distinct conic planes of $\mathcal{V}_2^4$ have exactly one point in common, and this common point belongs to $\mathcal{V}_2^4$. The tangent lines of the irreducible conics of $\mathcal{V}_2^4$ are called the tangents or tangent lines of $\mathcal{V}_2^4$. Since no point of the surface $\mathcal{V}_2^4$ is singular, all tangent lines of $\mathcal{V}_2^4$ at the point p of $\mathcal{V}_2^4$ are contained in a plane $\pi(p)$. This plane $\pi(p)$ is called the *tangent plane* of $\mathcal{V}_2^4$ at p. Since p is contained in exactly $q + 1$ irreducible conics of $\mathcal{V}_2^4$ and since no two conic planes through p have a line in common, the tangent plane $\pi(p)$ is the union of the $q+1$ tangent lines of $\mathcal{V}_2^4$ through p. Also $\mathcal{V}_2^4 \cap \pi(p) = \{p\}$. For any two distinct points p_1 and p_2 of $\mathcal{V}_2^4$, the tangent planes $\pi(p_1)$ and $\pi(p_2)$ have exactly one point in common. If $\mathcal{C}$ is a nonsingular conic on $\mathcal{V}_2^4$, if π is the plane of $\mathcal{C}$ and if $p' \in \mathcal{V}_2^4 \backslash \mathcal{C}$, then $\pi(p') \cap \pi = \emptyset$.

Suppose that q is odd. Then $\mathbf{PG}(5, q)$ admits a polarity that maps the set of all conic planes of $\mathcal{V}_2^4$ onto the set of all tangent planes of $\mathcal{V}_2^4$.

Suppose that q is even. Then the nuclei of the $q^2 + q + 1$ irreducible conics on $\mathcal{V}_2^4$ are the points of a plane $\mathbf{PG}(2, q)$, called the *nucleus* or *kernel* of $\mathcal{V}_2^4$. In Thas [180, 183], the following classification of good eggs is obtained.

Theorem 5.2.1 *Consider a TGQ $\mathbf{T}(n, 2n, q) = \mathbf{T}(\mathcal{O})$, q odd, with $\mathcal{O} = \mathcal{O}(n, 2n, q) = \{\pi, \pi_1, \ldots, \pi_{q^{2n}}\}$. If $\mathcal{O}$ is good at π, then we have one of the following.*

(a) *There exists a $\mathbf{PG}(3, q^n)$ in the extension $\mathbf{PG}(4n - 1, q^n)$ of the space $\mathbf{PG}(4n - 1, q)$ of $\mathcal{O}(n, 2n, q)$ which has exactly one point in common with the extension $\bar{\pi}_i$ of π_i to $\mathbf{GF}(q^n)$, with $i = 1, 2, \ldots, q^{2n}$. The set of these $q^{2n} + 1$ points is an elliptic quadric of $\mathbf{PG}(3, q^n)$ and $\mathbf{T}(\mathcal{O})$ is isomorphic to the classical GQ $\mathbf{Q}(5, q^n)$.*

(b) *We are not in Case (a) and there exists a $\mathbf{PG}(4, q^n)$ in $\mathbf{PG}(4n - 1, q^n)$ which intersects the extension $\bar{\pi}$ of π to $\mathbf{GF}(q^n)$ in a line M and which has exactly one point r_i in common with each space $\bar{\pi}_i$, with $i = 1, 2, \ldots, q^{2n}$. Let $\mathcal{W} = \{r_i \,\|\, i = 1, 2, \ldots, q^{2n}\}$ and let $\mathcal{M}$ be the set of all common points of M and the conics which contain exactly q^n points of $\mathcal{W}$. Then the set $\mathcal{W} \cup \mathcal{M}$ is the projection of a quadric Veronesean $\mathcal{V}_2^4$ from a point p in a conic plane of $\mathcal{V}_2^4$ onto a hyperplane $\mathbf{PG}(4, q^n)$; the point p is an exterior point of the conic of $\mathcal{V}_2^4$ in the conic plane. Also, the point-line dual of $\mathbf{T}(\mathcal{O})$ is isomorphic to the flock GQ $\mathcal{S}(\mathcal{F})$ with $\mathcal{F}$ a Kantor-Knuth flock, that is, $\mathcal{O}$ is a Kantor-Knuth egg.*

(c) *We are not in Cases (a) and (b) and there exists a $\mathbf{PG}(5, q^n)$ in $\mathbf{PG}(4n - 1, q^n)$ which intersects $\bar{\pi}$ in a plane μ and which has exactly one point r_i in common with each space $\bar{\pi}_i$, with $i = 1, 2, \ldots, q^{2n}$. Let $\mathcal{W} = \{r_i \,\|\, i = 1, 2, \ldots, q^{2n}\}$ and let $\mathcal{C}$ be the set of all common points of μ and the conics which contain exactly q^n points of $\mathcal{W}$. Then the set $\mathcal{W} \cup \mathcal{C}$ is a quadric Veronesean in $\mathbf{PG}(5, q^n)$.* ∎

Thas [180] contains a long technical proof of Theorem 5.2.1, buth with a weaker version of Case (b); Case (b) could be improved relying on Thas [183], see Section 9 of this paper. An alternative proof is contained in Lavrauw and Lunardon [96].

Remark 5.2.2 If we project the Veronesean $\mathcal{V}_2^4$ in $\mathbf{PG}(5, q^n)$ onto some $\mathbf{PG}(3, q^n) \subset \mathbf{PG}(5, q^n)$ from a line N that intersects $\mathcal{V}_2^4$ in two points of $\mathbf{PG}(5, q^{2n}) \setminus \mathbf{PG}(5, q^n)$ which are conjugate with respect to the extension $\mathbf{GF}(q^{2n})$ of $\mathbf{GF}(q^n)$, then we obtain an elliptic quadric of $\mathbf{PG}(3, q^n)$.

If the egg $\mathcal{O}(n, 2n, q) = \mathcal{O}$, with q odd, is good at its element π, then, by Corollary 5.1.4, the translation dual $\mathbf{T}(\mathcal{O}^*)$ of $\mathbf{T}(\mathcal{O})$ is the point-line dual of a flock GQ $\mathcal{S}(\mathcal{F})$ with elation point τ, where τ is the tangent space of $\mathcal{O}$ at π. The point (∞) of $\mathbf{T}(\mathcal{O}^*)$ is a translation line of $\mathcal{S}(\mathcal{F})$. We may assume that (∞) is the line $[\infty]$ of $\mathcal{S}(\mathcal{F})$ and then $\mathcal{F}$ is a semifield flock; see Section 4.7 and Payne and Rogers [125]. Conversely, if $\mathcal{F}$ is a semifield flock, then, by Section 4.7 and Corollary 5.1.4, $\mathcal{S}(\mathcal{F}) \cong \mathbf{T}(\mathcal{O})$ where $\mathcal{O}^*$ is good at its element τ which corresponds to the elation point of $\mathcal{F}$. If $\mathcal{F}$ is a Kantor-Knuth flock which is not linear, then $\mathcal{S}(\mathcal{F}) \cong \mathbf{T}(\mathcal{O}) \cong \mathbf{T}(\mathcal{O}^*)$ with $\mathcal{O}$ of Type (b) in Theorem 5.2.1; see Theorem 4.7.3. If $\mathcal{F}$ is a Ganley flock or the Penttila-Williams-Bader-Lunardon-Pinneri flock, then $\mathcal{S}(\mathcal{F}) \cong \mathbf{T}(\mathcal{O}^*)$ with $\mathcal{O}$ of Type (c) in Theorem 5.2.1. If the TGQ $\mathbf{T}(\mathcal{O})$ of order (s^2, s), with s odd, is good and the kernel has either size s or size $\sqrt{s}$, then it is clear that Case (c) cannot occur (as $n \in \{1, 2\}$) and so $\mathbf{T}(\mathcal{O})$ is isomorphic to the point-line dual of $\mathcal{S}(\mathcal{F})$, with $\mathcal{F}$ a Kantor-Knuth flock.

We end this section with considering the elements of a good egg $\mathcal{O}(n, 2n, q)$, with q odd, contained in a given $\mathbf{PG}(3n - 1, q)$ of the $\mathbf{PG}(4n - 1, q)$ containing $\mathcal{O}(n, 2n, q)$; we will heavily rely on Theorem 5.2.1.

Theorem 5.2.3 *Let* $\mathcal{O} = \{\pi, \pi_1, \pi_2, \ldots, \pi_{q^{2n}}\}$, *with* q *odd, be an egg in* $\mathbf{PG}(4n - 1, q)$ *which is good at* π. *If there is a subspace* $\mathbf{PG}(3n - 1, q)$ *of* $\mathbf{PG}(4n - 1, q)$ *which contains at least four elements of* $\mathcal{O}$, *but which does not contain* π, *then* $\mathcal{O}$ *is a Kantor-Knuth egg, that is, is of Type* (a) *or* (b) *in Theorem 5.2.1. If there is a subspace* $\mathbf{PG}(3n - 1, q)$ *which contains at least five elements of* $\mathcal{O}$, *but which does not contain* π, *then* $\mathcal{O}$ *is classical.*

Proof. Assume that the subspace $\mathbf{PG}(3n - 1, q)$ of $\mathbf{PG}(4n - 1, q)$ contains at least four elements of $\mathcal{O} \backslash \{\pi\}$, say π_i, with $i = 1, 2, 3, 4$. Now we rely on Theorem 5.2.1. We will use here the same notation. Assume, by way of contradiction, that $\mathcal{O}$ is not a Kantor-Knuth egg. Let $\mathcal{W} \cap \pi_i = \{r_i\}$, with $i = 1, 2, 3, 4$. We show that r_1, r_2, r_3, r_4 are not coplanar. Assume that these four points are contained in the plane $\mathbf{PG}(2, q^n)$. Then $\mathbf{PG}(2, q^n) \cap \mathcal{V}_2^4$, with $\mathcal{V}_2^4$ the quadric Veronesean $\mathcal{W} \cup \mathcal{C}$, is a conic $\mathcal{C}'$ which has a point r in common with the conic $\mathcal{C}$. It follows that π is a subspace of $\mathbf{PG}(3n - 1, q)$, a contradiction. Hence the points r_1, r_2, r_3, r_4 generate a $\mathbf{PG}(3, q^n)$. The space $\mathbf{PG}(3, q^n)$ contains a point m of the plane μ. The subspace generated by m and its conjugates belongs to the subspace generated by $\mathbf{PG}(3, q^n)$ and its conjugates, and so π has a nonempty intersection with $\mathbf{PG}(3n - 1, q)$. It follows that $\langle \pi, \pi_1, \pi_2 \rangle \cap \pi_i \neq \emptyset$, with $i = 3, 4$. As $\mathcal{O}$ is good at π, the $(3n - 1)$-dimensional space φ generated by π, π_1, π_2 contains exactly $q^n + 1$ elements of $\mathcal{O}$. If π_i, with $i \in \{3,4\}$, does not belong to φ, then any hyperplane containing φ and π_i contains at least $q^n + 2$ elements of $\mathcal{O}$, contradicting Theorem 3.8.1(i). Hence π_i is a subspace of φ, with $i = 3, 4$. Consequently $\varphi = \mathbf{PG}(3n - 1, q)$ contains π, a contradiction. So $\mathcal{O}$ is a Kantor-Knuth egg.

Next, assume that the subspace $\mathbf{PG}(3n - 1, q)$ of $\mathbf{PG}(4n - 1, q)$ contains at least five elements of $\mathcal{O}\backslash\{\pi\}$, say $\pi_1, \pi_2, \ldots, \pi_5$. We rely again on Theorem 5.2.1. Assume, by way of contradiction, that $\mathcal{O}$ is not classical. Then, by the first part of the proof, $\mathcal{O}$ is a nonclassical Kantor-Knuth egg. Let $\mathcal{W} \cap \pi_i = \{r_i\}$, with $i = 1, 2, \ldots, 5$. Clearly $\langle r_1, r_2, \ldots, r_5 \rangle \neq \mathbf{PG}(4, q^n)$, as otherwise $\mathcal{O}$ is contained in $\mathbf{PG}(3n - 1, q)$. Now suppose that $\langle r_1, r_2, \ldots, r_5 \rangle = \mathbf{PG}(3, q^n)$. Then, with the notation of Theorem 5.2.1, $M \cap \mathbf{PG}(3, q^n) \neq \emptyset$, and as in the previous section we find that $\pi \subset \mathbf{PG}(3n - 1, q)$, a contradiction. Consequently $r_1, r_2, \ldots, r_5$ are contained in a common plane $\mathbf{PG}(2, q^n)$. Clearly $\mathbf{PG}(2, q^n) \cap M = \emptyset$, as otherwise $\pi \subset \mathbf{PG}(3n - 1, q)$. The set $\mathcal{W} \cup \mathcal{M}$ is the projection of a quadric Veronesean $\mathcal{V}_2^4$ from a point p in a conic plane γ of $\mathcal{V}_2^4$ onto a hyperplane $\mathbf{PG}(4, q^n)$; the point p is an exterior point of the conic $\mathcal{C}'$ of $\mathcal{V}_2^4$ in γ. Let r_i be the projection of the point $r_i' \in \mathcal{V}_2^4$, with $i = 1, 2, \ldots, 5$. If $\langle r_1', r_2', \ldots, r_5' \rangle = \mathbf{PG}'(2, q^n)$, then $r_1', r_2', \ldots, r_5'$ are contained in a common conic of $\mathcal{V}_2^4$, hence $\mathcal{C}' \cap \mathbf{PG}'(2, q^n) \neq \emptyset$, and so $\mathbf{PG}(2, q^n) \cap M \neq \emptyset$, a contradiction. Consequently, $\langle r_1', r_2', \ldots, r_5' \rangle = \mathbf{PG}'(3, q^n)$, with $p \in \mathbf{PG}'(3, q^n)$. Clearly $\mathbf{PG}'(3, q^n) \cap \gamma = \{p\}$, as otherwise $\mathbf{PG}(2, q^n) \cap M \neq \emptyset$. As $\pi \not\subset \mathbf{PG}(3n - 1, q)$, no three points of $\{r_1', r_2', \ldots, r_5'\}$ are contained in a conic plane of $\mathcal{V}_2^4$. Let ζ be the bijection of a plane $\mathbf{PG}^*(2, q^n)$ onto $\mathcal{V}_2^4$ which maps the lines of $\mathbf{PG}^*(2, q^n)$ onto the conics of $\mathcal{V}_2^4$. Further, let $\mathcal{C}'^{\zeta^{-1}} = L, (r_i')^{\zeta^{-1}} = p_i$, with $i = 1, 2, \ldots, 5$. Then no three points of $\{p_1, p_2, \ldots, p_5\}$ are collinear. So $p_1, p_2, \ldots, p_5$ belong to just one conic $\mathcal{C}''$ of $\mathbf{PG}^*(2, q^n)$. It follows that $r_1', r_2', \ldots, r_5'$ generate a hyperplane of $\mathbf{PG}(5, q^n) \supset \mathcal{V}_2^4$; see page 117. So we have a contradiction, and conclude that $\mathcal{O}$ is classical.   ∎

Remark 5.2.4 (a) If $\mathcal{O}$ is a nonclassical Kantor-Knuth egg of $\mathbf{PG}(4n - 1, q)$, with q odd, then from Thas [180] it easily follows that there exist $(3n - 1)$-dimensional subspaces in $\mathbf{PG}(4n - 1, q)$ which contain exactly four elements of $\mathcal{O}$.

(b) Independently and in a completely different way, Lavrauw [94] also proved the last part of Theorem 5.2.3.

5.3 Coordinatization and Applications

We now introduce a coordinatization method of eggs and their duals which was first used by Payne in [118], and later on generalized by Lavrauw and Penttila [97].

5.3.1 Coordinatization of Eggs and Dual Eggs

The next theorem states the connection between additive q^n-clans and eggs, and is taken from [97]. We use the notation of §4.7.2.

Theorem 5.3.1 *The set* $C = \{A_t = \begin{pmatrix} t & y_t \\ 0 & z_t \end{pmatrix} \parallel t \in \mathbf{GF}(q^n)\}$ *is an additive* q^n*-clan if and only if the set* $\mathcal{O} = \{\pi(\lambda) \parallel \lambda \in \mathbf{GF}(q^n)^2 \cup \{\infty\}\}$, *with*

$$\pi(\lambda) = \{(t, -g_t(\lambda), -\lambda^{\delta_t}) \parallel t \in \mathbf{GF}(q^n)\}, \ \lambda \in \mathbf{GF}(q^n)^2,$$

and

$$\pi(\infty) = \{(0, t, (0,0)) \parallel t \in \mathbf{GF}(q^n)\},$$

where the triples defining $\pi(\lambda)$ *and* $\pi(\infty)$ *are considered as vectors of the* $4n$*-dimensional vectorspace over* $\mathbf{GF}(q)$, *is an egg of* $\mathbf{PG}(4n - 1, q)$, *where* $g_t(\lambda) = \lambda A_t \lambda^T$ *and* $\lambda^{\delta_t} = \lambda(A_t + A_t^T)$. *The tangent space at the element* $\pi(\lambda), \lambda \in \mathbf{GF}(q^n)^2$, *is*

$$\tau(\lambda) = \{(t, \beta\lambda^T + \lambda^{\delta_t}\lambda^T - g_t(\lambda), \beta) \parallel t \in \mathbf{GF}(q^n), \beta \in \mathbf{GF}(q^n)^2\},$$

and the tangent space at $\pi(\infty)$ *is*

$$\tau(\infty) = \{(0, t, \beta) \parallel t \in \mathbf{GF}(q^n), \beta \in \mathbf{GF}(q^n)^2\}.$$

■

For the rest of this section, it is convenient to write $\mathbb{F}$ instead of $\mathbf{GF}(q^n)$. If A_t is additive, so defines an egg in $\mathbf{PG}(4n - 1, q)$, we can write A_t as

$$A_t = \sum_{i=0}^{n-1} \begin{pmatrix} a_i & b_i \\ 0 & c_i \end{pmatrix} t^{q^i},$$

with $(a_0, a_1, \ldots, a_{n-1}) = (1, 0, \ldots, 0)$, for certain $b_i, c_i \in \mathbb{F}$. If an egg $\mathcal{O}$ is written in this form, we denote it by

$$\mathcal{O}(\bar{b}, \bar{c})$$

where $\bar{b} = (b_0, b_1, \ldots, b_{n-1})$ and $\bar{c} = (c_0, c_1, \ldots, c_{n-1})$. If $\mathcal{O}$ is as such, it is easy to explicitly write down the elements of the dual egg, and the tangent spaces to the latter. This was first done by Lavrauw and Penttila [97] in its full generality, based on observations of Payne in [118], who carried out the calculations for the eggs corresponding to the Roman TGQs.

We start with a lemma.

Lemma 5.3.2 *Let* tr *be the trace map from* $\mathbb{F}$ *to* $\mathbf{GF}(q)$. *Then*

$$\text{tr}\left(\sum_{i=0}^{n-1} \alpha_i t^{q^i}\right) = 0,$$

for all $t \in \mathbb{F}$ *if and only if*

$$\sum_{i=0}^{n-1} \alpha_i^{q^{n-1-i}} = 0.$$

Proof. Since the trace function is additive and $\text{tr}(x) = \text{tr}(x^q)$ for $x \in \mathbb{F}$, we have that

$$\text{tr}\left(\sum_{i=0}^{n-1} \alpha_i t^{q^i}\right) = \text{tr}\left[\left(\sum_{i=0}^{n-1} \alpha_i^{q^{n-1-i}}\right) t^{q^{n-1}}\right].$$

The fact that $\text{tr}(ax) = 0, \forall x \in \mathbb{F}$ implies that $a = 0$, ends the proof. $\blacksquare$

Theorem 5.3.3 *The elements of the translation dual* $\mathcal{O}^*(\bar{b}, \bar{c})$, *and the corresponding tangent spaces, of the egg* $\mathcal{O}(\bar{b}, \bar{c})$ *are given by*

$$\pi^*(\lambda) = \{(-g_t^*(\lambda), t, -\lambda t) \parallel t \in \mathbb{F}\}, \ \lambda \in \mathbb{F}^2,$$

$$\pi^*(\infty) = \{(t, 0, (0, 0)) \parallel t \in \mathbb{F}\},$$

$$\tau^*(\lambda) = \{(f^*(\beta, \lambda) + g_t^*(\lambda), t, \beta) \parallel t \in \mathbb{F}, \beta \in \mathbb{F}^2\}, \ \lambda \in \mathbb{F}^2,$$

$$\tau^*(\infty) = \{(t, 0, \beta) \parallel t \in \mathbb{F}, \beta \in \mathbb{F}^2\},$$

with

$$g_t^*(a, b) = \sum_{i=0}^{n-1} (a_i a^2 + b_i ab + c_i b^2)^{1/q^i} t^{1/q^i},$$

with $(a_0, a_1, \ldots, a_{n-1}) = (1, 0, \ldots, 0)$ and

$$f^*((a, b), (c, d)) = \sum_{i=0}^{n-1} (2a_i ac + b_i(ad + bc) + 2c_i bd)^{1/q^i}.$$

Proof. We calculate the vector space dual of $\pi(\lambda)$ and $\tau(\lambda)$ in $V(4n, q)$ with respect to the standard inproduct

$$((x, y, z, w), (x', y', z', w')) \mapsto \mathrm{tr}(xx' + yy' + zz' + ww'),$$

with tr the trace map from $\mathbb{F}$ onto $\mathbf{GF}(q)$. If (x, y, z, w) is an element of the vector space dual of $\tau(\lambda)$, then $\mathrm{tr}[xt + y(\beta\lambda^T + \lambda^{\delta_t}\lambda^T - g_t(\lambda)) + (z, w)\beta^T] = 0$, for all $t \in \mathbb{F}$ and all $\beta \in \mathbb{F}^2$. With $\lambda = (a, b)$ and $\beta = (c, d)$, we obtain

$$\mathrm{tr}[xt + y(ac + bd + \lambda^{\delta_t}\lambda^T - g_t(\lambda)) + zc + wd] = 0, \qquad (5.1)$$

for all $c, d, t \in \mathbb{F}$. When $t = 0$, this equation is satisfied for all $c, d \in \mathbb{F}$ if and only if $w = -by$ and $z = -ay$. Substituting this back into Equation (5.1), we obtain that $\mathrm{tr}[xt + y(\lambda^{\delta_t}\lambda^T - g_t(\lambda))] = 0$ for all $t \in \mathbb{F}$. Using the formulas for g_t and δ_t this is equivalent to

$$\mathrm{tr}[(x + y(a_0 a^2 + b_0 ab + c_0 b^2))t + \sum_{i=1}^{n-1} y(a_i a^2 + b_i ab + c_i b^2)t^{q^i}] = 0,$$

for all $t \in \mathbb{F}$. By Lemma 5.3.2, it follows that (x, y, z, w) is of the form

$$\left(-\sum_{i=0}^{n-1} (a_i a^2 + b_i ab + c_i b^2)^{1/q^i} t^{1/q^i}, t, -at, -bt\right),$$

for some $t \in \mathbb{F}$. The tangent spaces are handled in the same way. ∎

5.3.2 The Known Examples Revisited

Using Theorem 5.3.3, we now write down, for q odd, the known examples of good eggs, and their translation duals; for q even each known egg arises by field reduction either from an elliptic quadric of $\mathbf{PG}(3, q)$ or from a Tits ovoid. We only provide elements of type $\pi(\lambda)$ and $\pi^*(\lambda)$, $\lambda \in \mathbb{F} = \mathbf{GF}(q^n)$, as the elements $\pi(\infty)$ and $\pi^*(\infty)$ are all of the form

$$\{(0, t, 0, 0) \parallel t \in \mathbb{F}\} \text{ and } \{(t, 0, 0, 0) \parallel t \in \mathbb{F}\}$$

in this representation.

- KANTOR-KNUTH EGGS. In this case, the egg is isomorphic to its translation dual; see Payne [118]. Let q be odd, σ an automorphism of $\mathbb{F}$, and m a given nonsquare in $\mathbb{F}$. Then

$$\pi(\lambda) = \{(t, -a^2t + mb^2t^\sigma, -2at, 2bmt^\sigma) \mid\mid t \in \mathbb{F}\}$$

and

$$\pi^*(\lambda) = \{(-a^2t + (mb^2)^{\sigma^{-1}}t^{\sigma^{-1}}, t, -at, -bt) \mid\mid t \in \mathbb{F}\}.$$

- GANLEY AND ROMAN EGGS. For $q > 9$, the Ganley eggs are not isomorphic to their translation dual; see Payne [118]. They exist for $q = 3^h$, $h \geq 1$. Let m be a nonsquare in $\mathbb{F}$. Then

$$\pi(\lambda) = \{(t, -(a^2 - b^2m)t - abt^3 + b^2m^{-1}t^9, at - bt^3, -at^3$$
$$- b(mt + m^{-1}t^9)) \mid\mid t \in \mathbb{F}\}$$

and

$$\pi^*(\lambda) = \{(-(a^2 - b^2m)t - (ab)^{1/3}t^{1/3}$$
$$+ (b^2m^{-1})^{1/9}t^{1/9}, t, -at, -bt) \mid\mid t \in \mathbb{F}\}.$$

- THE PENTTILA-WILLIAMS-BADER-LUNARDON-PINNERI EGG AND ITS TRANSLATION DUAL. The Penttila-Williams-Bader-Lunardon-Pinneri egg is not isomorphic to its translation dual. Here $q = 3^5$. Then

$$\pi(\lambda) = \{(t, -a^2t - abt^3 + b^2t^{27}, at - bt^3, -at^3 - bt^{27}) \mid\mid t \in \mathbb{F}\}$$

and

$$\pi^*(\lambda) = \{(-a^2t - (ab)^{1/3}t^{1/3} + (b^2)^{1/27}t^{1/27}, t, -at, -bt) \mid\mid t \in \mathbb{F}\}.$$

5.3.3   A Geometric Connection between Semifield Flocks and Good Eggs

Let $\mathcal{F}$ be a semifield flock of the quadratic cone $\mathcal{K}$ with vertex v in $\mathbf{PG}(3, q^n)$, q odd. If one dualizes in $\mathbf{PG}(3, q^n)$, the q^n flock planes become the affine points of an n-dimensional affine space $\mathbf{AG}(n, q)$ over $\mathbf{GF}(q)$. The $(n - 1)$-dimensional space at infinity $\mathbf{PG}(n - 1, q)$ of $\mathbf{AG}(n, q)$ is contained in the set of internal points of the conic $\mathcal{C}$, that corresponds to the set of generators of $\mathcal{K}$, in the plane π_v corresponding to v. Over $\mathbf{GF}(q)$ the conic $\mathcal{C}$ becomes a pseudo-conic $\overline{\mathcal{C}}$ in the space $\mathbf{PG}(3n-1,q)$ which corresponds to π_v. Dualizing in $\mathbf{PG}(3n-1,q)$, the space $\mathbf{PG}(n-1,q)$ becomes a $(2n-1)$-space $\mathbf{PG}(2n-1,q)$ skew to the elements of the pseudo-conic $\overline{\mathcal{C}}'$ obtained by dualizing $\overline{\mathcal{C}}$.

Now suppose $\mathcal{O}$ is a good egg in $\mathbf{PG}(4n-1,q) = \langle \mathcal{O} \rangle$, q odd. Suppose $\mathcal{O}'$ is one of the pseudo-conics containing the good element π (see Theorem 5.1.9), and let $\pi' \in \mathcal{O} \setminus \mathcal{O}'$. Suppose τ' is the tangent space of $\mathcal{O}$ at π'; then $\tau' \cap \langle \mathcal{O}' \rangle = \beta$ is a $(2n-1)$-space in $\mathbf{PG}(3n-1,q) = \langle \mathcal{O}' \rangle$ skew to $\mathcal{O}'$. Now reverse the argument of the first paragraph to obtain a semifield flock.

In [95] Lavrauw proves that such a semifield flock is isomorphic to the semifield flock defining the good egg $\mathcal{O}$.

5.3.4 Further Characterizations of Good Eggs in Odd Characteristic

The following interesting theorems are due to Lavrauw [94].

Theorem 5.3.4 *Let $\mathcal{O}$ be an egg of $\mathbf{PG}(4n-1,q)$, q odd, which is good at π. Then the following properties are equivalent.*

(i) *$\mathcal{O}$ is classical.*

(ii) *There exists a triple (π', π'', π'''), with π', π'', π''' distinct elements of $\mathcal{O} \setminus \{\pi\}$ and where π' is not contained in the pseudo-conic $\mathcal{O}'$ of $\mathcal{O}$ in $\langle \pi, \pi'', \pi''' \rangle$, such that the tangent space of $\mathcal{O}$ at π' contains two distinct elements of the regular $(n-1)$-spread of $\langle \pi, \pi'', \pi''' \rangle$ which contains $\mathcal{O}'$.*

(iii) *Each triple (π', π'', π'''), with π', π'', π''' distinct elements of $\mathcal{O} \setminus \{\pi\}$ and where π' is not contained in the pseudo-conic $\mathcal{O}'$ of $\mathcal{O}$ in $\langle \pi, \pi'', \pi''' \rangle$, is such that the tangent space of $\mathcal{O}$ at π' contains two distinct elements of the regular $(n-1)$-spread of $\langle \pi, \pi'', \pi''' \rangle$ which contains $\mathcal{O}'$.*

Proof. Clearly (i) implies (iii), and (iii) implies (ii). It remains to show that (ii) implies (i).

So assume that (ii) is satisfied. By Corollary 5.1.4 and Section 4.7 the translation dual $\mathcal{O}^*$ of $\mathcal{O}$ defines a semifield flock $\mathcal{F}$. In the $(2n-1)$-dimensional space $\langle \pi, \pi'', \pi''' \rangle \cap \tau' = \xi$, with τ' the tangent space of $\mathcal{O}$ at π', a $(n-1)$-spread is induced by the regular $(n-1)$-spread defined by $\mathcal{O}'$ (the space ξ

is a line of the $\mathbf{PG}(2, q^n)$ defined by the regular $(n-1)$-spread). It follows that the semifield flock $\mathcal{F}$ is linear, and so $\mathcal{O}$ is classical. ∎

Theorem 5.3.5 *Let $\mathcal{O}$ be an egg of $\mathbf{PG}(4n-1, q)$, q odd, which is good at π. Then the following properties are equivalent.*

(i) *$\mathcal{O}$ is a Kantor-Knuth egg.*

(ii) *There exists a triple (π', π'', π'''), with π', π'', π''' distinct elements of $\mathcal{O}\backslash\{\pi\}$ and where π' is not contained in the pseudo-conic $\mathcal{O}'$ of $\mathcal{O}$ in $\langle \pi, \pi'', \pi''' \rangle$, such that the tangent space of $\mathcal{O}$ at π' contains an element ζ of the regular $(n-1)$-spread of $\langle \pi, \pi'', \pi''' \rangle$ which contains $\mathcal{O}'$.*

(iii) *Each triple (π', π'', π'''), with π', π'', π''' distinct elements of $\mathcal{O}\backslash\{\pi\}$ and where π' is not contained in the pseudo-conic $\mathcal{O}'$ of $\mathcal{O}$ in $\langle \pi, \pi'', \pi''' \rangle$, is such that the tangent space of $\mathcal{O}$ at π' contains an element ζ of the regular $(n-1)$-spread of $\langle \pi, \pi'', \pi''' \rangle$ which contains $\mathcal{O}'$.*

Proof. Assume that $\mathcal{O}$ is a Kantor-Knuth egg and let $\mathcal{F}$ be the corresponding flock. Let π', π'', π''' be distinct elements of $\mathcal{O}\backslash\{\pi\}$ where π' is not contained in the pseudo-conic $\mathcal{O}'$ of $\mathcal{O}$ in $\langle \pi, \pi'', \pi''' \rangle$. By Theorem 4.4.6 all planes of the elements of $\mathcal{F}$ contain a common point. It follows that $\langle \pi, \pi'', \pi''' \rangle \cap \tau' = \xi$, with τ' the tangent space of $\mathcal{O}$ at π', contains an element ζ of the regular $(n-1)$-spread defined by $\mathcal{O}'$. Hence (i) implies (iii). Since (iii) obviously implies (ii), it remains to show that (ii) implies (i).

So assume that (ii) is satisfied. The egg $\mathcal{O}^*$ defines a semifield flock $\mathcal{F}$. Also, all planes of the elements of $\mathcal{F}$ contain a common point. By Theorem 4.4.6 the flock $\mathcal{F}$ is a Kantor-Knuth flock, and so $\mathcal{O} \cong \mathcal{O}^*$ is a Kantor-Knuth flock. ∎

Remark 5.3.6 If $\mathcal{O}$ is not classical, then the element ζ in Theorem 5.3.5 defines an exterior point of the conic $\mathcal{C}'$ of $\mathbf{PG}(2, q^n)$ defined by the pseudo-conic $\mathcal{O}'$; see Theorem 4.4.6.

Theorem 5.3.7 *Let $\mathcal{O}$ be an egg in $\mathbf{PG}(4n-1, q)$, q odd, and let η be the subspace generated by three distinct egg elements.*

(i) *If $\mathcal{O}$ is a Kantor-Knuth egg, then any element π' of $\mathcal{O}$ is either contained in η or intersects η in at most one point.*

(ii) *If $\mathcal{O}$ is a Ganley-Payne egg, then any element π' of $\mathcal{O}$ is either contained in η or intersects η in at most a line.*

(iii) *If $\mathcal{O}$ is the translation dual of the Penttila-Williams-Bader-Lunardon-Pinneri egg, then any element π' of $\mathcal{O}$ is either contained in η or intersects η in at most a line.*

Proof. Using coordinates, see Lavrauw [94]. ∎

Chapter 6

Generalized Quadrangles, Nets and the Axiom of Veblen

Each regular point of a thick GQ S of order (s,t) defines a dual net $\mathcal{N}_x^D$. If $\mathcal{N}_x^D$ is not a dual affine plane and satisfies the Axiom of Veblen, then $\mathcal{N}_x^D$ is isomorphic to the "classical" dual net $\mathbf{H}_q^n$, for some q and n. In such a case strong properties for the GQ S can be deduced. In this chapter it is also shown that the existence of particular automorphisms of $\mathcal{N}_L^D$, where the regular line L is incident with some coregular point x of S, forces S to be a TGQ.

6.1 Generalized Quadrangles and the Axiom of Veblen

Consider a GQ $\mathbf{T}_3(\mathcal{O})$ of Tits, with $\mathcal{O}$ an ovoid of $\mathbf{PG}(3,q)$. Here $s = q$ and $t = q^2$. Then the point (∞) is coregular, that is, each line incident with (∞) is regular. It is an easy exercise to check that for each line incident with (∞) the corresponding dual net is isomorphic to $\mathbf{H}_q^3$; see Section 2.2 for the definition of $\mathbf{H}_q^3$. Hence for each line incident with the point (∞) the corresponding dual net satisfies the Axiom of Veblen. We now prove the converse. Theorem 6.1.1 and its corollary are due to Thas and Van Maldeghem [193].

Theorem 6.1.1 *Let $S = (\mathcal{P}, \mathcal{B}, \mathbf{I})$ be a GQ of order (s,t), with $s \neq t$, $s > 1$ and $t > 1$. If S has a coregular point x and if for each line L incident with x the corresponding dual net $\mathcal{N}_L^D$ satisfies the Axiom of Veblen, then S is isomorphic to a $\mathbf{T}_3(\mathcal{O})$ of Tits.*

Proof. Let L_1, L_2, L_3 be three lines no two of which are concurrent, let M_1, M_2, M_3 be three lines no two of which are concurrent, let $L_i \not\sim M_j$ if and only if $\{i, j\} = \{1, 2\}$ and assume that $x I L_1$. By 5.3.8 of Payne and Thas [128] it is sufficient to prove that for any line $L_4 \in \{M_1, M_2\}^{\perp}$ with $L_4 \not\sim L_i$, $i = 1, 2, 3$, there exists a line M_4 concurrent with L_1, L_2, L_4.

So let $L_4 \in \{M_1, M_2\}^{\perp}$ with $L_4 \not\sim L_i$, $i = 1, 2, 3$. Consider the line R containing $L_2 \cap M_2$ and concurrent with L_1. Further, consider the line R' containing $M_2 \cap L_4$ and concurrent with L_1. By the regularity of L_1 there is a line $R'' \in \{M_1, M_3\}^{\perp\perp}$ through the point $L_3 \cap M_2$. Clearly the lines L_1 and R'' are concurrent. So the line L_1 is concurrent with the lines R, R', R''; also the line M_2 is concurrent with the lines R, R', R''. By the regularity of L_1 the line R'' belongs to the line $\{R, R'\}^{\perp\perp}$ of the dual net $\mathcal{N}_{L_1}^D$ defined by L_1. Hence the lines $\{R, R'\}^{\perp\perp}$ and $\{M_1, M_3\}^{\perp\perp}$ of $\mathcal{N}_{L_1}^D$ have the element R'' in common. By the Axiom of Veblen, also the lines $\{M_1, R'\}^{\perp\perp}$ and $\{M_3, R\}^{\perp\perp}$ of $\mathcal{N}_{L_1}^D$ have an element M_4 in common. Consequently M_4 is concurrent with L_1, L_2, L_4. Now fro m 5.3.8 of Payne and Thas [128] it follows that $\mathcal{S}$ is isomorphic to a $\mathbf{T}_3(\mathcal{O})$ of Tits. ∎

As a corollary the following characterization of the classical GQ $\mathbf{Q}(5, s)$ is obtained.

Corollary 6.1.2 *Let $\mathcal{S}$ be a GQ of order (s, t) with $s \neq t, s > 1$ and $t > 1$.*

(i) *If s is odd, then $\mathcal{S}$ is isomorphic to the classical GQ $\mathbf{Q}(5, s)$ if and only if it has a coregular point x and if for each line L incident with x the corresponding dual net $\mathcal{N}_L^D$ satisfies the Axiom of Veblen.*

(ii) *If s is even, then $\mathcal{S}$ is isomorphic to the classical GQ $\mathbf{Q}(5, s)$ if and only if all its lines are regular and for at least one point x and all lines L incident with x the dual nets $\mathcal{N}_L^D$ satisfy the Axiom of Veblen.*

Proof. Let (x, L) be a flag of the GQ $\mathbf{Q}(5, s)$. By 3.2.4 of Payne and Thas [128] there is an isomorphism of $\mathbf{Q}(5, s)$ onto $\mathbf{T}_3(\mathcal{O})$, with $\mathcal{O}$ an elliptic quadric of $\mathbf{PG}(3, s)$, which maps x onto the point (∞). It follows that $\mathcal{N}_L^D$ satisfies the Axiom of Veblen.

Conversely, assume that the GQ $\mathcal{S}$ of order (s, t), with s odd, $s \neq t$, $s > 1$ and $t > 1$, has a coregular point x such that for each line L incident with x the dual net $\mathcal{N}_L^D$ satisfies the Axiom of Veblen. Then by Theorem 6.1.1 the GQ $\mathcal{S}$ is isomorphic to $\mathbf{T}_3(\mathcal{O})$. By Barlotti [8] and Panella [106] each ovoid

$\mathcal{O}$ of $\mathbf{PG}(3, s)$, with s odd, is an elliptic quadric. Now by Theorem 1.5.2(ii) we have $\mathcal{S} \cong \mathbf{T}_3(\mathcal{O}) \cong \mathbf{Q}(5, s)$.

Finally, assume that for the GQ $\mathcal{S}$ of order (s, t), with s even, $s \neq t$, $t > 1$, all lines are regular and that for at least one point x and all lines L incident with x the dual nets $\mathcal{N}_L^D$ satisfy the Axiom of Veblen. Then by Theorem 6.1.1 the GQ $\mathcal{S}$ is isomorphic to $\mathbf{T}_3(\mathcal{O})$. Since all lines of $\mathcal{S}$ are regular, by §2.1 we finally have $\mathcal{S} \cong \mathbf{T}_3(\mathcal{O}) \cong \mathbf{Q}(5, s)$. ∎

6.2 Translation Generalized Quadrangles and the Axiom of Veblen

This section is also taken from Thas and Van Maldeghem [193]. It explains the relationship between the Axiom of Veblen and good eggs.

Theorem 6.2.1 *Let $\mathcal{S}^{(x)}$ be a TGQ of order (s, s^2), $s \neq 1$, with base-point x. Then the dual net $\mathcal{N}_L^D$ defined by the regular line L, with $x \mathrm{I} L$, satisfies the Axiom of Veblen if and only if the egg $\mathcal{O}(n, 2n, q)$ which corresponds to $\mathcal{S}^{(x)}$ is good at its element π_i which corresponds to L.*

Proof. Assume that the dual net $\mathcal{N}_L^D$ satisfies the Axiom of Veblen. Let the egg $\mathcal{O}(n, 2n, q)$ correspond to $\mathcal{S}^{(x)}$ and let π_i correspond to L. We have $s = q^n$. The dual net has $q^n + 1$ points on a line and q^{2n} lines through a point. By Theorem 2.2.3 the dual net $\mathcal{N}_L^D$ is isomorphic to $\mathbf{H}_{q^n}^3$. Consider the TGQ $\mathbf{T}(n, 2n, q) \cong \mathcal{S}^{(x)}$ and let $\mathbf{PG}(3n, q)$ be a subspace skew to π_i in the projective space $\mathbf{PG}(4n, q)$ in which $\mathbf{T}(n, 2n, q)$ is defined. Let $\mathcal{O}(n, 2n, q) = \{\pi_0, \pi_1, \ldots, \pi_{q^{2n}}\}$, let $\langle \pi_i, \pi_j \rangle \cap \mathbf{PG}(3n, q) = \pi_j'$ for all $j \neq i$ (π_j' is $(n-1)$-dimensional), let $\mathbf{PG}(4n-1, q) \cap \mathbf{PG}(3n, q) = \mathbf{PG}(3n-1, q)$, with $\mathbf{PG}(4n-1, q)$ the space of $\mathcal{O}(n, 2n, q)$, and let $\tau_i \cap \mathbf{PG}(3n, q) = \tau_i'$, with τ_i the tangent space of $\mathcal{O}(n, 2n, q)$ at π_i (τ_i' is $(2n-1)$-dimensional). Then the dual net $\mathcal{N}_L^D$ is isomorphic to the following dual net $\mathcal{N}^D$:

- POINTS of $\mathcal{N}^D$ are the q^{2n} spaces π_j', $j \neq i$, and the q^{3n} points of $\mathbf{PG}(3n, q) \setminus \mathbf{PG}(3n-1, q)$;

- LINES of $\mathcal{N}^D$ are the q^{4n} subspaces of $\mathbf{PG}(3n, q)$ of dimension n which are not contained in $\mathbf{PG}(3n-1, q)$ and contain an element π_j', $j \neq i$, and

- INCIDENCE is the natural one.

Clearly the points π'_j, $j \neq i$, of $\mathcal{N}^D$ form a parallel class of points. Let M be a line of $\mathcal{N}^D$ incident with π'_j and let $\pi'_k \neq \pi'_j$, with $k \neq i \neq j$. As $\mathcal{N}^D \cong \mathbf{H}^3_{q^n}$ the elements π'_k and M of $\mathcal{N}^D$ generate a dual affine plane $\mathcal{A}^D$ in $\mathcal{N}^D$, and the plane $\mathcal{A}^D$ contains q^n points π'_l, $l \neq i$. Clearly the points of $\mathcal{A}^D$ not of type π'_l are the q^{2n} points of the subspace $\langle \pi'_k, M \rangle$ of $\mathbf{PG}(3n, q)$ which are not contained in $\mathbf{PG}(3n-1, q)$. Hence the q^n points of $\mathcal{A}^D$ of type π'_l are contained in $\langle \pi'_k, M \rangle \cap \mathbf{PG}(3n-1, q)$. It follows that these q^n elements π'_l are contained in a $(2n-1)$-dimensional space $\mathbf{PG}'(2n-1, q)$; also, they form a partition of $\mathbf{PG}'(2n-1, q) \backslash \tau'_i$. Consequently for any two elements $\pi'_l, \pi'_{l'}$, $l \neq i \neq l'$, the space $\langle \pi'_l, \pi'_{l'} \rangle$ contains exactly q^n elements π'_r, $r \neq i$. Hence for any two spaces π_l and $\pi_{l'}$ of $\mathcal{O}(n, 2n, q) \backslash \{\pi_i\}$, the $(3n-1)$-dimensional space $\langle \pi_i, \pi_l, \pi_{l'} \rangle$ contains exactly $q^n + 1$ elements of $\mathcal{O}(n, 2n, q)$. We conclude that $\mathcal{O}(n, 2n, q)$ is good at π_i.

Conversely, assume that $\mathcal{O}(n, 2n, q)$ is good at the element π_i which corresponds to L. As in the first part of the proof we project onto a $\mathbf{PG}(3n, q)$ and we use the same notation. Since $\mathcal{O}(n, 2n, q)$ is good at π_i, for any two elements $\pi'_l, \pi'_{l'}$, with $l \neq i \neq l'$, the space $\langle \pi'_l, \pi'_{l'} \rangle$ contains exactly q^n elements π'_r, $r \neq i$; these q^n elements form a partition of the points of $\langle \pi'_l, \pi'_{l'} \rangle$ which are not contained in τ'_i. If M, M' are distinct concurrent lines of $\mathcal{N}^D$, then it is easily checked that M and M' generate a dual affine plane $\mathcal{A}^D$ of order q^n in $\mathcal{N}^D$. As $\mathcal{A}^D$ satisfies the Axiom of Veblen, also $\mathcal{N}^D$ does. ∎

Corollary 6.2.2 *Let $\mathcal{S}^{(x)}$ be a TGQ of order (s, s^2), s odd and $s \neq 1$, with base-point x. If the dual net $\mathcal{N}^D_L$ defined by some regular line L, with $x I L$, satisfies the Axiom of Veblen, then $\mathcal{S}^{(x)}$ contains $s^3 + s^2$ classical subquadrangles $\mathbf{Q}(4, s)$ containing L.*

Proof.　This follows immediately from Theorem 6.2.1 and Theorem 5.1.9. ∎

6.3　Property (G) and the Axiom of Veblen

Property (G) for GQs of order (s, s^2) was introduced in Section 4.10. If the GQ $\mathcal{S}$ of order (s, s^2), $s \neq 1$, satisfies Property (G) at the line L, then we shall prove that L is regular and shall consider the dual net $\mathcal{N}^D_L$ in some detail.

Theorem 6.3.1 *Let $\mathcal{S} = (\mathcal{P}, \mathcal{B}, I)$ be a GQ of order (s, s^2), $s \neq 1$, satisfying Property (G) at the line L. Then L is regular.*

Proof. Let $S = (\mathcal{P}, \mathcal{B}, \mathrm{I})$ be a GQ of order (s, s^2), s even, satisfying Property (G) at the line L. Consider distinct points x and y incident with L, and let x be a center of the triad $\{y, z, u\}$. By Theorem 2.6.2, S has a subquadrangle S' of order s which contains $\{y, z, u\}^{\perp} \cup \{y, z, u\}^{\perp\perp}$. Let w be a point of S', with $w \nsim x$ and $w \nsim y$. Further, let r, r', r'' be distinct points of $\{x, w\}^{\perp'}$ (here "$\perp'$" denotes "perp" in S'), with $r\mathrm{I}L$. Then $\{r, r', r''\}^{\perp}$ and $\{r, r', r''\}^{\perp\perp}$ are contained in a subquadrangle of order s of S. Clearly the lines $xr, xr', xr'', wr, wr', wr''$ are common lines of S' and S''. By 2.3.1 of Payne and Thas [128], see also Section 1.1, the common points and lines of S' and S'' form a subquadrangle $S' \cap S''$ of order (s, t'), with $t' > 1$, of S''. Now Lemma 1.1.1 readily implies that $S' \cap S'' = S' = S''$. It follows that $\{r, r'\}^{\perp'\perp'} = \{r, r', r''\}^{\perp\perp}$, $\{r, r'\}^{\perp'} = \{r, r', r''\}^{\perp}$, and so $\{r, r'\}$ and $\{x, w\}$ are regular in S'. Hence x is regular in S'.

With the notation of the previous paragraph and interchanging the roles of x and y, we see that also y is regular in S'. Since in the previous paragraph y and r play the same role in S', also r is regular in S'. It follows immediately that all points of L are regular in S'. By Theorem 2.1.5 also the line L is regular in S'. Consider now a line M of S, with $M \nsim L$. Let $y \sim x'\mathrm{I}M$, $x \sim y'\mathrm{I}M$, and $x'' \in \{y, y'\}^{\perp} \setminus \{x, x'\}$. The subquadrangle $\overline{S}$ of order s defined by x, x', x'' contains L and M. Since L is regular in $\overline{S}$, the span $\{L, M\}^{\perp\perp}$ has size $s + 1$. We conclude that L is regular in S.

Next assume that s is odd, $s \neq 1$, and that the GQ $S = (\mathcal{P}, \mathcal{B}, \mathrm{I})$ of order (s^2, s) satisfies Property (G) at the point x. By Theorem 4.10.6 the GQ S is a flock GQ. Now we consider the construction of Knarr, see Section 4.9. If the point y is not collinear with x, that is, if y is a point of $\mathbf{PG}(5, s)$ not in x^ς, with ς the symplectic polarity used to construct S in $\mathbf{PG}(5, s)$, then $\{x, y\}^{\perp\perp}$ consists of the $s + 1$ points of the line $\langle x, y \rangle$ of $\mathbf{PG}(5, s)$. As $|\{x, y\}^{\perp\perp}| = s + 1$ the point x is regular. ∎

Remark 6.3.2 In the first part of the proof, where in the even case we show that x is regular in S', it is sufficient to assume that S satisfies Property (G) at the flag (x, L). By Theorem 4.10.7 it is sufficient in the odd case to assume that S satisfies Property (G) at some point pair $\{x, y\}$, with $x\mathrm{I}L\mathrm{I}y$. In fact, for all $s \neq 1$, the GQ S of order (s, s^2) is the point-line dual of a flock GQ. Hence L is an axis of symmetry (cf. Section 4.6), and consequently regular.

The following theorems of this section are also taken from Thas and Van Maldeghem [193].

Theorem 6.3.3 *Let* $\mathcal{S} = (\mathcal{P}, \mathcal{B}, I)$ *be a GQ of order* (s^2, s), *s even, satisfying Property* (G) *at the point* x. *Then the dual net* $\mathcal{N}_x^D$ *satisfies the Axiom of Veblen, and consequently* $\mathcal{N}_x^D \cong \mathbf{H}_s^3$.

Proof. Let $\mathcal{S} = (\mathcal{P}, \mathcal{B}, I)$ be a GQ of order (s^2, s), s even, satisfying Property (G) at the point x. By Theorem 6.3.1 the point x is regular. Let y be a point of the dual net $\mathcal{N}_x^D$, let A_1 and A_2 be distinct lines of $\mathcal{N}_x^D$ containing y, let B_1 and B_2 be distinct lines of $\mathcal{N}_x^D$ not containing y, and let $A_i \cap B_j \neq \emptyset$ for all $i, j \in \{1, 2\}$. Let $\{z\} = A_1 \cap B_1$ and let zIM, with $x \nmid M$. Further, let xIL, with $z \nmid L$, let u be the point of A_1 on L, and let u' be the point of B_1 on L. The line of $\mathcal{S}$ incident with u, respectively u', and concurrent with M is denoted by C, respectively C'; the line incident with z and x is denoted by N. Since $\mathcal{S}$ satisfies Property (G) at x, the triple $\{C, C', N\}$ is 3-regular. By Theorem 2.6.2 the lines of $\mathcal{S}$ concurrent with at least one line of $\{C, C', N\}^\perp$ and one line of $\{C, C', N\}^{\perp\perp}$, are the lines of a subquadrangle $\mathcal{S}'$ of order s of $\mathcal{S}$. As x is regular for $\mathcal{S}$ it is also regular for $\mathcal{S}'$. By Theorem 2.2.1 the point x in $\mathcal{S}'$ defines a projective plane π_x of order s. Clearly A_1, A_2, B_1, B_2 are lines of the projective plane π_x. Hence B_1 and B_2 intersect in π_x. Consequently $\mathcal{N}_x^D$ satisfies the Axiom of Veblen, and so $\mathcal{N}_x^D \cong \mathbf{H}_s^3$ by Theorem 2.2.3. ∎

Theorem 6.3.4 *Let* $\mathcal{S}^{(x)} = \mathbf{T}(\mathcal{O})$ *be a TGQ of order* (s, s^2), $s \neq 1$, *with base-point* x. *Then the dual net* $\mathcal{N}_L^D$ *defined by the regular line* L, *with* xIL, *satisfies the Axiom of Veblen if and only if the translation dual* $\mathcal{S}'^{(x')} = \mathbf{T}(\mathcal{O}^*)$ *of* $\mathcal{S}^{(x)}$ *satisfies Property* (G) *at the flag* (x', L'), *where* L' *corresponds to* L, *that is, where* L' *is the tangent space of* $\mathcal{O}$ *at* L. *In the even case,* $\mathcal{N}_L^D$ *satisfies the Axiom of Veblen if and only if* $\mathcal{S}^{(x)}$ *satisfies Property* (G) *at the flag* (x, L).

Proof. By Theorem 6.2.1 the dual net $\mathcal{N}_L^D$ satisfies the Axiom of Veblen if and only if $\mathcal{O} = \mathcal{O}(n, 2n, q)$ is good at the element L. By Theorem 5.1.2 the egg $\mathcal{O}$ is good at L if and only if $\mathbf{T}(\mathcal{O}^*)$ satisfies Property (G) at the flag (x', L'). By Theorem 5.1.5, for s even, $\mathbf{T}(\mathcal{O}^*)$ satisfies Property (G) at the flag (x', L') if and only if $\mathbf{T}(\mathcal{O})$ satisfies Property (G) at the flag (x, L). ∎

Theorem 6.3.5 *Let* $\mathcal{S}^{(x)}$ *be a TGQ of order* (s, s^2), *s odd and* $s \neq 1$, *with base-point* x. *If* xIL *and if the dual net* $\mathcal{N}_L^D$ *satisfies the Axiom of Veblen, then all lines concurrent with* L *are regular.*

Proof. Let N be concurrent with L, $x \nmid N$, and let the line M of $\mathcal{S}^{(x)}$ be nonconcurrent with N. From Theorem 5.1.9 easily follows that the lines

M, N are lines of a subquadrangle of $\mathcal{S}^{(x)}$ isomorphic to $\mathbf{Q}(4, s)$. Hence $\{M, N\}$ is a regular pair of lines. We conclude that the line N is regular in $\mathcal{S}^{(x)}$. ∎

6.4 Flock Generalized Quadrangles and the Axiom of Veblen

Let $\mathcal{F}$ be a *flock* of the quadratic cone $\mathcal{K}$ of $\mathbf{PG}(3, q)$. Then, by Theorem 4.6.3, with $\mathcal{F}$ corresponds a GQ $\mathcal{S}(\mathcal{F})$ of order (q^2, q). In Payne [118] it was shown that $\mathcal{S}(\mathcal{F})$ satisfies Property (G) at its point (∞); see Theorem 4.10.1. Hence by Theorem 6.3.1 we have the following result.

Theorem 6.4.1 *For any GQ $\mathcal{S}(\mathcal{F})$ of order (q^2, q) arising from a flock $\mathcal{F}$, the point (∞) is regular.* ∎

(Of course, we also know that (∞) is regular because it is a center of symmetry.)

The following theorem is taken from Thas and Van Maldeghem [193].

Theorem 6.4.2 *Consider the GQ $\mathcal{S}(\mathcal{F})$ of order (q^2, q) arising from the flock $\mathcal{F}$. If q is even, then the dual net $\mathcal{N}^D_{(\infty)}$ always satisfies the Axiom of Veblen and so $\mathcal{N}^D_\infty \cong \mathbf{H}^3_q$. If q is odd, then the dual net $\mathcal{N}^D_{(\infty)}$ satisfies the Axiom of Veblen if and only if $\mathcal{F}$ is a Kantor-Knuth flock.*

Proof. Consider the GQ $\mathcal{S}(\mathcal{F})$ of order (q^2, q) arising from the flock $\mathcal{F}$. Then $\mathcal{S}(\mathcal{F})$ satisfies Property (G) at the point (∞). Hence for q even, by Theorem 6.3.3, the dual net $\mathcal{N}^D_{(\infty)}$ satisfies the Axiom of Veblen, and consequently $\mathcal{N}^D_{(\infty)} \cong \mathbf{H}^3_q$.

Next, let q be odd. Suppose that $\mathcal{F}$ is a Kantor-Knuth flock. Then by Theorem 4.7.2 the point-line dual of $\mathcal{S}(\mathcal{F})$ is a TGQ, so is isomorphic to some $\mathbf{T}(\mathcal{O})$ with $\mathcal{O} = \mathcal{O}(n, 2n, q')$ and $q = q'^n$ and by Theorem 4.7.3, $\mathbf{T}(\mathcal{O}) \cong \mathbf{T}(\mathcal{O}^*)$. In Payne and Thas [129] it is shown that if $\mathcal{S}(\mathcal{F})$ is not classical, then the elation point (∞) of $\mathcal{S}(\mathcal{F})$ is fixed by the full automorphism group of $\mathcal{S}(\mathcal{F})$. Hence if $\mathcal{F}$ is not linear, then the point (∞) of $\mathcal{S}(\mathcal{F})$ corresponds to some line π of Type (b) of $\mathbf{T}(\mathcal{O})$, that is, to some element π of $\mathcal{O}$; if $\mathcal{F}$ is linear, so $\mathcal{S}(\mathcal{F})$ is classical, then we clearly may assume that (∞) corresponds to some element π of $\mathcal{O}$. Consequently $\mathbf{T}(\mathcal{O})$ satisfies Property (G) at π. By Theorem 5.1.2 the translation dual $\mathcal{O}^*$ of $\mathcal{O}$ is good at τ, with τ the tangent space of $\mathcal{O}$ at π. Hence, by Theorem 6.2.1, the dual net $\mathcal{N}_\tau^D$ which corresponds with the regular line τ of $\mathbf{T}(\mathcal{O}^*)$ satisfies the Axiom of Veblen. By Theorem 5.1.13 also the dual net $\mathcal{N}_\pi^D$ which corresponds with the regular line π of $\mathbf{T}(\mathcal{O})$ satisfies the Axiom of Veblen. It follows that the dual net $\mathcal{N}_{(\infty)}^D$ satisfies the Axiom of Veblen. Hence $\mathcal{N}_{(\infty)}^D \cong \mathbf{H}_q^3$.

Conversely, assume that q is odd and that the dual net $\mathcal{N}_{(\infty)}^D$ satisfies the Axiom of Veblen. Hence $\mathcal{N}_{(\infty)}^D \cong \mathbf{H}_q^3$. In the representation of Knarr, see Section 4.9, this dual net looks as follows:

- POINTS of $\mathcal{N}_{(\infty)}^D$ are the lines of $\mathbf{PG}(5, q)$ not containing $p = (\infty)$ but contained in one of the planes $pL_t = \pi_t$, with $L_t \in \mathcal{V}$ and $\mathcal{V}$ the BLT-set which corresponds to $\mathcal{F}$,

- LINES of $\mathcal{N}_{(\infty)}^D$ can be identified with the threedimensional subspaces of p^ξ not containing p (ξ is a symplectic polarity of $\mathbf{PG}(5, q)$), and

- INCIDENCE is inclusion.

By point-hyperplane duality in p^ξ, the net $\mathcal{N}_{(\infty)}$, which is the point-line dual of $\mathcal{N}_{(\infty)}^D$, is isomorphic to the following incidence structure:

- POINTS of $\mathcal{N}_{(\infty)}$ are the points of $p^\xi \setminus \delta$, with δ some threedimensional subspace of p^ξ,

- LINES of $\mathcal{N}_{(\infty)}$ are the planes of p^ξ not contained in δ, but containing one of the lines of a BLT-set $\mathcal{V}'$ in δ, where $\mathcal{V} \cong \mathcal{V}'$, and

- INCIDENCE is the natural one.

As the net $\mathcal{N}_{(\infty)}$ is isomorphic to the dual of $\mathbf{H}_q^3$, it is easily seen to be derivable; see e.g. De Clerck and Johnson [43]. Let $\mathcal{V}' = \{L_0', L_1', \ldots, L_q'\}$ be contained in the GQ $\mathbf{W}(q)'$. Then the line set $\tau = \{L_0', L_1'\}^{\perp'\perp'} \cup \{L_0', L_2'\}^{\perp'\perp'} \cup \cdots \cup \{L_0', L_q'\}^{\perp'\perp'}$, where "$\perp'$" denotes "perp" in $\mathbf{W}(q)'$, is a spread of $\mathbf{W}(q)'$; see e.g. Thas [172]. As $\mathcal{N}_{(\infty)}$ is derivable, by Biliotti and Lunardon [12] there are two distinct lines in δ intersecting the $q + 1$ lines of $\mathcal{V}'$. Hence the image of $\mathcal{V}'$ on the Klein quadric is contained in a threedimensional space, and so the q planes of the elements of the flock $\mathcal{F}$ all contain a common point. Now by Theorem 4.4.6, $\mathcal{F}$ is a Kantor-Knuth flock. ∎

Corollary 6.4.3 *Suppose that the* TGQ $\mathbf{T}(\mathcal{O})$*, with* $\mathcal{O} = \mathcal{O}(n, 2n, q)$ *and* q *odd, is the point-line dual of a flock GQ* $\mathcal{F}$*, where the point* (∞) *of* $\mathcal{S}(\mathcal{F})$ *corresponds to the element* π *of* $\mathcal{O}$*. Then* $\mathcal{O}$ *is good at the element* π *if and only if* $\mathcal{F}$ *is a Kantor-Knuth flock.*

Proof. Assume that $\mathcal{F}$ is a Kantor-Knuth flock. Then by Theorem 5.1.2 and Theorem 4.7.3 the egg $\mathcal{O}$ is good at π.

Conversely, assume that the egg $\mathcal{O}$ is good at π. Then by Theorem 6.2.1 the dual net $\mathcal{N}_\pi^D \cong \mathcal{N}_{(\infty)}^D$ satisfies the Axiom of Veblen. By Theorem 6.4.2 the flock $\mathcal{F}$ is a Kantor-Knuth flock. ∎

In Bader, Lunardon and Pinneri [4] a stronger version of Corollary 6.4.3 is obtained; the authors rely on Corollary 6.4.3, however.

Theorem 6.4.4 *Suppose that the* TGQ $\mathbf{T}(\mathcal{O})$ *with* $\mathcal{O} = \mathcal{O}(n, 2n, q)$ *and* q *odd, is the point-line dual of a flock GQ* $\mathcal{S}(\mathcal{F})$*, or, equivalently, assume that* $\mathcal{O}^*$ *is good at some element. Then* $\mathbf{T}(\mathcal{O}^*)$ *is the point-line dual of a flock GQ, or, equivalently,* $\mathcal{O}$ *is good at some element if and only if* $\mathcal{F}$ *is a Kantor-Knuth flock.*

Proof. Assume that $\mathcal{F}$ is a Kantor-Knuth flock. Then by Theorem 5.1.2 and Theorem 4.7.3 the eggs $\mathcal{O}$ and $\mathcal{O}^*$ are good.

Conversely, assume that the egg $\mathcal{O}$ is good at some element π. By Corollary 5.1.4 the GQ $\mathbf{T}(\mathcal{O}^*)$ is the point-line dual of a flock GQ with elation point τ, where τ is the tangent space of $\mathcal{O}$ at π. Hence $\mathbf{T}(\mathcal{O}^*)$ has an elation group with base-line τ. So G acts transitively on the lines, distinct from τ, incident with the point (∞) of $\mathbf{T}(\mathcal{O}^*)$. By Theorem 3.7.2, G is induced by a collineation group G' of $\mathbf{PG}(4n, q)$ (with $\mathbf{PG}(4n, q)$ the space in which $\mathbf{T}(\mathcal{O}^*)$ is constructed), which fixes $\mathcal{O}^*$, τ, and acts transitively on $\mathcal{O}^* \backslash \{\tau\}$. Hence there is a collineation group $\overline{G}$ of $\langle \mathcal{O} \rangle = \mathbf{PG}(4n - 1, q)$ fixing $\mathcal{O}$, π, and acting transitively on $\mathcal{O} \backslash \{\pi\}$. As $\mathbf{T}(\mathcal{O})$ is the point-line dual of a flock GQ, the GQ $\mathbf{T}(\mathcal{O})$ has an elation group with base-line $\pi' \in \mathcal{O}$. Hence there is a collineation group $\overline{\overline{G}}$ of $\mathbf{PG}(4n - 1, q)$ fixing $\mathcal{O}$, π', and acting transitively on $\mathcal{O} \backslash \{\pi'\}$. If $\pi \neq \pi'$, then the collineation group $\langle \overline{G}, \overline{\overline{G}} \rangle$ acts 2-transitively on $\mathcal{O}$. Consequently, $\mathcal{O}$ is good at each of its elements. Now by Theorem 5.1.12, $\mathcal{O}$ is classical, so the flock $\mathcal{F}$ is linear. If $\pi = \pi'$, then by Corollary 6.4.3, $\mathcal{F}$ is a Kantor-Knuth flock. ∎

6.5   Subquadrangles and the Axiom of Veblen

In this section, which is also taken from Thas and Van Maldeghem [193], we describe the impact of the Axiom of Veblen on the existence of subquadrangles of order s of GQs of order (s^2, s).

Theorem 6.5.1 *Let* $\mathcal{S} = (\mathcal{P}, \mathcal{B}, \mathrm{I})$ *be a GQ of order* (s, t), $s \neq 1 \neq t$, *having a regular point* x. *If* x *and any two points* y, z, *with* $y \not\sim x$ *and* $x \sim z \not\sim y$, *are contained in a proper subquadrangle* $\mathcal{S}'$ *of* $\mathcal{S}$ *of order* (s', t), *with* $s' \neq 1$, *then* $s' = t = \sqrt{s}$ *and the dual net* $\mathcal{N}_x^D$ *satisfies the Axiom of Veblen. It follows that* s *and* t *are prime powers, and that for each subquadrangle* $\mathcal{S}'$ *the projective plane* π_x *of order* t *defined by the regular point* x *of* $\mathcal{S}'$ *is Desarguesian.*

Conversely, if the dual net $\mathcal{N}_x^D$ *satisfies the Axiom of Veblen, then either*

(a) $s = t$, *or*

(b) $s = t^2$, s *and* t *are prime powers,* x *and any two points* y, z, *with* $y \not\sim x$ *and* $x \sim z \not\sim y$ *are contained in a subquadrangle* $\mathcal{S}'$ *of* $\mathcal{S}$ *of order* (t, t), *and the projective plane* π_x *of order* t *defined by the regular point* x *of* $\mathcal{S}'$ *is Desarguesian.*

Proof.   Let $\mathcal{S} = (\mathcal{P}, \mathcal{B}, \mathrm{I})$ be a GQ of order (s, t), $s \neq 1 \neq t$, having a regular point x.

First, assume that x and any two points y, z with $y \not\sim x$ and $x \sim z \not\sim y$, are contained in a proper subquadrangle $\mathcal{S}'$ of $\mathcal{S}$ of order (s', t), with $s' \neq 1$. As x is also regular for $\mathcal{S}'$, the GQ $\mathcal{S}'$ contains subquadrangles of order $(1, t)$. Then Lemma 1.1.1 implies $s' = t = \sqrt{s}$. By Theorem 2.2.1 the dual net $\mathcal{N}'^D_x$ arising from the regular point x of $\mathcal{S}'$, is a dual affine plane of order t. Hence $\mathcal{N}'^D_x$ satisfies the Axiom of Veblen. Now consider distinct lines A_1, A_2, B_1, B_2 of the dual net $\mathcal{N}_x^D$, where $A_1 \cap A_2 = \{z\}$, $z \notin B_1$, $z \notin B_2$, and $A_i \cap B_j \neq \emptyset$ for all $i, j \in \{1, 2\}$. Let $A_1 \cap B_1 = \{u\}$, $A_2 \cap B_2 = \{w\}$, and let $y \in \{u, w\}^{\perp} \backslash \{x\}$. Let $\mathcal{S}'$ be a subquadrangle of order t containing the points x, y, z of $\mathcal{S}$. Then A_1, A_2, B_1, B_2 are lines of the dual net $\mathcal{N}'^D_x$. As $\mathcal{N}'^D_x$ satisfies the Axiom of Veblen, we have $B_1 \cap B_2 \neq \emptyset$. It follows that the dual net $\mathcal{N}_x^D$ satisfies the Axiom of Veblen. Consequently $\mathcal{N}_x^D \cong \mathbf{H}_t^3$, and so s and t are prime powers. For any subquadrangle $\mathcal{S}'$ the dual net $\mathcal{N}'^D_x$ is a dual affine plane of order t, which is isomorphic to a dual affine plane of order t in $\mathbf{H}_t^3$. Hence $\mathcal{N}'^D_x$, and also the corresponding projective plane π_x, is Desarguesian.

Conversely, assume that the dual net $\mathcal{N}_x^D$ satisfies the Axiom of Veblen. Also, suppose that $s \neq t$, and so $s > t$ by Theorem 2.1.3. Then, by Theorem 2.2.3 and Corollary 2.2.4, we have $\mathcal{N}_x^D = \mathbf{H}_t^3$ and $s = t^2$, with s and t prime powers. Now consider any two points y, z, with $y \not\sim x$, $x \sim z \not\sim y$. As $\mathcal{N}_x^D \cong \mathbf{H}_t^3$ it is easily seen that z and $\{x, y\}^\perp$ generate a dual affine plane $\mathcal{A}$ of order t in $\mathcal{N}_x^D$. Let $A_1, A_2, \ldots, A_{t^2}$ be the lines of $\mathcal{A}$. Further, let $\mathcal{P}'$ be the point set of $\mathcal{S}$ consisting of the points of $A_1^\perp \cup A_2^\perp \cup \cdots \cup A_{t^2}^\perp$ and the points of $\mathcal{A}$. Clearly $\mathcal{P}'$ contains z and y, and $|\mathcal{P}'| = t^3 + t^2 + t + 1$. Further, any line of $\mathcal{S}$ incident with at least one point of $\mathcal{P}'$ either contains x or a point of $\mathcal{A}$; the set of all these lines is denoted by $\mathcal{B}'$. Also, any point incident with two distinct lines of $\mathcal{B}'$ belongs to $\mathcal{P}'$. Then by 2.3.1 of Payne and Thas [128], see also Section 1.1, $\mathcal{S}' = (\mathcal{P}', \mathcal{B}', \mathbf{I}')$, with $\mathbf{I}'$ the restriction of $\mathbf{I}$ to $(\mathcal{P}' \times \mathcal{B}') \cup (\mathcal{B}' \times \mathcal{P}')$ is a subquadrangle of $\mathcal{S}$ of order t. As in the first part of the proof one now shows that for any such subquadrangle $\mathcal{S}'$ the projective plane π_x defined by x is Desarguesian. ∎

Remark 6.5.2 Note that in the first part of the statement of Theorem 6.5.1, x need not be a regular point a priori; one easily deduces the regularity of x from the assumption about the subquadrangles, and Lemma 1.1.1.

6.6 Nets and Characterizations of Translation Generalized Quadrangles

First we give the definition of a $\mathcal{T}$-*net*, then we prove a characterization theorem for TGQs in terms of $\mathcal{T}$-nets, and finally several interesting corollaries will be obtained. This section is taken from Thas [185].

Let $\mathcal{N} = (\overline{\mathcal{P}}, \overline{\mathcal{B}}, \overline{\mathbf{I}})$ be a net of order k and degree r. Further, let R be a line of $\mathcal{N}$ and let $\mathcal{T}$ be the *parallel class* of $\overline{\mathcal{B}}$ containing R, that is, $\mathcal{T}$ consists of R and the $k - 1$ lines not concurrent with R. An automorphism θ of $\mathcal{N}$ is called a *transvection* with *axis* R if either $\theta = \mathbf{1}$ or if $\mathcal{T}$ is the set of all fixed lines of θ and R is the set of all fixed points of θ. The net $\mathcal{N}$ is a $\mathcal{T}$-*net* if for any two nonparallel lines $M, N \in \overline{\mathcal{B}} \backslash \mathcal{T}$ there is some transvection with axis belonging to $\mathcal{T}$ and mapping M onto N.

In particular, let $\mathcal{N} = (\overline{\mathcal{P}}, \overline{\mathcal{B}}, \overline{\mathbf{I}})$ be an affine plane of order k and let $\mathcal{D}$ be the corresponding projective plane. Then $\mathcal{N}$ is a $\mathcal{T}$-net if and only if the point z of $\mathcal{D}$ defined by $\mathcal{T}$ is a *translation point* of $\mathcal{D}$; see Hughes and Piper [75].

If $\mathcal{N}$ is the dual of $\mathbf{H}_q^n$, then it is easy to check that $\mathcal{N}$ is a $\mathcal{T}$-net for any parallel class $\mathcal{T}$.

Theorem 6.6.1 *Let $(\mathcal{S}^{(x)}, G)$ be a TGQ of order (s,t), $s \neq 1 \neq t$. Then for any line L incident with x, the dual net $\mathcal{N}_L^D$ defined by L is the dual of a $\mathcal{T}$-net $\mathcal{N}_L$, with $\mathcal{T}$ the parallel class of $\mathcal{N}_L$ defined by the point x.*

Proof. Let M, N be two distinct nonconcurrent lines of $\mathcal{S}$, concurrent with L but not incident with x. So M, N are nonparallel lines of the net $\mathcal{N}_L$, not belonging to the parallel class $\mathcal{T}$ defined by the point x. Let $R \in \{M, N\}^{\perp\perp}$, with $x \mathrm{I} R$, and let $L \mathrm{I} m \mathrm{I} M$, $L \mathrm{I} n \mathrm{I} N$. In $\mathcal{S}$ there is a symmetry $\tilde{\theta}$ about R with $m^{\tilde{\theta}} = n$. Then $M^{\tilde{\theta}} = N$. The symmetry $\tilde{\theta}$ induces an automorphism θ of the net $\mathcal{N}_L$ which fixes the parallel class $\mathcal{T}$ of $\mathcal{N}_L$ elementwise, and all points of $\mathcal{N}_L$ incident with the line R of $\mathcal{T}$; it is easily checked that no other elements of $\mathcal{N}_L$ are fixed. Also, $M^{\theta} = N$. Hence θ is a transvection with axis R of $\mathcal{N}_L$ mapping M onto N. It follows that $\mathcal{N}_L$ is a $\mathcal{T}$-net. ∎

Now we state the main theorem of this section.

Theorem 6.6.2 *Let $\mathcal{S} = (\mathcal{P}, \mathcal{B}, \mathrm{I})$ be a GQ of order (s,t), $s \neq 1 \neq t$, with coregular point x. If for at least one line L incident with x the dual net $\mathcal{N}_L^D$ is the dual of a $\mathcal{T}$-net $\mathcal{N}_L$ with $\mathcal{T}$ the parallel class of $\mathcal{N}_L$ defined by x, then $\mathcal{S}$ is a TGQ with base-point x.*

Proof. Let $\theta \neq 1$ be a transvection with axis R of the net $\mathcal{N}_L$, with $R \in \mathcal{T}$. Then θ fixes all lines of $\mathcal{S}$, distinct from L, incident with x and all sets $\{U, R\}^{\perp\perp}$, with $x \mathbf{I} U$, $U \sim L$. Now we prove that θ induces a symmetry γ about R of the GQ $\mathcal{S}$.

First, let r be a point of $\mathcal{S}$ not in $x^{\perp}$. Put $r \mathrm{I} U \mathrm{I} y \mathrm{I} L$ and $r \mathrm{I} M \mathrm{I} z \mathrm{I} R$. The line U^{θ} is concurrent with M; let $U^{\theta} \mathrm{I} r' \mathrm{I} M$. Now we put $r' = r^{\gamma}$; clearly $r' \neq r$ and distinct points not in $x^{\perp}$ are mapped by γ onto distinct points not in $x^{\perp}$. We show that $r \sim u$, $r \neq u$, $r \notin x^{\perp}$, $u \notin x^{\perp}$, implies that $r^{\gamma} \sim u^{\gamma}$, $r^{\gamma} \neq u^{\gamma}$. If $r \mathrm{I} A \sim L$, then $r^{\gamma} \mathrm{I} A^{\theta} \sim L$; if $u \mathrm{I} D \sim L$, then $u^{\gamma} \mathrm{I} D^{\theta} \sim L$.

Let $ru \sim L$. Then $A = D$, $A^{\theta} = D^{\theta}$, so $r^{\gamma} \sim u^{\gamma}$ with $r^{\gamma} \neq u^{\gamma}$. Next, let $ru \sim R$. Then also $r^{\gamma} \sim u^{\gamma}$, with $r^{\gamma} \neq u^{\gamma}$. Hence we assume $ru \not\sim L$ and $ru \not\sim R$. Let $ru \sim V$, $x \mathrm{I} V$ and $r u \mathrm{I} v \mathrm{I} V$. Let $r^{\gamma} \mathrm{I} W \mathrm{I} w \mathrm{I} V$. Assume, by way of contradiction, that $u^{\gamma} \mathbf{I} W$. The elements A, D, V are on a line of $\mathcal{N}_L^D$, so also $A^{\theta}, D^{\theta}, V^{\theta} = V$ are on a line of $\mathcal{N}_L^D$. So $W \sim D^{\theta}$. The lines R, ru, W are concurrent with V and with $r r^{\gamma}$. Consequently, as R is regular and $ru \sim uu^\gamma \sim R$, we have $uu^\gamma \sim W$. So a triangle with sides uu^γ, W, D^θ arises, clearly a contradiction. It follows that $u^\gamma IW$, and so $r^\gamma \sim u^\gamma$. Clearly $r^\gamma \neq u^\gamma$.

Now we define Z^γ for any line Z in $\mathcal{B}$. If xIZ, then we define $Z^\gamma = Z$. Next, assume $x \mathrel{\text{I}} Z$. Let r_1IZIr_2, $r_1 \neq r_2$, $r_1 \not\sim x \not\sim r_2$. Then $Z^\gamma = r_1^\gamma r_2^\gamma$. From the foregoing section follows that in this way Z^γ is well-defined. Clearly $Z^\gamma = Z$ for all lines $Z \sim R$. From the foregoing it also follows that for $x \mathrel{\text{I}} Z$ the lines Z and Z^γ are concurrent with a common line incident with x.

Let Z_1, Z_2 be two distinct lines incident with a common point $z \in x^\perp$. We will prove that Z_1^γ, Z_2^γ are incident with a common point $z' \in x^\perp$ with $z \sim z'$; also, $Z_1^\gamma \neq Z_2^\gamma$. If at least one of the lines Z_1, Z_2 is incident with x, this is obvious. So assume $x \mathrel{\text{I}} Z_i$, with $i = 1, 2$. If zIR, then clearly $Z_1^\gamma IzIZ_2^\gamma$ and $Z_1^\gamma \neq Z_2^\gamma$. Now let zIL. Then Z_1 and Z_2 are parallel lines of the net $\mathcal{N}_L$, so also Z_1^θ and Z_2^θ are parallel in $\mathcal{N}_L$. Hence $Z_1^\gamma = Z_1^\theta$, $Z_2^\gamma = Z_2^\theta$, L are concurrent in $\mathcal{S}$; also $Z_1^\gamma \neq Z_2^\gamma$. So let $z \mathrel{\text{I}} R$ and $z \mathrel{\text{I}} L$, say $zIVIx$. From the foregoing we have $Z_1^\gamma \sim V \sim Z_2^\gamma$. Assume, by way of contradiction, that $Z_1^\gamma \not\sim Z_2^\gamma$. Let $a_1IZ_1^\gamma$, $a_2IZ_2^\gamma$, $a_1 \mathrel{\text{I}} V \mathrel{\text{I}} a_2$, $a_1 \sim a_2$. Since γ defines a permutation of the set of all pairs consisting of distinct collinear points of $\mathcal{S}$ not in $x^\perp$, we have $a_1^{\gamma^{-1}} \sim a_2^{\gamma^{-1}}$, with $a_1^{\gamma^{-1}} IZ_1$ and $a_2^{\gamma^{-1}} IZ_2$, clearly a contradiction. Hence $Z_1^\gamma, Z_2^\gamma, V$ are concurrent in $\mathcal{S}$; also $Z_1^\gamma \neq Z_2^\gamma$.

Consider any point $z \in x^\perp$ and let Z_1IzIZ_2, with $Z_1 \neq Z_2$. Then, by definition, z^γ is the point incident with both Z_1^γ and Z_2^γ.

Now it is clear that γ is an automorphism of $\mathcal{S}$ which fixes all lines concurrent with R. Consequently, γ is a symmetry about R.

Let l_1 and l_2 be incident with L, $l_1 \neq x \neq l_2$ and $l_1 \neq l_2$. Now we choose two lines A_1, A_2, with l_1IA_1, l_2IA_2 and $R \in \{A_1, A_2\}^{\perp\perp}$. Then there is some transvection θ with axis R' of $\mathcal{N}_L$, for some $R' \in \mathcal{T}$, mapping A_1 onto A_2. As $(\{R', A_1\}^{\perp\perp})^\theta = \{R', A_1\}^{\perp\perp} = \{R', A_2\}^{\perp\perp}$, it follows that $R = R'$. For the corresponding symmetry γ about R we clearly have $l_1^\gamma = l_2$. Consequently, R is an axis of symmetry of $\mathcal{S}$.

Hence any line R incident with x, but distinct from L, is an axis of symmetry of $\mathcal{S}$. Now by Theorem 6.7.10 of K. Thas [206] $\mathcal{S}$ is a TGQ with base-point x. ∎

This theorem has several interesting and strong corollaries.

Corollary 6.6.3 *Let* $S = (\mathcal{P}, \mathcal{B}, \mathbf{I})$ *be a GQ of order* s, $s \neq 1$, *with coregular point* x. *If for at least one line* L *incident with* x *the corresponding projective plane* π_L *is a translation plane with as translation line the set of all lines of* S *incident with* x, *then* S *is a TGQ with base-point* x.

Proof. Follows immediately from Theorem 2.2.1, the first part of Section 6.6, and Theorem 6.6.2. ∎

Corollary 6.6.4 *Let* $S = (\mathcal{P}, \mathcal{B}, \mathbf{I})$ *be a GQ of order* s, $s \neq 1$, *with coregular point* x. *If for at least one line* L *incident with* x *the projective plane* π_L *is Desarguesian, then* S *is a TGQ with base-point* x. *If in particular* s *is odd, then* S *is isomorphic to the classical GQ* $\mathbf{Q}(4, s)$.

Proof. As any line of a Desarguesian plane is a translation line, by Corollary 6.6.3 the GQ S is a TGQ with base-point x. Now let s be odd. Then, by Theorem 3.9.6, we have $S \cong \mathbf{Q}(4, s)$. ∎

Corollary 6.6.5 *Let* $S = (\mathcal{P}, \mathcal{B}, \mathbf{I})$ *be a GQ of order* s, *with* $s \in \{5, 7\}$. *If* S *contains a coregular point, then* $S \cong \mathbf{Q}(4, s)$.

Proof. As there is a unique projective plane of order s, $s \in \{5, 7\}$, by Corollary 6.6.4 we have $S \cong \mathbf{Q}(4, s)$. ∎

A GQ of order 3 is isomorphic either to $\mathbf{Q}(4, 3)$ or to its dual $\mathbf{W}(3)$; see Section 1.8.

Corollary 6.6.6 *Let* $S = (\mathcal{P}, \mathcal{B}, \mathbf{I})$ *be a GQ of order* (s, s^2), $s \neq 1$, *with a regular line* L *for which the dual net* $\mathcal{N}_L^D$ *defined by* L *satisfies the Axiom of Veblen. If* L *is incident with a coregular point* x, *then* S *is a TGQ with base-point* x.

Proof. Immediate from Theorem 2.2.3, the first part of Section 6.6 and Theorem 6.6.2. ∎

Corollary 6.6.7 *Let* $S = (\mathcal{P}, \mathcal{B}, \mathbf{I})$ *be a GQ of order* (s, s^2), s *even, satisfying Property (G) at the line* L. *If* L *is incident with a coregular point* x, *then* S *is a TGQ with base-point* x.

Proof. Immediately from Theorem 6.3.3 and Corollary 6.6.6. ∎

Corollary 6.6.8 *Let* $\mathcal{S}(\mathcal{F})$ *be the* GQ *of order* (q, q^2), q *even, arising from a flock* $\mathcal{F}$. *If the point* (∞) *of* $\mathcal{S}(\mathcal{F})$ *is collinear with a regular point* x, *with* $(\infty) \neq x$, *then* $\mathcal{S}(\mathcal{F})$ *is isomorphic to the classical* GQ $\mathbf{H}(3, q^2)$ *arising from a nonsingular Hermitian variety of* $\mathbf{PG}(3, q^2)$.

Proof. The GQ $\mathcal{S}(\mathcal{F})$ is an EGQ with elation point (∞); see Section 4.6. It follows that the corresponding elation group acts transitively on the points distinct from (∞) which are incident with the line $L = (\infty)x$. Hence L is coregular. By Theorem 6.4.2 and Corollary 6.6.6, $\mathcal{S}(\mathcal{F})$ is a TGQ with base-line L. Now by Johnson's Theorem 5.1.11, $\mathcal{S}(\mathcal{F})$ is isomorphic to $\mathbf{H}(3, q^2)$. ∎

The last application of Theorem 6.6.2 is on subquadrangles.

Corollary 6.6.9 *Let* $\mathcal{S} = (\mathcal{P}, \mathcal{B}, \mathbf{I})$ *be a* GQ *of order* (s, t), $s \neq 1 \neq t$, *having a coregular point* x. *If for a fixed line* L, *with* $x \mathbf{I} L$, *and any two lines* M, N, *with* $M \nsim L$ *and* $L \sim N \nsim M$, *there is a proper subquadrangle* $\mathcal{S}'$ *of* $\mathcal{S}$ *of order* (s, t'), *with* $t' \neq 1$, *containing* L, M, N, *then* $t' = s = \sqrt{t}$, $\mathcal{S}$ *is a TGQ with base-point* x, *and* $\mathcal{S}'$ *is a TGQ with base-point* x. *For* s *even* $\mathcal{S}$ *satisfies Property* (G) *at the flag* (x, L) *and for* s *odd the translation dual* $\mathcal{S}^*$ *of* $\mathcal{S}$ *satisfies Property* (G) *at the flag* (x', L'), *with* x' *the base-point of* $\mathcal{S}^*$ *and* L' *the line of* $\mathcal{S}^*$ *corresponding to the line* L *of* $\mathcal{S}$. *For* s *odd the subquadrangles* $\mathcal{S}'$ *are isomorphic to* $\mathbf{Q}(4, s)$ *and* $\mathcal{S}^*$ *is isomorphic to the dual of a flock* GQ $\mathcal{S}(\mathcal{F})$. *If* s *is even and if* $\mathcal{S}^*$ *is isomorphic to the dual of a flock* GQ $\mathcal{S}(\mathcal{F})$, *then* $\mathcal{S} \cong \mathbf{Q}(5, s)$.

Proof. By Theorem 6.5.1 we have that $t' = s = \sqrt{t}$ and the dual net $\mathcal{N}_L^D$ satisfies the Axiom of Veblen. Hence by Corollary 2.2.4 $\mathcal{N}_L^D \cong \mathbf{H}_s^3$. By Corollary 6.6.6 the GQ $\mathcal{S}$ is a TGQ with base-point x. As $\mathcal{S}'$ contains x, $\mathcal{S}'$ is also a TGQ with base-point x. Let $\mathcal{S} \cong \mathbf{T}(\mathcal{O})$, with $\mathcal{O} = \mathcal{O}(n, 2n, q)$, and where $\pi \in \mathcal{O}$ corresponds to L. Now we show that the condition on the existence of subGQs implies that for any two spaces $\pi_i, \pi_j \in \mathcal{O} \backslash \{x\}$, with $i \neq j$, the $(3n - 1)$-dimensional space $\langle \pi, \pi_i, \pi_j \rangle$ contains exactly $q^n + 1$ elements of $\mathcal{O}$. Let $\mathbf{PG}(3n, q)$ be any $3n$-dimensional space in the $\mathbf{PG}(4n, q)$ containing $\mathbf{T}(\mathcal{O})$, which contains $\langle \pi, \pi_i, \pi_j \rangle$ but is not contained in $\mathbf{PG}(4n - 1, q) = \langle \mathcal{O} \rangle$. Next, let μ be an n-dimensonal space in $\mathbf{PG}(3n, q)$ containing π, with $\mu \nsubseteq \langle \pi, \pi_i, \pi_j \rangle$, let η be an n-dimensional space in $\mathbf{PG}(3n, q)$ containing π_i, with $\eta \nsubseteq \langle \pi, \pi_i, \pi_j \rangle$ and $\mu \cap \eta = \emptyset$, and let ρ be the n-dimensional space containing π_j and intersecting μ and η.

By the condition on the existence of subquadrangles, there is a subGQ of order $q^n = s$ of $\mathbf{T}(\mathcal{O})$ containing the lines π, η, μ, ρ. It easily follows that this subGQ contains all points of $\langle \pi, \pi_i, \rho \rangle \backslash \mathbf{PG}(4n - 1, q) = \mathbf{AG}(3n, q)$. Consequently the $q^n + 1$ lines of the subGQ incident with (∞) belong to a common $\mathbf{PG}(3n - 1, q)$, which clearly is $\langle \pi, \pi_i, \pi_j \rangle$. So $\mathcal{O}$ is good at π, and consequently, by Theorem 5.1.2, the translation dual $\mathcal{S}^*$ of $\mathcal{S}$ satisfies Property (G) at the flag (x', L'), with x' the base-point of $\mathcal{S}^*$ and L' the line of $\mathcal{S}^*$ corresponding to the line L of $\mathcal{S}$. So, by Theorem 5.1.5, for s even the GQ $\mathcal{S}$ satisfies Property (G) at the flag (x, L).

Now let s be odd. Then, by Theorem 5.1.9, all subquadrangles $\mathcal{S}'$ are isomorphic to $\mathbf{Q}(4, s)$. By Corollary 5.1.4 the GQ $\mathcal{S}^*$ is isomorphic to the point-line dual of a flock GQ $\mathcal{S}(\mathcal{F})$.

Finally, let s be even and assume that $\mathcal{S}^*$ is isomorphic to the point-line dual of a flock GQ $\mathcal{S}(\mathcal{F})$. Then, by Theorem 5.1.11 the GQ $\mathcal{S}^*$, and consequently also $\mathcal{S}$, is isomorphic to $\mathbf{Q}(5, s)$. ∎

Chapter 7

Ovoids and Subquadrangles

The first part of this chapter elaborates on ovoids of the classical GQ $\mathbf{Q}(4,q)$. In particular subtended ovoids and translation ovoids will be considered. The second part of Chapter 7 deals with subGQs of order q of TGQs of order (q, q^2). Also, there are sections on subquadrangles of flock GQs, on TGQs of order (q, q^2) having at least one subGQ of order q and on EGQs admitting certain subGQs.

7.1 Ovoids of $\mathbf{Q}(4,q)$

An *ovoid* $\mathcal{O}$ of the GQ $\mathbf{Q}(4,q)$ is a set of $q^2 + 1$ points, no two of which are collinear; equivalently, $\mathcal{O}$ has a point in common with each line of $\mathbf{Q}(4,q)$.

If the GQ $\mathbf{Q}(4,q)$ is contained in $\mathbf{PG}(4,q)$ as a nonsingular quadric $\mathbf{Q}$, and if the hyperplane $\mathbf{PG}(3,q)$ intersects the quadric $\mathbf{Q}$ in an elliptic quadric $\mathcal{O}$, then $\mathcal{O}$ is an ovoid of $\mathbf{Q}(4,q)$; such an ovoid is called *classical*. For q an even power of 2 no other ovoids of $\mathbf{Q}(4,q)$ are known. For $q = 2^{2h+1}$, with $h \geq 1$, one other type of ovoid is known; its projection from the nucleus n of $\mathbf{Q}$ onto a hyperplane not containing n, is the *Tits ovoid*; see Thas [158] (see also Subsection 12.3.2 below). In fact, in Thas [158] it is shown that with any ovoid of $\mathbf{Q}(4,q)$ with q even, corresponds, by projection, an ovoid of $\mathbf{PG}(3,q)$, and, conversely, with any ovoid of $\mathbf{PG}(3,q)$ with q even, corresponds an ovoid of $\mathbf{Q}(4,q)$. For q odd, Kantor [82] constructed two types of nonclassical ovoids. Starting from a Kantor-Knuth flock $\mathcal{F}$, the construction in Section 4.2 defines an ovoid $\mathcal{O}$ of $\mathbf{Q}(4,q)$. By Theorem 4.4.6 these ovoids of $\mathbf{Q}(4,q)$, q odd, are characterized by the property that they are the union of q conics which are mutually tangent at a common point.

Such an ovoid is classical if and only if the corresponding automorphism σ of $\mathbf{GF}(q)$ is the identity. An ovoid of $\mathbf{Q}(4, q)$ arising from a Kantor-Knuth flock will be called a *Kantor-Knuth ovoid* and will be denoted by $\mathcal{K}_1$. If $q = 3^{2h+1}$, with $h \geq 1$, then Kantor [82] proves that $\mathbf{Q}(4, q)$ has an ovoid which arises from the Ree-Tits ovoid on the nonsingular quadric in $\mathbf{PG}(6, q)$; such an ovoid is always nonclassical. An ovoid of $\mathbf{Q}(4, q)$ of this type will be called a *Kantor-Ree-Tits ovoid* and will be denoted by $\mathcal{K}_2$. Further, Penttila and Williams [131], with a slight assist from a computer, discovered a new ovoid of $\mathbf{Q}(4, 3^5)$. This ovoid will be called the *Penttila-Williams ovoid*. Finally we mention that for q odd one other class of ovoids of $\mathbf{Q}(4, q)$ is known; see Section 7.2.

With each ovoid $\mathcal{O}$ of $\mathbf{Q}(4, q)$ corresponds a spread of the dual GQ $\mathbf{W}(q)$. And with a spread of $\mathbf{W}(q)$ corresponds, by the standard argument, a translation plane π of order q^2; see e.g. Section 3.1 of Dembowski [45]. The plane π is Desarguesian if and only if $\mathcal{O}$ is classical. If $\mathcal{O}$ is of Tits type, then π is the Lüneburg plane; see Thas [158]. To an ovoid $\mathcal{K}_1$ corresponds a Knuth semifield plane; see Kantor [82].

7.2 Subquadrangles and Ovoids

Let $\mathcal{S}'$ be a subGQ of order (s, t') of the GQ $\mathcal{S}$ of order (s, t), with $t' < t$. If z is a point of $\mathcal{S}$ which is not contained in $\mathcal{S}'$, then the $1 + st'$ points of $\mathcal{S}'$ collinear with z form an ovoid $\mathcal{O}_z$ of $\mathcal{S}'$; see 2.2.1 of Payne and Thas [128] (or §1.1). Such an ovoid $\mathcal{O}_z$ of $\mathcal{S}'$ is called a *subtended ovoid*. For example, if we consider the subGQ $\mathbf{Q}(4, q)$ of the GQ $\mathbf{Q}(5, q)$, then $\mathcal{O}_z$ is an elliptic quadric on the quadric $\mathbf{Q}$ defining $\mathbf{Q}(4, q)$.

Consider an egg $\mathcal{O} = \mathcal{O}(n, 2n, q)$ and assume that $\mathcal{O}(n, 2n, q)$ is good at its element π. By Theorem 5.1.9 the TGQ $\mathbf{T}(\mathcal{O})$ contains exactly $q^{3n} + q^{2n}$ subGQs of order q^n containing the line π of $\mathbf{T}(\mathcal{O})$, and for q odd each of these subGQs is isomorphic to the classical GQ $\mathbf{Q}(4, q^n)$. Fix such a subGQ $\mathcal{S}' \cong \mathbf{Q}(4, q^n)$. Then by Lavrauw [93, 95] and Theorem 10.6.3 of K. Thas [206], up to isomorphism the subtended ovoid $\mathcal{O}_z$ of $\mathcal{S}'$ is independent of the choice of the point z. Further, Lavrauw [93, 95] proves that up to isomorphism the ovoid $\mathcal{O}_z$ is independent of the choice of the subGQ $\mathcal{S}'$.

Let $\mathcal{O}$ be an ovoid of $\mathbf{Q}(4, q)$, let $x \in \mathcal{O}$ and let L be a line of $\mathbf{Q}(4, q)$ incident with x. We call $\mathcal{O}$ a *translation ovoid with respect to the flag* (x, L) if there is an automorphism group $G_{x,L}$ of $\mathbf{Q}(4, q)$ which fixes $\mathcal{O}$, which fixes x linewise and L pointwise, and which acts transitively on the points of $(\mathcal{O} \setminus \{x\}) \cap y^\perp$ for each $y \neq x$ incident with L. We call $\mathcal{O}$ a *translation ovoid with respect to the point* x if $\mathcal{O}$ is a translation ovoid with respect to the flag (x, L), for each line L incident with x. In Bloemen, Thas and Van Maldeghem [14] it is shown that the ovoid $\mathcal{O}$ of $\mathbf{Q}(4, q)$ is a translation ovoid with respect to the point $x \in \mathcal{O}$ if and only if $\mathbf{Q}(4, q)$ admits an automorphism group of size q^2 which fixes $\mathcal{O}$ and all lines incident with x. It is easy to see that the ovoids $\mathcal{O}_z$ of $\mathcal{S}'$ in the previous paragraph are translation ovoids of $\mathcal{S}'$ with respect to the point (∞). For more on translation ovoids we refer to Bloemen, Thas and Van Maldeghem [14] and Lunardon [100].

Let $\mathcal{O}(n, 2n, q) = \mathcal{O}$ be a good egg in $\mathbf{PG}(4n-1, q)$, with q odd. By Corollary 5.1.4 and Theorem 4.7.2 the TGQ $\mathbf{T}(\mathcal{O}^*)$, with $\mathcal{O}^*$ the translation dual of $\mathcal{O}$, is the point-line dual of a flock GQ $\mathcal{S}(\mathcal{F})$, with $\mathcal{F}$ a semifield flock. Conversely, if $\mathcal{F}$ is a semifield flock, then $\mathcal{S}(\mathcal{F})$ is the point-line dual of a TGQ $\mathbf{T}(\mathcal{O}^*)$, where $\mathcal{O}$ is good (see Theorem 4.7.2). So with any semifield flock corresponds an ovoid of $\mathbf{Q}(4, s)$, which is a translation ovoid with respect to some point. Also, with any translation ovoid with respect to some point of $\mathbf{Q}(4, s)$ corresponds a semifield flock. For details we refer to Lunardon [100]. In the next section we will show how to construct, in a purely geometrically way, the semifield flock from the translation ovoid, and conversely.

In Thas [176] it was conjectured that the translation ovoids of $\mathbf{Q}(4, 3^h)$, $h > 2$, arising from the Ganley flocks (See Section 4.5) are new. This conjecture was proved by Thas and Payne [190], and the new ovoids are called *Thas-Payne ovoids*. Together with the ovoids of $\mathbf{Q}(4, q)$ mentioned in Section 7.1, these are the only known ovoids of the classical GQ $\mathbf{Q}(4, q)$.

7.3 Translation Ovoids and Semifield Flocks

The following construction is taken from Thas [181], see also Lunardon [100].

The plane $\mathbf{PG}(2, q^n)$, with $q = p^h$ and p prime, can be represented in $\mathbf{PG}(3n - 1, q)$ in such a way that points of the plane become $(n - 1)$-dimensional subspaces belonging to an $(n - 1)$-spread $\mathcal{T}$ of $\mathbf{PG}(3n - 1, q)$ and lines of the plane become $(2n - 1)$-dimensional subspaces of $\mathbf{PG}(3n - 1, q)$; see also Sections 3.6 and 4.5. In each of these $(2n - 1)$-dimensional subspaces corresponding to lines an $(n - 1)$-spread is induced by $\mathcal{T}$. Let $\mathcal{C}$ be a nonsingular conic of $\mathbf{PG}(2, q^n)$. With $\mathcal{C}$ corresponds a set $\mathcal{C}'$ of $q^n + 1$

$(n-1)$-dimensional subspaces of $\mathbf{PG}(3n-1,q)$. This set $\mathcal{C}'$ is a pseudo-conic of $\mathbf{PG}(3n-1,q)$.

Let π be a $(2n-1)$-dimensional subspace of $\mathbf{PG}(3n-1,q)$ having no point in common with the elements of $\mathcal{C}'$. Now we consider the GQ $\mathbf{T}_2(\mathcal{C}) \cong \mathbf{Q}(4,q^n)$ of Tits defined by $\mathcal{C}$. Then $\mathbf{T}_2(\mathcal{C}) \cong \mathbf{T}(\mathcal{C}')$, where $\mathcal{C}$ corresponds to $\mathcal{C}'$ and the points of Type (i) of $\mathbf{T}_2(\mathcal{C})$, which are the points not collinear with the point (∞) of $\mathbf{T}_2(\mathcal{C})$, correspond to the points of $\mathbf{PG}(3n,q) \setminus \mathbf{PG}(3n-1,q)$ where $\mathbf{PG}(3n,q)$ is the space containing $\mathbf{T}(\mathcal{C}')$; see Sections 3.6 and 4.5. Let $\widetilde{\pi}$ be a $2n$-dimensional subspace of $\mathbf{PG}(3n,q)$ containing π, but not contained in $\mathbf{PG}(3n-1,q)$. Then $(\widetilde{\pi} \setminus \mathbf{PG}(3n-1,q)) \cup \{(\infty)\}$ is an ovoid of the GQ $\mathbf{T}(\mathcal{C}')$, so defines an ovoid of $\mathbf{T}_2(\mathcal{C})$, and hence defines an ovoid of $\mathbf{Q}(4,q^n)$. Also, this ovoid of $\mathbf{Q}(4,q^n)$ is a translation ovoid with respect to the point which corresponds to the point (∞) of $\mathbf{T}(\mathcal{C}')$ (respectively, $\mathbf{T}_2(\mathcal{C})$). By Lunardon [100] any ovoid of $\mathbf{Q}(4,q^n)$ which is translation with respect to some point can be obtained in this way.

Now we dualize in $\mathbf{PG}(3n-1,q)$; choose e.g. a polarity in $\mathbf{PG}(3n-1,q)$ and consider the polar spaces of the subspaces of $\mathbf{PG}(3n-1,q)$. With $\mathcal{C}'$ corresponds a set $\hat{\mathcal{C}}'$ of q^n+1 $(2n-1)$-dimensional subspaces of $\mathbf{PG}(3n-1,q)$; $\hat{\mathcal{C}}'$ can be considered as a dual conic $\hat{\mathcal{C}}$ in a $\mathbf{PG}(2,q^n)$. With π corresponds an $(n-1)$-dimensional subspace $\hat{\pi}$ of $\mathbf{PG}(3n-1,q)$ which is skew to all elements of $\hat{\mathcal{C}}'$. By Section 4.5 this subspace $\hat{\pi}$ defines a semifield flock, and, conversely, any semifield flock can be obtained in this way.

Hence any ovoid $\mathcal{O}$ of $\mathbf{Q}(4,s)$ which is a translation ovoid with respect to some point, defines a semifield flock $\mathcal{F}_\mathcal{O}$, and, conversely, any semifield flock $\mathcal{F}$ of the quadratic cone $\mathcal{K}$ in $\mathbf{PG}(3,s)$ defines an ovoid $\mathcal{O}_\mathcal{F}$ of $\mathbf{Q}(4,s)$ which is a translation ovoid with respect to some point.

Let $\mathcal{F}$ be a semifield flock. Then the point-line dual of the GQ $\mathcal{S}(\mathcal{F})$ is a TGQ $\mathbf{T}(\mathcal{O}^*)$ of order (s,s^2), where the egg $\mathcal{O}$ is good. By Section 7.2, $\mathcal{O}$ defines a subtended ovoid $\mathcal{O}_z$ of the subquadrangle $\mathcal{S}' \cong \mathbf{Q}(4,s)$. By Lavrauw [93, 95] this ovoid $\mathcal{O}_z$ is isomorphic to the translation ovoid $\mathcal{O}_\mathcal{F}$ of $\mathbf{Q}(4,s)$, which corresponds to the flock $\mathcal{F}$.

7.4 Coordinates of the Known Nonclassical Ovoids of $\mathbf{Q}(4,q)$

Suppose that $\mathbf{Q}(4,q)$ has equation $X_2^2 = X_0X_1 + X_3X_4$. We now describe the known ovoids of $\mathbf{Q}(4,q)$ in coordinates.

First suppose that q is even.

TITS OVOIDS [219]. Define $\gamma : t \mapsto t^{2^{e+1}}$, with $t \in \mathbf{GF}(q)$ and $q = 2^{2e+1}$, $e \geq 1$. Then the *Tits ovoids* $\mathcal{TI}$ are described as follows (see also §12.3.2):

$$\mathcal{TI} = \{(0,0,0,1,0)\}$$
$$\cup \{(x,y,x^{\gamma/2+1} + y^{\gamma/2}, xy + x^{\gamma+2} + y^{\gamma}, 1) \parallel x,y \in \mathbf{GF}(q)\}.$$

Next suppose that q is odd.

KANTOR OVOIDS $\mathcal{K}_1$ [82]. Let n be a nonsquare of $\mathbf{GF}(q)$ and let $\sigma \neq 1$ be an arbitrary automorphism of $\mathbf{GF}(q)$. Then the Kantor ovoids $\mathcal{K}_1$ are given by

$$\mathcal{K}_1 = \{(0,0,0,1,0)\} \cup \{nx^{\sigma}, x, y, y^2 - nx^{\sigma+1}, 1) \parallel x,y \in \mathbf{GF}(q)\}.$$

KANTOR OVOIDS $\mathcal{K}_2$ [82]. Define $\phi : x \mapsto x^{3^h}$, with $x \in \mathbf{GF}(q)$ and $q = 3^{2h-1}$, $h \geq 2$. Then $\mathcal{K}_2$ is described as follows:

$$\mathcal{K}_2 = \{(0,0,0,1,0)\} \cup \{(x^{2\phi+3}+y^{\phi}, x, y, y^2-x^{2\phi+4}-xy^{\phi}, 1) \parallel x,y \in \mathbf{GF}(q)\}.$$

THAS-PAYNE OVOIDS $\mathcal{TP}$ [190]. Put $q = 3^h$, $h \geq 3$, and let n be a non-square of $\mathbf{GF}(q)$. Then $\mathcal{TP}$ is given by

$$\mathcal{TP} = \{(0,0,0,1,0)\} \cup \{(xn + (xn^{-1})^{1/9} + y^{1/3}, x, y, y^2$$
$$- x(xn + (xn^{-1})^{1/9} + y^{1/3}), 1) \in \mathbf{Q}(4,q) \parallel x,y \in \mathbf{GF}(q)\}.$$

If h is odd, -1 is a nonsquare, and then the ovoid can be written as

$$\mathcal{TP} = \{(0,0,0,1,0)\} \cup \{(-x - x^{1/9} + y^{1/3}, x, y, y^2 + x^2 + x^{10/9} - xy^{1/3}, 1)$$
$$\in \mathbf{Q}(4,q) \parallel x,y \in \mathbf{GF}(q)\}.$$

THE PENTTILA-WILLIAMS OVOID $\mathcal{PW}$ [131]. The Penttila-Williams ovoid $\mathcal{PW}$ of $\mathbf{Q}(4,3^5)$ is given by

$$\mathcal{PW} = \{(0,0,0,1,0)\} \cup \{(x^9 + y^{81}, x, y, y^2 - x^{10} - xy^{81}, 1) \parallel x,y \in \mathbf{GF}(3^5)\}.$$

7.5   Subquadrangles of $T(\mathcal{O})$, with $\mathcal{O}$ Good: the Even Case

First we discuss briefly the connection between GQs with a regular point and covers of the net associated with the regular point. In particular, we shall consider subGQs of order s of GQs of order (s, s^2) in this context.

Let $\mathcal{N} = (\mathcal{P}, \mathcal{B}, \mathrm{I})$ be a net of order k and degree r. An m-fold cover of $\mathcal{N}$ is a geometry $\overline{\mathcal{N}} = (\overline{\mathcal{P}}, \overline{\mathcal{B}}, \overline{\mathrm{I}})$ such that there exists a surjective map

$$\gamma : \overline{\mathcal{P}} \to \mathcal{P},$$

satisfying:

(i) for any $x \in \mathcal{P}$, $x^{\gamma^{-1}}$ consists of m pairwise noncollinear points;

(ii) for any pair $\{x, y\}$ of collinear points of $\mathcal{N}$, $\{x, y\}^{\gamma^{-1}}$ consists of m disjoint pairs of collinear points;

(iii) for any pair $\{x, y\}$ of noncollinear points of $\mathcal{N}$, $\{x, y\}^{\gamma^{-1}}$ is a set of pairwise noncollinear points of $\overline{\mathcal{N}}$;

(iv) for any line L of $\mathcal{N}$, if $\mathcal{P}_L = \{x \in \mathcal{P} \parallel x \mathrm{I} L\}$, then $\mathcal{P}_L^{\gamma^{-1}}$ consists of the points of m disjoint lines of $\overline{\mathcal{N}}$.

The map γ is called a *covering map*. A point or line of $\overline{\mathcal{N}}$ is said to be a *cover* of its image under γ.

Now let $\mathcal{S} = (\mathcal{P}, \mathcal{B}, \mathrm{I})$ be a GQ of order (s, t) with a regular point x, and let $\mathcal{N}_x$ be the associated net of order s and degree $t + 1$. The geometry $\overline{\mathcal{N}_x}$ with

- POINT SET $\mathcal{P} \setminus x^{\perp}$,

- LINE SET $\mathcal{B} \setminus \{L \in \mathcal{B} \parallel x \mathrm{I} L\}$ and

- INCIDENCE inherited from $\mathcal{S}$,

is a t-fold cover of $\mathcal{N}_x$. The covering map γ maps the point $y \in \mathcal{P} \setminus x^{\perp}$ to the point $\{x, y\}^{\perp}$ of $\mathcal{N}_x$. The cover has the additional property that for a nonincident point-line pair $\{y, L\}$ of $\overline{\mathcal{N}_x}$, either $y^{\gamma} = z^{\gamma}$ with $z \mathrm{I} L$ and $z \in \{x, y\}^{\perp\perp}$, or y is collinear with a unique point of $\overline{\mathcal{N}_x}$ incident with L. In general, given a net of order s and degree $t + 1$ with such a t-fold cover, it is possible to construct a GQ of order (s, t), which in the above case is a reconstruction of $\mathcal{S}$.

Suppose that $\mathcal{N}_x$ has a proper subnet $\mathcal{N}'_x$ of order s', with $s' \geq 2$, and degree $t + 1$. Then the subset $\mathcal{P}'$ of $\mathcal{P}$ consisting of x, the lines of $\mathcal{N}'_x$ and $y \in \mathcal{P} \setminus x^\perp$ such that $\{x, y\}^\perp$ is a point of $\mathcal{N}'_x$, is the point set of a subGQ $\mathcal{S}'$ of order (s', t) of $\mathcal{S}$. The lines of $\mathcal{S}'$ are the lines of $\mathcal{S}$ incident with at least one (and hence $s' + 1$) points of $\mathcal{P}'$. The point x is a regular point of $\mathcal{S}'$ with associated net $\mathcal{N}'_x$. It follows that $t = s'$, $t^2 = s$ and so the subnet $\mathcal{N}'_x$ is an affine plane of order t. For more details see K. Thas [200].

Let $\mathcal{S}$ be a GQ of order (q^2, q) with a regular point x such that the associated dual net $\mathcal{N}_x^D$ is isomorphic to $\mathbf{H}_q^3$. In Theorems 5.1.2, 6.2.1, 6.3.3 and 6.3.4 examples were given. In the next results, due to Brown and Thas [28], subGQs of such GQs will be considered.

Theorem 7.5.1 *Let* $\mathcal{S} = (\mathcal{P}, \mathcal{B}, \mathbf{I})$ *be a* GQ *of order* (q^2, q), $q \neq 1$, *with a regular point* x *such that the associated dual net* $\mathcal{N}_x^D$ *is isomorphic to* $\mathbf{H}_q^3$. *Then* x *is contained in the maximal number* $q^3 + q^2$ *subGQs of order* q. *If* $\mathcal{S}' = (\mathcal{P}', \mathcal{B}', \mathbf{I}')$ *is any subGQ of order* q *not containing* x, *then* $\mathcal{S}'$ *is isomorphic to the point-line dual of* $\mathbf{T}_2(\mathcal{O})$ *for some oval* $\mathcal{O}$ *of* $\mathbf{PG}(2, q)$.

Proof. Suppose that the net $(\mathbf{H}_q^3)^D \cong \mathcal{N}_x$ is constructed from the line L of $\mathbf{PG}(3, q)$; that is, by taking as POINTS the lines of $\mathbf{PG}(3, q)$ not meeting L, as LINES the points of $\mathbf{PG}(3, q) \setminus L$, and as INCIDENCE the incidence from $\mathbf{PG}(3, q)$. Then each plane π of $\mathbf{PG}(3, q)$ not containing L gives rise to an affine plane subnet of $(\mathbf{H}_q^3)^D$, which is the dual of π, with the point $\pi \cap L$ and the lines of π on $\pi \cap L$ removed. From the foregoing we see that π gives rise to a subGQ of order q. This gives $q^3 + q^2$ distinct subGQs of order q containing x, the maximal number by Lemma 5.1.8.

Now suppose that $\mathcal{S}' = (\mathcal{P}', \mathcal{B}', \mathbf{I}')$ is a subGQ of order q of $\mathcal{S}$, not containing x. The set $\mathcal{P} \setminus x^\perp$ is the point set of a q-fold cover of the net $\mathcal{N}_x \cong (\mathbf{H}_q^3)^D$ with covering map γ taking the point $y \in \mathcal{P} \setminus x^\perp$ to the point $\{x, y\}^\perp$ of $\mathcal{N}_x \cong (\mathbf{H}_q^3)^D$. The subGQ $\mathcal{S}'$ contains a unique line, say M, incident with x; see Section 2.2 of Payne and Thas [128]. The points of $\mathcal{S}$ incident with M, but distinct from x, define a parallel class of $(\mathbf{H}_q^3)^D$ the elements of which are contained in a plane of $\mathbf{PG}(3, q)$ containing L which we will denote by M^γ. The subGQ $\mathcal{S}'$ contains $q + 1$ points of M. These points define a set $\mathcal{O}$ of size $q + 1$ in the plane M^γ, and $\mathcal{O} \cap L = \emptyset$. Consider a line N of $\mathcal{S}'$ not concurrent with M. If $\mathcal{P}_N = \{z \in \mathcal{P} \parallel zIN\}$, then $\mathcal{P}_N^\gamma$ defines a point of $\mathbf{PG}(3, q)$ not in the plane M^γ. Further, since no two lines of $\mathcal{S}'$ are incident with a common point of $x^{\perp}$ not incident with M, it follows that γ gives a one-to-one correspondence between the q^3 lines of $\mathcal{S}'$ not concurrent with M and the q^3 points of $\mathbf{PG}(3,q) \setminus M^{\gamma}$. Each point of $\mathcal{S}'$ not incident with M is collinear with a unique point of M and so, under the map γ, defines a line of $\mathbf{PG}(3,q)$ meeting M^{γ} in a unique point of $\mathcal{O}$. Since no two lines of $\mathcal{S}'$ are concurrent in a point of $x^{\perp}$ not incident with M, it must also be the case that no two points of $\mathcal{S}'$, not incident with M, correspond under γ to the same line of $\mathbf{PG}(3,q)$. Thus γ gives a one-to-one correspondence between the set $\mathcal{P}' \setminus \mathcal{P}_M$, with $\mathcal{P}_M = \{z \in \mathcal{P} \parallel z \mathrm{I} M\}$, and the lines of $\mathbf{PG}(3,q)$ not in M^{γ} meeting M^{γ} in a point of $\mathcal{O}$. It is now a straightforward exercise to verify that $\mathcal{O}$ is an oval and that $\mathcal{S}'$ is isomorphic to the point-line dual of $\mathbf{T}_2(\mathcal{O})$.   ∎

We have an immediate corollary.

Corollary 7.5.2 *Let* $\mathcal{S} = (\mathcal{P}, \mathcal{B}, \mathrm{I})$ *be a GQ of order* (q^2, q), *q odd, with a regular point* x *such that the associated dual net* $\mathcal{N}_x^D$ *is isomorphic to* $\mathbf{H}_q^3$. *If* $\mathcal{S}' = (\mathcal{P}', \mathcal{B}', \mathrm{I}')$ *is any subGQ of order* q *not containing* x, *then* $\mathcal{S}'$ *is isomorphic to the classical GQ* $\mathbf{W}(q)$.

Proof. This follows from Theorem 7.5.1, Theorem 1.5.2(i), the remark which follows Theorem 1.5.2, and Theorem 1.4.1.   ∎

Suppose that $\mathcal{S} = (\mathcal{P}, \mathcal{B}, \mathrm{I})$ is a TGQ of order (s, s^2), $s \neq 1$, with $\mathcal{S} = \mathbf{T}(\mathcal{O})$, where $\mathcal{O}$ is good at its element π. By Theorem 6.2.1 the dual net $\mathcal{N}_{\pi}^D$ is isomorphic to $\mathbf{H}_s^3$. Suppose that $\mathbf{H}_s^3$ is constructed from the line L in $\mathbf{PG}(3,s)$. We identify $\mathbf{H}_s^3$ and $\mathcal{N}_{\pi}^D$, and assume that γ is the covering map from $\mathcal{P} \setminus x^{\perp}$ to the point set of $(\mathbf{H}_s^3)^D$.

Let G be the translation group of $\mathcal{S}$ about (∞). Then G has order s^4, fixes each line incident with (∞) and acts regularly on the elements of $\mathcal{P} \setminus x^{\perp}$. The group G induces an automorphism group G' of the dual net $\mathcal{N}_{\pi}^D = \mathbf{H}_s^3$. The identity mapping of $\mathcal{N}_{\pi}^D$ is induced by all symmetries of $\mathcal{S}$ about π, and so G' has size s^3. The group G' is induced by a collineation group of $\mathbf{PG}(3,s)$ fixing the line L; this collineation group will also be denoted by G'. If we denote the plane of $\mathbf{PG}(3,s)$ corresponding to the elements of $\mathcal{O} \setminus \{\pi\}$ by $(\infty)^{\gamma}$, then each element of G' fixes all points of $(\infty)^{\gamma}$, so is an axial collineation with axis $(\infty)^{\gamma}$. If $\theta \in G$ fixes a line of $\pi^{\perp}$ not incident with (∞), then it follows that θ is a symmetry about π. Hence G' acts regularly on the points of $\mathbf{PG}(3,s) \setminus (\infty)^{\gamma}$ and must be the group of elations of $\mathbf{PG}(3,s)$ with axis $(\infty)^{\gamma}$.

By Theorem 5.1.9 $\mathcal{S} = \mathbf{T}(\mathcal{O})$ contains exactly $s^3 + s^2$ subGQs of order s containing the line π. Now suppose that $\mathcal{S}' = (\mathcal{P}', \mathcal{B}', \mathbf{I}')$ is a subGQ of order s of $\mathcal{S}$ not containing the point (∞); then π cannot be a line of $\mathcal{S}'$. Consequently, by applying Theorem 7.5.1 it follows that $\mathcal{S}' \cong \mathbf{T}_2(\mathcal{O}')$, for some oval $\mathcal{O}'$ of $\mathbf{PG}(2, s)$, where $\mathbf{PG}(2, s)$ corresponds to the point m of $\mathcal{S}'$ which is incident with π; this plane $\mathbf{PG}(2, s)$ will be denoted by m^γ.

Lemma 7.5.3 *Let* $\mathcal{S} = (\mathcal{P}, \mathcal{B}, \mathbf{I})$ *be a TGQ of order* (s, s^2), $s \neq 1$, *with* $\mathcal{S} = \mathbf{T}(\mathcal{O})$, *where* $\mathcal{O}$ *is good at its element* π. *Suppose that* $\mathcal{S}' = (\mathcal{P}', \mathcal{B}', \mathbf{I}')$ *is a subGQ of order* s *of* $\mathcal{S}$ *not containing the point* (∞). *Then* $\mathcal{S}'$ *is isomorphic to the classical GQ* $\mathbf{Q}(4, s)$.

Proof. By the foregoing we have $\mathcal{S}' \cong \mathbf{T}_2(\mathcal{O}')$. Now we show that $\mathcal{O}'$ is a conic in m^γ, with m the unique point of $\mathcal{S}'$ incident with the line π. Let y be a point of $(\infty)^\gamma$ not on the line L and let φ be any nontrivial elation of $\mathbf{PG}(3, s)$ with axis $(\infty)^\gamma$ and center y. Let $\overline{\varphi}$ be an automorphism of $\mathcal{S}$ that induces φ in $\mathbf{PG}(3, s)$. Under γ the subGQ $\mathcal{S}'' = \mathcal{S}'^\varphi$ corresponds to $\mathbf{T}_2(\mathcal{O}'^\varphi)$, where $\mathcal{O}'^\varphi$ is an oval in the plane $(m^\gamma)^\varphi$. Under the action of the symmetry group about π the subGQ $\mathcal{S}''$ gives rise to s subGQs $\mathcal{S}'' = \mathcal{S}_1'', \mathcal{S}_2'', \ldots, \mathcal{S}_s''$, which intersect pairwise in the lines of $\mathcal{S}$ corresponding to the points of $\mathcal{O}'^\varphi$ and the points incident with these lines; see Sections 2.2 and 2.3 of Payne and Thas [128]. We now consider the intersection between $\mathcal{S}'$ and $\mathcal{S}_i''$. The only possible lines that lie in both $\mathcal{S}'$ and $\mathcal{S}_i''$, for some i, correspond to lines of $\mathbf{PG}(3, s)$ containing a point of $\mathcal{O}'$ and a point of $\mathcal{O}'^\varphi$. The subGQ $\mathcal{S}'$ contains $s^2 + 2s + 1$ such lines which are partitioned by the $\mathcal{S}_i''$. Now if $\mathcal{S}'$ and $\mathcal{S}_i''$ share such a line, then from Sections 2.2 and 2.3 of Payne and Thas [128] it easily follows that they share either $s + 1$ or $2s + 2$ such lines. It follows that there exists a $j \in \{1, 2, \ldots, s\}$ such that $\mathcal{S}'$ and $\mathcal{S}_j''$ share $2s + 2$ lines, that is, they meet in a grid. With this grid corresponds a regulus of $\mathbf{PG}(3, s)$, intersecting m^γ in $\mathcal{O}'$. It follows that $\mathcal{O}'$ must be a conic, and so $\mathcal{S}' \cong \mathbf{T}_2(\mathcal{O}') \cong \mathbf{Q}(4, s)$. ∎

We now prove that, when s is even, then, given the hypotheses of Lemma 7.5.3, the GQ $\mathcal{S}$ must be the classical GQ $\mathbf{Q}(5, s)$; this theorem is due to Brown and Thas [28]. Note that when s is odd the equivalent result is Theorem 7.6.2.

The proof of this theorem will make use of the following theorem by Thas and Payne [190] on subtended ovoids.

Theorem 7.5.4 *Let $S = (\mathcal{P}, \mathcal{B}, \mathbf{I})$ be a GQ of order (s, s^2), s even, having a subGQ S' isomorphic to $\mathbf{Q}(4, s)$. If in S' each ovoid $\mathcal{O}_x$ consisting of all the points collinear with a given point x of $S \setminus S'$ is an elliptic quadric, then S is isomorphic to $\mathbf{Q}(5, s)$.*   ∎

Remark 7.5.5 The equivalent theorem was proved in the odd case by Brown [22] and a proof valid for both s odd and even may be found in Brouns, Thas and Van Maldeghem [19] and in Brown [24].

We are now equipped to prove the classification theorem.

Theorem 7.5.6 *Let $S = (\mathcal{P}, \mathcal{B}, \mathbf{I})$ be a TGQ of order (s, s^2), s even, with $S = \mathbf{T}(\mathcal{O})$, where $\mathcal{O}$ is good at the element π. Suppose that $S' = (\mathcal{P}', \mathcal{B}', \mathbf{I}')$ is a subGQ of order s of S not containing the point (∞). Then S is isomorphic to the classical GQ $\mathbf{Q}(5, s)$.*

Proof. By Lemma 7.5.3, if m is the point incident with the line π and contained in S', then $S' \cong \mathbf{T}_2(\mathcal{C})$ where $\mathcal{C}$ is a conic in the plane m^γ of $\mathbf{PG}(3, s)$.

Let y be a point of $\mathcal{P} \setminus \mathcal{P}'$ such that $y \not\sim m$, and let $\mathcal{O}_y$ be the ovoid of S' subtended by y. We now show that $\mathcal{O}_y$ corresponds to an elliptic quadric of $\mathbf{Q}(4, s) \cong \mathbf{T}_2(\mathcal{C})$. Let φ be a nontrivial elation of $\mathbf{PG}(3, s)$ with axis $(\infty)^\gamma$ and center on the line L of $\mathbf{PG}(3, s)$. Let $\overline{\varphi}$ be an automorphism of S that induces φ in $\mathbf{PG}(3, s)$. By Sections 2.2 and 2.3 of Payne and Thas [128] there are three possibilities for $S' \cap S'^{\overline{\varphi}}$:

   (i) a point u, the $s + 1$ lines of S' and $S'^{\overline{\varphi}}$ incident with u and the points incident with these lines;

   (ii) an ovoid of S' and $S'^{\overline{\varphi}}$, and

   (iii) a grid.

Hence the subGQs S' and $S'^{\overline{\varphi}}$ can share either $0, 1, 2$ or $s + 1$ lines incident with the point m. Consequently, $\mathcal{C}$ and $\mathcal{C}^\varphi$ may share $0, 1, 2$ or $s + 1$ points. However, φ cannot fix $\mathcal{C}$ so the possible intersection numbers are $0, 1$ and 2. As any nontrivial elation of $\mathbf{PG}(3, s)$ has order 2, we may choose φ such that $|\mathcal{C} \cap \mathcal{C}^\varphi| = 2$, say $\mathcal{C} \cap \mathcal{C}^\varphi = \{z, z'\}$. By applying the symmetry group about π we may assume that y is contained in $S'^{\overline{\varphi}}$. As $|\mathcal{C} \cap \mathcal{C}^\varphi| = 2$, the subGQs $\mathcal{S}'$ and $\mathcal{S}'^{\overline{\varphi}}$ share exactly two lines incident with m, and so $\mathcal{S}' \cap \mathcal{S}'^{\overline{\varphi}}$ is a grid. Under the covering map γ, with the lines of the grid correspond z, z' and the $2s$ lines on z or z', not in m^γ, in a fixed plane δ of $\mathbf{PG}(3, s)$. Since the point y of $\mathcal{S}$ is not contained in $\mathcal{S}'$, but is contained in $\mathcal{S}'^{\overline{\varphi}}$, it follows that $y^\gamma \notin \delta$. Each of the $s + 1$ lines of $\mathcal{S}'^{\overline{\varphi}}$ incident with y is incident with a unique point of $\mathcal{S}' \cap \mathcal{S}'^{\overline{\varphi}}$. Looking in $\mathbf{PG}(3, s)$ at the $\mathbf{T}_2(\mathcal{C})$ and $\mathbf{T}_2(\mathcal{C}^\varphi)$ models for $\mathcal{S}'$ and $\mathcal{S}'^{\overline{\varphi}}$, respectively, we see that this set $\mathcal{D}$ of $s + 1$ points must contain the $s - 1$ points of the projection of $\mathcal{C}^\varphi \setminus \{z, z'\}$ from y^γ onto δ, that is, $s - 1$ points of a conic (completed to a conic by adding z and z'). The remaining two points of $\mathcal{D}$ are the planes spanned by y^γ and the tangent lines of $\mathcal{C}$ at z and z'; these tangent lines are also the tangent lines of $\mathcal{C}^\varphi$ at z and z'. Under the isomorphism from $\mathbf{T}_2(\mathcal{C})$ to $\mathbf{Q}(4, s)$ the images of these $s + 1$ points of $\mathbf{T}_2(\mathcal{C})$ are the $s + 1$ points of a conic, so under the isomorphism from $\mathbf{Q}(4, s)$ to $\mathbf{W}(s)$ the corresponding $s + 1$ points of $\mathbf{W}(s)$ are also the $s + 1$ points of a conic $\mathcal{C}'$; see Brown [23] for an explicit isomorphism from $\mathbf{T}_2(\mathcal{C})$ to $\mathbf{W}(s)$. To $\mathcal{O}_y$ corresponds an ovoid $\mathcal{O}'_y$ of $\mathbf{W}(s)$. By Thas [158] $\mathcal{O}'_y$ is also an ovoid of the projective space $\mathbf{PG}(3, s)$ which contains $\mathbf{W}(s)$. As $\mathcal{O}'_y$ contains a conic, by Brown [23], $\mathcal{O}'_y$ is an elliptic quadric of $\mathbf{PG}(3, s)$. Hence to $\mathcal{O}_y$ corresponds an elliptic quadric on $\mathbf{Q}(4, s)$.

Now suppose that $y \in \mathcal{P} \setminus \mathcal{P}'$ such that $y \sim m$ with $y \neq m$. Again let $\mathcal{O}_y$ be the ovoid of $\mathcal{S}'$ subtended by y. Let N be any line of $\mathcal{S}$ incident with y but not with m. Then N meets $\mathcal{S}'$ in a unique point, n say. Each of the $s - 1$ points incident with N, but distinct from y and n, subtends an ovoid of $\mathcal{S}'$ which corresponds to an elliptic quadric on $\mathbf{Q}(4, s)$. Since the s points incident with N, but distinct from n, subtend s ovoids that partition the points of $\mathcal{S}'$ not collinear with n, it follows that the ovoid $\mathcal{O}_y$ is determined by the $s - 1$ ovoids subtended by the points incident with N, but distinct from y and n. Looking in the $\mathbf{Q}(4, s)$ or $\mathbf{T}_2(\mathcal{C})$ model of $\mathcal{S}'$, we see that with $\mathcal{O}_y$ corresponds an elliptic quadric of $\mathbf{Q}(4, s)$.

Since each point of $\mathcal{P} \setminus \mathcal{P}'$ subtends an ovoid in $\mathcal{S}'$, which defines an elliptic quadric in $\mathbf{Q}(4, s)$, it follows from Theorem 7.5.4 that $\mathcal{S} \cong \mathbf{Q}(5, s)$. ∎

Note that a similar proof may be possible in the case when s is odd. However, given that the characterizations of an elliptic quadric of $\mathbf{Q}(4, s)$ (as ovoid of the latter) are more involved for s odd than in the case where s is even, any such proof will be more complicated and probably much longer than that given in Theorem 7.6.2.

Now we will consider subGQs of order s of TGQs of order (s, s^2), for any

$s \neq 1$, where $\mathcal{S}'$ does contain the point (∞). The following theorem is folklore.

Theorem 7.5.7 Let $\mathcal{S} = (\mathcal{P}, \mathcal{B}, \mathrm{I})$ be a TGQ of order (s, s^2), $s \neq 1$, with $\mathcal{O} = \mathcal{O}(n, 2n, q)$, and assume that $\mathcal{S}' = (\mathcal{P}', \mathcal{B}', \mathrm{I}')$ is a subGQ of order s containing the point (∞) of $\mathcal{S}$. Then $\mathcal{S}' = \mathbf{T}(\mathcal{O}')$, with $\mathcal{O}' = \mathcal{O}'(n, n, q)$ a pseudo-oval on $\mathcal{O}$.

Proof. Assume $\mathcal{S}$ admits a subquadrangle $\mathcal{S}' = (\mathcal{P}', \mathcal{B}', \mathrm{I}')$ of order s containing the point (∞). Let $L_0, L_1, \ldots, L_s$ be the $s + 1$ lines of $\mathcal{S}$ incident with (∞) and contained in $\mathcal{S}'$, say $L_i = \pi_i$ with $\pi_i \in \mathcal{O}$, $i = 0, 1, \ldots, s$. If $\mathbf{PG}(n, q) = M$ is a line of $\mathcal{S}'$ concurrent with L_0 and not containing (∞), then $\langle \mathbf{PG}(n, q), \pi_i \rangle \setminus \mathbf{PG}(4n - 1, q)$, with $\langle \mathcal{O} \rangle = \mathbf{PG}(4n - 1, q)$ and $i \in \{1, 2, \ldots, s\}$, is a $2n$-dimensional affine space $\mathbf{AG}^{(i)}(2n, q)$ consisting entirely of points of $\mathcal{S}'$. In this way arise the $s^3 - s^2$ points of $\mathcal{S}'$ not collinear with (∞) or with $L_0 \cap M$, together with the s points of M not incident with L_0. Remark that the points of $\mathbf{AG}^{(i)}(2n, q)$, $i \in \{1, 2, \ldots, s\}$, are the points not collinear with (∞) of a grid $\mathcal{S}^{(i)}$. Now let $N = \mathbf{PG}'(n, q)$ be a line of $\mathcal{S}'$ concurrent with L_1, not concurrent with M and not containing (∞). As N contains a point of the grid $\mathcal{S}^{(i)}$, the space $\mathbf{PG}'(n, q)$ has a point in common with the space $\mathbf{AG}^{(i)}(2n, q)$, $i = 2, 3, \ldots, s$. Letting vary $\mathbf{PG}'(n, q)$, we see that the $s - 1$ affine spaces $\mathbf{AG}^{(i)}(2n, q)$, $i = 2, 3, \ldots, s$, are contained in $\langle \pi_1, \mathbf{AG}^{(2)}(2n, q) \rangle = \mathbf{PG}(3n, q)$. Clearly also $\mathbf{AG}^{(1)}(2n, q)$ is contained in $\mathbf{PG}(3n, q)$. It easily follows that the points of $\mathcal{S}'$ not collinear with (∞) are the points of the affine space $\mathbf{AG}(3n, q) = \mathbf{PG}(3n, q) \setminus \mathbf{PG}(4n - 1, q)$, that $\pi_0, \pi_1, \ldots, \pi_s$ are contained in a common $\mathbf{PG}(3n - 1, q)$, and that $\mathcal{S}' = \mathbf{T}(\mathcal{O}')$, with $\mathcal{O}' = \{\pi_0, \pi_1, \ldots, \pi_s\}$. ∎

Remark 7.5.8 Several (more general) versions of Theorem 7.5.7 can be found in K. Thas [197], see also [203].

7.6 Subquadrangles of $\mathbf{T}(\mathcal{O})$, with $\mathcal{O}$ Good: the Odd Case

The theorems of this section are due to Brown and Thas [29].

Theorem 7.6.1 Let $\mathcal{S} = (\mathcal{P}, \mathcal{B}, \mathrm{I})$ be a TGQ of order (s, s^2), s odd and $s \neq 1$, with $\mathcal{S} = \mathbf{T}(\mathcal{O})$ and $\mathcal{O}$ good at some element π. If $\mathcal{S}'$ is a subGQ of order s of $\mathcal{S}$ containing the point (∞), then $\mathcal{S}' \cong \mathbf{Q}(4, s)$ and either $\mathcal{S}'$ is one of the $s^3 + s^2$ (classical) subGQs of order s containing the line π of $\mathcal{S}$, or $\mathcal{S} \cong \mathbf{Q}(5, s)$.

Proof. By Theorem 5.1.9 the GQ contains exactly $s^3 + s^2$ subGQs of order s containing the line π of $\mathcal{S}$; each of them is isomorphic to the classical GQ $\mathbf{Q}(4, s)$.

Assume $\mathcal{S}$ admits a subGQ $\mathcal{S}' = (\mathcal{P}', \mathcal{B}', \mathbf{I}')$ of order s containing the point (∞) but not containing the line π. If $\mathcal{S} = \mathbf{T}(\mathcal{O})$, with $\mathcal{O} = \mathcal{O}(n, 2n, q)$, then $\mathcal{S}' = \mathbf{T}(\mathcal{O}')$, with $\mathcal{O}' = \{\pi_0, \pi_1, \dots, \pi_s\}$, $\pi_i \in \mathcal{O}$ for $i = 0, 1, \dots, s$, and $\pi_0, \pi_1, \dots, \pi_s$ contained in a common $\mathbf{PG}(3n - 1, q)$; see the proof of Theorem 7.5.7. Now by Theorem 5.2.3 and the uniqueness of the GQ of order $(3, 9)$ the egg $\mathcal{O}$ is classical, and so $\mathcal{S} \cong \mathbf{Q}(5, s)$. ∎

Theorem 7.6.2 *Let $\mathcal{S} = (\mathcal{P}, \mathcal{B}, \mathbf{I})$ be a TGQ of order (s, s^2), s odd and $s \neq 1$, with $\mathcal{S} = \mathbf{T}(\mathcal{O})$ and $\mathcal{O}$ good at some element π. If $\mathcal{S}'$ is a subGQ of order s of $\mathcal{S}$ not containing the point (∞), then $\mathcal{S} \cong \mathbf{Q}(5, s)$.*

Proof. By Corollary 5.1.4, $\mathbf{T}(\mathcal{O})$ is the translation dual of the point-line dual of a flock GQ. If π is the good element of $\mathcal{O}$, then, by Section 3 of K. Thas [205] (see also K. Thas [206] or Chapter 10) every point of the line π is a translation point of the TGQ $\mathcal{S}$. One of these translation points, say m, belongs to $\mathcal{S}'$. As there is an automorphism of $\mathcal{S}$ mapping m onto (∞), there is a subGQ $\mathcal{S}''$ of order s which contains the point (∞). Hence, by Theorem 7.6.1, $\mathcal{S} \cong \mathbf{Q}(5, s)$. ∎

7.7 Subquadrangles: Remaining Cases and Some Applications

Each known GQ of order (s, s^2), $s \neq 1$, is either the point-line dual of a flock GQ, or a TGQ $\mathbf{T}(\mathcal{O})$, with either $\mathcal{O}$ or $\mathcal{O}^*$ good. In this section we will determine the subquadrangles of order s of flock GQs and of TGQs $\mathbf{T}(\mathcal{O})$, with $\mathcal{O}^*$ good. As most of the proofs are quite long, just a few of them will actually be given. Finally, we will mention some interesting characterization theorems, which are related to the determination of these subquadrangles.

The next theorem is due to O'Keefe and Penttila [92], but we give here another and very short proof.

Theorem 7.7.1 *If $\mathcal{S}(\mathcal{F})$ is a flock GQ of order (q^2, q), q even, with center (∞), then $\mathcal{S}(\mathcal{F})$ contains exactly $q^3 + q^2$ subGQs of order q containing the point (∞).*

Proof. Payne [118] proved that $|\{L,M,N\}^{\perp\perp}| = q+1$ for all triads $\{L,M,N\}$, with $(\infty)\mathrm{I}L$, for which there is a line U such that $(\infty)\mathrm{I}U$ and $\{L,M,N\} \subseteq U^{\perp}$; see also Theorem 4.10.1. That is, the set $\{L,M,N\}$ is 3-regular. Now let $\{L,M,N\}$ be such a triad. If $\mathcal{B}'$ is the set of lines of $\mathcal{S}(\mathcal{F})$ incident with points of the form $X \cap Y$, with $X \in \{L,M,N\}^{\perp}$ and $Y \in \{L,M,N\}^{\perp\perp}$, and $\mathcal{P}'$ is the set of points which are incident with at least two distinct lines of $\mathcal{B}'$, then endowed with the induced incidence I', $\mathcal{S}' = (\mathcal{P}',\mathcal{B}',\mathrm{I}')$ is a subGQ of order q of $\mathcal{S}(\mathcal{F})$, see Theorem 2.6.2. Now a standard counting argument yields $q^3 + q^2$ of such subGQs. By Lemma 5.1.8, $\mathcal{S}(\mathcal{F})$ contains exactly $q^3 + q^2$ subGQs of order q containing the point (∞). ∎

Now we mention, without proof, a strong characterization theorem for $\mathbf{Q}(5,s)$ on which we will rely in the proof of Theorem 7.7.3. Theorem 7.7.2 and 7.7.3 are due to Brown and Thas [28].

Theorem 7.7.2 *Let $\mathcal{S} = (\mathcal{P},\mathcal{B},\mathrm{I})$ be a GQ of order (s,s^2), $s \neq 1$, and assume one of the following conditions is satisfied:*

(i) *s is odd and $\mathcal{S}$ satisfies Property (G) at the pair $\{x,y\}$ with $x \sim y \neq x$ and $L = xy$, or*

(ii) *s is even, $\mathcal{S}$ is the point-line dual of a flock GQ and L is the corresponding base-line of $\mathcal{S}$.*

If at least one triad $\{z_1,z_2,z\}$, with $z\mathrm{I}L$ and where $\{z_1,z_2,z\}^{\perp}$ does not contain a point incident with L, is 3-regular, then $\mathcal{S}$ is classical. ∎

Theorem 7.7.3 *Let $\mathcal{S}$ be a nonclassical flock GQ of order (q^2,q), q even, with base-point (∞). Then any subGQ of $\mathcal{S}$ of order q contains (∞).*

Proof. Let $\mathcal{S}^D$ be the point-line dual of a nonclassical flock GQ of order (q^2,q), q even. The base-line of $\mathcal{S}^D$ will be denoted by L. Suppose, by way of contradiction, that $\mathcal{S}'$ is a subGQ of order q of $\mathcal{S}^D$ which does not contain L. Let x be the point of $\mathcal{S}'$ incident with L an let $y\mathrm{I}L$, with $x \neq y$. Further, let $\mathcal{O}_y$ be the ovoid of $\mathcal{S}'$ consisting of the $q^2 + 1$ points of $\mathcal{S}'$ collinear with y. Also, let $z \in \mathcal{O}_y \setminus \{x\}$, let M be a line of $\mathcal{S}'$ incident with x, let N be a line of $\mathcal{S}'$ incident with z, and let $M \not\sim N$. Let $M\mathrm{I}u \sim z$ and $N\mathrm{I}w \sim x$. Then by Theorem 4.10.1 and Theorem 2.6.2 there is a subGQ $\mathcal{S}''$ of order q of $\mathcal{S}^D$ containing y,u,w,x,z. From Lemma 1.1.1 easily follows that $\mathcal{S}' \cap \mathcal{S}''$ is a grid containing the lines M, N. Hence the pair $\{M, N\}$ is regular for any line N of S' not concurrent with M. Consequently M is a regular line of S', so in S any line incident with x is regular. Now by Theorem 2.1.5 the point x is regular in S'. If r is a point of S' not collinear with x, then $\{x, r\}^{\perp'}$ and $\{x, r\}^{\perp'\perp'}$ contain $q + 1$ points of S'. Let $\{x, r\}^{\perp'} = \mathcal{A}$ and $\{x, r\}^{\perp'\perp'} = \overline{\mathcal{A}}$. If $r' \in \overline{\mathcal{A}} \setminus \{x, r\}$, then $\{x, r, r'\}$ is 3-regular in S and $\{x, r, r'\}^{\perp} = \mathcal{A}$ does not contain a point incident with L. By Theorem 7.7.2, the GQ S^D is classical, a contradiction. ∎

Theorem 7.7.4 *Let S be a flock GQ of order (q^2, q), q odd, with base-point (∞). If S' is a subGQ of order q of S containing the point (∞), then S is a Kantor-Knuth semifield flock GQ. Hence $S' \cong \mathbf{W}(q)$ and either $S \cong \mathbf{H}(3, q^2)$ or S' is one of the $q^3 + q^2$ subGQs of order q containing the point (∞).* ∎

Remark 7.7.5 The proof of this theorem is quite long; the last part of the statement follows from Theorem 7.6.1.

Corollary 7.7.6 *Let S be a flock GQ of order (q^2, q), q odd, with base-point (∞). If the net $\mathcal{N}_{(\infty)}$ defined by the regular point (∞) of S contains at least one proper subnet having the same degree as $\mathcal{N}_{(\infty)}$, then S is a Kantor-Knuth semifield flock GQ.*

Proof. In K. Thas [200] it is shown that the subnet is an affine plane of order q. From the proof of Theorem 6.5.1 then follows that S has a subGQ of order q containing the point (∞), and so by Theorem 7.7.4 we are done. ∎

Theorem 7.7.7 *Let S be a nonclassical flock GQ of order (q^2, q), q odd, with base-point (∞). Then any subGQ of S of order q contains (∞).* ∎

Remark 7.7.8 The proof of Theorem 7.7.7 is long and technical. Central in this proof is the construction of Knarr; see Section 4.9.

As a byproduct of the proof of Theorem 7.7.7 we obtain the counterpart of Corollary 6.6.8; no proof will be given.

Theorem 7.7.9 *If $S = (\mathcal{P}, \mathcal{B}, \mathrm{I})$ is a flock GQ of order (q^2, q), with q odd, then the point-line dual of S is a TGQ if and only if S has a regular point x, with $(\infty) \neq x \sim (\infty)$, where (∞) is the elation point of S.* ∎

Next we will consider subGQs of TGQs $\mathbf{T}(\mathcal{O}^*)$ of order (s, s^2), where the translation dual $\mathcal{O}^*$ of $\mathcal{O}$ is good.

If s is even, then the egg $\mathcal{O}(n, 2n, q)$ is good if and only if $\mathcal{O}^*(n, 2n, q)$ is good; see Corollary 5.1.6.

If $\mathcal{O}^*(n, 2n, q)$, q odd, is good, then, by Corollary 5.1.4 the TGQ $\mathbf{T}(\mathcal{O})$ is isomorphic to the point-line dual of a flock GQ.

So from Theorems 7.5.6, 7.5.7, 7.7.4 and 7.7.7 we have the following result.

Theorem 7.7.10 *Let $\mathcal{S} = \mathbf{T}(\mathcal{O})$, with $\mathcal{O} = \mathcal{O}(n, 2n, q)$, be a TGQ such that the translation dual $\mathcal{O}^*$ of $\mathcal{O}$ is good at the tangent space τ of $\mathcal{O}$ at $\pi \in \mathcal{O}$, and assume that $\mathcal{S}$ contains a subGQ $\mathcal{S}'$ of order $q^n = s$. If q is even and $\mathcal{S}'$ does not contain the translation point (∞), then $\mathcal{S} \cong \mathbf{Q}(5, s)$; if q is even and $\mathcal{S}'$ contains the translation point (∞), then $\mathcal{S}' = \mathbf{T}(\mathcal{O}')$, with $\mathcal{O}' = \mathcal{O}'(n, n, q)$ a pseudo-oval on $\mathcal{O}$. If q is odd and $\mathcal{S}'$ does not contain the translation point (∞), then $\mathcal{S} \cong \mathbf{Q}(5, s)$; if q is odd and $\mathcal{S}'$ contains the translation point (∞), then $\mathcal{S}$ is a Kantor-Knuth semifield flock GQ, $\mathcal{S}' \cong \mathbf{Q}(4, s)$ and either $\mathcal{S} \cong \mathbf{Q}(5, s)$ or $\mathcal{S}'$ is one of the $s^3 + s^2$ subGQs of order s containing the point (∞).* ∎

7.8 Translation Generalized Quadrangles with One Classical Subquadrangle

Recall Brown's result on ovoids in $\mathbf{PG}(3, q)$ and its equivalent statement in the theory of TGQs.

Theorem 7.8.1 (Brown [23])   (i) *If an ovoid of $\mathbf{PG}(3, q)$, q even, has a conic section, the ovoid must be an elliptic quadric.*

  (ii) *If a $\mathbf{T}_3(\mathcal{O})$ of order (q, q^2), where $\mathcal{O}$ is an ovoid of $\mathbf{PG}(3, q)$, q even, has a classical subGQ of order q containing the point (∞), then $\mathbf{T}_3(\mathcal{O}) \cong \mathbf{Q}(5, q)$, that is, $\mathbf{T}_3(\mathcal{O})$ arises from a nonsingular elliptic quadric in $\mathbf{PG}(5, q)$.*

Proof.   See Brown [23].   ∎

We note that the only classical GQs of order q are the GQ $\mathbf{Q}(4, q)$, which arises from a nonsingular quadric in $\mathbf{PG}(4, q)$, and the symplectic quadrangle $\mathbf{W}(q)$. However, if q is even, then by Theorem 1.4.1, $\mathbf{Q}(4, q) \cong \mathbf{W}(q)$.

Brown and Lavrauw [26] generalized Theorem 7.8.1 by obtaining a similar result for generalized ovoids.

We first prove a lemma.

Lemma 7.8.2 *Every* $(2n-1)$-*space* α *in* $\mathbf{PG}(3n-1,q)$, *q even, which is skew to all elements of a pseudo-conic, is the span of two elements of the regular* $(n-1)$-*spread induced in* α *by the pseudo-conic.*

Proof. Let α be a $(2n-1)$-space skew to all elements of the pseudo-conic $\mathcal{O}$ in $\pi = \mathbf{PG}(3n-1,q)$. After dualizing, we obtain an $(n-1)$-space α^* that is disjoint from all elements of a dual pseudo-conic $\mathcal{O}^*$. Embed π in a $3n$-dimensional space $\mathbf{PG}(3n,q)$, and embed α^* in an n-space $\mathbf{PG}(n,q) \subseteq \mathbf{PG}(3n,q)$, where $\mathbf{PG}(n,q) \not\subseteq \pi$. Considering $\mathbf{AG}(3,q^n)$ as an $\mathbf{AG}(3n,q)$, and identifying $\mathbf{AG}(3n,q)$ with $\mathbf{PG}(3n,q) \setminus \pi$, we obtain the following geometrical structure in the projective completion $\mathbf{PG}(3,q^n)$ of $\mathbf{AG}(3,q^n)$. With $\mathcal{O}^*$ corresponds a dual conic $\overline{\mathcal{O}}$ in the plane $\mathbf{PG}(2,q^n) = \mathbf{PG}(3,q^n) \setminus \mathbf{AG}(3,q^n)$, and with $\mathbf{PG}(n,q) \setminus \alpha^*$ corresponds a point set $\mathcal{F}^*$ of size q^n in $\mathbf{AG}(3,q^n)$ with the property that the line of $\mathbf{PG}(3,q^n)$ joining any two distinct points of $\mathcal{F}^*$ has no point in common with the union of the elements of $\overline{\mathcal{O}}$. So $\mathcal{F}^*$ is a flock of the dual conic $\overline{\mathcal{O}}$. By Section 4.5, the corresponding flock $\mathcal{F}$ of the quadratic cone is a semifield flock, and so, by Theorem 4.5.1, the flock $\mathcal{F}$ is linear. Hence $\mathcal{F}^*$ is an affine line. So α^* is an element of the regular $(n-1)$-spread of π defined by $\mathcal{O}^*$, and hence α contains q^n+1 elements of the regular $(n-1)$-spread of π defined by $\mathcal{O}$. ∎

We are ready to obtain the result of Brown and Lavrauw. We state their result in terms of translation generalized quadrangles.

Theorem 7.8.3 *If* $\mathcal{S} = \mathbf{T}(\mathcal{O})$ *is a TGQ of order* (q^n, q^{2n}), *where* $\mathcal{O}$ *is a generalized ovoid of* $\mathbf{PG}(4n-1,q)$, *q even, which has a classical subGQ* $\mathcal{S}'$ *of order* q *containing the point* (∞), *then* $\mathbf{T}(\mathcal{O}) \cong \mathbf{Q}(5,q^n)$.

Proof. Suppose $\mathcal{O}' \subseteq \mathcal{O}$ is the pseudo-conic on $\mathcal{O}$ which corresponds to $\mathcal{S}'$; see Theorem 7.5.7. For x a point of $\mathcal{S} \setminus \mathcal{S}'$, let $\mathcal{O}_x$ be the ovoid of $\mathcal{S}'$ subtended by x. First suppose that x is a point of Type (ii) of $\mathbf{T}(\mathcal{O})$ — so x is a subspace of dimension $3n$ meeting $\mathbf{PG}(4n-1,q)$ in the tangent space at some egg element. Let $\mathbf{T}(\mathcal{O})$ be contained in $\mathbf{PG}(4n,q)$ and let $\mathbf{PG}(3n,q) \subset \mathbf{PG}(4n,q)$ be the $3n$-space containing $\mathcal{O}'$ that yields $\mathbf{T}(\mathcal{O}') = \mathcal{S}'$. Then $\mathcal{O}_x$ consists of the point (∞) plus the q^{2n} points contained in

$(x \cap \mathbf{PG}(3n, q)) \setminus \mathbf{PG}(3n-1, q)$, with $\mathbf{PG}(3n-1, q)$ the space containing $\mathcal{O}'$. The $(2n-1)$-dimensional space $x \cap \mathbf{PG}(3n-1, q) = \mathbf{PG}(2n-1, q)$ is skew to the pseudo-conic $\mathcal{O}'$, so applying Lemma 7.8.2, we have that it is the span of two elements of the regular $(n-1)$-spread of $\mathbf{PG}(2n-1, q)$ induced by the pseudo-conic. Now represent $\mathcal{S}'$ over $\mathbf{GF}(q^n)$, so as a $\mathbf{T}_2(\mathcal{C}) \cong \mathbf{Q}(4, q^n)$ with $\mathcal{C}$ a conic of $\mathbf{PG}(2, q^n)$. Then the ovoids of $\mathbf{T}_2(\mathcal{C})$ consisting of the point (∞) of $\mathbf{T}_2(\mathcal{C})$ together with the affine points of a plane skew to $\mathcal{C}$ correspond to the elliptic quadrics of $\mathbf{Q}(4, q^n)$ containing a fixed point. As a subtended ovoid can be subtended by at most two distinct points (by Section 2.5), and as there are $q^{2n}(q^n - 1)/2$ elliptic quadrics on $\mathbf{Q}(4, q^n)$ containing a given point, it follows that each of the $q^{2n}(q^n - 1)$ points of $\mathcal{S} \setminus \mathcal{S}'$ of Type (ii) subtends a doubly subtended classical ovoid.

Now let y be a point of $\mathcal{S} \setminus \mathcal{S}'$ not collinear with (∞), and let $\mathcal{O}_y$ be the corresponding subtended ovoid of $\mathcal{S}'$. We now work in the $\mathbf{T}_2(\mathcal{C})$ model of $\mathcal{S}'$; assume $\mathbf{T}_2(\mathcal{C})$ is contained in $\mathbf{PG}(3, q^n)$. As $y \not\sim (\infty)$, we have that $\mathcal{O}_y = \mathcal{A} \cup \{\pi_p \parallel p \in \mathcal{C}\}$, where $\mathcal{A}$ is a set of $q^{2n} - q^n$ points of Type (i) of $\mathbf{T}_2(\mathcal{C})$ and π_p is a point of Type (ii) of $\mathbf{T}_2(\mathcal{C})$ (a plane containing $p \in \mathcal{C}$). If a plane π of $\mathbf{PG}(3, q^n)$ contains no point of $\mathcal{C}$, then $(\pi \setminus \mathbf{PG}(2, q^n)) \cup \{(\infty)\}$ is a classical ovoid subtended by two points, x and x' of $\mathcal{S} \setminus \mathcal{S}'$, by the first part of the proof. If y is collinear with x or x', then $\pi \cap \mathcal{A}$ is a single point. If y is collinear with neither x nor x', then $\{x, x', y\}$ is a triad of $\mathcal{S}$. Hence $|\pi \cap \mathcal{A}| = q^n + 1$, as this is the number of centers of $\{x, x', y\}$. Next, suppose that π contains a unique point p of $\mathcal{C}$. If $\pi = \pi_p \subseteq \mathcal{O}_y$, then π contains no point of $\mathcal{A}$. If $\pi \neq \pi_p$, then the q^n lines of π incident with p and not in the plane of $\mathcal{C}$ are lines of $\mathbf{T}_2(\mathcal{C})$, and so contain precisely one point of $\mathcal{A}$. Whence $|\pi \cap \mathcal{A}| = q^n$. Next, suppose that π contains two distinct points, p and p', of $\mathcal{C}$. Of the $q^n + 1$ projective lines in π incident with p, one is contained in $\mathbf{PG}(2, q^n)$, one in π_p and $q^n - 1$ are lines of $\mathbf{T}_2(\mathcal{C})$ containing a unique point of $\mathcal{A}$. Hence $|\pi \cap \mathcal{A}| = q^n - 1$. Finally, if $\pi = \mathbf{PG}(2, q^n)$, then π contains no point of $\mathcal{A}$.

Consider the set of points of $\mathbf{PG}(3, q^n)$ defined by $\overline{\mathcal{O}_y} = \mathcal{A} \cup \mathcal{C}$. By the above, the plane intersections with $\overline{\mathcal{O}_y}$ have size 1 or $q^n + 1$, and a straightforward count shows that $\overline{\mathcal{O}_y}$ is an ovoid of $\mathbf{PG}(3, q^n)$. Further, since $\overline{\mathcal{O}_y}$ contains $\mathcal{C}$, it is an elliptic quadric by Theorem 7.8.1. Hence $\mathcal{O}_y$ is a classical ovoid of $\mathcal{S}' \cong \mathbf{Q}(4, q^n)$. So every point of $\mathcal{S}$ outside $\mathcal{S}'$ subtends an elliptic quadric in $\mathcal{S}'$. By Theorem 7.5.4 the theorem now follows. ∎

Corollary 7.8.4 *An egg in $\mathbf{PG}(4n-1, q)$, q even, contains a pseudo-conic if and only if it is classical.* ∎

Further on in this chapter, we will improve these results by showing that an elation generalized quadrangle of order (q, q^2), q even, with a subGQ $\mathbf{Q}(4, q)$ containing the elation point, arises from a nonsingular elliptic quadric in $\mathbf{PG}(5, q)$. This is taken from K. Thas [208]. The theorem is a corollary of a structural result (cf. Proposition 7.9.2 and Corollary 7.9.3) independent of the characteristic. Another corollary generalizes also a result of Brown and Lavrauw [27].

7.9 Elation Generalized Quadrangles with a Subquadrangle

Recall that a *center of transitivity* x of a GQ $\mathcal{S}$ is a point for which there is an automorphism group of $\mathcal{S}$ fixing x linewise and acting transitively on the points noncollinear with x.

The next lemma is well-known (it can, for instance, be found in [197]).

Lemma 7.9.1 *Let $\mathcal{S}$ be a GQ of order (s, s^2), $s > 1$, and suppose that $\mathcal{S}'$ is a subGQ of order s. Furthermore, assume that $x \in \mathcal{S}'$ is a center of transitivity of $\mathcal{S}$. Then x is also a center of transitivity of $\mathcal{S}'$, and there are s distinct subGQs of order s which all contain the same lines through x, and of which the sets of points not collinear with x partition the set of points of $\mathcal{S}$ not collinear with x.*

Proof. Let y, y', $y \neq y'$, in $\mathcal{S}'$ both be not collinear with x. Let θ be a whorl of $\mathcal{S}$ about x mapping y onto y'. If $\mathcal{S}'^{\theta} \neq \mathcal{S}'$, then $\mathcal{S}' \cap \mathcal{S}'^{\theta}$ must be a subGQ of order $(s, 1)$ (by Lemma 1.1.1), a contradiction. Whence $\mathcal{S}' \cap \mathcal{S}'^{\theta} = \mathcal{S}'$ and it follows that x is a center of transitivity for $\mathcal{S}'$. The lemma now easily follows. ∎

Proposition 7.9.2   (i) *Let $\mathcal{S}^{(x)}$ be an EGQ of order (q, q^2), $q > 1$, containing a subGQ $\mathcal{S}'$ of order q which has an axis of symmetry L incident with x. Then L is an axis of symmetry in $\mathcal{S}^{(x)}$.*

 (ii) *Let $\mathcal{S}^{(x)}$ be a GQ of order (q, q^2), $q > 1$, with center of transitivity x, containing a subGQ $\mathcal{S}'$ of order q which has an axis of symmetry L incident with x. Then L is an axis of symmetry in $\mathcal{S}^{(x)}$.*

Proof. (i) Let $\mathcal{S}'$ be as in the statement of the proposition, and suppose G is the group of elations of $\mathcal{S}$ about x. Then there are q elements in the orbit $\mathcal{S}'^G$ by Lemma 7.9.1, and by that same lemma they precisely intersect in the $q+1$ lines of $\mathcal{S}'$ which are incident with x, together with the points on these lines. Whence the elements of $\mathcal{S}'^G$ partition the lines of $\mathcal{S}$ not incident with x and concurrent with one of the $q+1$ lines on x in $\mathcal{S}'$, and also the points of $\mathcal{S}$ not collinear with x. Let L be as in the statement of the proposition, and suppose $M \sim L \neq M$ is not incident with x. Then there is precisely one element, denoted $\mathcal{S}'_M$, in $\mathcal{S}'^G$ which contains M, and no other element of $\mathcal{S}'^G$ contains a point of $M \setminus \{L \cap M\}$. Take an arbitrary element θ of G_M. Then θ fixes $\mathcal{S}'_M$. By Theorem 3.2.1, θ induces a symmetry about L in $\mathcal{S}'_M$, so that θ fixes all lines of $\mathcal{S}'_M$ through each point of L. We obtain the following crucial property for G:

(S) *For each point $y \mathrm{I} L$, $y \neq x$, each element of G fixes a constant number*
    $c = aq$, $a \in \mathbb{N}$, of lines incident with y and different from L.

We emphasize the fact that this constant is the same for each point on L different from x.

Suppose that $\theta \neq 1$ is as above (θ fixes M). Suppose that θ fixes some point $z \sim x$ which is not on L. Then Theorem 3.2.1 implies that the fixed elements structure of θ is a thick subGQ of order (q, q^2), and so $\theta = 1$, contradiction. So θ fixes $f = q+1$ points, namely the points of L. Suppose that z is a point not collinear with x for which $z^\theta \sim z \neq z^\theta$; then $zz^\theta \sim L$ (otherwise $z, z^\theta, \mathrm{proj}_L z$ are the points of a triangle). It follows that the number of points of $\mathcal{S}$ which are mapped by θ onto a collinear point different from itself, is $g = q^3 + q^2 c$. Applying Theorem 3.3.1, we obtain

$$(q^2+1)(q+1) + q^3 + q^3 a \equiv q^3 + 1 \mod q^2 + q.$$

Hence $a + 1 \equiv 0 \mod q+1$, so that $c = aq \geq q^2$, hence $c = q^2$. It follows that θ is a symmetry about L of $\mathcal{S}$.

(ii)   By Bose and Shrikhande [18], see also Section 2.5, we have that for each $z \nsim x$, $\{z, x\}^{\perp\perp} = \{z, x\}$. Then by Theorem 3.2.2(4), there is a normal subgroup $G' \trianglelefteq G$ (where G is as in Part (i) of the proof) for which $\mathcal{S}^{(x)}$ is an EGQ with elation group G'. Now apply (i).   ∎

Corollary 7.9.3   (i)   Let $\mathcal{S}^{(x)}$ be an EGQ of order (q, q^2), $q > 1$, containing a subGQ $\mathcal{S}'$ of order q which has at least one axis of symmetry L incident with x. Then $\mathcal{S}^{(x)}$ is a TGQ for the translation point x.

(ii) Let $\mathcal{S}^{(x)}$ be a GQ of order (q, q^2), $q > 1$, with center of transitivity x, containing a subGQ $\mathcal{S}'$ of order q which has at least one axis of symmetry L incident with x. Then $\mathcal{S}^{(x)}$ is a TGQ for the translation point x.

Proof. By the proof of Proposition 7.9.2, we only have to obtain (i). Suppose that $\mathcal{S}^{(x)}$ is as assumed in (i). By Lemma 7.9.1, one derives that $\mathcal{S}'$ is an EGQ with elation point x (the stabilizer of $\mathcal{S}'$ in the elation group of $\mathcal{S}^{(x)}$ is an elation group of $\mathcal{S}'$). Say this elation group of $\mathcal{S}'$ is G, and suppose $L \mathrm{I} x$ is an axis of symmetry in $\mathcal{S}'$. Then clearly the full group of symmetries (of $\mathcal{S}'$) about L is a (normal) subgroup of G. By Theorem 3.3.18, it follows that $\mathcal{S}'$ is a TGQ with translation point x, so that each line incident with x in $\mathcal{S}'$ is an axis of symmetry of $\mathcal{S}'$. By Proposition 7.9.2, each of these lines is also an axis of symmetry of $\mathcal{S}$. Now the theorem follows from Theorem 3.3.17. ∎

Remark 7.9.4 (a) Let $\mathcal{S}^{(x)}$ be as in (i) or (ii) of Proposition 7.9.2, and suppose there is a subGQ $\mathcal{S}'$ of order q which has a regular line L incident with x. Then by Theorem 3.3.13, L is an axis of symmetry of $\mathcal{S}'$. So by Proposition 7.9.2, L is an axis of symmetry of $\mathcal{S}^{(x)}$. Whence Corollary 7.9.3 can also be restated in this way (and the same holds for some of the following results).

(b) Let $\mathcal{S}^{(x)}$ be an EGQ of order (s, s^2), $s > 1$, containing the s distinct subGQs $\mathcal{S}_1, \mathcal{S}_2, \ldots, \mathcal{S}_s$ of order s which share a line L incident with some point x of $\mathcal{S}$, and which also share the same $s + 1$ lines incident with x. Suppose L is regular in all these subGQs. Suppose that there is some line $M \sim L \neq M$ which is not incident with x so that there is a group of automorphisms H of $\mathcal{S}$ which fixes x linewise, and acts transitively on the points of M different from $L \cap M$. Then there is an $i \in \{1, 2, \ldots, s\}$ so that $M \in \mathcal{S}_i$. By Theorem 2.3.10 of [206], L is an axis of symmetry in $\mathcal{S}_i$, and a similar argument as in the proof of Proposition 7.9.2 leads to the fact that L is an axis of symmetry in $\mathcal{S}$.

Not much is known about EGQs which are also dual EGQs, and which are nonclassical. We will prove some specific results in Section 12.2. In O'Keefe and Penttila [90], some results on certain TGQs which are dual EGQs can be found. In Thas and K. Thas [192], see Chapter 8, a general theory is initiated about TGQs which are also dual TGQs. Also, if $\mathcal{S}$ is a thick nonclassical GQ with elation point x and elation line L, then by K. Thas and Van Maldeghem [215], or Van Maldeghem [231], one may assume that $x \mathrel{I} L$. (If $x \mathrel{\not I} L$, it follows easily that each point of $\mathcal{S}$ is an elation point, and each line an elation line, and then [231] — see also Corollary 11.7.3 — implies that the GQ is classical or dual classical.) Here, we obtain a result on flock GQs which are also dual EGQs.

Theorem 7.9.5 *Let $\mathcal{S} = \mathcal{S}(\mathcal{F})$ be a flock generalized quadrangle of order (q^2, q), q even, which is also a dual EGQ. Then $\mathcal{S}$ is classical, i.e., $\mathcal{S}(\mathcal{F}) \cong \mathbf{H}(3, q^2)$.*

Proof. We look at the situation in the point-line dual $\mathcal{S}(\mathcal{F})^D$ of $\mathcal{S}(\mathcal{F})$. Then the line $[\infty]$ which corresponds to the point (∞) of $\mathcal{S}(\mathcal{F})$ is contained in $q^3 + q^2$ subGQs of order q; see Theorem 7.7.1. Suppose $\mathcal{S}'$ is such a subGQ. We may assume that the line $[\infty]$ is incident with an elation point x, so that by Lemma 7.9.1, $\mathcal{S}'$ is an EGQ with elation point x. The line $[\infty]$ is an axis of symmetry of $\mathcal{S}(\mathcal{F})^D$, cf. Section 4.6, so that it also is an axis of symmetry of $\mathcal{S}'$. Now by Corollary 7.9.3 we have that $\mathcal{S}$ is also a TGQ with translation point x. The conclusion follows from Theorem 4.5.1. ∎

The following theorem is a nice corollary of Corollary 7.9.3.

Theorem 7.9.6   (i) *Let $\mathcal{S}^{(x)}$ be an EGQ of order (q, q^2), $q > 1$, containing a subGQ $\mathcal{S}'$ of order q which is a TGQ with translation point x. Then $\mathcal{S}^{(x)}$ is a TGQ with translation point x.*

  (ii) *Let $\mathcal{S}^{(x)}$ be a GQ of order (q, q^2), $q > 1$, with center of transitivity x, containing a subGQ $\mathcal{S}'$ of order q which is a TGQ with translation point x. Then $\mathcal{S}^{(x)}$ is a TGQ with translation point x.*

Proof. Each line of $\mathcal{S}'$ incident with x is an axis of symmetry of $\mathcal{S}'$. ∎

The next theorem generalizes the result of Brown and Lavrauw to EGQs.

Theorem 7.9.7 *Let $\mathcal{S}^{(x)}$ be an EGQ of order (q, q^2), q even, containing a classical subGQ $\mathcal{S}'$ of order q containing x. Then $\mathcal{S}^{(x)} \cong \mathbf{Q}(5, q)$.*

Proof. The only classical GQ of order q, q even, is $\mathbf{Q}(4, q) \cong \mathbf{W}(q)$, and this is a TGQ for each of its points. By Theorem 7.9.6, $\mathcal{S}^{(x)}$ is a TGQ. Now apply Theorem 7.8.3. ∎

Let $\mathcal{O}$ be a generalized oval in $\mathbf{PG}(3n-1,q) = \langle \mathcal{O} \rangle$, q even. Then by Theorem 3.9.1 the $q^n + 1$ tangent spaces to the elements of $\mathcal{O}$ meet in an $(n-1)$-dimensional subspace η of $\mathbf{PG}(3n-1,q)$. It is easily seen that for any $\pi \in \mathcal{O}$, $\mathcal{O}' = (\mathcal{O} \setminus \{\pi\}) \cup \{\eta\}$ is again a generalized oval. If $\mathcal{O}$ is a generalized conic, that is, if $\mathbf{T}(\mathcal{O}) \cong \mathbf{Q}(4,q^n)$, then $\mathcal{O}'$ is called a *pointed generalized conic* or a *generalized pointed conic*.

We will not prove the next result in the present monograph.

Theorem 7.9.8 *Let $\mathcal{S}^{(\infty)} = \mathbf{T}(\mathcal{O})$ be a TGQ of order (q^n, q^{2n}), q even, containing a subGQ $\mathbf{T}(\mathcal{O}')$ through (∞), where $\mathcal{O}'$ is a pointed generalized conic in $\mathbf{PG}(3n-1,q)$. Then either $q^n = 4$ and $\mathcal{S}^{(\infty)} \cong \mathbf{Q}(5,4)$, or $q^n = 8$ and $\mathcal{S}^{(\infty)} \cong \mathbf{T}_3(\mathcal{O}'')$ with $\mathcal{O}''$ a Suzuki-Tits ovoid of $\mathbf{PG}(3,8)$.*

Proof. See Brown and Lavrauw [27]. ∎

We now generalize to EGQs.

Theorem 7.9.9 *Let $\mathcal{S}^{(\infty)}$ be an EGQ of order (q^n, q^{2n}), q even, containing a subGQ $\mathbf{T}(\mathcal{O})$, where $\mathcal{O}$ is a pointed generalized conic in $\mathbf{PG}(3n-1,q)$, having (∞) as translation point. Then either $q^n = 4$ and $\mathcal{S}^{(\infty)} \cong \mathbf{Q}(5,4)$, or $q^n = 8$ and $\mathcal{S}^{(\infty)} \cong \mathbf{T}_3(\mathcal{O}')$ with $\mathcal{O}'$ a Suzuki-Tits ovoid of $\mathbf{PG}(3,8)$.*

Proof. This theorem follows immediately from Theorem 7.9.6 and Theorem 7.9.8. ∎

Chapter 8

Translation Generalized Ovals

In this chapter, we first introduce new objects called "translation generalized ovals" and "translation generalized ovoids", and make a thorough study of these objects. We then obtain several new characterizations of the $T_2(\mathcal{O})$ of Tits and the classical generalized quadrangle $Q(4, q)$ in even characteristic, including the complete classification of 2-transitive generalized ovals for the even case. We also prove that translation generalized ovoids do not exist.

8.1 Translation Generalized Ovoids and Translation Generalized Ovals

Let θ be an involution in the projective general linear automorphism group $\mathbf{PGL}_{k+1}(q)$ of the projective space $\mathbf{PG}(k, q)$, with q even. Then the set of all fixed points of θ is a subspace $\mathbf{PG}(r, q)$ of $\mathbf{PG}(k, q)$, called the *axis* of θ. The fixed hyperplanes for θ are all hyperplanes containing some subspace $\mathbf{PG}(k - r - 1, q)$ of $\mathbf{PG}(k, q)$, which is called the *center* of θ. Also, $\mathbf{PG}(k - r - 1, q) \subseteq \mathbf{PG}(r, q)$ and so $k \leq 2r + 1$. For any point $x \notin \mathbf{PG}(r, q)$ we have $xx^\theta \cap \mathbf{PG}(k - r - 1, q) \neq \emptyset$. If π_1 and π_2 are mutually skew $(k - r + 1)$-dimensional subspaces of $\mathbf{PG}(k, q)$, which are skew to a given $\mathbf{PG}(r, q) \subseteq \mathbf{PG}(k, q)$, then there is exactly one involution θ of $\mathbf{PG}(k, q)$ with axis $\mathbf{PG}(r, q)$ mapping π_1 onto π_2; the center of θ is the subspace $\langle \pi_1, \pi_2 \rangle \cap \mathbf{PG}(r, q)$. Further, all involutions of $\mathbf{PG}(k, q)$ with given axis $\mathbf{PG}(r, q)$ form an elementary abelian group.

For more details on involutions of finite projective spaces we refer to Chapter 16 of [139]. Notice that in [139] the term "fundamental space" is used instead of "axis", and the term "fundamental axis" instead of "center".

Let $\mathcal{O}$ be a generalized ovoid in $\mathbf{PG}(2n + m - 1, q)$, $n \neq m$, respectively a generalized oval in $\mathbf{PG}(3n - 1, q)$, with q even. Then $\mathcal{O}$ is a *translation generalized ovoid*, respectively a *translation generalized oval*, with *axis* the tangent space $\mathbf{PG}^{(i)}(n+m-1, q)$ at $\mathbf{PG}^{(i)}(n-1, q) \in \mathcal{O}$, if there is a group of involutions of $\mathbf{PGL}_{2n+m}(q)$ with axis $\mathbf{PG}^{(i)}(n + m - 1, q)$, fixing $\mathcal{O}$ and acting regularly on $\mathcal{O} \setminus \{\mathbf{PG}^{(i)}(n - 1, q)\}$.

If $n = m = 1$, then a translation generalized oval is just called a *translation oval*; if $2n = m = 2$, it is called a *translation ovoid*. All translation ovals of $\mathbf{PG}(2, q)$, $q = 2^h$, were determined by Payne in [107]; up to recoordinatization, they are always of the form

$$\{(1, t, t^{2^i}) \parallel t \in \mathbf{GF}(q)\} \cup \{(0, 0, 1)\},$$

where i is fixed in $\{1, 2, \ldots, h - 1\}$ and $(h, i) = 1$.

Suppose $\mathcal{S} = (\mathcal{P}, \mathcal{B}, \mathbf{I})$ is a TGQ of order q^n with translation point (∞). Let $x \not\sim (\infty)$ be a regular point. Then each point of $\mathcal{P} \setminus (\infty)^\perp$ is regular. Whence each point of $\mathcal{S}$ is regular, and $\mathcal{S} \cong \mathbf{W}(q^n)$ by Theorem 2.2.2.

Theorem 8.1.1 (Structure of Translation Generalized Ovoids and Ovals)

(i) *A translation generalized ovoid in $\mathbf{PG}(2n + m - 1, q)$, q even, does not exist.*

(ii) *Let $\mathcal{O}$ be a nonclassical generalized oval in $\mathbf{PG}(3n - 1, q)$, with q even. Then the following are equivalent.*

  (a) *$\mathcal{S} = \mathbf{T}(\mathcal{O})$ contains a regular point x different from the translation point (which is necessarily collinear with (∞));*

  (b) *the point-line dual $\mathcal{S}^D$ of $\mathcal{S}$ is a TGQ (with as base-point the line $x(\infty)$ of $\mathcal{S}$);*

  (c) *$\mathcal{O}$ is a translation generalized oval with axis the tangent space of $\mathcal{O}$ at its element $x(\infty)$.*

Proof. Let $\mathcal{O}$ be a nonclassical generalized oval in $\mathbf{PG}(3n-1, q)$. Suppose x is a regular point different from (∞). Then by the observation preceding the theorem, x is collinear with (∞). Hence each point $y\mathbf{I}x(\infty)$, $y \neq (\infty)$, is regular. By Corollary 3.3.9, also the point (∞) is regular. The translations of $\mathbf{T}(\mathcal{O})$ fixing the point x, induce elations, with as axis the set of points of $x(\infty)$, of the projective plane π_x defined by the regular point x of $\mathcal{S}$. Hence π_x is a translation plane having as translation line the set of points on $x(\infty)$. Then by Thas [185], $\mathcal{S}^D$ is a TGQ where the base-point is the line $x(\infty)$ of $\mathcal{S}$. So $\mathcal{S}$ contains a regular point x different from the translation point if and only $x \sim (\infty)$ and $\mathcal{S}^D$ is a TGQ with as base-point the line $x(\infty)$ of $\mathcal{S}$.

Suppose that $\mathcal{S}^D$ is a TGQ with as base-point the line $x(\infty)$ of $\mathcal{S}$. Then the translation group of $\mathcal{S}^D$ has a subgroup of size q^n acting regularly on the lines, distinct from $x(\infty)$, incident with (∞). Remark that all translations of $\mathcal{S}^D$ are involutions. Further, let $x(\infty) = \pi$ and let τ be the tangent space of $\mathcal{O}$ at π. We have $\pi^\theta = \pi$, $\tau^\theta = \tau$ and $\eta^\theta = \eta$, with η the kernel of $\mathcal{O}$. Clearly the axis of θ is contained in τ, as otherwise elements of $\mathcal{O} \setminus \{\pi\}$ would be fixed by θ. If $\pi_1 \in \mathcal{O} \setminus \{\pi\}$, then $\langle \pi_1, \pi_1^\theta \rangle \cap \tau = \pi'$ belongs to the center of θ, so is a subspace of the axis of θ. Let $u \in \pi'$. The unique line containing u and intersecting η and π is fixed by θ, and it follows that also π and η belong to the axis of θ. Hence τ is the axis of θ and π' is the center of θ. Hence $\mathcal{O}$ is a translation generalized oval with as axis the tangent space of $\mathcal{O}$ at its element $x(\infty)$.

Finally, suppose that $\mathcal{O}$ is a translation generalized oval with axis the tangent space τ of $\mathcal{O}$ at π, where π is the line $x(\infty)$. Let $G(\mathcal{O})$ be the group of involutions with axis τ which acts regularly on $\mathcal{O} \setminus \{\pi\}$. Embed $\mathbf{PG}(3n - 1, q) = \langle \mathcal{O} \rangle$ in $\mathbf{PG}(3n, q)$. Let η be an arbitrary $2n$-dimensional subspace of $\mathbf{PG}(3n, q)$ which is not contained in $\mathbf{PG}(3n - 1, q)$, and which contains τ. Then there is a group G' of involutions of $\mathbf{PG}(3n, q)$ with axis η which induces $G(\mathcal{O})$ in $\mathbf{PG}(3n - 1, q)$; the corresponding automorphism group of $\mathbf{T}(\mathcal{O})$ will also be denoted by G'. The axis γ is a point of $\mathbf{T}(\mathcal{O})$ and G' fixes $\gamma^\perp$ pointwise. As $|G'| = q^n$, the point γ is a center of symmetry, and so is a regular point by Theorem 3.3.4. Part (ii) of the theorem is proved.

Now suppose that $\mathcal{O}$ is a translation generalized ovoid in $\mathbf{PG}(2n+m-1, q)$, so $n < m$, with axis τ (τ being the tangent space of $\mathcal{O}$ at the element π). Then by the final part of the proof of Part (ii), each $(n + m)$-dimensional subspace of $\mathbf{PG}(2n + m, q) \supseteq \mathbf{PG}(2n + m - 1, q)$ which is not contained in $\mathbf{PG}(2n + m - 1, q)$ and which contains τ, is a regular point of $\mathbf{T}(\mathcal{O})$. Hence $s \geq t$, so $n \geq m$, contradiction. The theorem is proved. ∎

Remark 8.1.2 (On the Notion of "Translation Generalized Oval/Ovoid")

Let $\mathcal{O}$ be a generalized ovoid, respectively generalized oval, in $\mathbf{PG}(2n + m - 1, q)$, and let $\pi \in \mathcal{O}$. Suppose τ is the tangent space of $\mathcal{O}$ at π. Suppose there is a subgroup G of $\mathbf{P\Gamma L}_{2n+m}(q)$ that fixes τ pointwise and that acts regularly on $\mathcal{O} \setminus \{\pi\}$. We will call this Property (*) for the moment. Then by an argument similar to the one in the final part of the proof of Theorem 8.1.1, each $(n + m)$-dimensional subspace of $\mathbf{PG}(2n + m, q) \supseteq \mathbf{PG}(2n + m - 1, q)$ which is not contained in $\mathbf{PG}(2n + m - 1, q)$ and which contains τ, is a regular point of $\mathbf{T}(\mathcal{O})$. So $n = m$. It easily follows that also (∞) is regular, and so, by Theorem 2.1.5, q is even. (Alternatively by 1.5.2(i) of [128] it follows that q is even and so, by Theorem 2.1.5, (∞) is regular.) Let $LI(\infty)$ be the line of regular points of $\mathbf{T}(\mathcal{O})$ that corresponds to the space τ of $\mathbf{PG}(3n - 1, q)$. Then in the same way as in the first part of the proof of Theorem 8.1.1, the dual $\mathbf{T}(\mathcal{O})^D$ of $\mathbf{T}(\mathcal{O})$ is a TGQ with as base-point the line L. It follows that G is an elementary abelian group, so that each nontrivial element of G is an involution. Hence a generalized oval, respectively ovoid, satisfies Property (*) if and only if it is a translation generalized oval, respectively ovoid. This is the reason why we only defined translation generalized ovals and ovoids in even characteristic.

Corollary 8.1.3 (see also Chapter 12 **of [128] and K. Thas [206])**   *Let $\mathcal{O}$ be a nonclassical oval in $\mathbf{PG}(2, q)$, q even. Then $\mathcal{S} = \mathbf{T}_2(\mathcal{O})$ contains a regular point x different from the translation point if and only if $x \sim (\infty)$ and $\mathcal{S}^D$ is a $\mathbf{T}_2(\mathcal{O})$ with base-point the point corresponding to $x(\infty)$, if and only if $\mathcal{O}$ is a translation oval with axis the tangent space of $\mathcal{O}$ at its element $x(\infty)$.*   ∎

Remark 8.1.4 Tits [221] defines a *translation ovoid* ("*ovoïde à translation*") $\mathcal{O}$ of $\mathbf{PG}(n, \mathbb{K})$, $n = 2, 3$ and $\mathbb{K}$ a (not necessarily finite) field, in a slightly different way; here $\mathcal{O}$ is a set of points such that for each point $x \in \mathcal{O}$ the union of the tangent lines of $\mathcal{O}$ at x is a hyperplane of $\mathbf{PG}(n, \mathbb{K})$, called the *tangent space* of $\mathcal{O}$ at x. He demands that *for each tangent space T to $\mathcal{O}$, there is a group of elations with axis T, fixing $\mathcal{O}$ globally, and acting regularly on the points of $\mathcal{O}$ not incident with T.* Amongst many other results, he then proves that $\mathbf{PG}(n, \mathbb{K})$, $n = 2, 3$, contains a translation ovoid if and only if the characteristic of $\mathbb{K}$ is 2 and there is a subfield $\mathbb{L}$ of $\mathbb{K}$ for which $(n - 1)[\mathbb{K} : \mathbb{L}] \leq [\mathbb{L} : \mathbb{L}^2]$. Suppose that $\mathbb{K}$ is finite, i.e., $\mathbb{K}$ is isomorphic to $\mathbf{GF}(2^r)$ for some r. If $n = 3$, then clearly no such subfield exists, so $\mathbf{PG}(3, 2^r)$ does not contain translation ovoids in the sense of Tits. If $n = 2$, then the only possibility is that $\mathbb{L} = \mathbb{K} = \mathbf{GF}(2^r)$, and in this case the translation ovoid is a conic of $\mathbf{PG}(2, \mathbb{K})$.

Let $\mathcal{S} = \mathbf{T}(\mathcal{O})$ be a TGQ of order s, with $s = q^n$ even. Let $\mathcal{O} \subseteq \mathbf{PG}(3n - 1, q)$, and write $\mathcal{O} = \{\pi, \pi_1, \ldots, \pi_{q^n}\}$. Suppose η is the *nucleus* of $\mathcal{O}$, that is, the common $(n-1)$-dimensional space of all tangent spaces of $\mathcal{O}$. Consider the $(2n-1)$-space $\tau = \langle \pi, \eta \rangle$ (that is, the tangent space of $\mathcal{O}$ at π). For each $i \in \{1, 2, \ldots, q^n\}$, the set

$$\{\pi, \eta\} \cup \{\langle \pi_i, \pi_j \rangle \cap \tau \parallel i \neq j\}$$

is an $(n-1)$-spread of τ, denoted $\mathcal{T}_i$.

Note that if $\mathcal{O}$ is a translation generalized oval with axis τ, then all the $\mathcal{T}_i$'s coincide. It is also important to note that if all the $\mathcal{T}_i$'s coincide, we have that if γ is an element of that spread distinct from π and η, and if $j \in \{1, 2, \ldots, q^n\}$, then $\langle \gamma, \pi_j \rangle$ contains *precisely one other* element π_r of $\mathcal{O}$ (and hence is disjoint from π_k if $k \neq j, r$). Suppose that all the $(n-1)$-spreads $\mathcal{T}_i$ coincide. We say that $\mathcal{O}$ is *projective at* τ if the following property holds:

> Let γ be an element of $\mathcal{T} = \mathcal{T}_i$ (for all i), where $\pi \neq \gamma \neq \eta$, and let $j \neq k$ be in $\{1, 2, \ldots, q^n\}$ such that $\langle \gamma, \pi_j \rangle \neq \langle \gamma, \pi_k \rangle$. By the above, there are elements $\pi_{j'}$, $j' \neq j$, and $\pi_{k'}$, $k' \neq k$, so that $\pi_{j'} \subseteq \langle \gamma, \pi_j \rangle$ and $\pi_{k'} \subseteq \langle \gamma, \pi_k \rangle$. Then $\langle \pi_j, \pi_k \rangle \cap \langle \pi_{k'}, \pi_{j'} \rangle$ is an element of $\mathcal{T}$.

Remark 8.1.5 Each translation oval $\mathcal{O}$ of $\mathbf{PG}(2, s)$ with axis L, s even, has the (in that case trivial) property that the $\mathcal{T}_i$s coincide, and also is projective at L. It should be noticed that for a generalized oval the fact that the $\mathcal{T}_i$s coincide, does not imply that $\mathcal{O}$ is projective; this follows from Theorem 8.1.6 below, and the existence of ovals in $\mathbf{PG}(2, q)$, q even, which are not translation ovals.

Suppose all the $\mathcal{T}_i$s coincide, and that the generalized oval $\mathcal{O}$ is projective at $\langle \pi, \eta \rangle = \tau$. Let π_j and π_k be arbitrary distinct elements of $\mathcal{O} \setminus \{\pi\}$, and suppose that $\langle \pi_j, \pi_k \rangle \cap \tau = \tau_{jk}$. Let θ_{jk} be the involution with axis τ, center τ_{jk}, and which maps π_j onto π_k. Then θ_{jk} fixes $\mathcal{O}$ globally, and sends π_j onto π_k. If then G is the group consisting of all the involutions θ_{jk}, $j \neq k$, and $j, k \in \{1, 2, \ldots, q^n\}$, together with the identity, then it is clear that G is a group of involutions with axis τ, acting regularly on $\mathcal{O} \setminus \{\pi\}$. Whence $\mathcal{O}$ is a translation generalized oval.

So we have the following theorem.

Theorem 8.1.6 *Let $\mathcal{O}$ be a generalized oval in $\mathbf{PG}(3n - 1, q)$, q even, and use the above notation. Then $\mathcal{O}$ is a translation generalized oval with axis τ if and only if all the $\mathcal{T}_i$s coincide and $\mathcal{O}$ is projective at τ.* ∎

Note that for $n = 1$, this is the well-known geometrical characterization of translation ovals in $\mathbf{PG}(2, q)$ (q even).

8.2 Note on the Definition of Translation Generalized Oval/Ovoid

In this section, we make the following useful observation.

Theorem 8.2.1 *A generalized oval $\mathcal{O}$ of $\mathbf{PG}(3n - 1, q)$, q even, is a translation generalized oval with axis τ, where τ is the tangent space of $\mathcal{O}$ at $\pi \in \mathcal{O}$, if and only if there is a subgroup of involutions of $\mathbf{PGL}_{3n}(q)$ which fixes π and acts regularly on the remaining elements of $\mathcal{O}$.*

Proof. Let G be a group of involutions as stated. Let θ be a nontrivial element of G. Suppose η is the kernel of $\mathcal{O}$. Let Π be an arbitrary n-dimensional subspace of $\mathbf{PG}(3n - 1, q)$ containing π but not contained in τ, and suppose χ is a $(2n - 1)$-dimensional subspace of $\mathbf{PG}(3n - 1, q)$ containing η and meeting τ in η. First of all, one notes that $\Pi^\theta \neq \Pi$. For, suppose that this is not the case. Let π' be the unique element of $\mathcal{O} \setminus \{\pi\}$ intersecting Π. Then $\pi'^\theta = \pi'$, clearly a contradiction.

For each $p \in \chi$, put $p^\alpha = \langle p, \pi \rangle^\theta \cap \chi$. Then α is a linear involutory automorphism of χ. As α does not fix any point of $\chi \setminus \eta$, it follows that α must fix each point of η (as the dimension of η is $n - 1$). But θ and α have the same action on η, so that θ must fix η pointwise. By the same argument, θ also fixes π pointwise, so that τ is fixed pointwise by θ.

Now suppose $\zeta \in \mathcal{O} \setminus \{\pi\}$. Then $\langle \zeta, \zeta^\theta \rangle \cap \langle \pi, \eta \rangle$ belongs to the center, so that the center has dimension at least $n - 1$. So the dimension of the axis is at most $2n - 1$, from which it follows that τ *is the axis of θ. The theorem readily follows.*   ∎

From the proof of Theorem 8.2.1 we have the following useful observation.

Observation 8.2.2 *Let $\mathcal{O}$ be a generalized oval of $\mathbf{PG}(3n-1, q)$, q even, and let τ be the tangent space of $\mathcal{O}$ at the element $\pi \in \mathcal{O}$. Suppose that there is an involution in $\mathbf{PGL}_{3n}(q)$ which stabilizes $\mathcal{O}$, which fixes π and does not fix any of the remaining elements of $\mathcal{O}$. Then θ is an involution with axis τ.*   ∎

Remark 8.2.3 It should be remarked that Theorem 8.2.1 is not true for generalized ovoids. For, consider the GQ $\mathbf{Q}(5, q^n)$, q even, and suppose p

is a point of $\mathbf{Q}(5, q^n)$. Then p is a translation point, and $\mathrm{Aut}(\mathbf{Q}(5, q^n))_p$ contains a subgroup H which acts naturally as $\mathbf{PSL}_2(q^{2n})$ on the set of lines incident with p. In particular, for any line $L \mathrm{I} p$, the stabilizer of L in H contains a normal elementary abelian 2-subgroup which acts regularly on the lines incident with p and distinct from L. So, if $\mathcal{O}$ is the generalized ovoid in $\mathbf{PG}(4n - 1, q)$ corresponding to the TGQ $\mathbf{Q}(5, q^n)^{(p)}$, and τ is the tangent space of $\mathcal{O}$ at $\pi \in \mathcal{O}$, where π corresponds to L, then there is a subgroup G of involutions of $\mathbf{PGL}_{4n}(q)$ which fixes π and acts regularly on the remaining elements of $\mathcal{O}$. But G can not fix τ pointwise by Theorem 8.1.1.

8.3 Characterizations of the $\mathbf{T}_2(\mathcal{O})$ of Tits

Not many characterizations of the $\mathbf{T}_2(\mathcal{O})$ of Tits, $\mathcal{O}$ an oval in $\mathbf{PG}(2, s)$ with s even, are known. It is the goal of this section to provide new characterizations of these objects relying on Theorem 3.9.7 and the theory of translation generalized ovals.

Theorem 8.3.1 *Let* $\mathcal{S}^{(\infty)} = \mathbf{T}(\mathcal{O})$ *be a TGQ of order s with base-point* (∞). *Suppose there is some line* $L \mathrm{I} (\infty)$ *for which the associated projective plane* π_L *is Desarguesian. Suppose that* $x \neq (\infty)$ *is a regular point, $x \nmid L$. Then* $\mathcal{O}$ *is a regular translation generalized oval, i.e.* $\mathcal{S} \cong \mathbf{T}_2(\mathcal{O}')$ *of Tits for some translation oval* $\mathcal{O}'$ *of* $\mathbf{PG}(2, s)$, *and conversely.*

Proof. First suppose that x is not collinear with (∞). Then since $\mathcal{S}^{(\infty)}$ is a TGQ, each point not collinear with (∞) is regular. It follows that each point is regular, and then $\mathcal{S}^{(\infty)} \cong \mathbf{W}(s)$ by Theorem 2.2.2. Also, s is even by Theorem 2.1.5. The theorem follows.

Now suppose $x \sim (\infty)$. By 1.5.2(i) of [128] s is even (alternatively, all points of the line $x(\infty)$ are regular, so s is even by Theorem 2.1.5). Then Theorem 8.1.1 implies that $\mathcal{O}$ is a translation generalized oval with axis the tangent of $\mathcal{O}$ at $x(\infty)$. The result now follows from Theorems 8.1.6 and 3.9.7.

The converse is obvious. ∎

The following theorem is very general.

Theorem 8.3.2 *Let $S = S^{(\infty)} = \mathbf{T}(\mathcal{O})$ be a TGQ of order s, s even, with base-point (∞). Suppose there is a line $L\mathrm{I}(\infty)$ for which the associated projective plane π_L is Desarguesian. Suppose also that there is a line $M \neq L$ which is incident with (∞), and an involution θ of $\mathrm{Aut}(S)_{(\infty)}$ which fixes M pointwise and which does not fix any of the lines incident with (∞) and different from M. Then $S \cong \mathbf{T}_2(\mathcal{O}')$ of Tits for some oval $\mathcal{O}'$ of $\mathbf{PG}(2,s)$.*

Proof. Let π, respectively π', be the element of $\mathcal{O}$ corresponding to L, respectively M, and let τ' be the tangent space of $\mathcal{O}$ at π'. Suppose θ is as in the theorem. Then by Observation 8.2.2, θ fixes τ' pointwise and does not fix any element of $\mathcal{O} \setminus \{\pi'\}$. So π and π^θ induce the same spread $\mathcal{T}$ in τ'. Moreover, since π_L is Desarguesian, we have that $\mathcal{T}$ is regular. Now apply Theorem 3.9.7. ∎

Corollary 8.3.3 *Let $S = S^{(\infty)} = \mathbf{T}(\mathcal{O})$ be a TGQ of order s, s even, with base-point (∞). Suppose there is a line $L\mathrm{I}(\infty)$ for which the associated projective plane π_L is Desarguesian. Suppose that there is a nontrivial symmetry about some point $x \,\mathrm{I}\, L$. Then $S \cong \mathbf{T}_2(\mathcal{O}')$ of Tits for some oval $\mathcal{O}'$ of $\mathbf{PG}(2,s)$.*

Proof. If x is not collinear with (∞) then one observes that each point of S is a translation point, so that each point is regular. The corollary then follows from Theorem 2.2.2 and Theorem 1.5.2(i).

If $x \sim (\infty)$, we consider a nontrivial involution in the group of symmetries generated by the given one, and then the corollary follows from Theorem 8.3.2. ∎

8.4   A Characterization of Translation Generalized Ovals

Let $\mathcal{O} = \{\pi, \pi_1, \ldots, \pi_{q^n}\}$ be a generalized oval in $\mathbf{PG}(3n-1,q)$, with $q = 2^h$ even, and let τ be the tangent of $\mathcal{O}$ at π. Now we define a point-line incidence structure $\mathbf{A}(\mathcal{O})$ as follows:

- POINTS are the elements of $\mathcal{O} \setminus \{\pi\}$;

- LINES are the pairs $\{\pi_i, \pi_j\}$, with $i, j \in \{1, 2, \ldots, q^n\}$ and $i \neq j$;

- INCIDENCE is containment.

Hence $\mathbf{A}(\mathcal{O})$ is the complete graph with vertex set $\mathcal{O} \setminus \{\pi\}$. Further, two lines $\{\pi_i, \pi_j\}$ and $\{\pi_k, \pi_l\}$ are called *parallel* if either $\{\pi_i, \pi_j\} = \{\pi_k, \pi_l\}$ or $\langle \pi_i, \pi_j \rangle \cap \tau = \langle \pi_k, \pi_l \rangle \cap \tau$.

Theorem 8.4.1 *The incidence structure* $\mathbf{A}(\mathcal{O})$ *provided with parallelism is isomorphic to the* hn-*dimensional affine space* $\mathbf{AG}(hn, 2)$ *over* $\mathbf{GF}(2)$ *if and only if* $\mathcal{O}$ *is a translation generalized oval with axis* τ.

Proof. Assume that $\mathbf{A}(\mathcal{O})$ provided with parallelism is isomorphic to $\mathbf{AG}(hn, 2)$. If $\{\pi_i, \pi_j\}$ is a line of $\mathbf{A}(\mathcal{O})$ and $\pi_k \notin \{\pi_i, \pi_j\}$, then there is exactly one line of $\mathbf{A}(\mathcal{O})$ containing π_k and parallel to $\{\pi_i, \pi_j\}$. Also, if the distinct lines $\{\pi_i, \pi_j\}$ and $\{\pi_k, \pi_l\}$ are parallel, then also $\{\pi_i, \pi_k\}$ and $\{\pi_j, \pi_l\}$ are parallel. Hence, by Theorem 8.1.6, $\mathcal{O}$ is a translation generalized oval with axis τ.

Conversely, assume that $\mathcal{O}$ is a translation generalized oval with axis τ. Then, by Theorem 8.1.6, the q^n $(n-1)$-spreads of τ defined by the respective elements of $\mathcal{O} \setminus \{\pi\}$ coincide, and further $\mathcal{O}$ is projective at π. Now by [98] the point-line incidence structure $\mathbf{A}(\mathcal{O})$ provided with parallelism is isomorphic to $\mathbf{AG}(hn, 2)$.   ∎

Remark 8.4.2 Let $\mathbf{PG}(hn, 2)$ be the projective completion of the affine space $\mathbf{A}(\mathcal{O})$. Then the points at infinity of $\mathbf{A}(\mathcal{O})$, that is, the points of $\mathbf{PG}(hn, 2)$ not in $\mathbf{A}(\mathcal{O})$, can be identified with the elements of $\mathcal{T} \setminus \{\pi, \eta\}$, where $\mathcal{T}$ is the common $(n-1)$-spread of τ defined by the elements of $\mathcal{O} \setminus \{\pi\}$ and η is the kernel of $\mathcal{O}$. Any line of $\mathbf{PG}(hn-1, 2) = \mathbf{PG}(hn, 2) \setminus \mathbf{A}(\mathcal{O})$ is of the form $\{\alpha, \beta, \gamma\}$, with $\alpha = \langle \pi_i, \pi_j \rangle \cap \langle \pi_k, \pi_l \rangle$, $\beta = \langle \pi_i, \pi_k \rangle \cap \langle \pi_j, \pi_l \rangle$, $\gamma = \langle \pi_i, \pi_l \rangle \cap \langle \pi_j, \pi_k \rangle$ and $\pi_i, \pi_j, \pi_k, \pi_l$ distinct elements of $\mathcal{O} \setminus \{\pi\}$.

8.5 Classification of 2-Transitive Generalized Ovals in Even Characteristic

We call a generalized oval $\mathcal{O}$ in $\mathbf{PG}(3n-1, q)$ 2-*transitive* if there is a subgroup of $\mathbf{P\Gamma L}_{3n}(q)$ which stabilizes $\mathcal{O}$ and acts 2-transitively on its elements. Note that by Thas and K. Thas [191], cf. Theorem 3.7.2, this is equivalent to asking that the automorphism group of $\mathbf{T}(\mathcal{O})$, which we may demand to fix the translation point (∞) (as otherwise $\mathbf{T}(\mathcal{O}) \cong \mathbf{Q}(4, q^n)$), acts 2-transitively on the lines incident with (∞).

We now classify the 2-transitive generalized ovals in even characteristic, without relying on the classification of finite simple groups. We will look

at the equivalent problem for the corresponding TGQs. From the theorem
will follow which groups arise.

Theorem 8.5.1 *Let $\mathcal{S} = \mathcal{S}^{(\infty)} = \mathbf{T}(\mathcal{O})$ be the TGQ of order (q^n, q^n), q even,
which arises from a 2-transitive generalized oval $\mathcal{O}$ in $\mathbf{PG}(3n - 1, q)$. Then
$\mathcal{S} \cong \mathbf{Q}(4, q^n)$, and so $\mathcal{O}$ is classical.*

Proof. Since q is even, the point (∞) of $\mathbf{T}(\mathcal{O})$ is regular. The plane $\pi_{(\infty)}^D$,
which is the dual of $\pi_{(\infty)}$, is a translation plane of order q^n, where the
parallel classes correspond to the lines of $\mathcal{S}$ incident with (∞). We suppose
that $\mathrm{Aut}(\mathcal{S})$ fixes (∞) for reasons of conveniency (otherwise $\mathcal{S} \cong \mathbf{Q}(4, q^n)$).
Then $\mathrm{Aut}(\mathcal{S})$ acts 2-transitively on the parallel classes of $\pi_{(\infty)}^D$. We have
the following possibilities by Schulz [136] and Czerwinsky [42], see also
Lüneburg [101] (Chapter VI):

(a) $\pi_{(\infty)}^D$ is Desarguesian, and $\mathrm{Aut}(\mathcal{S})$ has a subgroup K which induces
   $\mathbf{PSL}_2(q^n)$ on the line at infinity of $\pi_{(\infty)}^D$;

(b) $\pi_{(\infty)}^D$ is a Lüneburg plane (so that n is even, and h odd, with $q = 2^h$
   — cf. also the next chapter), and $\mathrm{Aut}(\mathcal{S})$ has a subgroup K which
   induces the Suzuki group $\mathbf{Sz}(\sqrt{q^n})$ on the line at infinity of $\pi_{(\infty)}^D$.

We take K in such a way that the translation group G of $\mathcal{S}$ is contained in
K.

Let α be an involution of K which fixes the element $\pi \in \mathcal{O}$ but not all
elements of $\mathcal{O}$; then π is the only element of $\mathcal{O}$ which is fixed by α. We
regard α as being an element of $\mathbf{P\Gamma L}_{3n}(q)$. Since h is odd, α is a linear
involution. By Observation 8.2.2, α is an involution with axis τ, the latter
being the tangent space of $\mathcal{O}$ at π. As $\mathbf{PSL}_2(q^n)$ and $\mathbf{Sz}(\sqrt{q^n})$ both are
generated by their involutions (a fact that can be proved, for instance, by
using their representations given in the next chapter), we have proved that
the nucleus η of $\mathcal{O}$ is fixed pointwise by K (regarded as a subgroup of
$\mathbf{PGL}_{3n}(q)$).

Let z' be a point of $\mathcal{S}$ not collinear with (∞). Then since $\mathcal{S}$ is a TGQ, it
is easy to see that $K_{z'} = K'$ still induces $\mathbf{PSL}_2(q^n)$, respectively $\mathbf{Sz}(\sqrt{q^n})$,
on the lines incident with (∞). Let $LI(\infty)$ be arbitrary, and consider a
Sylow 2-subgroup S_2 of K_L'. As S_2 — seen as a subgroup of $\mathbf{PGL}_{3n}(q)$ —
fixes η pointwise, S_2 fixes each line of $\mathcal{S}$ incident with $\mathrm{proj}_L z'$ (and S_2 acts
sharply transitively on the set of lines incident with (∞) and different from

L). If we interpret S_2 as an automorphism group of π_L, it thus follows that S_2 also must fix L pointwise (since in a finite projective plane any central collineation is also axial, and conversely).

The group $G_{\mathrm{proj}_{L}z'}$ has order q^{2n} and fixes L pointwise, so together with S_2, it generates a group H such that, if N is the group of symmetries about L, H/N acts sharply transitively on the set of spans $\{U, V\}^{\perp\perp}$ of nonconcurrent lines $U, V \in L^{\perp}$, and fixes L pointwise. So H/N is a translation group of π_L^D of order q^{2n}, and hence it is elementary abelian. In particular $S_2 \cong S_2 N/N$ also is. So $\mathbf{PGL}_{3n}(q)_{\mathcal{O}}$ has a subgroup of involutions that fixes π and acts sharply transitively on $\mathcal{O} \setminus \{\pi\}$.

By Theorem 8.2.1, we conclude that $\mathcal{O}$ is a translation generalized oval with axis τ. So any point incident with L is regular. As L was an arbitrary line incident with (∞), it follows that each point of $(\infty)^{\perp}$ is regular, so that each point of $\mathcal{S}$ is regular (cf. 1.3.6(iv) of [128]). It follows that $\mathcal{S} \cong \mathbf{Q}(4, q^n)$ by Theorem 2.2.2, so that Case (b) cannot occur. Hence by Theorem 1.5.2 and §3.6, $\mathcal{O}$ is classical.   ∎

Chapter 9

Moufang Sets and Translation Moufang Sets

In this chapter, we gather some results from group theory, that we will need later on. The central notion is the notion of a *Moufang set*. This can be viewed as a generalization of a one-dimensional translation group.

We begin without any finiteness restriction, but later on, when dealing with classification, we will have to consider only finite Moufang sets.

9.1 Definition and General Results

A *Moufang set (on the set X)* is a system $(X, (U_x)_{x \in X})$ consisting of a set X, $|X| > 2$, and a family of groups U_x of permutations of X indexed by X itself and satisfying the following conditions.

(MS1) U_x fixes $x \in X$ and is sharply transitive on $X \setminus \{x\}$.

(MS2) In the full permutation group of X, each U_x normalizes the set of subgroups $\{U_y \parallel y \in X\}$ (that is, it acts on the latter by conjugation).

The groups U_x will be called *root groups*. The elements of U_x are often called *root elations*. If U_x is abelian for some $x \in X$, then for all $x \in X$ and we call the Moufang set a *translation Moufang set*.

If $(X, (U_x)_{x \in X})$ is a Moufang set, and $Y \subseteq X$, then Y, $|Y| > 2$, induces a *sub Moufang set* if for each $y \in Y$, the stabilizer $(U_y)_Y$ acts sharply transitively on $Y \setminus \{y\}$. In this case $(Y, ((U_y)_Y)_{y \in Y})$ is a Moufang set.

The group S generated by the U_x, for all $x \in X$, is called the *little projective group* of the Moufang set. A permutation of X that normalizes the set of subgroups $\{U_y \parallel y \in X\}$, is called an *automorphism* of the Moufang set. The set S of all automorphisms of the Moufang set is a group G, called the *full projective group* of the Moufang set. Any group H, with $S \leq H \leq G$, is called a *projective group* of the Moufang set. We have the following easy lemma.

Lemma 9.1.1   (i) *The little projective group S of a Moufang set $(X, (U_x)_{x \in X})$ acts doubly transitively on the set X.*

  (ii) *A permutation group H (acting on X) is a projective group if and only if $U_x \trianglelefteq H_x$, for every $x \in X$.*

Proof.   (i) This follows immediately from the fact that every point stabilizer S_x acts transitively on $X \setminus \{x\}$, since $U_x \leq S_x$.

(ii)   Since any automorphism preserves the set $\{U_x \parallel x \in X\}$ under conjugation, it is clear that $U_x \trianglelefteq H_x$, for all $x \in X$, if H is a projective group. Conversely, if H is a group of permutations of X with $U_x \trianglelefteq H_x$, for all $x \in X$, then clearly $S \leq H$. Now let $h \in H$ and $x \in X$ both be arbitrary. By (i), there exists $g \in S$ with $x^h = x^g$. Since $hg^{-1} \in H_x$, we have $U_x^{hg^{-1}} = U_x$, and so $U_x^h = U_x^g = U_{x^g}$. Hence h normalizes the set of root groups and is therefore an automorphism.   ∎

We will sometimes refer to a *proper* Moufang set for the case in which the little projective group is not sharply doubly transitive, and *improper* otherwise.

In the finite case, there is a complete classification of Moufang sets. In the infinite case, there is little theory known, and we are still far from a classification. Only examples are studied in the literature. We will prove some elementary general properties (some of them do not appear in the literature, however), and then we describe the finite 2-transitive examples which are not sharply 2-transitive (but we do not prove the classification).

Since Moufang sets deal with sharply transitive actions, we first review the basics of those actions. We supply a proof for completeness' sake, but these proofs and different ones may be found in every text book on basic group theory.

A group element g is said to act *freely* on a set S, if the group generated by g does not contain any nontrivial element that fixes $x \in S$. A group acts *freely* on S if only the identity fixes some $x \in S$.

Proposition 9.1.2 *Let (G, X) be a sharply transitive group.*

(i) *Then we can identify X with G and the action of $g \in G$ on the element $x \in X = G$ is given by right multiplication xg.*

(ii) *Suppose some permutation u commutes with every element of $G = X$. Then u acts freely on X and the action of u is given by left multiplication with some element $h \in G$ (so u maps every $x \in X$ to hx).*

(iii) *If a transitive permutation group H centralizes G, then H acts sharply transitively on X and there is an isomorphism $\varphi : H \to G$ such that the action of $h \in H$ on $G = X$ is given by $h : x \mapsto h^{-\varphi}x$. If in particular G is abelian, then $H = G$.*

(iv) *Suppose the sharply transitive permutation groups H and G normalize each other. Then either G and H have a nontrivial intersection, or G and H centralize each other.*

Proof. (i) Take any fixed element $x \in X$ and identify each element $g \in G$ with $x^g \in X$. Then $(x^g)^h$ gets identified with gh, and (i) follows.

(ii) Now put $X = G$. If $[u, G] = \{\mathrm{id}\}$, then every element of $\langle u \rangle$ centralizes G, so we only have to show that u acts fixed point freely if it is nontrivial. Let e be the identity in G and suppose that u maps e to $h \in G$. If $g \in G$ is arbitrary, then $g^u = (eg)^u = (e^u)g = hg$ (the second equality uses the fact that u commutes with g), hence u maps g to hg.

(iii) By (ii), H acts freely on X and hence sharply transitively, and there is a bijection $\theta : H \to G$ such that the action of $h \in H$ on $X = G$ is given by left multiplication with h^θ. If $h_1, h_2 \in H$, and $g \in G$, then $(h_1 h_2)^\theta g = g^{h_1 h_2} = (g^{h_1})^{h_2} = h_2^\theta h_1^\theta g$, so $(h_1 h_2)^\theta = h_2^\theta h_1^\theta$. Putting $\varphi : H \to G : h \mapsto (h^\theta)^{-1}$, we see that φ is an isomorphism and so we can write $h^\theta = (h^{-1})^\varphi = h^{-\varphi}$.

(iv) If G and H intersect trivially, then from $[g, h] = g^{-1}h^{-1}gh \in H^g h \subseteq H$ and $g^{-1}h^{-1}gh \in g^{-1}G^h = G$ we deduce that $[g, h]$ is trivial, for all $g \in G$ and all $h \in H$. ∎

By (iii) of the previous proposition, and its proof, it follows that for every sharply transitive permutation group G acting on a set X, there is a unique sharply transitive permutation group H acting on X and centralizing G in $\mathbf{Sym}(X)$. We say that the action of H is *opposite* the action of G and denote H by G^{opp}.

Note that, if G is abelian, then the two actions are obviously the same.

We now get back to the Moufang sets. A first question that can be asked is: do we really need all the root groups in the definition? This is one answer.

Proposition 9.1.3 *Let* X *(*$|X| > 2$*) be a set and let* $a, b \in X$ *be distinct. Let* U_a *and* U_b *be two permutation groups acting on* X *such that* U_a *fixes* a *and acts sharply transitively on* $X \setminus \{a\}$*, while* U_b *fixes* b *and acts sharply transitively on* $X \setminus \{b\}$*. Then* U_a *and* U_b *are root groups of at most one common Moufang set. Also, they are root groups of a Moufang set if and only if for each* $u \in U_a^{\times}$*,* $u \neq \mathrm{id}$*, there exists* $v \in U_b$ *such that* $U_a^v = U_b^u$*, if and only if* U_a *is normal in* G_a*,* U_b *is normal in* G_b*, and* U_a *is conjugate to* U_b *in* $G := \langle U_a, U_b \rangle$*.*

Proof. Suppose first that $(X, (U_x)_{x \in X})$ is a Moufang set. For given $u \in U_a \setminus \{\mathrm{id}\}$, define $v \in U_b$ by $a^v = b^u$. Then (MS2) implies that $U_a^v = U_{a^v} = U_{b^u} = U_b^u$. By the doubly transitivity of G, it maps U_a under conjugation to U_b. Also, since G normalizes the U_x, the stabilizer G_a of a must normalize U_a, hence $U_a \trianglelefteq G_a$ and likewise $U_b \trianglelefteq G_b$.

Now suppose that for each $u \in U_a^{\times}$ there exists $v \in U_b$ such that $U_a^v = U_b^u$. Clearly $a^v = b^u$. This motivates the following notation. For each $x \in X \setminus \{a, b\}$, denote by u_x the unique element of U_a mapping b to x and let v_x be the unique element of U_b mapping a to x. Define $U_x := U_a^{v_x} = U_b^{u_x}$. Then we claim that $(X, (U_x)_{x \in X})$ is a Moufang set. Indeed, (MS1) is obvious, so consider (MS2). We have to show that, for all $x \in X$, the group U_x permutes by conjugation the groups U_y, $y \in X$. For $x \in \{a, b\}$, this is trivial. If $x \notin \{a, b\}$, then we can write an arbitrary element of U_x as $v_x^{-1} u v_x$, with $u \in U_a$, and since each of v_x^{-1}, u and v_x preserves by conjugation $\{U_y \parallel y \in X\}$, also $v_x^{-1} u v_x$ does. It is now also clear that U_a and U_b are root groups of at most one common Moufang set.

Now suppose that U_a is normal in G_a, U_b is normal in G_b, and U_a is conjugate to U_b in $G = \langle U_a, U_b \rangle$. Let $u \in U_a^{\times}$ be arbitrary, and let $v \in U_b$ be such that $b^u = a^v$. Then $a^{vu^{-1}} = b$. Since U_a and U_b are conjugate in G, there exists $g \in G$ such that $U_a^g = U_b$. Clearly $a^g = b$, as a and b are unambiguously determined by U_a and U_b, respectively. This implies that $vu^{-1}g^{-1} \in G_a$, and so, by assumption, $U_a^{vu^{-1}g^{-1}} = U_a$, hence $U_a^v = U_a^{gu} = U_b^u$. The previous paragraph completes the proof of the proposition. ∎

The condition that for each $u \in U_a^\times$ there exists $v \in U_b$ such that $U_a^v = U_b^u$ is used by Timmesfeld [216] to define rank one groups. But a rank one group has the additional condition that all root groups are nilpotent, and we do not want this restriction.

For a set X, elements $a, b \in X$ (with $|X| > 2$), and permutation groups U_a, U_b of X such that U_a (respectively, U_b) fixes a (respectively, b) and acts sharply transitively on $X \setminus \{a\}$ (respectively, $X \setminus \{b\}$), we call (X, U_a, U_b) a *Moufang triple* if, for $G = \langle U_a, U_b \rangle$, U_a is normal in G_a, U_b is normal in G_b, and U_a is conjugate to U_b in G.

We can use the previous proposition to recognize easily sub Moufang sets.

Corollary 9.1.4 *Let* $\mathcal{M} = (X, (U_x)_{x \in X})$ *be a Moufang set. Suppose that* $Y \subseteq X$, *with* $|Y| \geq 3$, *and consider two distinct elements* $\infty, O \in Y$. *Let* V_∞ *and* V_O *be the permutation groups induced on* Y *by the stabilizers* $(U_\infty)_Y$ *and* $(U_O)_Y$, *respectively. Then* (Y, V_∞, V_O) *is a Moufang triple — and hence induces a sub Moufang set of* $\mathcal{M}$ *— if and only if* V_∞ *and* V_O *act transitively on* $Y \setminus \{\infty\}$ *and* $Y \setminus \{O\}$, *respectively.*

Proof. It is clear that the stated transitivity condition is necessary. We now show that it is also sufficient.

First we observe that the group H generated by V_∞ and V_O acts doubly transitively on Y. Let $h \in H$ be such that $\infty^h = O$ and $O^h = \infty$. Then h can be written as product of elements of $V_\infty \cup V_O$, and reading these elements as belonging to $U_\infty \cup U_O$, we see that there is some element $g \in \langle U_\infty, U_O \rangle$ such that $G/Y = h$. It follows that V_∞^h is the restriction to Y of $(U_\infty)_Y^g$, and hence this must be equal to V_O. Hence V_∞ and V_O are conjugate in H. Similarly one shows that V_∞ and V_O are normal in H_∞ and H_O, respectively. The assertion now follows directly from Proposition 9.1.3. ∎

Proposition 9.1.3 tells us something about the possibility of two root groups to be contained in the same Moufang set. Now we want to look at the situation where we have two Moufang sets acting on the same set X and sharing at least one root group, and we want to find conditions under which these Moufang sets are the same. More precisely, we have the following result.

Theorem 9.1.5 *Let $(X, (U_x)_{x \in X})$ be a Moufang set, and let $a, b \in X$ be distinct. Moreover, let W_b be a permutation group acting on X, fixing b and acting sharply transitively on $X \setminus \{b\}$. Suppose that (X, U_a, W_b) is a Moufang triple. If $U_b \cap W_b$ is nontrivial, or if U_b and W_b normalize each other, then $W_b = U_b$. If U_b and W_b centralize each other, then $U_b = W_b$ and the Moufang set is a translation Moufang set.*

Proof. Let $\mathcal{M}$ be the Moufang set determined by U_a and U_b, and let $\mathcal{N} = (X, (W_x)_{x \in X})$ be the one determined by U_a and W_b, with $U_a = W_a$. First suppose that $v \in U_b \cap W_b$ is nontrivial. Then $W_{a^v} = W_a^v = U_a^v = U_{a^v}$ is a root group in both $\mathcal{M}$ and $\mathcal{N}$, and since these Moufang sets are both determined by U_a and U_{a^v}, they must coincide. Hence $U_b = W_b$.

Now suppose that U_b centralizes W_b. Then the action of W_b is opposite the action U_b. Take arbitrary elements $u \in U_a$, $v \in U_b$ and $w \in W_b$. Then $[v, w] = \mathrm{id}$, hence $[v^u, w^u] = \mathrm{id}$ and we see that the action of W_{b^u} is opposite the action of U_{b^u}. If $|X| = 3$, then clearly $U_b = W_b$ is commutative and the result follows. Hence we may assume $|X| \geq 4$, so that we can take two distinct elements c, d in $X \setminus \{a, b\}$. We then see that $\mathcal{N}$ is determined by the permutation groups W_c and W_d the actions of which are opposite those of U_c and U_d, respectively.

By the doubly transitivity of the little projective group $G(\mathcal{M})$ of $\mathcal{M}$, there is a permutation $g \in G(\mathcal{M})$ interchanging a with b. Since $[U_b, W_b] = \{\mathrm{id}\}$, we also have $[U_b^g, W_b^g] = \{\mathrm{id}\}$. Hence, if we denote U_a^{opp} by V_a, then, since $U_b^g = U_{b^g} = U_a$, and since W_b^g acts sharply transitively on $X \setminus \{a\}$, we obtain $W_b^g = V_a$. If $c^{g^{-1}} = c'$ and $d^{g^{-1}} = d'$, then similarly, one easily shows that $W_{c'}^g = W_c$ and $W_{d'}^g = W_d$. Hence the conjugate $\mathcal{N}^g$ of $\mathcal{N}$ contains the root groups W_c and W_d and hence coincides with $\mathcal{N}$. Comparing the root groups fixing b, we see that $U_b = W_b$, and $[U_b, W_b] = \{\mathrm{id}\}$ implies that U_b is abelian. Hence $\mathcal{M} = \mathcal{N}$ is a translation Moufang set.

Finally suppose that U_b and W_b normalize each other. If they share a nontrivial permutation of X, then by the first part of the proof $U_b = W_b$. If they intersect trivially, then by Proposition 9.1.2(iv) they centralize each other, and the second part of our proof implies again $U_b = W_b$. ∎

9.2 Finite Moufang Sets

In this section, our aim is to comment on the following classification result of finite Moufang sets.

Theorem 9.2.1 *Let $\mathcal{M} = (X, (U_x)_{x \in X})$ be a finite Moufang set with little projective group S and full projective group G. Then one of the following cases occurs.*

(2T) *S acts sharply doubly transitively on X, there is a prime number p and a positive integer n such that $|X| = p^n$, and S contains a normal sharply transitive subgroup N, which is an elementary abelian p-group.*

(Ch) *S is a Chevalley group, there exists a prime number p and a positive integer n such that $|X| = p^n + 1$, and U_x is a p-group of nilpotency class at most 3. We have one of the following cases:*

- *$S \cong \mathbf{PSL}_2(p^n)$, $p^n \geq 4$, is simple and X is the projective line $\mathbf{PG}(1, p^n)$;*

- *n is a multiple of 3, $S \cong \mathbf{PSU}_3(p^{n/3})$, $p^n \geq 27$, is simple and X is the Hermitian unital $\mathcal{U}_H(p^{n/3})$;*

- *$p = 2$, $n = 2n'$ is even, n' is odd, $S \cong \mathbf{Sz}(2^{n'})$, $n' \geq 3$, is simple and X is the Suzuki-Tits ovoid $\mathcal{O}_{ST}(2^{n'})$;*

- *$p = 3$, $n = 3n'$, n' is odd, $S \cong \mathbf{R}(3^{n'})$, $n' \geq 1$, is simple for $n' \geq 3$, and X is the Ree-Tits unital $\mathcal{U}_R(3^{n'})$; if $n' = 1$, then $\mathbf{R}(3) \cong \mathbf{P\Gamma L}_2(8)$ has a simple subgroup of index 3, which coincides with the commutator subgroup of S.*

In all cases, we have that G is the full automorphism group of S. Also, the root groups are precisely the Sylow p-subgroups of S.

∎

We will not prove this theorem here, but the interested reader can find proofs in Shult [143] and Hering, Kantor and Seitz [69]. We will content ourselves with some comments on this theorem, and with an explicit definition of the Moufang sets appearing in (Ch) above.

It should be noted that in the papers [143, 69], the term "split BN-pair of rank 1" is used instead of "Moufang set".

Concerning the sharply doubly transitive case, it follows from Frobenius' Theorem that there is a normal sharply transitive subgroup N acting on X. The group U_x, for arbitrary $x \in X$, clearly acts transitively on $N^\times$ by conjugation, hence all elements in N have the same order and N is a p-group. Since every p-group has a nontrivial center, and since U_x acts transitively on $N^\times$, we see that N must be elementary abelian. It follows that $|X|$ is a prime power. The standard example here acts on a finite field $\mathbf{GF}(q)$, for some prime power q, and the actions are given by

$$\mathbf{GF}(q) \to \mathbf{GF}(q) : x \mapsto ax + b,$$

with $a \in \mathbf{GF}(q) \setminus \{0\}$ and $b \in \mathbf{GF}(q)$. The normal sharply transitive subgroup consists of the maps with $a = 1$.

A second class of examples is given by a slight modification of the previous example. If q is odd and a perfect square, then we retain the previous maps with a a square in $\mathbf{GF}(q) \setminus \{0\}$, and substitute the other maps with the maps $\mathbf{GF}(q) \to \mathbf{GF}(q) : x \mapsto ax^{\sqrt{q}} + b$, with a a nonsquare in $\mathbf{GF}(q)$. We call it the *nonstandard example related to* $\mathbf{GF}(q)$.

We now turn to the examples under (Ch), the *Chevalley type groups*.

As a general remark, we would like to point out that in the following descriptions we are primarily interested in the finite case, but we choose the notation in such a way that also the general (infinite) case is covered, disregarding nevertheless noncommutative fields.

9.2.1   The Case $\mathbf{PSL}_2(q)$

This is the prototype of all Moufang sets. Let $\mathbb{K}$ be any (finite) field, and denote by $\mathbf{PG}(1, \mathbb{K})$ the projective line over $\mathbb{K}$. So we may identify $\mathbf{PG}(1, \mathbb{K})$ with the set of all 1-dimensional subspaces of a given 2-dimensional vector space $\mathbb{K} \times \mathbb{K}$. The elements of $\mathbf{PG}(1, \mathbb{K})$ can be written as (a, b), with $a, b \in \mathbb{K}$, $(a, b) \neq (0, 0)$ and (a, b) identified with (ca, cb) for all $c \in \mathbb{K}^\times$. We set $O = (0, 1)$ and $\infty = (1, 0)$. We define U_O as the (multiplicative) group of matrices $(k)_O := \begin{pmatrix} 1 & k \\ 0 & 1 \end{pmatrix}$, $k \in \mathbb{K}$, U_∞ consists of the matrices $(k)_\infty := \begin{pmatrix} 1 & 0 \\ k & 1 \end{pmatrix}$, for all $k \in \mathbb{K}$, and the action of a matrix M on an element (a, b) is by right multiplication $(a, b)M$ (conceiving (a, b) as a (2×1)-matrix). It is easy to calculate that $U_\infty^{(k)_O} = U_O^{(k^{-1})_\infty}$, hence by Proposition 9.1.3, we indeed have a Moufang set acting on $\mathbf{PG}(1, \mathbb{K})$. Since U_∞ is abelian, we have a translation Moufang set. We denote this Moufang set by $\mathcal{M}(\mathbf{PG}(1, \mathbb{K}))$. The little projective group is equal to $\mathbf{PSL}_2(\mathbb{K})$ and is simple if $|\mathbb{K}| \geq 4$. For $|\mathbb{K}| = 2$, we have the unique Moufang set on 3 points, and for $|\mathbb{K}| = 3$, we obtain the unique Moufang set on 4 points. Both are improper Moufang sets, isomorphic to the standard examples related to $\mathbf{GF}(3)$ and $\mathbf{GF}(4)$, respectively.

9.2.2 The Case $\mathrm{PSU}_3(q)$

Let $\mathbb{K}$ be a field having some quadratic Galois extension $\mathbb{F}$. This means that $\mathbb{F}$ admits some field involution σ and the elements of $\mathbb{F}$ fixed by σ are precisely those of the subfield $\mathbb{K}$. In order to treat all characteristics at the same time, we introduce the following notation.

For an arbitrary $i \in \mathbb{F} \setminus \mathbb{K}$, we can write any element $k \in \mathbb{F}$ as

$$\frac{k^\sigma i - k i^\sigma}{i - i^\sigma} + \frac{k - k^\sigma}{i - i^\sigma} i,$$

with both $(k i^\sigma - k^\sigma i)(i^\sigma - i)^{-1}$ and $(k - k^\sigma)(i - i^\sigma)^{-1}$ in $\mathbb{K}$. So $\mathbb{F}$ can be regarded as a 2-dimensional vector space over $\mathbb{K}$ and the map $\mathbb{F} \to \mathbb{K} : x \mapsto x + x^\sigma$ is $\mathbb{K}$-linear, nontrivial, hence surjective with 1-dimensional kernel. We denote the inverse image of the element $k \in \mathbb{K}$ by $\mathbb{K}^{(k)}$.

Now we define our set X. It is the set of all projective points (x_0, x_1, x_2), up to a multiplicative nonzero factor, of the projective plane $\mathbf{PG}(2, \mathbb{F})$ satisfying the equation $x_0 x_2^\sigma + x_0^\sigma x_2 = x_1 x_1^\sigma$. We can write

$$X = \{(1,0,0)\} \cup \{(k, x, 1) \;\|\; x \in \mathbb{F}, k \in \mathbb{K}^{(x x^\sigma)}\},$$

and also

$$X = \{(0,0,1)\} \cup \{(1, x, k) \;\|\; x \in \mathbb{F}, k \in \mathbb{K}^{(x x^\sigma)}\}.$$

We set $O = (0,0,1)$ and $\infty = (1,0,0)$. The group of collineations $U_\infty = \{(x, k)_\infty \;\|\; x \in \mathbb{F}, k \in \mathbb{K}^{(x x^\sigma)}\}$, where $(x, k)_\infty$ is the collineation of $\mathbf{PG}(2, \mathbb{F})$ defined by the linear transformation with matrix

$$\begin{pmatrix} 1 & 0 & 0 \\ x^\sigma & 1 & 0 \\ k & x & 1 \end{pmatrix},$$

acts sharply transitively on $X \setminus \{\infty\}$ (on the right!) and fixes ∞. Likewise, the group $U_O = \{(x, k)_O \;\|\; x \in \mathbb{F}, k \in \mathbb{K}^{(x x^\sigma)}\}$, where $(x, k)_O$ is the collineation of $\mathbf{PG}(2, \mathbb{F})$ defined by the linear transformation with matrix

$$\begin{pmatrix} 1 & x & k \\ 0 & 1 & x^\sigma \\ 0 & 0 & 1 \end{pmatrix},$$

fixes O and acts sharply transitively on $X \setminus \{O\}$. It is also easy to calculate that, if $k \in \mathbb{K}^{(x x^\sigma)}$, then $k^{-1} \in \mathbb{K}^{((x/k)(x/k)^\sigma)}$, and also that

$$U_O^{(x,k)_\infty} = U_\infty^{(x/k, 1/k)_O}.$$

By Proposition 9.1.3, we obtain a Moufang set, which we denote by $\mathcal{M}(\mathbf{H}(2, \mathbb{F}, \sigma))$, or when $|\mathbb{K}| = q$, briefly by $\mathcal{M}(\mathbf{H}(2, q^2))$, since in this case σ is uniquely determined and given by $\sigma : x \mapsto x^q$. We call it a *Hermitian Moufang set*.

The little projective group of $\mathcal{M}(\mathbf{H}(2, \mathbb{F}, \sigma))$ is the unitary group $\mathbf{PSU}_3(\mathbb{F}, \sigma)$, denoted just $\mathbf{PSU}_3(q)$ when $|\mathbb{K}| = q$. It is a simple group whenever $|\mathbb{K}| \geq 3$ and has order $q^3(q^3 + 1)(q^2 - 1)/(q + 1, 3)$. Note that $|X| = q^3 + 1$ in this case. For $|\mathbb{K}| = 2$, $\mathbf{PSU}_3(2)$ is isomorphic to the sharply doubly transitive nonstandard example related to $\mathbf{GF}(9)$, and hence is not simple (but solvable).

There is a little proposition that can be proved now.

Proposition 9.2.2 *Let* $U_\infty = \{(x, k)_\infty \ \| \ x \in \mathbb{F}, k \in \mathbb{K}^{(xx^\sigma)}\}$ *be, as above, a root group of the Moufang set* $\mathcal{M}(\mathbf{H}(2, \mathbb{F}, \sigma))$, *with* $|\mathbb{F}| > 4$. *Then* U_∞ *is nilpotent of class* 2 *and the following subgroups of* U_∞ *coincide:*

    (i) the center $Z(U_\infty)$;

    (ii) the derived group $[U_\infty, U_\infty]$;

    (iii) the subgroup of elements in U_∞ *that induce a central collineation in* $\mathbf{PG}(2, \mathbb{F})$ *(necessarily with center* $(1, 0, 0)$*).*

If $|\mathbb{K}| = q$, *then* $|U_\infty| = q^3$ *and* $|Z(U_\infty)| = q$.

Proof. It is easy to calculate that all these subgroups coincide with $\{(0, k)_\infty \ \| \ k \in \mathbb{K}^{(0)}\}$. The other assertions are obvious. ∎

9.2.3 The Case $\mathbf{Sz}(q)$

Let $\mathbb{K}$ be a field of characteristic 2, and denote by $\mathbb{K}^2$ its subfield of all squares. Suppose that $\mathbb{K}$ admits some *Tits endomorphism* θ, i.e., the endomorphism θ is such that it maps x^θ to x^2, for all $x \in \mathbb{K}$. If $|\mathbb{K}| = 2^n$, then n must be odd, say $n = 2e + 1$, and θ maps x to $x^{2^{e+1}}$, for all $x \in \mathbb{K}$. Let $\mathbb{K}^\theta$ denote the image of $\mathbb{K}$ under θ. In the finite case $\mathbb{K}^\theta = \mathbb{K}$. Let L be a subspace of the vector space $\mathbb{K}$ over $\mathbb{K}^\theta$, such that $\mathbb{K}^\theta \subseteq L$ (this implies that $L \setminus \{0\}$ is closed under taking multiplicative inverse). In the finite case $L = \mathbb{K}$. We also assume that L generates $\mathbb{K}$ as a ring. We now define the *Suzuki-Tits Moufang set* $\mathcal{M}(\mathbf{Sz}(\mathbb{K}, L, \theta))$.

Let X be the following set of points of $\mathbf{PG}(3, \mathbb{K})$, where the coordinates with respect to some given basis are:

$$X = \{(1,0,0,0)\} \cup \{(a^{2+\theta} + aa' + a'^{\theta}, 1, a', a) \parallel a, a' \in L\}.$$

One now calculates that

$$X = \{(0,1,0,0)\} \cup \{(1, a^{2+\theta} + aa' + a'^{\theta}, a, a') \parallel a, a' \in L\}.$$

We set $\infty = (1,0,0,0)$ and $O = (0,1,0,0)$.

Let $(x, x')_\infty$ be the collineation of $\mathbf{PG}(3, \mathbb{K})$ determined by

$$(x_0\ x_1\ x_2\ x_3) \mapsto (x_0\ x_1\ x_2\ x_3) \begin{pmatrix} 1 & 0 & 0 & 0 \\ x^{2+\theta} + xx' + x'^{\theta} & 1 & x' & x \\ x & 0 & 1 & 0 \\ x^{1+\theta} + x' & 0 & x^{\theta} & 1 \end{pmatrix},$$

and let $(x, x')_O$ be the collineation of $\mathbf{PG}(3, \mathbb{K})$ determined by

$$(x_0\ x_1\ x_2\ x_3) \mapsto (x_0\ x_1\ x_2\ x_3) \begin{pmatrix} 1 & x^{2+\theta} + xx' + x'^{\theta} & x & x' \\ 0 & 1 & 0 & 0 \\ 0 & x^{1+\theta} + x' & 1 & x^{\theta} \\ 0 & x & 0 & 1 \end{pmatrix}.$$

Define the groups

$$U_\infty = \{(x, x')_\infty \parallel x, x' \in L\} \text{ and } U_O = \{(x, x')_O \parallel x, x' \in L\}.$$

Both groups U_∞ and U_O act on X, as an easy computation shows (for U_O use the second description of X above), and they act sharply transitively on $X \setminus \{(1,0,0,0)\}$ and $X \setminus \{(0,1,0,0)\}$, respectively. Moreover, one can check that $(U_O)^{(x,x')_\infty} = (U_\infty)^{(y,y')_O}$, with

$$y = \frac{x'}{x^{2+\theta} + xx' + x'^{\theta}} \text{ and } y' = \frac{x}{x^{2+\theta} + xx' + x'^{\theta}}.$$

It follows from Proposition 9.1.3 that we obtain a Moufang set $\mathcal{M}(\mathbf{Sz}(\mathbb{K}, L, \theta))$, called a *Suzuki-Tits Moufang set*. The group $\mathbf{Sz}(\mathbb{K}, L, \theta)$ is the *Suzuki group* generated by U_∞ and U_O. If $|\mathbb{K}| = q$, then we denote it by $\mathbf{Sz}(q)$. One has $|\mathbf{Sz}(q)| = q^2(q^2 + 1)(q - 1)$ and $|X| = q^2 + 1$. All Suzuki groups are simple groups unless $|\mathbb{K}| = 2$. The group $\mathbf{Sz}(2)$ is a sharply doubly transitive standard Moufang set related to $\mathbf{GF}(5)$.

In order to understand better the structure of U_∞, we can identify each point $(a^{2+\theta} + aa' + a'^\theta, 1, a', a)$ of $X \setminus \{\infty\}$ with the ordered pair (a, a'). Then the unique element of U_∞ that maps $(0,0)$ to (b, b') is given by

$$(b, b')_\infty : (a, a') \mapsto (a + b, a' + b' + ab^\theta).$$

The root group U_∞ is hence given abstractly by the set $\{(a, a')_\infty \,\|\, a, a' \in L\}$ with operation $(a, a')_\infty \oplus (b, b')_\infty = (a + b, a' + b' + ab^\theta)_\infty$. We can now state the following proposition.

Proposition 9.2.3 *Let U_∞ be as above (a root group of the Moufang set $\mathcal{M}(\mathbf{Sz}(\mathbb{K}, L, \theta)))$, and suppose $|U_\infty| > 4$. Then U_∞ is nilpotent of class 2 and the following subgroups of U_∞ coincide:*

(i) *the center $Z(U_\infty)$;*

(ii) *the derived group $[U_\infty, U_\infty]$;*

(iii) *the subset of elements in U_∞ which fix at least two points of $\mathbf{PG}(3, \mathbb{K})$;*

(iv) *the subset of elements of order 1 and 2 of U_∞.*

If $|\mathbb{K}| = q$, then $|U_\infty| = q^2$ and $|Z(U_\infty)| = q$.

Proof. The equality of the subgroups under (i), (ii) and (iv) follows from an elementary and easy computation, using the abstract description of U_∞ above. For (iii) we consider the matrix representation of U_∞ and one easily calculates that $(0, a')_\infty$ fixes the line spanned by $(1, 0, 0, 0)$ and $(0, 0, 1, 0)$ pointwise. If $a \neq 0$, then $(a, a')_\infty$ only fixes $(1, 0, 0, 0)$, as the reader can verify for himself.   ∎

Remark 9.2.4 The Moufang set $\mathcal{M}(\mathbf{Sz}(\mathbb{K}, L, \theta))$ can also be defined as the set of absolute points of a polarity in a certain generalized quadrangle. We explain this in detail further on.

9.2.4 The Case $\mathbf{R}(q)$

Let $\mathbb{K}$ be a field of characteristic 3, and denote by $\mathbb{K}^3$ its subfield of all third powers. In the finite case, $\mathbb{K}^3$ is just $\mathbb{K}$. Suppose that $\mathbb{K}$ admits some Tits endomorphism θ, i.e., the endomorphism θ is such that it maps x^θ to x^3, for all $x \in \mathbb{K}$. In the finite case, $|\mathbb{K}|$ is a power of 3 with odd exponent, say

$|\mathbb{K}| = 3^{2e+1}$, and θ maps x to $x^{3^{e+1}}$. Let $\mathbb{K}^\theta$ denote the image of $\mathbb{K}$ under θ. In the finite case, again necessarily $\mathbb{K}^\theta = \mathbb{K}$. We now define the *Ree-Tits Moufang set* $\mathcal{M}(\mathbf{R}(\mathbb{K}, \theta))$.

For $a, a', a'' \in \mathbb{K}$, we put

$$f_1(a, a', a'') = -a^{4+2\theta} - aa''^\theta + a^{1+\theta}a'^\theta + a''^2 + a'^{1+\theta} - a'a^{3+\theta} - a^2 a'^2,$$
$$f_2(a, a', a'') = -a^{3+\theta} + a'^\theta - aa'' + a^2 a',$$
$$f_3(a, a', a'') = -a^{3+2\theta} - a''^\theta + a^\theta a'^\theta + a'a'' + aa'^2.$$

Let X be the following set of points of $\mathbf{PG}(6, \mathbb{K})$, where the coordinates with respect to some given basis are:

$$X = \{(1, 0, 0, 0, 0, 0, 0)\}$$
$$\cup \{(f_1(a, a', a''), -a', -a, -a'', 1, f_2(a, a', a''), f_3(a, a', a'')) \parallel a, a', a'' \in \mathbb{K}\}.$$

A tedious calculation shows that

$$X = \{(0, 0, 0, 0, 1, 0, 0)\}$$
$$\cup \{(1, f_2(a, a', a''), f_3(a, a', a''), a'', f_1(a, a', a''), -a', -a) \parallel a, a', a'' \in \mathbb{K}\}.$$

We set $\infty = (1, 0, 0, 0, 0, 0, 0)$ and $O = (0, 0, 0, 0, 1, 0, 0)$.

Let $(x, x', x'')_\infty$ be the collineation of $\mathbf{PG}(6, \mathbb{K})$ determined by

$$(x_0\ x_1\ x_2\ x_3\ x_4\ x_5\ x_6) \mapsto (x_0\ x_1\ x_2\ x_3\ x_4\ x_5\ x_6)$$

$$\times \begin{pmatrix}
1 & 0 & 0 & 0 & 0 & 0 & 0 \\
p & 1 & 0 & -x & 0 & x^2 & -x'' - xx' \\
q & x^\theta & 1 & x' - x^{1+\theta} & 0 & r & s \\
x'' & 0 & 0 & 1 & 0 & x & -x' \\
f_1(x, x', x'') & -x' & -x & -x'' & 1 & f_2(x, x', x'') & f_3(x, x', x'') \\
x' - x^{1+\theta} & 0 & 0 & 0 & 0 & 1 & -x^\theta \\
x & 0 & 0 & 0 & 0 & 0 & 1
\end{pmatrix},$$

with

$$p = x^{3+\theta} - x'^\theta - xx'' - x^2 x',$$
$$q = x''^\theta + x^\theta x'^\theta + x'x'' - xx'^2 - x^{2+\theta}x' - x^{1+\theta}x'' - x^{3+2\theta},$$
$$r = x'' - xx' + x^{2+\theta},$$
$$s = x'^2 - x^{1+\theta}x' - x^\theta x'',$$

and let, with the same notation, $(x, x', x'')_O$ be the collineation of $\mathbf{PG}(6, \mathbb{K})$ determined by

$$(x_0\ x_1\ x_2\ x_3\ x_4\ x_5\ x_6) \mapsto (x_0\ x_1\ x_2\ x_3\ x_4\ x_5\ x_6)$$

$$\times \begin{pmatrix} 1 & f_2(x, x', x'') & f_3(x, x', x'') & x'' & f_1(x, x', x'') & -x' & -x \\ 0 & 1 & -x^\theta & 0 & x' - x^{1+\theta} & 0 & 0 \\ 0 & 0 & 1 & 0 & x & 0 & 0 \\ 0 & -x & x' & 1 & -x'' & 0 & 0 \\ 0 & 0 & 0 & 0 & 1 & 0 & 0 \\ 0 & x^2 & -x'' - xx' & x & p & 1 & 0 \\ 0 & r & s & x^{1+\theta} - x' & q & x^\theta & 1 \end{pmatrix}.$$

For the computations, it can be interesting to note that $(x, x', x'')_O = (x, x', x'')^g_\infty$, with g the collineation of $\mathbf{PG}(6, \mathbb{K})$ determined by

$$(x_0, x_1, x_2, x_3, x_4, x_5, x_6) \mapsto (x_4, x_5, x_6, -x_3, x_0, x_1, x_2).$$

Define the groups

$$U_\infty = \{(x, x', x'')_\infty \parallel x, x', x'' \in \mathbb{K}\} \text{ and } U_O = \{(x, x', x'')_O \parallel x, x', x'' \in \mathbb{K}\}.$$

The groups U_∞ and U_O both act on X, as an easy computation shows (for U_O use the second description of X above), and they act sharply transitively on $X \setminus \{(1, 0, 0, 0, 0, 0, 0)\}$ and $X \setminus \{(0, 0, 0, 0, 1, 0, 0)\}$, respectively. Moreover, it can be checked that $(U_O)^{(x, x', x'')_\infty} = (U_\infty)^{(y, y', y'')_O}$, with

$$y = -\frac{f_3(x, x', x'')}{f_1(x, x', x'')},$$

$$y' = -\frac{f_2(x, x', x'')}{f_1(x, x', x'')},$$

$$y'' = -\frac{x''}{f_1(x, x', x'')}.$$

Hence, by Proposition 9.1.3 we obtain a Moufang set, which we denote by $\mathcal{M}(\mathbf{R}(\mathbb{K}, \theta))$ and call a *Ree-Tits Moufang set*. Its little projective group, generated by U_∞ and U_O, is the *Ree group* $\mathbf{R}(\mathbb{K}, \theta)$. If $|\mathbb{K}| = q$, then we denote the corresponding Ree group by $\mathbf{R}(q)$. In this case we have $|\mathbf{R}(q)| = q^3(q^3 + 1)(q - 1)$. All Ree groups are simple groups unless $q = 3$, in which case $\mathbf{R}(3) \cong \mathbf{P\Gamma L}_2(8)$. Hence $[\mathbf{R}(3), \mathbf{R}(3)] \cong \mathbf{PSL}_2(8)$ is simple, but not sharply doubly transitive, because it has order 504.

Following 7.7.7 of [232], we can now define U_∞ abstractly as follows. Define

$$(a, a', a'') := (f_1(a, a', a''), -a', -a, -a'', 1, f_2(a, a', a''), f_3(a, a', a'')).$$

Then the unique element of U_∞ that maps $(0,0,0)$ to (x, x', x'') is given by

$$(x, x', x'')_\infty : (a, a', a'') \mapsto (a+x, a'+x'+ax^\theta, a''+x''+ax'-a'x-ax^{1+\theta}).$$

The root group U_∞ can now be defined as the set $\{(a, a', a'')_\infty \| a, a', a'' \in \mathbb{K}\}$ with operation

$$(a, a', a'')_\infty \oplus (b, b', b'')_\infty = (a+b, a'+b'+ab^\theta, a''+b''+ab'-a'b-ab^{1+\theta})_\infty.$$

We have the following proposition.

Proposition 9.2.5 *Let U_∞ be as above (a root group of the Moufang set $\mathcal{M}(\mathbf{R}(\mathbb{K}, \theta)))$. Then U_∞ is nilpotent of class 3 and the center $Z(U_\infty)$ coincides with the second central derived group $[U_\infty, U_\infty]_{[2]}$. Also the following subgroups of U_∞ coincide:*

(i) *the derived group $[U_\infty, U_\infty]$;*

(ii) *the subset of elements in U_∞ which fix at least two points of $\mathbf{PG}(6, \mathbb{K})$;*

(iii) *the subset of elements of order 3 of U_∞.*

Also, the derived group $[U_\infty, U_\infty]$ is abelian, hence U_∞ is solvable of length 2. Finally, if $|\mathbb{K}| = q$, then $|U_\infty| = q^3$, $|[U_\infty, U_\infty]| = q^2$ and $|Z(U_\infty)| = q$.

Proof. The coinciding of $Z(U_\infty)$ and $[U_\infty, U_\infty]_{[2]}$ is readily verified using the above abstract description of U_∞. Likewise, the equality of the subgroups under (i) and (iii) follows from an elementary and easy computation. For (ii) we consider the matrix representation of U_∞ and one easily calculates that $(0, x', x'')_\infty$ fixes the plane spanned by $(1,0,0,0,0,0,0)$, $(0,0,0,0,0,0,1)$ and $(0, x'^2, 0, -x'x'', 0, x''^2 - x'^{1+\theta}, 0)$ pointwise. If $x \neq 0$, then $(x, x', x'')_\infty$ only fixes $(1,0,0,0,0,0,0)$, as the reader can verify for himself. ■

Remark 9.2.6 The Moufang set $\mathcal{M}(\mathbf{R}(\mathbb{K}, \theta))$ can also be defined as the set of absolute points of a polarity in a certain generalized hexagon, see 7.7 of [232].

9.2.5 Sub Moufang Sets

We end this chapter with describing which finite Moufang sets are sub Moufang sets of other finite ones.

First consider $\mathcal{M}(\mathbf{PG}(1, q))$. Clearly, if $\mathbf{GF}(q')$ is a subfield of $\mathbf{GF}(q)$, then by considering a projective subline over $\mathbf{GF}(q')$ of $\mathbf{PG}(1, q)$, we obtain a sub Moufang set isomorphic to $\mathcal{M}(\mathbf{PG}(1, q'))$. Up to isomorphism there are no other sub Moufang sets of $\mathcal{M}(\mathbf{PG}(1, q))$. Note that the improper standard Moufang sets related to any field cannot be isomorphic to sub Moufang sets of $\mathcal{M}(\mathbf{PG}(1, q))$ unless they are isomorphic to $\mathcal{M}(\mathbf{PG}(1, 2))$ or $\mathcal{M}(\mathbf{PG}(1, 3))$, although they are somehow related to affine lines, which can be viewed as subgeometries of projective lines.

Next consider $\mathcal{M}(\mathbf{H}(2, q^2))$. Again, $\mathcal{M}(\mathbf{H}(2, q'^2))$ is isomorphic to a sub Moufang set if and only if $\mathbf{GF}(q'^2)$ is a subfield of $\mathbf{GF}(q^2)$ but not of $\mathbf{GF}(q)$. But here, also $\mathcal{M}(\mathbf{PG}(1, q))$ is a sub Moufang set, together with all its sub Moufang sets mentioned in the previous paragraph. Indeed, with the notation of Subsection 9.2.2, we restrict X to the projective points with second coordinate 0, i.e., to the set $\{(1, 0, 0)\} \cup \{(k, 0, 1) \parallel k + k^\sigma = 0\}$. Also, we restrict U_∞ to the elements $(0, k)_\infty$, with $k + k^\sigma = 0$ and with U_O to the elements $(0, k)_O$, with $k + k^\sigma = 0$. There are no other sub Moufang sets up to isomorphism.

The only sub Moufang sets of the Suzuki-Tits Moufang set $\mathcal{M}(\mathbf{Sz}(q))$ are isomorphic to the Suzuki-Tits Moufang sets $\mathcal{M}(\mathbf{Sz}(q'))$, where $\mathbf{GF}(q')$ is a subfield of $\mathbf{GF}(q)$. No projective lines to be found here!

Finally, concerning the Ree-Tits Moufang set $\mathcal{M}(\mathbf{R}(q))$, we have the sub Moufang sets over a subfield, $\mathcal{M}(\mathbf{R}(q'))$, with $\mathbf{GF}(q')$ a subfield of $\mathbf{GF}(q)$, and also $\mathcal{M}(\mathbf{PG}(1, q))$, together with all its sub Moufang sets. This can be seen as follows. With the notation of Subsection 9.2.4, we restrict X to the point set $Y = \{(x'^{1+\theta}, -x', 0, 0, 1, x'^\theta, 0) \parallel x' \in \mathbb{K}\} \cup \{(1, 0, 0, 0, 0, 0, 0)\}$. Then one easily sees that the commutative groups $\{(0, x', 0)_\infty \parallel x' \in \mathbb{K}\}$ and $\{(0, x', 0)_O \parallel x' \in \mathbb{K}\}$ act sharply transitively on $Y \setminus \{\infty\}$ and $Y \setminus \{O\}$, respectively. Corollary 9.1.4 now implies that we indeed obtain a sub Moufang set, which is readily seen to be isomorphic to $\mathcal{M}(\mathbf{PG}(1, q))$. There are no other sub Moufang sets up to isomorphism.

Remark 9.2.7 Let $\mathcal{M}(\mathbf{R}(q))$ be a Ree-Tits Moufang set on the set X. Recall that $|X| = q^3 + 1$. Let $\mathcal{B}$ be the collection of subsets of size $q + 1$ on which a sub Moufang set isomorphic to $\mathcal{M}(\mathbf{PG}(1, q))$ exists. Then the space $\mathcal{U}_R = (X, \mathcal{B})$ is a *unital*, i.e., a system of points and blocks such that every two points are contained in a unique block, and, for some natural number q, the number of points is equal to $q^3 + 1$, and the number of points on each block is $q + 1$. This unital is called the *Ree-Tits unital*.

The Hermitian curve $\mathbf{H}(2, q^2)$ is also a unital, if endowed with all secants. In fact, this *Hermitian unital* can also be constructed as follows. Let $\mathcal{M}(\mathbf{H}(2, q^2))$ be a Hermitian Moufang set on the set X. Let $\mathcal{B}$ be the collection of subsets of size $q + 1$ on which a sub Moufang set isomorphic to $\mathcal{M}(\mathbf{PG}(1, q))$ exists. Then $\mathcal{U}_H = (X, \mathcal{B})$ is that unital.

Chapter 10

Configurations of Translation Points

Let $\mathcal{S}^{(p)}$ be a TGQ with translation point p. Then an easy exercise yields that we have the following possibilities for its configuration of translation points:

(a) p is the only translation point;

(b) there is a line of translation points (incident with p);

(c) every point is a translation point.

If we are in Case (c), then by Theorems 10.5.3 and 10.7.1 below, and Theorem 2.2.2, the TGQ is isomorphic to one of $\mathbf{Q}(4, q)$, $\mathbf{Q}(5, q)$ for some q. Often in the existing literature, one speaks of "the translation point" of a given nonclassical TGQ S. So in view of the hypothetical Class (b), the question arises whether there are nonclassical examples with a configuration of translation points as described in (b).

In the papers [86, 201, 202] and the monograph [206], a more general class of GQs was studied, namely the "span-symmetric generalized quadrangles" ("SPGQs"). The defining property is that they have nonconcurrent axes of symmetry. The main result of Kantor [86] and K. Thas [201] is that an SPGQ of order s ($s > 1$) necessarily is isomorphic to $\mathbf{Q}(4, s)$ (and then s is a prime power). So since each SPGQ of order s ($s > 1$) is classical, each nonclassical TGQ of order s has a unique, and whence well-defined, translation point.

Surprisingly there are nonclassical examples of TGQs of order (s, s^2), $s > 1$ and s an odd prime power, with distinct translation points, so TGQs for which the notion "translation point" is not uniquely defined! This was first noted in [205]. From the paper [202], a classification of all GQs in the Class (b) could then be deduced.

In this chapter, we will describe several aspects of the theory of SPGQs (which have direct implications for TGQs), state the main results, along with a number of lemmas (without proofs) by which we will point out the general ideas of the proofs of these main results. Some lemmas have an elementary (combinatorial) proof, and sometimes a sketch of a proof will be given. We will then elaborate on the consequences for the theory of TGQs.

Also, we will comment on the classification of TGQs which lie in (c), in particular on the proof of K. Thas [199] which does not use the deep group theory of Fong and Seitz [59, 60].

10.1 Span-Symmetric Generalized Quadrangles

Suppose S is a GQ of order (s, t), $s, t \neq 1$, and suppose L and M are distinct nonconcurrent axes of symmetry; then it is easy to see by transitivity that every line of $\{L, M\}^{\perp\perp}$ is an axis of symmetry, and S is called a *span-symmetric generalized quadrangle* (SPGQ) with *base-span* $\{L, M\}^{\perp\perp}$. Note that $|\{L, M\}^{\perp\perp}| = s + 1$, as L and M are regular lines by Theorem 3.3.4.

Let S be a span-symmetric generalized quadrangle of order (s, t), $s, t \neq 1$, with base-span $\{L, M\}^{\perp\perp}$. Throughout this chapter, we will use the following notation.

First of all, the base-span will always be denoted by $\mathcal{L}$. The group which is generated by all the symmetries about the lines of $\mathcal{L}$ is G, and sometimes we will call it the *base-group*. This group clearly acts 2-transitively on the lines of $\mathcal{L}$, and fixes every line of $\mathcal{L}^{\perp}$. The set of all the points which are on lines of $\{L, M\}^{\perp\perp}$ is denoted by Ω (of course, Ω is also the set of points on the lines of $\{L, M\}^{\perp}$; we have that $|\{L, M\}^{\perp}| = |\{L, M\}^{\perp\perp}| = s + 1$). We will refer to $\Gamma = (\Omega, \mathcal{L} \cup \mathcal{L}^{\perp}, \mathtt{I}')$, with $\mathtt{I}'$ being the restriction of $\mathtt{I}$ to $(\Omega \times (\mathcal{L} \cup \mathcal{L}^{\perp})) \cup ((\mathcal{L} \cup \mathcal{L}^{\perp}) \times \Omega)$, as being *the base-grid*.

The substructure of fixed elements of an element of the base-group is given by the following theorem. Recall that a *partial spread* of a GQ is a set of mutually nonconcurrent lines.

Theorem 10.1.1 *Let S be an SPGQ of order (s, t), $s \neq 1 \neq t$, with base-span $\mathcal{L}$ and base-group G. If $\theta \neq 1$ is an element of G, then the substructure $S_{\theta} = (\mathcal{P}_{\theta}, \mathcal{B}_{\theta}, \mathtt{I}_{\theta})$ of elements fixed by θ must be given by one of the following.*

(i) $\mathcal{P}_\theta = \emptyset$ and $\mathcal{B}_\theta$ is a partial spread containing $\mathcal{L}^\perp$.

(ii) There is a line $L \in \mathcal{L}$ for which $\mathcal{P}_\theta$ is the set of points incident with L, and $M \sim L$ for each $M \in \mathcal{B}_\theta$ ($\mathcal{L}^\perp \subseteq \mathcal{B}_\theta$).

(iii) $\mathcal{B}_\theta$ consists of $\mathcal{L}^\perp$ together with a subset $\mathcal{B}'$ of $\mathcal{L}$; $\mathcal{P}_\theta$ consists of those points incident with lines of $\mathcal{B}'$.

(iv) $\mathcal{S}_\theta$ is a subGQ of order (s, t') with $s \leq t' < t$. This forces $t' = s$ and $t = s^2$.

Proof.   See 10.7.1 of [128].   ∎

We also mention the next result.

Theorem 10.1.2 *If $\mathcal{S}$ is an SPGQ of order (s, t), $s, t \neq 1$ and $t < s^2$, with base-group G, then G acts regularly on the set of $(s + 1)s(t - 1)$ points of $\mathcal{S}$ which are not on any line of $\mathcal{L}$.*

Proof.   See 10.7.2 of [128].   ∎

10.2   Groups Admitting a 4-Gonal Basis

Let $\mathcal{S}$ be an SPGQ of order $s \neq 1$ with base-span $\mathcal{L}$, and put $\mathcal{L} = \{L_0, L_1, \ldots, L_s\}$. The group of symmetries about L_i is denoted by G_i, $i = 0, 1, \ldots, s$. One notes the following properties (see [114] and 10.7.3 of [128]):

(1) the groups $G_0, G_1, \ldots, G_s$ form a complete conjugacy class in G, and are all of order s, $s \geq 2$;

(2) $G_i \cap N_G(G_j) = \{\mathbf{1}\}$ for $i \neq j$;

(3) $G_i G_j \cap G_k = \{\mathbf{1}\}$ for i, j, k distinct, and

(4) $|G| = s^3 - s$.

We say that G is a group with a 4-*gonal basis* $\mathfrak{B} = \{G_0, G_1, \ldots, G_s\}$ if these four conditions are satisfied.

It is possible to recover the SPGQ $\mathcal{S}$ of order s, $s > 1$, from the base-group G starting from 4-gonal bases, in the following way, see [114, 128].

Suppose G is a group of order $s^3 - s$ with a 4-gonal basis $\mathfrak{B} = \{G_0, G_1, \ldots, G_s\}$, and let $G_i^* = N_G(G_i)$ for $i = 0, 1, \ldots, s$. Define a point-line incidence structure $\mathcal{S}_\mathfrak{B} = (\mathcal{P}_\mathfrak{B}, \mathcal{B}_\mathfrak{B}, \mathrm{I}_\mathfrak{B})$ as follows.

- $\mathcal{P}_\mathfrak{B}$ consists of two kinds of POINTS.

  (a) Elements of G.

  (b) Right cosets of the G_i^*s.

- $\mathcal{B}_\mathfrak{B}$ consists of three kinds of LINES.

  (i) Right cosets of G_i, $0 \le i \le s$.

  (ii) Sets $M_i = \{G_i^* g \parallel g \in G\}$, $0 \le i \le s$.

  (iii) Sets $L_i = \{G_j^* g \parallel G_i^* \cap G_j^* g = \emptyset, 0 \le j \le s, j \ne i\} \cup \{G_i^*\}$, $0 \le i \le s$.

- INCIDENCE. $\mathrm{I}_\mathfrak{B}$ is the natural incidence: a line $G_i g$ of Type (i) is incident with the s points of Type (a) contained in it, together with that point $G_i^* g$ of Type (b) containing it. The lines of Types (ii) and (iii) are already described as sets of those points with which they are to be incident.

Then $\mathcal{S}_\mathfrak{B} = (\mathcal{P}_\mathfrak{B}, \mathcal{B}_\mathfrak{B}, \mathrm{I}_\mathfrak{B})$ is a GQ of order s which is span-symmetric for the base-span $\{L_0, L_1\}^{\perp\perp}$ [114, 128]. Also, if $\mathcal{S}$ is an SPGQ of order s, $s \ne 1$, with base-span $\mathcal{L}$ and base-group G, and where $\mathfrak{B}$ is the corresponding 4-gonal basis, then $\mathcal{S} \cong \mathcal{S}_\mathfrak{B}$ [114, 128].

So we have the following theorem, which was first noted by Payne.

Theorem 10.2.1 *An SPGQ of order $s \ne 1$ with given base-span $\mathcal{L}$ is canonically equivalent to a group G of order $s^3 - s$ with a 4-gonal basis $\mathfrak{B}$.*

Proof. See 10.7.8 of [128]. ∎

It is also important to recall the following from [128].

Theorem 10.2.2 *Let $\mathcal{S}$ be an SPGQ of order $s \ne 1$, with base-span $\mathcal{L}$. Then every line of $\mathcal{L}^\perp$ is an axis of symmetry.*

Proof. See 10.7.9 of [128]. ∎

So for any two distinct lines U and V of $\mathcal{L}^\perp$, $\mathcal{S}$ is also an SPGQ with base-span $\{U, V\}^{\perp\perp}$. The corresponding base-group will be denoted by $G^\perp$. It should be emphasized that this property only holds for SPGQs of order s, $s > 1$, as will follow from later observations.

Now suppose G is a group of order $s^3 - s$, $s > 1$, where s is a power of the prime p, and suppose that G has a 4-gonal basis $\mathfrak{B} = \{G_0, G_1, \ldots, G_s\}$. Since the groups G_i all have order s, all these groups are Sylow p-subgroups in G. Since $\mathfrak{B}$ is a complete conjugacy class, this means that every Sylow p-subgroup of G is contained in $\mathfrak{B}$, and hence G has *exactly* $s + 1$ Sylow p-subgroups. We have proved the following theorem.

Theorem 10.2.3 *Suppose G is a group of order $s^3 - s$, $s > 1$, with s a prime power. Then G can have at most one 4-gonal basis. In particular, if G has a 4-gonal basis, then it is unique.* ∎

As a corollary we obtain the following fundamental result, see [206].

Theorem 10.2.4 *Suppose $\mathcal{S}$ is an SPGQ of order (s, t), $1 < s \leq t < s^2$. Then $\mathcal{S}$ is isomorphic to the classical GQ $\mathbf{Q}(4, s)$ if and only if the base-group is isomorphic to $\mathbf{SL}_2(s)$.*

Proof. Suppose that the base-group G is isomorphic to $\mathbf{SL}_2(s)$; then s is a power of a prime and hence by Theorem 10.2.3, $\mathbf{SL}_2(s)$ has at most one 4-gonal basis. As $t < s^2$, G acts regularly on the points of $\mathcal{S}$ not in Ω by Theorem 10.1.2, so $(s + 1)s(t - 1) = (s + 1)s(s - 1)$, and $s = t$. Now suppose L and M are nonconcurrent lines of $\mathbf{Q}(4, s)$. Then L and M are axes of symmetry, and hence $\mathbf{Q}(4, s)$ is span-symmetric for the base-span $\{L, M\}^{\perp\perp}$. In this case, the base-group is isomorphic to $\mathbf{SL}_2(s)$, see e.g. [114], which proves that $\mathbf{SL}_2(s)$ indeed has a 4-gonal basis for each prime power s, that is necessarily unique by Theorem 10.2.3. Hence, by Theorem 10.2.1, there is only one GQ (up to isomorphism) which can arise from $\mathbf{SL}_2(s)$ using 4-gonal bases. ∎

10.3 SPGQs and Moufang Sets

By the following theorem, the base-group of an SPGQ is essentially known.

Theorem 10.3.1 *Suppose S is an SPGQ of order (s,t), $s,t \neq 1$, with base-span $\mathcal{L}$ and base-group G. Let N be the kernel of the action of G on the lines of $\mathcal{L}$. Then G/N acts as a sharply 2-transitive group on $\mathcal{L}$, or is isomorphic (as a permutation group) to one of the following:*

 (i) $\mathbf{PSL}_2(s)$;

 (ii) *the Ree group* $\mathbf{R}(\sqrt[3]{s})$;

 (iii) *the Suzuki group* $\mathbf{Sz}(\sqrt{s})$;

 (iv) *the unitary group* $\mathbf{PSU}_3(\sqrt[3]{s})$,

each with its natural action of degree $s + 1$.

Proof.   The group G (and hence also G/N) is doubly transitive on $\mathcal{L}$, and for every $L \in \mathcal{L}$ the full group of symmetries about L, which acts regularly on $\mathcal{L} \setminus \{L\}$, is a normal subgroup of the stabilizer of L in G/N. This means that $(G/N, \mathcal{L})$ is a Moufang set. Theorem 9.2.1 provides the above list of possibilities for G/N, noting that G is generated by the normal subgroups mentioned above.   ∎

10.4   Basic Structural Lemmas

Let S be an SPGQ of order (s,t), with $s \neq 1 \neq t$. We keep using the standard notation introduced earlier.

Suppose that $s \leq t < s^2$ for the rest of this section.

The following set of lemmas can be found in Kantor [86] and K. Thas [201]. One can also use [206] as a general reference.

Lemma 10.4.1 *The group G is a perfect group if $(G/N, \mathcal{L})$ is not sharply 2-transitive, and if $G/N \not\cong \mathbf{R}(3)$.*

Proof.   See Lemma 7.6.2 of [206].   ∎

Lemma 10.4.2 *N is a subgroup of the center of G.*

Proof. Let H be a full group of symmetries about any line L of $\mathcal{L}$. Then H and N normalize each other, so $N \cap H = \{1\}$ implies that they centralize each other. The lemma now follows as G is generated by the symmetries about the lines of $\mathcal{L}$. ∎

Lemma 10.4.3 *We have that G/N either acts naturally as $\mathbf{PSL}_2(s)$ on $\mathcal{L}$, or sharply 2-transitively.*

Proof. See Lemma 7.6.5 of [206]. ∎

The proof of the following lemma uses the theory of universal central extensions of (perfect) groups.

Lemma 10.4.4 *If G/N acts as $\mathbf{PSL}_2(s)$, then $G \cong \mathbf{SL}_2(s)$ and $\mathcal{S} \cong \mathbf{Q}(4, s)$.*

Proof. See Lemma 7.6.6 of [206]. ∎

It remains to consider the case where G/N acts sharply 2-transitively.

First suppose that G/N acts sharply 2-transitively on the lines of $\mathcal{L}$, and that $s = t$.

Every line of $\mathcal{L}^\perp$ is an axis of symmetry by Theorem 10.2.2.

Since the lines of $\mathcal{L}^\perp$ are also axes of symmetry, we can assume that the base-group $G^\perp$ corresponding to these lines also acts sharply 2-transitively on $\mathcal{L}^\perp$ — otherwise $G^\perp$ is isomorphic to $\mathbf{SL}_2(s)$, and then $\mathcal{S}$ is classical by Theorem 10.2.4. Hence G and $G^\perp$ contain normal central subgroups N and $N^\perp$ which act trivially on the points of Ω, both of order $s - 1$ (where Ω is the set of points on the lines of the base-span). Also, G and $G^\perp$ act regularly on the points of $\mathcal{S}$ not in Ω by Theorem 10.1.2.

Let x be a point and L a line of a projective plane Π. Then Π is said to be (x, L)-*transitive* if the group of all collineations of Π with center x and axis L acts transitively on the points, distinct from x and not on L, of any line through x. The following theorem is a step in the Lenz-Barlotti classification of finite projective planes, see e.g. [45, 145]; it states that the Lenz-Barlotti Class **III.2** is empty.

Theorem 10.4.5 (Yaqub [239]) *Let Π be a finite projective plane, containing a nonincident point-line pair (x, L) for which Π is (x, L)-transitive, and* *assume that* Π *is* (y, xy)-*transitive for every point* y *on* L. *Then* Π *is Desarguesian.*   ∎

As every axis of symmetry L of a GQ is a regular line, there is a projective plane π_L canonically associated with L (by Theorem 2.2.1) if the GQ has order s ($s \neq 1$). Using Theorem 10.4.5, one then obtains the following result.

Theorem 10.4.6 *Suppose that S is an SPGQ of order s, where $s \neq 1$, with base-group G and base-span $\mathcal{L}$. Also, let N be the kernel of the action of G on the lines of $\mathcal{L}$, and suppose that G/N acts as a sharply 2-transitive group on the lines of $\mathcal{L}$. Then S is isomorphic to $\mathbf{Q}(4, 2)$ or $\mathbf{Q}(4, 3)$.*

Proof.   See Theorem 7.7.2 of [206].   ∎

Now suppose that $s < t < s^2$ for the rest of the section.

There is a purely combinatorial way to get rid of the sharply 2-transitive case for $s < t < s^2$, as follows. In the papers [86] and [201], Maschke's Theorem on complete reducibility of representations was used for this purpose.

Let $s < t < s^2$, and suppose G acts sharply 2-transitively on $\mathcal{L}$. If $x \in \mathcal{S} \setminus \Omega$, then x^N is a set of $t - 1$ points which are all collinear with each point of $x^\perp \cap \Omega = X$. By 1.4.1 of [128] we have $s(t - 2) \leq s^2$ and so $t \leq s + 2$. As $t > s$ and $t \neq s+1$, we have $t = s+2$. Since there are nontrivial symmetries about lines of $\mathcal{S}$, Theorem 3.3.2 implies that s is even.

The paper [86] (see Chapter 7 of [206]) contains a group theoretical treatment for the case $s < t < s^2$, using Maschke's Theorem, cf. Gorenstein [64], which was also needed in [206] to handle the case $t = s - 2$ with s even. We now give a combinatorial solution also for that case, which completes the combinatorial approach already largely obtained in [206].

As G/N is a sharply 2-transitive group, it contains a normal elementary abelian p-group A of size $s + 1$, with $s + 1 = p^h$ for some h (this follows from the fact that $(G/N, \mathcal{L})$ is a Frobenius group). Call A' the subgroup of G for which $A'/N = A$; as A is normal in G/N, it now follows easily that A' is normal in G.

Let x be a point of $\mathcal{S} \setminus \Omega$; then there are two distinct lines incident with x that do not hit Γ, so that the total number of such lines is $s(s + 1)2$. So the stabilizer in G of a line M which does not contain a point of Γ has size at least $(s+1)/2$. Since $s+1 = p^h$ with p odd, it follows that the size of the stabilizer is exactly $s+1$ (recall that G acts regularly on the points of $\mathcal{S} \setminus \Omega$), and that the action of the stabilizer on the points of M is regular. As A' is a Sylow p-subgroup of G, we have that $G_M \leq A'$, and $|M^{A'}| = s+1$. As $N \leq Z(G)$ acts transitively on the lines of $M^{A'}$, we have that G_M fixes each line in $M^{A'}$.

If two distinct lines of $M^{A'}$ would intersect, then any element of G_M fixes their common point, clearly a contradiction. Hence $M^{A'}$ is a partial spread of the GQ. The set of all points incident with lines of $M^{A'}$ is an orbit R of A'. If $m \mathrm{I} M$ and $\tilde{M}$ is the second line incident with m which does not contain a point of Ω, then similarly R is the set of all points incident with lines of $\tilde{M}^{A'}$. So R is the point set of a grid having as lines $M^{A'} \cup \tilde{M}^{A'}$. It easily follows that M^G is a partial spread of $\mathcal{S}$ and that $M^G \cup \mathcal{L}^{\perp}$ is a spread of $\mathcal{S}$. Consider any $i \in \{0, 1, \ldots, s\}$, take any point $m \mathrm{I} M$ and suppose O is the unique line on m meeting $L_i \in \mathcal{L}$. As any orbit R' of A' is the point set of a grid, it contains at most one point of O. Hence O has a unique point in common with every orbit of A'. Take any $\theta \in G_i$ and $\beta \in G_M$. Then $M^{\theta\beta}$ is concurrent with O^{β} and belongs to the spread $M^G \cup \mathcal{L}^{\perp}$, so $M^{\theta\beta} = M^{\theta}$. It easily follows that G_M fixes every line of the spread, while acting sharply transitively on the points of each of these lines. By definition, this means that $\mathcal{L}^{\perp} \cup M^G$ is a *spread of symmetry*.

Before proceeding, we need a nice construction, first observed by Payne, of GQs of order $(s-1, s+1)$ from (thick) GQs of order s with a regular point.

The Payne Derivative

Let s be a regular point of the GQ $\mathcal{S} = (\mathcal{P}, \mathcal{B}, \mathrm{I})$ of order s, $s > 1$. Define an incidence structure $\mathcal{P}(\mathcal{S}, x) = \mathcal{S}' = (\mathcal{P}', \mathcal{B}', \mathrm{I}')$ as follows:

- The POINT SET $\mathcal{P}'$ is the set $\mathcal{P} \setminus x^{\perp}$.

- The LINES of $\mathcal{B}'$ are of two types:

    - the elements of Type (a) are the lines of $\mathcal{B}$ which are not incident with x;

    - the elements of Type (b) are the hyperbolic lines $\{x, y\}^{\perp\perp}$ where $y \not\sim x$.

- INCIDENCE I' is containment (regarding a line of $\mathcal{S}$ as a set of points).

Then S' is a GQ of order $(s - 1, s + 1)$, which we call the *Payne derivative* of S with respect to x.

Theorem 10.4.7 (De Soete and Thas [49]) *Let S be a GQ of order $(s, s + 2)$, with $s > 1$, and suppose S has a spread of symmetry T. Let H be the group of automorphisms of S that fix T linewise. Then there exists a GQ S' of order $s + 1$ with center of symmetry x, such that*

$$S \cong \mathcal{P}(S', x).$$

Moreover, T arises from the set of hyperbolic lines containing x, and H is induced by the group of symmetries about x. ∎

So there exists a GQ S' of order $s + 1$ with center of symmetry (∞) such that $\mathcal{P}(S', (\infty)) \cong S$, and $\mathcal{L}^{\perp} \cup M^G = T'$ arises from the hyperbolic lines containing (∞), while G_M is induced by the symmetries about (∞). Moreover, as G preserves T', a result of De Winter and K. Thas [50] states that G is induced by an automorphism group of S' that fixes (∞).

Now consider a nontrivial symmetry ϕ of S about some fixed line L_j of the base-span $\mathcal{L}$. Then L_j is a line of Type (a) in S', and ϕ is induced by a collineation of S' that fixes (∞), every point on L_j (as a line of S'), and is a whorl about y with $y \mathrm{I} L_j$, $y \not\sim (\infty)$. This contradicts Theorem 3.2.1, and ends the combinatorial proof of the sharply 2-transitive case. ∎

10.5   Classification of SPGQs of Order (s, t), $1 < s \le t < s^2$

The classification of SPGQs of order (s, t) with $1 \ne s \le t < s^2$ can now be completed.

Theorem 10.5.1 *Let S be an SPGQ of order (s, t), where $1 < s \le t < s^2$. Then $s = t$ is a prime power and S is isomorphic to $\mathbf{Q}(4, s)$.*

Proof.   Adopt the notation $G, N, G/N, \mathcal{L}$, etc. from above. By Lemma 10.4.3, G/N either acts as a sharply 2-transitive group on $\mathcal{L}$, or as $\mathbf{PSL}_2(s)$. If G/N acts as $\mathbf{PSL}_2(s)$, then $G \cong \mathbf{SL}_2(s)$, $s = t$ and S is classical by Lemma 10.4.4. If G/N acts sharply 2-transitively on $\mathcal{L}$, then $S \cong \mathbf{Q}(4, 2)$ or $\mathbf{Q}(4, 3)$. Whence the result. ∎

This leads to the complete classification of groups having a 4-gonal basis.

Theorem 10.5.2 *A finite group is isomorphic to* $\mathbf{SL}_2(s)$ *for some* s *if and only if it has a* 4*-gonal basis.* ∎

It also follows that the order of an SPGQ is essentially known.

Theorem 10.5.3 *Let* $\mathcal{S}$ *be an* SPGQ *of order* (s, t), $s, t \neq 1$. *Then* $t \in \{s, s^2\}$. ∎

10.6 SPGQs of Order (s, s^2)

In [202], the author proved that if $\mathcal{S}$ is a span-symmetric generalized quadrangle of order (s, s^2), $s \neq 1$ and s odd, with base-span $\mathcal{L}$, then $\mathcal{S}$ contains $s + 1$ subquadrangles, all isomorphic to the classical GQ $\mathbf{Q}(4, s)$, which mutually intersect in the base-grid $\Gamma = (\Omega, \mathcal{L} \cup \mathcal{L}^{\perp}, \mathbf{I}')$. Also, the base-group G acts freely on the points of $\mathcal{S} \setminus \Omega$ and $G \cong \mathbf{SL}_2(s)$. (Note that $|G| = |\mathbf{SL}_2(s)| = s^3 - s$.)

Already since the beginning of the study of SPGQs (late 1970's), it was believed (and posed as a problem to prove) that SPGQs of order (s, s^2), $s > 1$, always have ("many") classical subGQs of order s, all passing through the base-grid. The solution of that problem in the odd case by K. Thas [202] involved a proof that was largely valid in the even case. In Chapter 7 of [206], the problem was completely solved for the general case. The fact that an SPGQ of order (s, s^2) has $s + 1$ subGQs isomorphic to $\mathbf{Q}(4, s)$ in this position endows the GQ with a lot of structure, and this observation is crucial for many applications in theory of SPGQs. For instance, Kantor used this to classify what he calls "grid-symmetric generalized quadrangles" — these are SPGQs for which the perp of the base-span also is a base-span, see [87]. Other applications will be obtained further on.

Here, we give a brief account of the proof. The reader is referred to [206] for further details.

For the rest of this section, let $\mathcal{S}$ *be an* SPGQ *of order* (s, t), $s \neq 1 \neq t$, *with base-grid* $\Gamma = (\Omega, \mathcal{L} \cup \mathcal{L}^{\perp}, \mathbf{I}')$ *and base-group* G, *and let* L_i, G_j, *etc. be as before.*

If p is a point which is not an element of Ω, and U is a line through p which meets Ω in a certain point $q \mathbf{I} L_k$ of Γ, then clearly every point on U which is different from q is a point of the G-orbit which contains p. We will use this observation freely throughout this chapter.

The proof of the next lemma was first obtained in K. Thas [202].

Lemma 10.6.1 *Suppose S is an SPGQ of order (s,t), $s \neq 1 \neq t$, with base-grid Γ and base-group G. Then G has size at least $s^3 - s$.*

Proof. If $s = t$, then we already know that G has order $s^3 - s$, so suppose $s \neq t$ (so that $t = s^2$ by Theorem 10.5.3).

Set $\mathcal{L} = \{L_0, L_1, \ldots, L_s\}$ and suppose G_i is the group of symmetries about L_i for all feasible i. Suppose p is a point of S not in Ω, and consider the following $s+1$ lines $N_i := \text{proj}_p L_i$. If Λ is the G-orbit which contains p, then every point of N_i not in Ω is also a point of Λ. Now consider an arbitrary point $q \neq p$ on N_0 which is not on a line of $\mathcal{L}$. Then again every point of $\text{proj}_q L_i$, $i \neq 0$, not in Ω is a point of Λ. Hence we have the following inequality:

$$|\Lambda| \geq 1 + (s+1)(s-1) + (s-1)^2 s, \qquad (10.1)$$

from which it follows that $|\Lambda| \geq s^3 - s^2 + s$. Now fix a line U of $\mathcal{L}^\perp$. Every line of S which meets this line and which contains a point of Λ is completely contained in $\Lambda \cup \Omega$. Also, G acts transitively on the points of U. Suppose k is the number of lines through a (= every) point of U that are completely contained in $\Lambda \cup \Omega$ (as point sets). If we count in two ways the number of point-line pairs (u, M) for which $u \in \Lambda, M \sim U$ and $u \mathbf{I} M$, then it follows that

$$k(s+1)s = |\Lambda| \geq s(s^2 - s + 1)$$
$$\implies k \geq \frac{s^2 - s + 1}{s+1} = s - 2 + \frac{3}{s+1},$$

and hence, since $k \in \mathbb{N}$, we have that $k \geq s - 1$. Thus $|\Lambda| \geq s^3 - s$ and so also $|G|$.   ∎

In the rest of this section, we can suppose that $s > 3$; if $s = 2$, then $S \cong \mathbf{Q}(5,2)$ by Section 1.8. In that case, there are $s + 1 = 3$ subGQs of order 2, isomorphic to $\mathbf{Q}(4,2)$ and mutually intersecting in the base-grid. The base-group stabilizes each of these subGQs (as it is generated by symmetries), and acts freely on the points of S not in Ω; whence $G \cong \mathbf{SL}_2(2)$. Similar properties hold for the case $s = 3$.

For $\sqrt{t} = s > 3$, the strategy of handling the situation is essentially the same as that followed in Section 10.4. The aim again is to prove that G is a perfect central extension of G/N, when $(G/N, \mathcal{L})$ is not sharply 2-transitive. The latter case has to be handled separately (in an entirely different way than in the case $t < s^2$). Then, one eventually comes to the following result, using Lemma 10.6.1.

Proposition 10.6.2 *Suppose that S is an SPGQ of order (s, s^2), $s > 3$, with base-span $\mathcal{L} = \{U, V\}^{\perp\perp}$, base-group G and base-grid $\Gamma = (\Omega, \mathcal{L} \cup \mathcal{L}^{\perp}, \mathrm{I}')$. Furthermore, let N be the kernel of the action of G on $\mathcal{L}$. Then G/N is isomorphic to $\mathbf{PSL}_2(s)$. Also, G acts freely on $S \setminus \Omega$.*

Proof. See Section 7.10 of [206], in particular the Lemmas 7.10.2 — 7.10.7. ∎

We know that G acts freely on the points of $S \setminus \Omega$. Assume that G has order $(s + 1)s(s - 1)$. Let Λ be an arbitrary G-orbit in $S \setminus \Omega$, and fix a line W of $\mathcal{L}^{\perp}$. By the fact that G acts freely on the points of $S \setminus \Omega$, that $|G| = (s+1)s(s-1)$ and that G acts transitively on the points incident with W, we have that any point on W is incident with exactly $s - 1$ lines of S which are completely contained in Λ except for the point on W which is in Ω, and that every point of Λ is incident with a line which meets W (recall that G is generated by groups of symmetries). Now define the following incidence structure $S' = S'(\Gamma) = (\mathcal{P}', \mathcal{B}', \mathrm{I}')$.

- LINES. The elements of $\mathcal{B}'$ are the lines of S' and they are essentially of two types:

  (1) the lines of $\mathcal{L}^{\perp} \cup \mathcal{L}^{\perp\perp}$;

  (2) the lines of S which contain a point of Λ and a point of Ω.

- POINTS. The elements of $\mathcal{P}'$ are the points of the incidence structure and they are just the points of $\Omega \cup \Lambda$.

- INCIDENCE. Incidence I' is the induced incidence.

Then S' is an SPGQ of order s. By Theorem 10.5.1, S' is isomorphic to the GQ $\mathbf{Q}(4, s)$.

Theorem 10.6.3 *Suppose S is an* SPGQ *with base-span $\mathcal{L}$ and of order (s, s^2), $s \neq 1$. Assume that G has size $s^3 - s$. Then there exist $s + 1$ subquadrangles of order s which are all isomorphic to $\mathbf{Q}(4, s)$, so that they mutually intersect in the base-grid Γ.*

Proof. From each G-orbit in $S \setminus \Omega$ arises a subGQ of order s which is isomorphic to $\mathbf{Q}(4, s)$, and there are exactly $s + 1$ such distinct G-orbits. ∎

The Main Structure Theorem for SPGQs of Order (s, s^2)

We are ready to prove the main result for SPGQs of order (s, s^2) [206, Chapter 7].

Theorem 10.6.4 *Suppose S is an* SPGQ *of order (s, s^2), $s \neq 1$. Then S contains $s + 1$ subquadrangles, all isomorphic to the classical GQ $\mathbf{Q}(4, s)$, which mutually intersect in the base-grid Γ. Also, the base-group G acts freely on $S \setminus \Omega$, $|G| = s^3 - s$ and $G \cong \mathbf{SL}_2(s)$.*

Proof. The fact that G acts freely on $S \setminus \Omega$ implies that G is a perfect group, and also that $G/N \cong \mathbf{PSL}_2(s)$ if N is the kernel of the action of G on $\mathcal{L}$; see [206]. Moreover, N is contained in the center of G (recall the proof of Lemma 10.4.2). Thus, G is a perfect central extension of $\mathbf{PSL}_2(s)$, and this leads to the fact that $G \cong \mathbf{SL}_2(s)$ and that $|G| = s^3 - s$; see again [206]. The result follows from Theorem 10.6.3. ∎

10.7 Translation Generalized Quadrangles with a Line of Translation Points

Although it was conjectured in 1983 by Payne that every SPGQ of order (s, s^2), $s > 1$, is isomorphic to $\mathbf{Q}(5, s)$, see PROBLEM 26 of [115], such a similar result as in the case $t < s^2$ cannot hold.

For, let $\mathcal{K}$ be the quadratic cone with equation $X_0 X_1 = X_2^2$ of $\mathbf{PG}(3, q)$, q odd. Then the q planes π_t with equation

$$tX_0 - mt^\sigma X_1 + X_3 = 0,$$

where $t \in \mathbf{GF}(q)$, m a given nonsquare in $\mathbf{GF}(q)$ and σ a given field automorphism of $\mathbf{GF}(q)$, define a Kantor-Knuth semifield flock $\mathcal{F}$ of $\mathcal{K}$; see

Section 4.5. In [116], Payne noted that the dual Kantor-Knuth flock generalized quadrangles are span-symmetric (this observation is rather hidden in [116]), and this infinite class of generalized quadrangles contains nonclassical examples; $\mathcal{F}$ is not a linear flock (and then the GQ is nonclassical) if and only if σ is not the identity. Moreover, every nonclassical dual Kantor-Knuth flock GQ contains a line L for which every line which meets L is an axis of symmetry (see [206]). Hence there is some line each point of which is a translation point. In order to initialize a theory for thick span-symmetric generalized quadrangles of order (s, s^2) with $s \neq 1$, and to answer several important questions in a broader context, K. Thas studied the following problem in [202, 205, 206].

PROBLEM. *Classify all thick generalized quadrangles with distinct translation points.*

In the papers [202], [205] and [206, Chapter 7], this program was carried out completely. Before coming to a synthesis of the main result, cf. Theorem 10.7.2 below, we need the next theorem.

Theorem 10.7.1 (Fong and Seitz [59, 60]; Kantor [85]; K. Thas [199])
If S is a GQ of order (s, s^2), $s > 1$, each line of which is an axis of symmetry, then S is isomorphic to $\mathbf{Q}(5, s)$.

Proof (Taken from K. Thas [199]). Let U and V be any two nonconcurrent lines of S. Then S is an SPGQ with base-span $\{U, V\}^{\perp\perp}$ and by Theorem 10.6.4, $\{U, V\}$ is contained in $s + 1$ subGQs of order s. Now 5.3.5 of [128] implies the theorem. ∎

Theorem 10.7.2 *Suppose S is a generalized quadrangle of order (s, t), $s \neq 1 \neq t$, with two distinct collinear translation points. Then we have the following possibilities:*

(i) *$s = t$, s is a prime power and $S \cong \mathbf{Q}(4, s)$;*

(ii) *$t = s^2$, s is even, s is a prime power and $S \cong \mathbf{Q}(5, s)$;*

(iii) *$t = s^2$, $s = q^n$ with q odd, where $\mathbf{GF}(q)$ is the kernel of the TGQ $S = S^{(\infty)}$ with (∞) an arbitrary translation point of S, $q \geq 4n^2 - 8n + 2$ and S is the point-line dual of a flock GQ $S(\mathcal{F})$ where $\mathcal{F}$ is a Kantor-Knuth flock;*

(iv) $t = s^2$, $s = q^n$ with q odd, where $\mathbf{GF}(q)$ is the kernel of the TGQ $S = S^{(\infty)}$ with (∞) an arbitrary translation point of S, $q < 4n^2 - 8n + 2$ and S is the translation dual of the point-line dual of a flock GQ $S(\mathcal{F})$ for some flock $\mathcal{F}$.

If a thick GQ S has two noncollinear translation points, then S is isomorphic to one of $\mathbf{Q}(4, s)$, $\mathbf{Q}(5, s)$.

Conversely, if $S = \mathbf{T}(\mathcal{O})$ is a good thick TGQ of order (s, s^2) in odd characteristic, then S has a line of translation points.

Proof.   Let S be as assumed. If S contains two noncollinear translation points, then by Theorems 10.5.1, 10.5.3 and 10.7.1 S is isomorphic to one of $\mathbf{Q}(4, s)$, $\mathbf{Q}(5, s)$.

Suppose that S contains two collinear distinct translation points u and v. Then for $s = t$, the statement follows immediately from Theorem 10.5.1. If $s \neq t$, then $t = s^2$ by Theorem 10.5.3. By transitivity, every point of $L := uv$ is a translation point, and hence every line of $L^\perp$ is an axis of symmetry. If we fix some base-span $\mathcal{L}$ for which $L \in \mathcal{L}^\perp$, then by Theorem 10.6.4 there are $s + 1$ classical subGQs of order s which mutually intersect precisely in the base-grid $\Gamma = (\Omega, \mathcal{L} \cup \mathcal{L}^\perp, \mathbf{I}')$ (with the usual notation). Fix an arbitrary translation point $(\infty)\mathbf{I}L$, and consider the TGQ $S^{(\infty)} = \mathbf{T}(\mathcal{O})$ with base-point (∞). Then by Theorem 7.5.7, $\mathcal{O}$ is good at its element π which corresponds to L. If s is even, the result follows by Theorem 7.8.3.

Now let s be odd. Then Theorem 5.1.2 implies that the translation dual $\mathbf{T}(\mathcal{O}^*)$ of $\mathbf{T}(\mathcal{O})$ satisfies Property (G) at the flag $((\infty)', L')$, where $(\infty)'$ corresponds to (∞), and L' to L. If we now apply Theorem 4.10.6, then it follows that $\mathbf{T}(\mathcal{O}^*)$ is the point-line dual of a flock generalized quadrangle $S(\mathcal{F})$. The theorem now follows from Theorem 5.1.14.

Conversely, suppose that $S = \mathbf{T}(\mathcal{O})$ is a good thick TGQ of order (s, s^2), s odd, with $\pi \in \mathcal{O}$ a good element. Let x be any point of $\mathbf{T}(\mathcal{O})$ incident with π. Then by Theorem 6.2.1 and Theorem 6.3.5, x is coregular. By Corollary 6.6.9, we conclude that x is a translation point.

The theorem is proved.   ∎

Corollary 10.7.3 *A GQ of order (s, t), $s \neq 1 \neq t$ and s even, has two distinct collinear translation points if and only if S is isomorphic to $\mathbf{Q}(4, s)$ or $\mathbf{Q}(5, s)$.*   ∎

Corollary 10.7.4 *Suppose $\mathcal{S} = \mathbf{T}(\mathcal{O})$ is a TGQ of order (q^n, q^m), where q is odd if $n = m$. Suppose (∞) is the base-point of the TGQ. Let $\mathcal{S}^* = \mathbf{T}(\mathcal{O}^*)$ be its translation dual with base-point $(\infty)'$. Then we have one of the following possibilities:*

(i) $|\mathrm{Aut}(\mathcal{S})| = |\mathrm{Aut}(\mathcal{S}^*)|$;

(ii) $|\mathrm{Aut}(\mathcal{S})| = (q^n + 1)|\mathrm{Aut}(\mathcal{S}^*)|$ *with $n \neq m$;*

(iii) $|\mathrm{Aut}(\mathcal{S}^*)| = (q^n + 1)|\mathrm{Aut}(\mathcal{S})|$ *with $n \neq m$.*

Proof. The $\mathrm{Aut}(\mathcal{S})$-orbit of (∞), with $\mathcal{S} = (\mathcal{P}, \mathcal{B}, \mathrm{I})$, is either $\mathcal{P}$, a line, or $\{(\infty)\}$. If $n = m$ with q odd and if the $\mathrm{Aut}(\mathcal{S})$-orbit of (∞) is not $\{(\infty)\}$, then by 1.3.6 (iv) of [128] each line of $\mathcal{S}$ is regular, so $\mathcal{S} \cong \mathbf{Q}(4, q^n)$ by Theorem 2.2.2, hence $\mathcal{S} \cong \mathcal{S}^*$. If every point of $\mathcal{S}$ is a translation point, then $\mathcal{S}$ is classical, $\mathcal{S} \cong \mathcal{S}^*$, and so $\mathrm{Aut}(\mathcal{S}) \cong \mathrm{Aut}(\mathcal{S}^*)$. In all other cases we have one of (i),(ii),(iii) by Theorem 3.11.1. ∎

The following corollary, which shows that all possibilities in Corollary 10.7.4 occur, was deduced in K. Thas [205] in the context of TGQs arising from flocks.

Corollary 10.7.5 (1) *Suppose $\mathcal{S} = \mathbf{T}(\mathcal{O})$ is a TGQ of order (s, s^2), s even, and let (∞) be the base-point. Let $\mathcal{S}^* = \mathbf{T}(\mathcal{O}^*)$ be the translation dual with base-point $(\infty)'$. Then $|\mathrm{Aut}(\mathcal{S})| \cong |\mathrm{Aut}(\mathcal{S}^*)|$.*

(2) *Suppose $\mathcal{S} = \mathbf{T}(\mathcal{O})$ is a TGQ of order (s, s^2), s odd, which arises from a flock but which is not the point-line dual of a Kantor-Knuth semifield flock GQ. Let (∞) be the base-point of $\mathbf{T}(\mathcal{O})$. Let $\mathcal{S}^* = \mathbf{T}(\mathcal{O}^*)$ be the translation dual of $\mathbf{T}(\mathcal{O})$ with base-point $(\infty)'$. Then $|\mathrm{Aut}(\mathcal{S}^*)| = (s+1)|\mathrm{Aut}(\mathcal{S})|$.*

Proof. Suppose we are in Case (1). Then by Theorem 10.7.2, both $\mathcal{S}$ and $\mathcal{S}^*$ either have precisely one translation point, or all their points are as such. The result follows from Theorem 3.11.1.

If we are in Case (2), then $\mathcal{S}^*$ is the translation dual of the point-line dual of a flock GQ — hence $\mathcal{S}^*$ is good, and so by Theorem 10.7.2 it has a line of translation points. By Theorem 10.7.2 and Theorem 6.4.4, $\mathcal{S}$ has only one translation point, and the result follows from Theorem 3.11.1. ∎

10.8   On the Classification of Translation Generalized Quadrangles

Combining the obtained results, we arrive at the following theorem.

Theorem 10.8.1 *Let* $S = \mathbf{T}(\mathcal{O})$ *be a TGQ of order* (q^n, q^m) *with translation point* (∞). *Then we have the following possibilities.*

(i) $n = m$, *and either* $S \cong \mathbf{Q}(4, q^n)$, *or*

    (a) S *has one translation point, q is even, and* $\mathrm{Aut}(S) = \mathrm{Aut}(S)_{(\infty)}$;

    (b) S *has one translation point, q is odd, and if u and v are arbitrary points of* $\mathbf{T}(\mathcal{O})$ *and* $\mathbf{T}(\mathcal{O}^*)$, *not collinear with* (∞) *and* $(\infty)'$ *(where* $(\infty)'$ *is the translation point of* $S^* = \mathbf{T}(\mathcal{O}^*))$, *respectively, then*

$$\mathrm{Aut}(S)_u = [\mathrm{Aut}(S)_{(\infty)}]_u \cong \mathrm{Aut}(S^*)_v = [\mathrm{Aut}(S^*)_{(\infty)'}]_v.$$

    Also, $|\mathrm{Aut}(S)| = |\mathrm{Aut}(S^*)|$.

(ii) $n < m$, *and then we have one of the following cases.*

    (a) $S \cong \mathbf{Q}(5, q^n)$.

    (b) S *and* S^* *are not classical, have two distinct collinear translation points, and then q is odd, $m = 2n$, and* $S \cong S^* \cong S(\mathcal{F})^D$, *where* $\mathcal{F}$ *is a Kantor-Knuth semifield flock.*

    (c) S *and* S^* *are not classical, S is not the point-line dual of a Kantor-Knuth semifield flock GQ, has a line of translation points, and then $m = 2n$, q is odd, and S is the translation dual of the point-line dual of a flock GQ. In this case we have that* $|\mathrm{Aut}(S)| = (q^n + 1)|\mathrm{Aut}(S^*)|$. *Moreover, we have that*

$$[\mathrm{Aut}(S)_{(\infty)}]_u \cong \mathrm{Aut}(S^*)_v = [\mathrm{Aut}(S^*)_{(\infty)'}]_v,$$

    where (∞) *is an arbitrary translation point on the line of translation points* $[\infty]$ *of S, where* $(\infty)'$ *is the translation point of S^*, and where u and v are arbitrary points of* $\mathbf{T}(\mathcal{O})$ *and* $\mathbf{T}(\mathcal{O}^*)$, *not collinear with* (∞) *and* $(\infty)'$, *respectively.*

    (d) S *and* S^* *are not classical, S^* is not the point-line dual of a Kantor-Knuth semifield flock GQ, has a line of translation points, and then*

$m = 2n$, q is odd, and $\mathcal{S}^*$ is the translation dual of the point-line dual of a flock GQ. Similarly as in Case (ii) (c), we have that $|\mathrm{Aut}(\mathcal{S}^*)| = (q^n + 1)|\mathrm{Aut}(\mathcal{S})|$. Also,

$$[\mathrm{Aut}(\mathcal{S}^*)_{(\infty)'}]_v \cong \mathrm{Aut}(\mathcal{S})_u = [\mathrm{Aut}(\mathcal{S})_{(\infty)}]_u,$$

where $(\infty)'$ is an arbitrary translation point on the line of translation points $[\infty]'$ of $\mathcal{S}^*$, where (∞) is the translation point of $\mathcal{S}$, and where u and v are arbitrary points of $\mathbf{T}(\mathcal{O})$ and $\mathbf{T}(\mathcal{O}^*)$, not collinear with (∞) and $(\infty)'$, respectively.

(e) $\mathcal{S}$ and $\mathcal{S}^*$ both have precisely one translation point, and then $\mathrm{Aut}(\mathcal{S}) = \mathrm{Aut}(\mathcal{S})_{(\infty)}$ and $\mathrm{Aut}(\mathcal{S}^*) = \mathrm{Aut}(\mathcal{S}^*)_{(\infty)'}$, where the notation is obvious. Furthermore, $|\mathrm{Aut}(\mathcal{S})| = |\mathrm{Aut}(\mathcal{S}^*)|$, and

$$\mathrm{Aut}(\mathcal{S})_u \cong \mathrm{Aut}(\mathcal{S}^*)_v,$$

where u and v are arbitrary points of $\mathbf{T}(\mathcal{O})$ and $\mathbf{T}(\mathcal{O}^*)$, not collinear with (∞) and $(\infty)'$, respectively.

If $H = \mathrm{Aut}(\mathcal{S})_{(\infty)}$ is the stabilizer of (∞) in $\mathrm{Aut}(\mathcal{S})$, then we have that $H \cong T \rtimes H_x$, with T the translation group of $\mathcal{S}^{(\infty)}$ and x any point not collinear with (∞).

Proof. First suppose that $n = m$. Then by Theorem 10.7.2, $\mathcal{S}$ has one translation point if $\mathcal{S}$ is not classical. Part (i) then follows from Theorem 3.11.1.

Next suppose that $n < m$, and that $\mathcal{S}$ is not isomorphic to $\mathbf{Q}(5, q^n)$. If $\mathcal{S} = \mathbf{T}(\mathcal{O})$ has two distinct translation points, then they are necessarily collinear, and from Theorem 10.7.2 it follows that q is odd and that $(\mathcal{S}^*)^D$ is a flock GQ $\mathcal{S}(\mathcal{F})$, or, equivalently, $\mathcal{O}$ is good at some element π (cf. Corollary 5.1.4). If $\mathcal{S}^*$ also contains distinct translation points, then Theorem 6.4.4 implies that $\mathcal{F}$ is a Kantor-Knuth semifield flock, and hence $\mathcal{S}^* \cong \mathcal{S}$ by Theorem 4.7.3. Suppose $\mathcal{S} \not\cong \mathcal{S}^*$. Then $\mathcal{S}^*$ has one translation point and Part (c) follows from Theorems 3.11.1, 10.7.2 and Corollary 10.7.4. Part (d) is obtained from Part (c) by interchanging the role of $\mathcal{S}$ and $\mathcal{S}^*$. If we are not in one of the previous cases, then we are in Case (e) by Theorem 3.11.1. The final part of the statement now follows from Theorem 3.12.1. ∎

Remark 10.8.2 Based on the theory of SPGQs, and in particular TGQs with more than one translation point, K. Thas has described a detailed conjectural classification program (called "Blueprint" in *loc. cit.*) for all TGQs in the final chapter of [206]. (The Blueprint first appeared in [204].)

In the next chapter, we will consider GQs which are TGQs with respect to every point without any finiteness assumption. We will develop our ideas in the more general context of Moufang GQs.

Chapter 11

Moufang Quadrangles with a Translation Point

In this chapter, we study generalized quadrangles which are translation with respect to every point. In fact, these are a special case of the class of so-called *Moufang quadrangles*. We will formulate and prove a lot of equivalent, but seemingly much weaker, conditions than the Moufang Condition. From time to time, and when appropriate, we apply these results to translation Moufang GQs. We also show a basic correspondence between general Moufang GQs and translation Moufang GQs, namely, that every Moufang GQ contains a full or ideal thick translation Moufang GQ.

All these results will be shown without any finiteness assumption. This guarantees that all arguments be purely geometric, or combinatorial group theoretic. In the finite case, other proofs exist (but not necessarily shorter ones!) making use of deeper group theoretic theorems such as the classification of Moufang sets and of irreducible split BN-pairs of rank 2 (which are not available in the infinite case).

11.1 Notation

Two nonintersecting lines L, M will sometimes be called *opposite*; likewise two noncollinear points x, y will sometimes be called *opposite*.

For every point x and line L, we denote the set of elements incident with x and L by $[x]$ and $[L]$, respectively. With this notation, the mapping $\rho_{L,M} : [L] \to [M] : x \mapsto \operatorname{proj}_M x$ is a bijection for every line M opposite L. We shall use this map below, so we treat $\rho_{L,M}$ as standard notation.

Let S be a finite GQ of order (s, t), $s, t \geq 2$. A subquadrangle of order (s, t'), for some $t' \in \mathbb{N}$, will be called a *full* subGQ; a subquadrangle of order (s', t), for some $s' \in \mathbb{N}$, will be called an *ideal* subGQ. In general, a *full* subquadrangle S' of a (not necessarily finite) generalized quadrangle S is a subquadrangle with the property that every point of S incident with some line of S' belongs to S'. Dually, one defines in general an *ideal* subquadrangle.

Now let S be a thick GQ of order (s, t), let S' be a full subGQ of order (s, t'), $t' < t$, and let S'' be a full subGQ of S' of order (s, t''), $t'' < t'$. Then $t = s^2$, $t' = s$ and $t'' = 1$; see Lemma 1.1.1. In particular, S'' does not admit a proper full subGQ.

We start with some general lemmas on subGQs and collineation groups, some of which are trivial in the finite case. We gather these in the next section. In Section 11.3 we introduce a lot of conditions on the collineation group of a GQ S and its action on S. All these conditions assume a transitive collineation group with a so-called *Tits system*, and therefore we study this situation in Section 11.4. In the subsequent sections, we prove the promised equivalences of conditions we introduced, and show that every Moufang GQ contains a subGQ which is a TGQ.

11.2   Some General Elementary Lemmas

The first lemma is a characterization of GQs, grids and dual grids.

Lemma 11.2.1 *Let $S = (\mathcal{P}, \mathcal{B}, \mathrm{I})$ be an incidence structure satisfying the following properties.*

(i) *Each element is incident with at least 2 elements.*

(ii) *There exist a point x and a line L not incident with x.*

(iii) *For each $x \in \mathcal{P}$ and each $L \in \mathcal{B}$, with x not incident with L, there exist a unique point y and a unique line M for which $x \mathrm{I} M \mathrm{I} y \mathrm{I} L$.*

Then S is either a generalized quadrangle, or a grid, or a dual grid.

Proof.   Suppose, by way of contradiction, that there are two points x_1, x_2 both incident with two distinct lines L_1, L_2. Then every line incident with x_1 must be incident with x_2, as otherwise we contradict (iii). Dually, every point on any line incident with both x_1, x_2 must be incident with every line through both x_1, x_2. Conditions (i) and (ii) imply that there exists a line L not incident with any of the points on L_1. Then the line through x_1 meeting L must also be incident with x_2, but it contains a point — namely, the point on L — not incident with L_1, a contradiction.

Suppose now that S is not a grid, nor a dual grid. Left to show is that there are a constant number of points on a line and, dually, a constant number of lines through a point. If all lines are incident with exactly two points, then it is easy to see that we obtain a dual grid; likewise, if all points are incident with exactly two lines, then we have a grid. So we may suppose that there is a line L with three points, and a point x incident with three lines. Condition (iii) readily implies that opposite lines carry the same number of points, and opposite points are incident with the same number of lines. We claim that every line $M \sim L$, $M \neq L$, is opposite some line $K \not\sim L$. Indeed, let x_1, x_2, x_3 be three points incident with L, and let $M_i \mathrm{I} x_i$ be arbitrary but distinct from L, $i = 1, 2, 3$. We show that there is a line opposite both L and M_1. Let $y_j \mathrm{I} M_j$, $j = 2, 3$, $y_j \neq x_j$. It is clear that x cannot be collinear with all four of x_2, x_3, y_2, y_3. But since x_2 and x_3 are opposite y_3 and y_2, respectively, at least one of y_2, y_3 is incident with at least three lines. One of these three lines does not meet L nor M_1. The claim is proved. Now M_1 is incident with the same number of points as L, and the lemma is proved. ∎

We refer to Condition (iii) of the previous lemma as the *Main Axiom of GQs*.

For thick GQs, we have the following alternative definition.

Lemma 11.2.2 *Let $S = (\mathcal{P}, \mathcal{B}, \mathrm{I})$ be an incidence structure satisfying the following properties.*

(i′) *Some point is incident with at least three lines, and some line is incident with at least three points.*

(ii′) *There exists an ordinary quadrangle in S, i.e., there are four points $x_1, x_2, x_3, x_4 \in \mathcal{P}$ with $x_1 \sim x_2 \sim x_3 \sim x_4 \sim x_1$, where $x_1 \not\sim x_3$ and $x_2 \not\sim x_4$.*

(iii) *For each $x \in \mathcal{P}$ and each $L \in \mathcal{B}$, with x not incident with L, there exist a unique point y and a unique line M for which $x \mathrm{I} M \mathrm{I} y \mathrm{I} L$.*

Then S is a thick generalized quadrangle.

Proof. By (iii), the lines x_1x_2 and x_3x_4, and the lines x_2x_3 and x_4x_1 in (ii′) do not meet. Let L be any line of S. If L is incident with one of the points x_i, $i \in \{1,2,3,4\}$, say with x_1, then we can apply (iii) to the pair $\{x_3, L\}$ and obtain a second point on L. If L is not incident with one of $x_1, \ldots, x_4$, then we can apply (iii) to L and each of $x_1, \ldots, x_4$. It is easy to see that this gives rise to at least two distinct points on L. Similarly, one shows that each point of S is incident with at least two lines. Hence we have proved (i) of the previous lemma. As (ii) of that lemma is an immediate consequence of (ii′), we conclude that S is either a thick GQ, or a grid or dual grid. But the latter two do not satisfy (i′).

The lemma is proved.   ∎

The next lemma is evidently true in the finite case, comparing orders.

Proposition 11.2.3 *Let $S = (\mathcal{P}, \mathcal{B}, \mathrm{I})$ be a thick GQ and let S' be a thick subGQ of S. If S' contains at least one line L such that every point in S on L is also a point of S', then S' is a full subGQ. Also, if S' is both full and ideal in S, then it coincides with S.*

Proof. If M is a line in S' opposite L, then the above map $\rho_{L,M}$ is a bijection in both S and S'; hence all points of M in S belong to S'. The first result now follows by repeatedly applying this observation and noting that the graph $(\mathcal{B}, \not\sim)$ is connected (it even has diameter 2).

For the second assertion, let z be an arbitrary point of S, and let L be a line of S'. If $z\mathrm{I}L$, then z belongs to S' since S' is a full subGQ. Suppose now $z \not\mathrm{I} L$. Since S' is full in S, the point $\mathrm{proj}_L z$ belongs to S'. Since S' is ideal in S, the line $\mathrm{proj}_z L \mathrm{I} \mathrm{proj}_L z$ belongs to S'. Since S' is full in S, the point z must now belong to S'. Similarly all lines of S belong to S'.

The proposition is proved.   ∎

The connection between collineations and subGQs will become apparent from the following proposition.

Proposition 11.2.4 *Let φ be a collineation of the GQ S. Let $\mathcal{P}'$ ($\mathcal{B}'$) be the set of invariant points (lines) of S under the action of φ. If $\mathcal{P}'$ contains two opposite points of S, and $\mathcal{B}'$ contains two opposite lines of S, then $\mathcal{P}'$ and $\mathcal{B}'$ are the point and line set, respectively, of either a subGQ S' of S, or a grid or dual grid.*

Proof. Let I' be the restriction of I to the sets $\mathcal{P}'$ and $\mathcal{B}'$. According to Lemma 11.2.1, it suffices to show that $\mathcal{S}' = (\mathcal{P}', \mathcal{B}', I')$ is an incidence structure in which every element is incident with at least two elements, in which there is a point of $\mathcal{P}'$ not incident with some line of $\mathcal{B}'$, and in which the Main Axiom of GQs holds. The latter is clear since, if a point x is fixed, and a line L not through x is fixed, then $\mathrm{proj}_x L$ and $\mathrm{proj}_L x$ are uniquely determined by x and L and therefore fixed. We now show that every line of $\mathcal{S}'$ is incident with at least two points of $\mathcal{S}'$.

Let L be any line of $\mathcal{S}'$. By assumption, there are two opposite points x_1, x_2 in $\mathcal{S}'$. Hence the points $\mathrm{proj}_L x_i$, $i = 1, 2$, belong to $\mathcal{S}'$. If $\mathrm{proj}_L x_1 \neq \mathrm{proj}_L x_2$, then we are done. Suppose now that $\mathrm{proj}_L x_1 = \mathrm{proj}_L x_2$. By assumption there are also two opposite lines in $\mathcal{S}'$. Consequently not all elements of $\mathcal{B}'$ contain the point $\mathrm{proj}_L x_1$. Let $M \in \mathcal{B}'$ be such that it is not incident with $\mathrm{proj}_L x_1$. If M meets L in another point, then we are done, so we assume that M is opposite L in $\mathcal{S}'$. Then the line M cannot be concurrent with both $\mathrm{proj}_{x_1} L$ and $\mathrm{proj}_{x_2} L$, since a triangle would appear. We may assume without loss of generality that M is not concurrent with $\mathrm{proj}_{x_1} L$. Since x_1 and $\mathrm{proj}_L x_1$ are collinear, the Main Axiom of GQs implies that $y_1 := \mathrm{proj}_M x_1$ and $y_2 := \mathrm{proj}_M (\mathrm{proj}_L x_1)$ are different points, belonging to $\mathcal{P}'$. Hence also $\mathrm{proj}_L y_1$ and $\mathrm{proj}_L x_1 = \mathrm{proj}_L (\mathrm{proj}_M (\mathrm{proj}_L x_1))$ are two distinct points on L belonging to $\mathcal{S}'$. Hence every line of $\mathcal{S}'$ is incident with at least two points of $\mathcal{S}'$ and dually, every point of $\mathcal{S}'$ is on at least two lines of $\mathcal{S}'$. Now it is also clear that $\mathcal{S}'$ contains a point x and a line L which are not incident. $\blacksquare$

Here is an immediate useful corollary.

Corollary 11.2.5 *Let φ be a collineation in a thick GQ $\mathcal{S}$ fixing an ordinary quadrangle, fixing all points on a certain line of that quadrangle, and fixing all lines through a certain point (vertex) of that quadrangle. Then φ is the identity.*

Proof. By the previous proposition, φ fixes pointwise a subGQ which is by assumption thick. By Proposition 11.2.3, this subGQ coincides with $\mathcal{S}$. Hence φ fixes everything and is the identity. $\blacksquare$

We end this section with two general combinatorial results. The first one is folklore, the second one is an observation of Cuypers, see [21].

Lemma 11.2.6 *Let $S = (\mathcal{P}, \mathcal{B}, I)$ be a thick GQ and let $x \in \mathcal{P}$ be arbitrary. Then the graph (Γ, E) with vertices the points of S opposite x, and adjacency given by collinearity is connected.*

Proof. Let y, z be two points opposite x. If some point in $\{y, z\}^{\perp}$ is not collinear with x, then y and z belong to the same connected component of Γ. If all points of $\{y, z\}^{\perp}$ belong to $x^{\perp}$, then replacing y by any point y' collinear with y and opposite x yields a pair $\{y', z\}$ for which $\{y', z\}^{\perp}$ is not contained in $x^{\perp}$. Now y is adjacent to y', which is in the same connected component as z. ∎

Lemma 11.2.7 *Let $S = (\mathcal{P}, \mathcal{B}, I)$ be a thick GQ not of order $(2, 2)$. Let $\{x, L\}$ be a flag of S, with $x \in \mathcal{P}$ and $L \in \mathcal{B}$. Define the following graph (Γ, E). The vertex set Γ is the union of the set of points of S opposite x and the set of lines of S opposite L. The edge set E is the set of flags of S whose elements are both contained in Γ. Then the graph (Γ, E) is connected.*

Proof. It is clearly enough to show that every two vertices of (Γ, E) that correspond with points of S can be connected by a path in (Γ, E). So let y, z be two points of S opposite x. Since the order of S is unequal $(2, 2)$, we may assume, without loss of generality, that every point of S is incident with at least four lines. Since there is exactly one line through each point opposite x which is not opposite L, we see that the valency of y in (Γ, E) is at least 3. Similarly for z. We may obviously assume that y and z cannot be joined by a path of length 2 or 4 in (Γ, E). This means that y and z are not incident with a line opposite L, and that the intersection point (if it exists) of any line through y opposite L with any line through z opposite L is collinear with x.

We now assume first that $\mathrm{proj}_L y = \mathrm{proj}_L z$, but $\mathrm{proj}_y L \neq \mathrm{proj}_z L$. Since there is one line through each of y, z not opposite L, we see that there are at least three points x_1, x_2, x_3 not on L collinear with all of x, y, z. Now let y' be a point collinear with y, $y' \neq y$, such that $x_3 I y y'$ and such that y' is opposite x (hence belonging to Γ). Clearly, y and y' can be connected inside (Γ, E) (by a path of length 2). If y' is collinear with z, then y', z, x_3 are the vertices of a triangle, contradiction. So y' is not collinear with z, and at least one of the lines $L_i := \mathrm{proj}_{y'} z x_i$, $i = 1, 2$, say L_1, is not concurrent with L (since they cannot be equal). Also, the intersection point z' of L_1 and $z x_1$ is opposite x. We now have the path $(z, z z', z', L_1, y', y y', y)$ of length 6 connecting z with y.

If, still under the assumption that $\text{proj}_L y = \text{proj}_L z$, we have $\text{proj}_y L = \text{proj}_z L$, and so yz is a line, then we apply the previous paragraph to a point y^* opposite x, collinear with $\text{proj}_L y$, and not on yz. We obtain that y and y^* are connected in (Γ, E), and y^* and z are connected in (Γ, E), hence y and z are connected by a path in (Γ, E).

We now assume that $\text{proj}_L y \neq \text{proj}_L z$. Then the line $M := \text{proj}_z(\text{proj}_y L)$ is opposite L and the point $\text{proj}_M y$ is opposite x. Hence we have a path from z to $\text{proj}_M z$ inside (Γ, E). But by the foregoing, we also have a path from $\text{proj}_M y$ to y, hence from z to y.

The lemma is proved. ∎

If $\mathcal{S}$ is a GQ of order $(2, 2)$, then it is easy to show that (Γ, E) of the previous lemma has exactly two connected components, each being an ordinary quadrangle in $\mathcal{S}$.

11.3 The Moufang Property and Analogues

We introduce some more notation. Let $\mathcal{S} = (\mathcal{P}, \mathcal{B}, \text{I})$ be a GQ. An *apartment* is an ordinary quadrangle in $\mathcal{S}$. It consists of four points and four lines. A *root* is a set of five "consecutive" elements of an apartment. Hence a root contains either two points x_1, x_2 and three lines L_1, L_2, L_3, with $L_1 \text{I} x_1 \text{I} L_2 \text{I} x_2 \text{I} L_3$, or dually, it contains three points and two lines. It will be useful to distinguish between these dual notions. Therefore, we will call the former a *root*, and the latter (dually) a *dual root*. A root without its extremal lines will be called a *panel*, more precisely, the *interior (panel)* of the root. Similar but dual definitions for *dual panels*. So a panel is a set of two distinct collinear points, together with the joining line.

Note that the Main Axiom of GQs implies that every pair of flags (a *flag* being an incident point-line pair) of $\mathcal{S}$ is contained in at least one apartment of $\mathcal{S}$.

We are now ready to define the Moufang Property for generalized quadrangles.

Let $\mathcal{S}$ be a generalized quadrangle with full collineation group $G = \text{Aut}(\mathcal{S})$. Let π be a panel. Then we say that π *has the Moufang Property* (or, equivalently, π is a *Moufang panel*) if for some root α with interior π, the group $G^{[\pi]}$ of collineations (called *root elations* and, dually, *dual root elations*) fixing every element incident with any element of π acts transitively on the set of apartments containing α. Let $\pi = \{x, y, L\}$, with $x, y \in \mathcal{P}$ and $L \in \mathcal{B}$. Let M be any line through x distinct from L. Suppose π is a Moufang panel, and let α be a root with interior π such that $G^{[\pi]}$ acts transitively on the set of apartments containing α. Let u, u' be two points on M distinct from x. Suppose the unique line of α incident with y and unequal L is N. Then N is opposite M and we may consider the distinct points $u_1 = \text{proj}_N u$ and $u_1' = \text{proj}_N u'$. Using the Main Axiom of GQs, it is quite easy to see that u_1 and u_1' determine unique apartments Σ and Σ', respectively, containing α. Hence there is a collineation $\theta \in G^{[\pi]}$ mapping Σ to Σ', and hence mapping u_1 to u_1'. But since M is fixed under θ by assumption, θ maps u to u', and hence, for every line M' through y distinct from L, θ maps the (unique) apartment containing π, u and M' to the (unique) apartment containing π, u' and M'. We have shown that the definition of Moufang panel is independent of the root involved in that definition.

We can even say a little more. Suppose that $\theta \in G^{[\pi]}$, with π as above. Suppose that θ leaves an apartment through π invariant. Then by Corollary 11.2.5, θ is the identity. Hence, in general, $G^{[\pi]}$ acts freely on the set of apartments through α, with α a root with interior π, and it acts sharply transitively on that set precisely when π is a Moufang panel.

If every panel and every dual panel of the GQ S has the Moufang Property, then we say that S has the Moufang Property, or that S is a Moufang GQ. If every panel is a Moufang panel, or if every dual panel is a Moufang dual panel, then we say that S is a half Moufang GQ. If S is a Moufang GQ, then the group generated by all root elations and dual root elations is called the little projective group of S.

Note that a symmetry about a line L is automatically a root elation, and it will sometimes be called an axial root elation (with axis L). Dually we define central root elations (with center some point x). It is now clear that every line of a GQ in which every point is a translation point, is an axis of symmetry, and hence the GQ is half Moufang.

Let $\{x, L\}$ be a flag of the GQ S. Let MIx and yIL such that y is not incident with M. As above, it is easy to see with the aid of Corollary 11.2.5, that the group $G^{[x,L]}$ of collineations fixing all lines through x and fixing all points on L acts freely on the set of apartments containing $\{x, y, L, M\}$. We call the flag $\{x, L\}$ a Moufang flag if the group $G^{[x,L]}$ acts transitively (and hence sharply transitively) on the set of apartments containing $\{x, y, L, M\}$. As before, one shows easily that this definition is independent of the chosen line M through x, $M \neq L$, and of the chosen point y on L, $y \neq x$.

We see that the definition of a Moufang flag is a self-dual one, hence there is no need to introduce something like a "dual Moufang flag". If every flag of the GQ S is a Moufang flag, then we call S *3-Moufang*, where the number 3 refers to the length of the sequence (y, L, x, M) as a path in the incidence graph of S (which is the graph $(\mathcal{P} \cup \mathcal{B}, \mathtt{I})$). The 3-Moufang Condition is a completely self-dual one, and so one might think that a notion of "half 3-Moufang" cannot be defined. But actually, one can take a kind of greatest common divisor of the 3-Moufang Condition and the half Moufang Condition to obtain the following definition of a half 3-Moufang generalized quadrangle.

Let $\{x, L\}$ be a flag of the GQ S. Let $z\mathtt{I}M\mathtt{I}x$ and $K\mathtt{I}y\mathtt{I}L$ such that $z \neq x$, $K \neq L$ and y is not incident with M. As before, the group $G_z^{[x,L]}$ acts freely on the set of apartments containing $\{x, y, z, L, M\}$, and the group $G_K^{[x,L]}$ acts freely on the set of apartments containing $\{x, y, K, L, M\}$. When we have transitivity for each z, then in the first case, we say that the flag $\{x, L\}$ is *half 3-Moufang at x*, while in the second case we talk about *half 3-Moufang at L*. The GQ S is called *half 3-Moufang* if either every flag $\{x, L\}$ of S is half 3-Moufang at x, or every flag $\{x, L\}$ of S is half 3-Moufang at L.

Now let x be any point of the GQ S. Recall that if the group $G^{[x]}$ of collineations fixing all lines through x acts transitively on the set of points of S opposite x (here, this is not necessarily a regular action!), then we say that x is a *center of transitivity*. Dually, we define an *axis of transitivity*. If all points are centers of transitivity, and all lines are axes of transitivity, then we say that S is a *2-Moufang* GQ. If either all points are centers of transitivity, or all lines are axes of transitivity, then we call S a *half 2-Moufang* GQ.

Let $\{x, L\}$ again be a flag of the GQ S, with x a point and L a line. Another flag $\{y, M\}$ is called *opposite* $\{x, L\}$ if the point y is opposite x and the line M is opposite L. If the group $G_{x,L}$ of all collineations of S fixing the flag $\{x, L\}$ acts transitively on the set of flags opposite $\{x, L\}$, then we say that $\{x, L\}$ is a *transitive flag*. If all flags are transitive, then we say that S is a *1-Moufang* GQ, or, equivalently, that S satisfies the *Tits Condition*, or that S is a *Tits* GQ. The last two names are motivated by their group theoretic counterparts, the *Tits systems*, see below. If S satisfies the Tits Condition, then the corresponding collineation group G acts transitively on the set of ordered pairs of opposite flags of S.

For convenience, we shall sometimes also refer to the (half) 4-Moufang Condition for a (half) Moufang GQ.

Let $\mathcal{S} = (\mathcal{P}, \mathcal{B}, \mathbf{I})$ be a Tits GQ with full collineation group G. Let $\{x, L\}$ be any flag in $\mathcal{S}$, with $x \in \mathcal{P}$ and $L \in \mathcal{B}$. Set $B := G_{x,L}$. Let Σ be an apartment containing $\{x, L\}$. Define $N := G_\Sigma$, the (setwise) stabilizer of Σ, and set $H = B \cap N$. Then H is precisely the group of collineations fixing Σ elementwise. Hence $H \trianglelefteq N$. Now, if there exists a nilpotent normal subgroup U of B such that $UH = B$, then we say that $\mathcal{S}$ has the *Fong-Seitz Property*. A *Fong-Seitz* GQ is a Tits GQ with this additional group theoretic property. From group-theoretic point of view this is a natural condition. Note that, a priori, nothing guarantees uniqueness of U, but one can actually show that it is unique (see De Medts, Haot, Tent and Van Maldeghem [48]); in the finite case it is not so difficult to see that U is the unique Sylow p-subgroup of B, when G contains the little projective group of the corresponding Moufang GQ — as we will later prove, every Fong-Seitz GQ is a Moufang GQ. Hence B is always a split extension of the nilpotent group U, and $B \cong U : H$.

All Moufang GQs have been classified by Tits and Weiss in the monograph *Moufang Polygons* [229]. Technically, the classification was completed in 1997. In the finite case, the classification of Moufang GQs follows from the classification of Fong-Seitz GQs by Fong and Seitz [59, 60]. They show that the Moufang Condition implies the Fong-Seitz Property, and then they use entirely group theoretic methods to classify the Fong-Seitz GQs (in fact, they classify all Fong-Seitz *generalized polygons*, but the part of quadrangles is the largest by far). The proof is highly nontransparent from a geometric point of view, and it is approximately 100 pages long. Corollary 11.6.7 and Theorem 11.7.4 reduce in a completely geometric and elementary combinatorial group theoretic way the Fong-Seitz Condition to the Moufang Condition (including the infinite case!), and hence combined with the classification theorem of Tits and Weiss (mentioned above), this classifies all Fong-Seitz GQs. In the finite case, one can also appeal to the alternative geometric classification of Moufang quadrangles by Payne and Thas [128, Chapter 9], Kantor [85] and K. Thas [199], see also Thas [188]. All this replaces the bulk of the papers of Fong and Seitz [59, 60] by elementary methods, and generalizes it at the same time to the infinite case.

In this chapter, we essentially prove that all the conditions introduced above, except for the Tits Condition and the half 2-Moufang Condition, are equivalent. In particular, this will imply that a GQ in which every point is a translation point is a Moufang GQ. Also, we will show that every Moufang GQ has a thick subGQ which is a TGQ. For all this, we will not need finiteness, so we prove things in the most general setting.

Concerning the Tits Condition, it can be proved, using the classification of finite simple groups, that also this condition is, in the finite case, equivalent with the Moufang Condition; see Buekenhout and Van Maldeghem [33]. In the infinite case, there are a lot of counter examples, the most striking ones being without doubt those constructed by Tent [150]. But we will not use the classification of finite simple groups in this book, and so we will not prove the result of Buekenhout and Van Maldeghem mentioned above. We refer to work of K. Thas [209, 210, 211] on the classification of finite Tits GQs without the classification of finite simple groups.

Concerning the half 2-Moufang Condition, it can be shown that it is also equivalent to the Moufang Condition, at least in the finite case. This has been done by K. Thas and Van Maldeghem [215], but the proof of that result is beyond the scope of this book.

11.4 Tits Generalized Quadrangles and Tits Systems

As we saw in Chapter 3, a 4-gonal family is the group theoretic equivalent of the notion of an EGQ. Likewise, we will axiomatize group theoretically the notion of a Tits GQ. This gives rise to *groups with a BN-pair*, also called *Tits systems*. This notion was introduced by Tits [218, 220] for spherical buildings, and we specialize it here to the class of GQs.

At this stage, it is convenient to introduce a *generalized distance* between flags of $\mathcal{S}$ as follows. Let $F = \{x, L\}$ and $F' = \{x', L'\}$ with $x, x' \in \mathcal{P}$ and $L, L' \in \mathcal{B}$ be two flags of $\mathcal{S}$. Then $\delta(F, F') = z \in \{-3, -2, -1, 0, 1, 2, 3, 4\}$, where $|z| = \frac{1}{2}(d(x, x') + d(L, L'))$ and if $|z| \neq 4$, then z has the same sign as $d(x', x) + d(L', x) - d(x', L) - d(L', L)$ (distances are measured in the incidence graph of $\mathcal{S}$ and the sign of 0 is "+"). Note that $\delta(F, F') = -\delta(F', F)$ if $\delta(F, F')$ is even but not equal to 4; otherwise $\delta(F, F') = \delta(F', F)$.

In the next proposition we will be dealing with double cosets. It is convenient to think of these as unions of ordinary cosets. For instance, the double coset HgK for subgroups H, K of some group G and $g \in G$ consists of the union of all left cosets hgK for h ranging over H, or of all right cosets Hgk for k ranging over K. Note that, in contrast with ordinary cosets, double cosets need not contain an equal number of elements. For instance, $|HgH| > |HH|$ whenever g does not normalize H.

We will also consider a quotient group G/N as consisting of cosets of N in G. With this convention, it makes sense to multiply elements of the quotient group with elements or subgroups of G; it is just the ordinary multiplication of subsets in a group.

Proposition 11.4.1 *Let $\mathcal{S} = (\mathcal{P}, \mathcal{B}, I)$ be a thick Tits generalized quadrangle and let G be a collineation group of $\mathcal{S}$ such that G acts transitively on the set of ordered pairs of opposite flags of $\mathcal{S}$. Let $F = \{x, L\}$ be any flag in $\mathcal{S}$, with $x \in \mathcal{P}$ and $L \in \mathcal{B}$, and let Σ be an apartment containing F. Define $B := G_{x,L}$, $N := G_\Sigma$ and $H = B \cap N$ as above. Then $N \cap (G_x \setminus G_L)$ and $N \cap (G_L \setminus G_x)$ are nonempty. Choose arbitrarily s_x and s_L, respectively, in these sets. Then (G, B, N) and s_x, s_L satisfy the following properties.*

(BN1) $G = \langle B, N \rangle$.

(BN2) $H \trianglelefteq N$ *and* $W := N/H = \langle s_x H, s_L H \rangle$ *is isomorphic to* **Dih**$_8$*, the dihedral group of order* 8.

(BN3) $BsBwB \subseteq BwB \cup BswB$, *for all* $w \in N$ *and* $s \in \{s_x, s_L\}$.

(BN4) $sBs \neq B$, *for* $s \in \{s_x, s_L\}$.

Also, B does not contain any nontrivial normal subgroup of G and H coincides with

$$\bigcap_{w \in N} w^{-1} B w.$$

Proof. Let $F' = \{x', L'\}$, $x' \in \mathcal{P}$, $L' \in \mathcal{B}$, be the flag in Σ that is opposite F. Also, let $M \neq L$ and $M' \neq L'$ be the "second" lines in Σ incident with x and x', respectively. By the Tits Property there is some element $s_x \in G_x$ mapping the pair $(\{x, L\}, \{x', L'\})$ to the pair $(\{x, M\}, \{x', M'\})$. Clearly s_x stabilizes Σ and belongs to $N \cap (G_x \setminus G_L)$. Similarly we find $s_L \in N \cap (G_L \setminus G_x)$. We may treat the elements s_x and s_L as arbitrary elements in $N \cap (G_x \setminus G_L)$ and $s_L \in N \cap (G_L \setminus G_x)$, respectively.

Remark that the Tits Property is equivalent with G being transitive on the set of flags of $\mathcal{S}$, together with requiring that B acts transitively on the set of apartments containing F.

We now show that N acts transitively on the flags of Σ. By flag transitivity, there is a collineation g mapping any flag F' of Σ onto F, and by our previous remark, there is a collineation $g' \in B$ mapping Σ^g back to Σ. Then $gg' \in N$ and maps F' onto F.

We now show (BN1). Let $g \in G$ be arbitrary. Let Σ' be an apartment of $\mathcal{S}$ containing the flags F and F^g. By the Tits Property there is some $g_1 \in B$

mapping Σ' to Σ. Since N acts transitively on the flags of Σ, we can find $w \in N$ with $(F^{gg_1})^w = F$. Hence $gg_1 w \in B$ and so $G = BNB$.

We turn to (BN2). The action of N on Σ is clearly isomorphic to a subgroup of the natural action of $\mathbf{Dih}_8$ on a square. By our remark above, this action in fact coincides with that natural action. Since $\mathbf{Dih}_8$ acts sharply transitively on the "flags" of the corresponding square, we see that $H = B \cap N$ is the elementwise stabilizer of Σ, and hence a normal subgroup of the setwise stabilizer N. Moreover, H is the kernel of the permutation group (N, Σ) (Σ seen as a set of flags) and as such $N/H \cong \mathbf{Dih}_8$.

Returning to our proof above of (BN1), one sees that, if we define $w_z \in N/H$ as the coset of H in N whose elements map F onto the flag of Σ at generalized distance z, then an arbitrary element $g \in G$ that maps F onto a flag at generalized distance z from F belongs to $Bw_z B$ (which defines a unique double coset of B since $H \leq B$). Conversely, it is easy to see that every element of $Bw_z B$ maps F to a flag F' at generalized distance $\delta(F, F') = z$ from F. Hence G is the disjoint union of the double cosets $Bw_z B$, for $z \in \{-3, -2, -1, 0, 1, 2, 3, 4\}$.

Now let F' be a flag with $\delta(F, F') = i$. If $i < 0$, then one easily sees that $\delta(F, F'^{s_L}) = 1 - i$. If $i > 0$, then either $\delta(F, F'^{s_L}) = 1 - i$, or $\delta(F, F'^{s_L}) = i$. Now, one checks that $w_{1-i} = w_i s_L = w_i w_1$. This shows $BwBs_L B \subseteq BwB \cup Bws_L B$, for all $w \in W$. Taking inverses, (BN3) for $s = s_L$ follows. Similarly, (BN3) is also true for $s = s_x$ (and $s_x H = w_{-1}$).

Regarding (BN4), we may choose a $g \in B$ which does not fix the unique point incident with L contained in Σ and distinct from x. This is possible by the thickness assumption. For this g, the image of F under $s_L g s_L$ clearly differs from F (and lies at generalized distance 1 from F). Hence $s_L B s_L \neq B$. Similarly $s_x B s_x \neq B$.

If $K \trianglelefteq G$ is contained in B, then $K = K^g \trianglelefteq B \cap B^g$, for every $g \in G$. Hence K stabilizes every flag F^g, for g ranging over G, and so K fixes every flag (by flag transitivity of G) and must therefore be trivial.

Since $w^{-1}Bw$ is the stabilizer of the flag F^w, and F^w ranges over all flags of Σ as w ranges over N, every element of $\bigcap_{w \in N} w^{-1}Bw$ belongs to N and hence to H. Conversely, every element of H fixes every flag of Σ and hence belongs to $w^{-1}Bw$, for all $w \in N$.

The proposition is proved.   ∎

The group W is called the *Weyl group* of (G, B, N), and the elements s_x and s_L *representatives of the standard generators of W.*

We now would like to prove a converse of the previous proposition. To that aim, we define the notion of a group with a BN-pair of Type $\mathbf{B}_2$, also called a Tits system of Type $\mathbf{B}_2$. Since we will not encounter any other types of such groups, we will shortly talk about groups with a BN-pair and Tits systems.

Let G be a group, and let B and N be two subgroups of G. Set $H = B \cap N$. Then (G, B, N) is called a *Tits system*, or (B, N) is called a *BN-pair in G*, or G is called a *group with BN-pair* (B, N), if there exist elements s_x and s_L of G, with $\{s_x^2, s_L^2\} \subseteq H$, such that (BN1) up to (BN4) of Proposition 11.4.1 hold. The group W in (BN2) will be called the *Weyl group* and the cosets $s_x H$ and $s_L H$ the *standard generators of W*.

If $\mathcal{S}$ is a Tits GQ, and (G, B, N) is as in Theorem 11.4.1, then we call (G, B, N) a *natural Tits system associated with $\mathcal{S}$*.

If (G, B, N) is a Tits system with Weyl group W and if s_x and s_L are corresponding representatives of the standard generators of W, then we define the following incidence structure $\mathcal{S}_{G,B,N} = (\mathcal{P}, \mathcal{B}, \mathbf{I})$. Define $P_x = \langle B, B^{s_x} \rangle$ and $P_L = \langle B, B^{s_L} \rangle$.

- POINTS. The elements of $\mathcal{P}$ are the right cosets of P_x.

- LINES. The elements of $\mathcal{B}$ are the right cosets of P_L.

- INCIDENCE. For $g, h \in G$, the point $P_x g$ is incident with the line $P_L h$ if $P_x g \cap P_L h \neq \emptyset$ (and in this case we may choose $g = h$).

The group G acts (on the right) as a collineation group on $\mathcal{S}_{G,B,N}$, and it is flag transitive since every flag can be written as $\{P_x g, P_L g\}$ (and is the image under g of the "standard" flag $\{P_x, P_L\}$). If we denote the point P_x by x, then the point $P_x g$ can be written as x^g. Similarly, we write P_L as L, and every line can be written as L^g, for some $g \in G$.

Before continuing, we prove a lemma, which will later lead to the *Bruhat Decomposition*.

First a definition. Let W be the Weyl group of some Tits system, and let s_x, s_L be representatives of the standard generators. Then the minimum number of factors needed to write an element $w \in W$ as a product of $s_x H$ and $s_L H$ is called the *length* of w, and denoted $\ell(w)$. Since W is a dihedral group of order 8, we have $\ell(w) \in \{0, 1, 2, 3, 4\}$, for an arbitrary element w of W.

Lemma 11.4.2 *If (G, B, N) is a Tits system with Weyl group W, then for $w, w' \in W$, one has $BwB = Bw'B$ if and only if $w = w'$.*

Proof. Clearly, if $w = w'$, then $BwB = Bw'B$. Now assume that $w \neq w'$ and assume that $BwB = Bw'B$. First suppose that $w = s$ is a standard generator of W. Note that $B \subseteq BsBsB$, since $sBsB$ contains $sHsH = H \leq B$. Then (BN3) tells us that $B \subseteq BsBsB = BsBw'B \subseteq Bsw'B \cup Bw'B$. Hence either $sw' \subseteq B$ or $w' \subseteq B$, implying that either $w' = s$ or w' is the identity. The former contradicts our hypothesis, the latter implies $BsB = B$, implying s is the identity, again a contradiction.

So we may assume that $1 < \ell(w) \leq \ell(w')$. Let $w = sw''$, with s a standard generator, and $\ell(w'') = \ell(w) - 1$. Then $Bsw''B = Bw'B$ implies the existence of $b, b' \in B$ with $sw'' = bw'b'$, hence $w'' = s^{-1}bw'b' \in BsBw'B \subseteq Bsw'B \cup Bw'B$. So either $Bw''B = Bsw'B$, or $Bw''B = Bw'B$. An easy induction on the length of w implies now that either $w'' = sw'$, contradicting $w = sw''$ as s is an involution, or $w'' = w'$, contradicting $\ell(w') \geq \ell(w) > \ell(w'')$.

The lemma is proved. ∎

Now let $s \in \{s_x, s_L\}$ and note that (BN3) implies $BsBsB \subseteq BsB \cup B$. Since (BN4) says that $BsBsB \neq B$, we deduce that at least one element w of BsB belongs to $BsBsB$, and hence multiplying w at the left and at the right with B, we obtain $BsB \subseteq BsBsB$. We now see that $BsBsB = BsB \cup B$. It follows that $Bs_xB \cup B = Bs_xBs_xB = P_x$ and $Bs_LB \cup B = Bs_LBs_LB = P_L$.

We now claim that G is generated by P_x and P_L. Indeed, by the previous paragraph, $s_xH \subseteq P_x$ and $s_LH \subseteq P_L$. Also, (BN2) implies that N is generated by s_xH and s_LH, and so $N \leq \langle P_x, P_L \rangle$. Hence, since $B \leq P_x \cap P_L$, we obtain $\langle B, N \rangle \leq \langle P_x, P_L \rangle$ and (BN1) completes the proof of our claim.

Next we claim that $P_x \cap P_L = B$. Indeed, since $s_x \notin B$, the double cosets Bs_xB and B are distinct, and likewise Bs_LB and B are distinct. Also, Lemma 11.4.2 implies that $Bs_xB \neq Bs_LB$. The claim is now clear.

As a consequence, we may identify the flags of $\mathcal{S}_{G,B,N}$ with the right cosets of B. Denote by Σ the subgeometry induced by the set of flags

$$\{B, Bs_x, Bs_L, Bs_xs_L, Bs_Ls_x, Bs_xs_Ls_x, Bs_Ls_xs_L, Bs_xs_Ls_xs_L\}.$$

Just as we showed $Bs_xB \neq Bs_LB$ above using (BN2) and (BN3), the same conditions also guarantee that those eight flags are different. Notice that (using the equality $Bs_x s_L s_x s_L = Bs_L s_x s_L s_x$, which follows from $W \cong$ Dih_8)

$$B \subseteq P_x \cap P_L,$$

$$Bs_L \subseteq P_x s_L \cap P_L,$$

$$Bs_x s_L \subseteq P_x s_L \cap P_L s_x s_L,$$

$$Bs_L s_x s_L \subseteq P_x s_L s_x s_L \cap P_L s_x s_L,$$

$$Bs_x s_L s_x s_L \subseteq P_x s_L s_x s_L \cap P_L s_x s_L s_x,$$

$$Bs_x s_L s_x \subseteq P_x s_L s_x \cap P_L s_x s_L s_x,$$

$$Bs_L s_x \subseteq P_x s_L s_x \cap P_L s_x,$$

$$Bs_x \subseteq P_x \cap P_L s_x.$$

Hence Σ induces a quadrangle in $\mathcal{S}_{G,B,N}$, if we can show that the point x is not collinear with $x^{s_L s_x s_L}$ and dually, that the line L is not concurrent with $L^{s_x s_L s_x}$ (and then it follows that x^{s_L} is not collinear with $x^{s_L s_x}$ by applying s_L to both points and remarking that $Bs_L s_x s_L s_L \subseteq Bs_L s_x H = BH s_L s_x = Bs_L s_x$, and likewise for the lines L^{s_x} and $L^{s_x s_L}$). Now

$$P_x s_L s_x s_L = Bs_x Bs_L s_x s_L \cup Bs_L s_x s_L$$

$$\subseteq Bs_x s_L s_x s_L B \cup Bs_L s_x s_L B,$$

which follows from multiplying the right hand side of the first line at the right with B and then applying (BN3) with $s = s_x$ and $w = s_L s_x s_L$.

Any line $P_L g$ incident with P_x shares a coset of B with P_x. The cosets of B in P_x are of the form B and $Bs_x b$, with $b \in B$. Hence we may put $g = s_x b$ (leaving the more trivial case of $g \in B$ to the reader). Hence the coset $P_L g$ is equal to $Bs_x b \cup Bs_L Bs_x b$, and hence belongs to the union of the double cosets $Bs_x B \cup Bs_L s_x B$ (indeed, (BN3) implies that $Bs_L Bs_x B \subseteq Bs_L s_x B \cup Bs_x B$; taking inverses of the left hand side, applying (BN3) again, and then again taking inverses, we deduce $Bs_L Bs_x B \subseteq Bs_L s_x B \cup Bs_L B$. So $Bs_L Bs_x B \subseteq Bs_L s_x B$ by taking intersections, using Lemma 11.4.2). This union should now contain a right coset of B that is also contained in one of the double cosets $Bs_x s_L s_x s_L B$ or $Bs_L s_x s_L B$. This implies, however, that one of these double cosets coincides with one of $Bs_x B$ or $Bs_L s_x B$, contradicting Lemma 11.4.2.

The dual argument is similar, and so we conclude that Σ is a quadrangle. It is easy to see that s_x and s_L preserve Σ and act as generating reflections of the dihedral action of N on Σ.

Now we claim that $G = BNB$. Since $G = \langle B, N \rangle$, it suffices to show that every element $bnb'n'b''$, with $b, b', b'' \in B$ and $n, n' \in N$, belongs to BNB. Noticing that every element of N can be written as a finite product of s_x and s_L times an element of $H \leq B$, the claim easily follows from (BN3) using induction on the length $\ell(n)$ of n.

Together with Lemma 11.4.2, this gives rise to the Bruhat Decomposition, which states that

$$G = \overset{\cdot}{\bigcup_{w \in W}} BwB.$$

Consider any flag F^g. This flag corresponds to the coset Bg. By the foregoing paragraphs we can write $g = b'nb$, with $b, b' \in B$ and $n \in N$. Hence $Bg = Bnb$ and so b^{-1} fixes F and maps F^g into Σ.

It now follows rather easily that the geometry $\mathcal{S}_{G,B,N}$ does not contain any digon $\{x_1, x_2, L_1, L_2\}$, with $x_1, x_2 \in \mathcal{P}$ and $L_1, L_2 \in \mathcal{B}$ with $x_i I L_j$, $i, j \in \{1, 2\}$. Indeed, by flag transitivity, we may assume $x_1 = x$ and $L_1 = L$. By the previous paragraph, we may also assume that the flag $\{x_2, L_2\}$ is contained in Σ, a contradiction.

Now suppose that $\mathcal{S}_{G,B,N}$ contains a triangle $x_1 I L_3 I x_2 I L_1 I x_3 I L_2 I x_1$, with the x_i points and the L_i lines, $i = 1, 2, 3$. We may again assume $x_1 = x$ and $L_2 = L$, and by the previous paragraph, we may assume that the flag $\{x_2, L_1\}$ belongs to Σ. Since Σ is a quadrangle, we must have $x_2 = x^{s_L s_x s_L}$. But then L_3 is incident with both x and $x^{s_L s_x s_L}$, contradicting the fact that these points are not even collinear.

Now let x' be any point not incident with some line L'. We may assume $x' = x$ and $L' \in \Sigma$. It follows that x' is incident with at least two lines (the lines through x in Σ), that L' is incident with at least two points, and that there is a flag $\{y, M\}$ with $x' I M I y I L'$. The flag $\{y, M\}$ is unique by the previous paragraph. Hence we have shown that $\mathcal{S}_{G,B,N}$ is a GQ (note that by flag transitivity, it cannot be a grid or a dual grid without order).

The group G acts as a collineation group, by multiplication at the right, on $\mathcal{S}_{G,B,N}$. Since every element of the kernel K of that action must in particular fix B, we see that K is a normal subgroup of G contained in B. Conversely, if $K' \trianglelefteq G$, and $K' \leq B$, then, since the stabilizer of the flag Bg, $g \in G$, clearly coincides with B^g, the group K' stabilizes every flag of $\mathcal{S}_{G,B,N}$, and hence belongs to the kernel K. So K is the biggest normal subgroup of G contained in B, and as such equal to

$$K = \bigcap_{g \in G} B^g.$$

The group G/K acts faithfully on $\mathcal{S}_{G,B,N}$ and the stabilizer of the flag F is B/K.

Now we determine the stabilizer of Σ. Since N stabilizes Σ and acts transitively on the flags of Σ, it suffices to determine the elementwise stabilizer S of Σ, and then NS is the global stabilizer of Σ. So suppose some $b \in B$ stabilizes Σ. Since Σ is uniquely determined by the opposite flags F and $F^{s_x s_L s_x s_L}$, this is equivalent with $b \in B$ fixing $F^{s_x s_L s_x s_L}$. Hence $S = B \cap B^{s_x s_L s_x s_L}$. Note that then b automatically belongs to B^w, for every $w \in W$. Hence we can also write

$$S = \bigcap_{w \in W} B^w.$$

We have proved the following theorem.

Theorem 11.4.3 *Let (G, B, N) be a Tits system with Weyl group W. Then the geometry $\mathcal{S}_{G,B,N}$ defined above is a generalized quadrangle satisfying the Tits Condition. Setting*

$$K = \bigcap_{g \in G} B^g \text{ and } S = \bigcap_{w \in W} B^w,$$

G/K acts naturally and faithfully by right translation on $\mathcal{S}_{G,B,N}$. Also, B is the stabilizer of a unique flag F and NS is the stabilizer of a unique apartment containing F, and the triple $(G/K, B/K, NS/K)$ is a natural Tits system associated with $\mathcal{S}_{G,B,N}$. ∎

In the literature, the Tits system (G, B, N) is called *saturated* precisely when $N = NS$, with S as above. Replacing N by NS, every Tits system is "equivalent" to a saturated one.

It is now clear that the notion of a Fong-Seitz quadrangle is equivalent to the notion of a saturated Tits system (G, B, N) with the property that there exists a normal nilpotent subgroup U of B with $B = UH$. If we do not insist on (G, B, N) being saturated, then we must require that $B = US$, with S as above. In any case, we call such Tits systems *split*.

11.5 Properties of Moufang Quadrangles

In this section, we will show that the Moufang Property implies all properties that we introduced in Section 11.3.

From now on until the end of this chapter we always assume that the GQs we are dealing with are thick, except if explicitly mentioned otherwise.

First we prove some easy assertions. They belong to folklore, but an explicit reference is Van Maldeghem [232], Chapter 6.

Proposition 11.5.1 (i) *Every Moufang GQ is a 3-Moufang GQ.*

(ii) *Every half Moufang GQ is a half 3-Moufang GQ.*

(iii) *Every 3-Moufang GQ is a 2-Moufang GQ.*

(iv) *Every half 3-Moufang GQ is a half 2-Moufang GQ.*

(v) *Every half 2-Moufang GQ is a Tits GQ.*

(vi) *Every k-Moufang GQ is always a half k-Moufang GQ, $k = 2, 3, 4$.*

Proof. Let $S = (\mathcal{P}, \mathcal{B}, I)$ be a GQ with full collineation group G. Let $\pi := \{x_1, L_1, x_2\}$ and $\pi' := \{L_0, x_1, L_1\}$ be a panel and a dual panel, respectively, with $x_1, x_2 \in \mathcal{P}$, $L_0, L_1 \in \mathcal{B}$.

First suppose that π and π' are Moufang. Then the groups $G^{[\pi]}$ and $G^{[\pi']}$ generate the group $G^{[x_1, L_1]}$ which acts transitively on the set of apartments containing $\pi \cup \pi'$. Hence the flag $\{x_1, L_1\}$ is a Moufang flag. This shows (i).

Next, suppose that π is Moufang. Then the flag $\{x_1, L_1\}$ is clearly half 3-Moufang at L_1. This shows (ii).

Suppose now that the flags $\{x_1, L_1\}$ and $\{x_2, L_1\}$ are Moufang. Then the group generated by $G^{[x_1, L_1]}$ and $G^{[x_2, L_1]}$ clearly acts transitively on the set of apartments containing x_1, L_1, x_2. This shows (iii).

We now show (iv). More exactly, we assume that for every line $L I x_1$, the flag $\{x_1, L\}$ is half 3-Moufang at L, and we show that the group $G^{[x_1]}$ acts transitively on the set of points opposite x_1. If y_1, y_2 are two distinct collinear points opposite x_1, then the line M joining y_1 and y_2 meets some line L' incident with x_1. The element of $G^{[x_1, L']}$ mapping y_1 to y_2 evidently belongs to $G^{[x_1]}$, and (iv) follows from Lemma 11.2.6.

Suppose now that S is half 2-Moufang, and that all points are centers of transitivity. Since the stabilizer of a dual panel $\{L_0, x_1, L_1\}$ acts transitively on the panels $\{x_2, L_2, x_3\}$, with $L_1 I x_2 I L_2 I x_3$, $x_1 \neq x_2 \neq x_3$ and $L_1 \neq L_2$, using the group $G^{[x_1]}$, we see that G acts transitively on the set of flags of

S, and that the stabilizer of the flag $\{x_1, L_0\}$ acts transitively on the set of flags opposite it. This proves (v).

Finally, (vi) is trivial.   ∎

The Moufang Property is inherited by every subGQ. In the finite case, this means that classical and dual classical GQs only have classical and dual classical subGQs. Also this is folklore, but we present a proof.

Proposition 11.5.2 *Any thick subquadrangle $S' = (\mathcal{P}', \mathcal{B}', \mathrm{I})$ of a Moufang quadrangle $S = (\mathcal{P}, \mathcal{B}, \mathrm{I})$ is a Moufang quadrangle.*

Proof. Let α be a (dual) root in S' with interior (dual) panel π, and let Σ and Σ' be two apartments in S' containing α. Let u be the (dual) root elation with respect to α which maps Σ onto Σ'. Then the intersection S'' of S' with its image S'^u under u clearly satisfies the assumptions of Lemma 11.2.2, and hence is a thick subquadrangle in both S' and S'^u. Moreover, looking at the elements incident with one of the elements of π, we see that S'' is full and ideal in both S' and S'^u. Lemma 11.2.3 implies that S'' coincides with both S' and S'^u, hence $S' = S'^u$ and u is a (dual) root elation in S'. It is now straightforward to conclude that S' is a Moufang GQ.   ∎

We will now show that a Moufang GQ is a Fong-Seitz GQ, see also (33.4) of Tits and Weiss [229].

Theorem 11.5.3 *Every thick Moufang generalized quadrangle is a Fong-Seitz generalized quadrangle.*

Proof. By Proposition 11.5.1, we already know that a Moufang GQ is a Tits GQ. So let S be a Moufang GQ and take a flag $\{x, L\}$, then we only have to construct a nilpotent normal subgroup U of the stabilizer B of $\{x, L\}$ such that $UH = B$, with H the intersection of B with the stabilizer N of some apartment Σ of S containing x and L. The construction is quite easy, and also the fact that $UH = B$ will be easily proved. But the proof that U is nilpotent will need some arguments.

Fix an apartment Σ containing $\{x, L\}$. We now give new names to the elements of Σ. We put $\Sigma = \{x_1, x_2, \ldots, x_8\}$, where we read the subscripts modulo 8, with $x_i \mathrm{I} x_{i+1}$, for all $i \in \mathbb{Z} \mod 8$, and where $x_2 = x$ and $x_3 = L$. Hence, for convenience, we abandon here the notational convention of capital letters for lines and small letters for points. Instead, we have even index for points and odd index for lines. The group of (dual) root elations related to the (dual) panel $\{x_{i-1}, x_i, x_{i+1}\}$ will be denoted U_i. We put $U = \langle U_1, U_2, U_3, U_4 \rangle$.

First we show that $U = U_1 U_2 U_3 U_4$. Let $u_i \in U_i$ and $u_{i+1} \in U_{i+1}$ be arbitrary. Since u_i fixes all elements incident with x_{i-1}, and u_{i+1} fixes x_{i-1}, we see that $[u_i, u_{i+1}]$ fixes all elements incident with x_{i-1}. Similarly, $[u_i, u_{i+1}]$ fixes all elements incident with x_{i+2}. Both u_i and u_{i+1} fix all elements incident with either x_i or x_{i+1}, hence also $[u_i, u_{i+1}]$ does. We conclude that $[u_i, u_{i+1}]$ fixes Σ and an ideal and full subGQ, hence $[u_i, u_{i+1}] = \mathrm{id} = [u_{i+1}, u_i]$.

Now let $u_{i-1} \in U_{i-1}$ and $u_{i+1} \in U_{i+1}$. As above we see that $[u_{i-1}, u_{i+1}]$ fixes all elements incident with x_{i-1}, with x_i and with x_{i+1}. Hence $[u_{i-1}, u_{i+1}] \in U_i$. Similarly, $[U_{i+1}, U_{i-1}] \leq U_i$.

Finally, let $u_1 \in U_1$ and $u_4 \in U_4$. Then $u := [u_1, u_4]$ fixes, as before, all elements incident with both x_2 and x_3. Put $x_5' = x_5^u$. Since x_4 is fixed by u, x_5' is a line through x_4. By the Moufang Condition, there exists a dual root elation $u_2 \in U_2$ mapping x_5 to x_5'. Likewise, there exists a root elation $u_3 \in U_3$ mapping x_8 to x_8^u. The collineation $u u_3^{-1} u_2^{-1}$ now fixes all elements incident with x_2 or x_3, and fixes x_8 and x_5, hence Σ, and so has to be the identity by Corollary 11.2.5. It follows that $u = u_2 u_3 = u_3 u_2 \in U_2 U_3 = U_3 U_2$. Similarly, $[U_4, U_1] \leq U_2 U_3$.

Now note that $b^{-1}a = a[a, b]b^{-1}$, for arbitrary group elements. Hence, if $u_i, u_i' \in U_i$, for $i \in \{1, 2, 3, 4\}$ arbitrary, then, putting $[u_1^{-1} u_1', u_4] = u_2'' u_3''$, we obtain

$$(u_1 u_2 u_3 u_4)^{-1}(u_1' u_2' u_3' u_4') = u_3^{-1} u_1^{-1} u_1' u_2'' u_3'' u_4^{-1} u_2^{-1} u_2' u_3' u_4', \qquad (11.1)$$
$$= u_1^{-1} u_1' [u_1^{-1} u_1', u_3] u_2'' u_2^{-1} u_2' u_3^{-1} u_3''$$
$$\times [u_2^{-1} u_2', u_4] u_3' u_4^{-1} u_4', \qquad (11.2)$$

which clearly belongs to $U_1 U_2 U_3 U_4$. Hence $U_1 U_2 U_3 U_4$ is a subgroup of U. On the other hand, since $U_i \leq U_1 U_2 U_3 U_4$, for $i = 1, 2, 3, 4$, we also have $U \leq U_1 U_2 U_3 U_4$. Hence $U = U_1 U_2 U_3 U_4$.

Note that, since U_{i+1} and U_i centralize each other, $U_i U_{i+1}$ is a group, for all $i \in \mathbb{Z} \mod 8$. Also, the group U_{i-1} normalizes $U_i U_{i+1}$, so $U_{i-1} U_i U_{i+1}$ is a group, too, for all $i \in \mathbb{Z} \mod 8$.

We further claim that, if we write an element u of U as $u = u_1 u_2 u_3 u_4$, where $u_i \in U_i$, $i = 1, 2, 3, 4$, then this is unique. Indeed, suppose also $u =$

$u_1' u_2' u_3' u_4'$, with $u_i' \in U_i$, $i = 1, 2, 3, 4$. First we show that, if $a_1 a_2 a_3 a_4 = \mathrm{id}$, with $a_i \in U_i$, $i = 1, 2, 3, 4$, then $a_i = \mathrm{id}$ for all $i \in \{1, 2, 3, 4\}$. Indeed, all of a_1, a_2, a_3 fix x_1, and a_4 fixes x_1 only if $a_4 = \mathrm{id}$. Hence $a_4 = \mathrm{id}$. Similarly, $a_3 = \mathrm{id}$ by considering the action on x_8. Similarly $a_2 = a_1 = \mathrm{id}$. Now from Equation (11.2), we deduce that $\mathrm{id} = u^{-1} u = (u_1 \ldots u_4)^{-1} u_1' \ldots u_4'$ implies that $u_1^{-1} u_1' = \mathrm{id}$, hence $u_1 = u_1'$. This implies $[u_1^{-1} u_1', u_3] = [u_1^{-1} u_1', u_4] = \mathrm{id}$, and so $u_2'' = u_3'' = \mathrm{id}$, and now it follows rather quickly that $u_2 = u_2'$, $u_3 = u_3'$ and $u_4 = u_4'$. Hence the claim is proved.

So we know exactly what an element of U looks like. To prove nilpotency, we want to show in particular that the U_i are nilpotent. That is our next task. We will show that each U_i is nilpotent of class at most 2, and that one of U_1 and U_2 is abelian.

If both U_1 and U_2 are abelian, then there is nothing to prove, since by the Tits Property the groups U_{2i} are conjugate amongst themselves, and also the groups U_{2i+1} are mutually conjugate, for all $i \in \mathbb{Z} \mod 4$. So we may assume that for instance U_1 is not abelian. We first claim that every element of $V_1 := [U_1, U_1]$ fixes every line meeting x_1. Indeed, let x be any point on x_1 (distinct from x_8 and from x_2). Let $v_1 = [u_1, u_1']$ be an element of V_1. Let $u_3 \in U_3$ be such that $x_8^{u_3} = x$. Then $u_1^{u_3}$ fixes all lines through x and through x_2, and fixes all points on x_1. Moreover, it has the same action on the set of points incident with x_3 as u_1. Hence $[u_1^{u_3}, u_1'] v_1^{-1}$ fixes x_3 pointwise. But it belongs to U_1, hence $v_1 = [u_1^{u_3}, u_1']$, and the latter fixes all lines through x. The claim is proved. Next we show that $[V_1, U_1]$ is trivial. Let $u_1 \in U_1$ and $v_1 \in V_1$ be arbitrary. Consider any nontrivial element u_4 in U_4, then the conjugate $u_1^{u_4}$ has the same action on the set of points incident with x_3 as u_1. Hence $[u_1, v_1]$ has the same action on that point set as $[u_1^{u_4}, v_1]$. But the latter fixes all lines concurrent with x_1 (since v_1 does, and $u_1^{u_4}$ fixes x_1), and fixes all points on $x_1^{u_4}$, which is not equal to x_1. This implies that $[u_1^{u_4}, v_1]$ is the identity, and hence so is $[u_1, v_1]$. We have shown that U_1 is nilpotent of class 2. Since V_1 fixes $x_1^\perp$ elementwise and U_3 stabilizes $x_1^\perp$ globally, the commutator $[V_1, U_3]$ fixes $x_1^\perp$ elementwise and hence is trivial, since it is also a subgroup of U_2. Hence, by conjugating,

$$[V_{2i+1}, U_{2i-1}] = [V_{2i+1}, U_{2i+3}] = \mathrm{id}, \qquad (11.3)$$

for all integers i modulo 4. Let $u_1 \in U_1^\times$ and $u_4 \in U_4^\times$ be arbitrary. Then $[u_1, u_4] = u_1^{-1} u_1^{u_4}$. The latter expression shows that, if we fix u_4, then, since u_1 fixes x_1 pointwise, the action of $[u_1, u_4]$ on the points of x_1 is determined by $u_1^{u_4}$. But $U_1^{u_4}$ acts transitively on the points of x_1 distinct from x_2, hence we see that, writing $[u_1, u_4] = u_2 u_3$, with $u_i \in U_i$, $i = 2, 3$, we may choose, for each u_4, an arbitrary u_3, and then some (unique) $u_1 \in U_1$ and $u_2 \in U_2$ exist for which $[u_1, u_4] = u_2 u_3$. Since u_3 and $u_1^{u_4}$ have the same action on the points of x_1, we see that $u_3 = u_1^{u_4 u}$, for the unique $u \in U_8 U_1$ that maps $x_8^{u_4}$ to x_4. Since now V_1 and V_3 are conjugate (as U_1 and U_3 are), we deduce that $u_3 \in V_3$ if and only if $u_1 \in V_1$. Dually, we may fix u_1, choose u_2 and get unique u_3 and u_4 with $[u_1, u_4] = u_2 u_3$. We now apply this as follows. We choose $v_1 \in V_1$ nontrivial, but fixed. We let u_2 and u_2' be arbitrary but nontrivial in U_2. Our aim is to show that $[u_2, u_2'] = \text{id}$. We consider the unique $u_3 \in U_3$ and $u_4 \in U_4$ for which $[v_1, u_4] = u_2 u_3$. We now first claim that $[u_2 u_3, u_2'] = \text{id}$. Indeed, $[u_2 u_3, u_2'] = [[v_1, u_4], u_2']$, and this is equal to $[u_4, v_1] u_2'^{-1} v_1^{-1} u_4^{-1} v_1 u_4 u_2'$. First notice that, by a simple computation and the fact that U_4 centralizes U_3, we have $[u_2', u_4^{-1}] = [u_2', u_4^{-1}]^{u_4} = [u_2', u_4]^{-1}$. Secondly, remember that also V_1 centralizes U_3. Hence we have

$$
\begin{aligned}
v_1 u_4 u_2' &= u_2' v_1 u_2'^{-1} u_4 u_2' u_4^{-1} u_4 \\
&= u_2' v_1 [u_2', u_4^{-1}] u_4 \\
&= u_2' [u_2', u_4^{-1}] v_1 u_4
\end{aligned}
\tag{11.4}
$$

and also

$$
\begin{aligned}
v_1^{-1} u_4^{-1} u_2' &= v_1^{-1} u_2' [u_2', u_4] u_4^{-1} \\
&= u_2' v_1^{-1} u_4^{-1} [u_2', u_4].
\end{aligned}
\tag{11.5}
$$

So one calculates

$$
\begin{aligned}
[u_2 u_3, u_2'] &= [[v_1, u_4], u_2'] \\
&= [u_4, v_1] u_2'^{-1} v_1^{-1} u_4^{-1} v_1 u_4 u_2' \tag{11.6} \\
&= [u_4, v_1] u_2'^{-1} v_1^{-1} u_4^{-1} u_2' [u_2', u_4^{-1}] v_1 u_4 \tag{11.7} \\
&= [u_4, v_1] u_2'^{-1} u_2' v_1^{-1} u_4^{-1} [u_2', u_4] [u_2', u_4^{-1}] v_1 u_4 \tag{11.8} \\
&= [u_4, v_1] v_1^{-1} u_4^{-1} v_1 u_4 \tag{11.9} \\
&= [u_4, v_1] [v_1, u_4] = \text{id}, \tag{11.10}
\end{aligned}
$$

where we use Equation (11.4) to go from Equation (11.6) to Equation (11.7), and Equation (11.5) to go from Equation (11.7) to Equation (11.8). Our claim is proved. But now $\text{id} = [u_2 u_3, u_2'] = [u_2, u_2']^{u_3} [u_3, u_2'] = [u_2, u_2']^{u_3}$, implying that u_2 and u_2' commute. Hence U_2 is abelian if U_1 is not.

Hence from now on, we may assume, without loss of generality, that U_2 is abelian, regardless of U_1 being abelian or not. Note that the Tits Property implies that U_{2i} is abelian, for all i. We now show that $[U_2, U_4]$ centralizes U_1 (and also U_5 by the same token). Indeed, let $u_i \in U_i$ be arbitrary, $i = 1, 2, 4$, then we have (using the fact that u_2 centralizes $[U_4, U_1] \leq U_3 U_2$ because U_2 is abelian)

$$[[u_2, u_4], u_1] = [u_2^{-1} u_4^{-1} u_2 u_4, u_1] = [u_4^{-1}, u_1]^{u_2 u_4} [u_4, u_1]$$
$$= [u_4^{-1}, u_1]^{u_4} [u_4, u_1] = \text{id}.$$

Hence by a shift of the indices, we obtain that $[U_{2i}, U_{2i+2}]$ centralizes both U_{2i+3} and U_{2i-1}. We will denote the centralizer of U_{2i-1} and U_{2i+3} in U_{2i+1} by W_{2i+1}, for all i. Note that we have just shown that $[U_{2i}, U_{2i+2}] \leq W_{2i+1}$, and that Equation (11.3) implies that $V_{2i+1} \leq W_{2i+1}$. Now, we claim that $W_1 \trianglelefteq U_1$. Indeed, Let $w \in W_1$ and $u \in U_1$ be arbitrary, and let $v \in U_3$. Then $v^{(w^u)} = ((uvu^{-1})^w)^u = u^{-1} u^w v u^{-w} u = [u, w] v [w, u]$, and the latter is equal to v since $v \in U_3$ commutes with $[u, w] \in U_2$, and since $[w, u]^{-1} = [u, w]$. The claim is proved.

Before making the final calculations, we remark that $U_1 U_2 U_3 \trianglelefteq U$ (and not just a subgroup); this follows from

$$(u_1 u_2 u_3)^{u_4} = u_1 u_2 u_3 [u_1 u_2 u_3, u_4] = u_1 u_2 u_3 [u_1, u_4]^{u_2 u_3} [u_2, u_4]^{u_3} \in U_1 U_2 U_3,$$

since $[u_1, u_4]^{u_2 u_3} \in (U_2 U_3)^{u_2 u_3} = U_2 U_3$ and $[u_2, u_4]^{u_3} \in U_3^{u_3} = U_3$, for all $u_i \in U_i$, $i = 1, 2, 3, 4$. Similarly $U_2 U_3 U_4 \trianglelefteq U$. Hence also $U_2 U_3 \trianglelefteq U$. We also remark that, defining V_1 as the trivial group $[U_1, U_1]$ in case that U_1 is abelian, we always have that $[V_1, U_3]$ is trivial. Also, since $[V_1, U_4] \leq U_2 V_3$ if U_1 is nonabelian, as shown above, and since this remains true if U_1 is abelian (and V_1 is trivial in this case), we see that $V_1^{u_4} \leq V_1 U_2 V_3$ (and the latter is indeed a group), for all $u_4 \in U_4$.

Now we take $u := u_1 u_2 u_3 u_4$ and $u' := u_1' u_2' u_3' u_4'$, with $u_i, u_i' \in U_i$, $i = 1, 2, 3, 4$, arbitrary, and we calculate $[u, u']$. Note that, in general,

$$[ab, cd] = [a, d]^b [a, c]^{db} [b, d] [b, c]^d. \tag{11.11}$$

Assuming, as above, that U_2 and U_4 are abelian, and taking into account that u_2 and u_2' centralize each other and that they centralize U_1, and hence that $[u_1 u_2, u_1' u_2'] = [u_1, u_1']$, and similarly $[u_3 u_4, u_3' u_4'] = [u_3, u_3']$, we calculate:

$$[u, u'] = [u_1 u_2, u_3' u_4']^{u_3 u_4} [u_1 u_2, u_1' u_2']^{u_3' u_4' u_3 u_4} [u_3 u_4, u_3' u_4'] [u_3 u_4, u_1' u_2']^{u_3' u_4'},$$
$$= [u_1 u_2, u_3' u_4']^{u_3 u_4} [u_1, u_1']^{u_3' u_4' u_3 u_4} [u_3, u_3'] [u_3 u_4, u_1' u_2']^{u_3' u_4'}. \tag{11.12}$$

We now examine each factor of Equation (11.12). We will often use the following fact. Since for $u_i \in U_i$, and $u_j \in U_j$, one has $u_j[u_j, u_i] = u_j^{u_i}$,

$$U_j^{u_i} \subseteq U_j[U_j, U_i]. \tag{11.13}$$

Firstly,

$$[u_1u_2, u_3'u_4'] = [u_1, u_4']^{u_2}[u_1, u_3']^{u_4'u_2}[u_2, u_4'][u_2, u_3']^{u_4'},$$

and so this belongs to $(U_2U_3)^{u_2}.U_2^{u_4'u_2}.U_3 = U_2U_3.U_2U_3.U_3 = U_2U_3$. Hence $[u_1u_2, u_3'u_4']^{u_3u_4} \in U_2U_3^{u_3u_4}$. The latter equals $U_2^{u_3u_4}U_3^{u_3u_4} = U_2^{u_4'}U_3$, which is in $U_2[U_2, U_4]U_3 = U_2U_3$ by (11.13) above. Secondly, $[u_1, u_1']^{u_3'u_4'u_3u_4} = [u_1, u_1']^{u_4'u_3u_4}$, since $[V_1, U_3]$ is trivial. We further deduce that $[u_1, u_1']^{u_4'u_3u_4} \in (V_1U_2V_3)^{u_3u_4} \leq V_1U_2V_3$ (again using (11.13)). Thirdly, $[u_3, u_3'] \in V_3$. Finally, $[u_3u_4, u_1'u_2']^{u_3'u_4'} \in U_2U_3$, similarly to the first factor. Consequently, $[u, u'] \in V_1U_2U_3$. We now examine $[U, V_1U_2U_3]$. With the above notation, this means that we must evaluate $[u_1u_2u_3u_4, v_1'u_2'u_3']$, with $v_1' \in V_1$ arbitrary. We redefine u' as $v_1'u_2'u_3'$. We can do that evaluation by putting $u_1' = v_1'$ and $u_4' = $ id in Equation (11.12). We obtain, keeping into account that u_2 centralizes U_1 and U_3, hence $[u_1u_2, u_3'] = [u_1, u_3']$, and that v_1' centralizes U_1, hence $[u_1, v_1'] = $ id,

$$[u, u'] = [u_1, u_3']^{u_3u_4}[u_3, u_3'][u_3u_4, v_1'u_2']^{u_3'}. \tag{11.14}$$

We examine each factor of Equation (11.14). Firstly, $[u_1, u_3']^{u_3u_4} \in U_2W_3$. Secondly, $[u_3, u_3'] \in V_3 \leq W_3$. Finally, applying Equation (11.11), noting $[u_3, v_1'] = [u_3, u_2'] = $ id, and remarking that $[u_4, u_2']$ commutes with u_3' since both u_4 and u_2' do, we compute (also using (11.13)), $[u_3u_4, v_1'u_2']^{u_3'} = [u_4, u_2'][u_4, v_1']^{u_3'} \in W_3.U_2V_3 = U_2W_3$. We have shown that for the second central derivative of U holds $[U, U]_{[2]} \leq [U, U_2U_3] \leq U_2W_3$. So, in order to have an estimate for the third central derivative $[U, U]_{[3]}$, we consider an arbitrary $w_3' \in W_3$ and redefine $u' = u_2'w_3'$. We can now substitute id for v_1' and w_3' for u_3' in Equation (11.14) and calculate:

$$[u, u'] = [u_1, w_3']^{u_3u_4}[u_3, w_3'][u_3u_4, u_2']^{w_3'},$$
$$= [u_3, w_3'][u_3u_4, u_2']^{w_3'}. \tag{11.15}$$

Both factors of the right hand side of Equation (11.15) are contained in W_3. So $[U, U]_{[3]} \leq W_3$. But W_3 by definition centralizes U_1, and additionally centralizes U_2 and U_4. Hence we see that, putting $u' = w_3'$, and hence substituting id for u_2' in Equation (11.15),

$$[u, u'] = [u_3, w_3'], \tag{11.16}$$

and so $[U, U]_{[4]} \leq V_3$. Choosing now $v'_3 \in V_3$ arbitrarily, we see that v'_3 centralizes U_1, U_2, U_3 and U_4, hence $v'_3 \in Z(U)$ and so $[U, U]_{[5]}$ is trivial. This shows that U is nilpotent (and of class at most 5).

We now show that $UH = B$. Take an arbitrary element b of B. Put $x'_6 = x_6^b$ and $x'_7 = x_7^b$. We claim that there is an element of U mapping $\{x_6, x_7\}$ onto $\{x'_6, x'_7\}$. Indeed, let $u_1 \in U_1$ be the unique root elation mapping x_4 onto $x'_4 := \text{proj}_{x_3} x'_6$. Dually, let u_4 be the unique dual root elation of U_4 mapping x_1 onto $x'_1 := \text{proj}_{x_2} x'_7$. Then $u_1 u_4$ maps $\{x_4, x_1\}$ onto $\{x'_4, x'_1\}$. Let $u_3 \in U_3$ be such that it maps $x_8^{u_1 u_4} I x'_1$ onto $x'_8 := \text{proj}_{x'_1} x_2$. Finally, let $u_2 \in U_2$ be such that $u_2^{u_4}$ (which fixes $x'_1 = x_1^{u_4}$ pointwise) maps $x_5^{u_1 u_4 u_3} I x'_4$ to $x'_5 := \text{proj}_{x'_4} x'_7$, or equivalently, $x_7^{u_1 u_4 u_3}$ to x'_7. Then we see that $u := u_1 u_4 u_3 u_2^{u_4} \in U$ maps $\{x_6, x_7\}$ onto $\{x'_6, x'_7\}$. Our claim is proved. But now $bu^{-1} \in H$, so $b \in HU$, hence $HU = B$, or, taking inverses, $UH = B$.

In order to show that $U \trianglelefteq B = UH$, it suffices to show that $U_i^h = U_i$, for all i, and for all $h \in H$. But this is trivial.

This completes the proof of Theorem 11.5.3. ∎

We now prove a lemma, connecting the notion of symmetry about a line to the notion of a nontrivial line span in Moufang quadrangles.

Lemma 11.5.4 *Let S be a Moufang quadrangle. Then a root elation $u \in U_1$ is a symmetry about x_1 if and only if x_5^u belongs to the span $\{x_1, x_5\}^{\perp\perp}$.*

Proof. If u is a symmetry, then x_5^u belongs to $\{x_1, x_5\}^{\perp\perp}$ because u fixes all lines concurrent with x_1, hence also $\{x_1, x_5\}^\perp$. Since x_5 meets every line of $\{x_1, x_5\}^\perp$, also x_5^u does.

Conversely, suppose $x_5^u \in \{x_1, x_5\}^{\perp\perp}$. Then it is easy to see that all lines of $\{x_1, x_5\}^\perp$ are fixed by u. Let x'_7 be such a line, $x'_7 \neq x_3$, and denote by x'_0 the intersection of x_1 and x'_7. Let $u' \in G^{[x'_0, x_1, x_2]}$ be such that $x_5^{uu'} = x_5$. Then $uu' \in G^{[x_1, x_2]}$ fixes the quadrangle determined by x_1, x_3, x_5, x'_7, and hence by Corollary 11.2.5 $uu' = \text{id}$. Since $u' \in G^{[x'_0]}$, also $u \in G^{[x'_0]}$. Since x'_0 was essentially arbitrary on x_1, we now see that u is a symmetry about x_1. ∎

We will now show that every Moufang GQ contains a full or ideal subTGQ.

Theorem 11.5.5 *Every Moufang generalized quadrangle contains a thick full or ideal translation subquadrangle.*

Proof. We use the same notation and conventions as in the proof of Theorem 11.5.3 above. In particular, we assume that U_2 is abelian.

Suppose that for some $i \in \mathbb{Z} \mod 8$, an element u_i of $U_i^\times$ commutes with all elements of U_{i+2}, and let $u \in U_{i+2}$ be arbitrary. All elements incident with x_{i-1}^u are fixed by both $u_i[u_i, u]$ (since the latter equals u_i^u, and u_i fixes all elements incident with x_{i-1}) and $[u_i, u]$ (since it is the identity). Hence u_i fixes all elements incident with x_{i-1}^u. Since u was arbitrary and U_{i+2} acts transitively on the elements incident with x_i distinct from x_{i+1}, it follows that u_i fixes $x_i^\perp$ elementwise. Hence u_i is a symmetry about x_i.

Hence, if W_3 is trivial, then, since $[U_2, U_4] \leq W_3$, all elements of U_2 commute with all elements of U_4, and the previous paragraph implies that U_2 consists entirely of symmetries about x_2. By transitivity, every point is a center of symmetry, and every line is a translation line. Hence $\mathcal{S}$ is a TGQ.

Now suppose that W_3 is not trivial. Then by definition every element of W_3 commutes with all elements of U_1 and so, every element of W_3 is a symmetry about x_3. We now redefine W_{2i+1} as the subgroup of U_{2i+1} containing all symmetries about x_{2i+1}, for all $i \in \mathbb{Z}$. Our assumption implies that none of W_{2i+1} is trivial, and Lemma 11.5.4 implies that they are all conjugate to one another. We also remark that $[V_2, V_4] \leq W_3$.

We define the following point set in $x_2^\perp$:

$$\Omega = (\{x_2, x_6\}^\perp)^{W_3} \cup (\{x_2, x_6\}^\perp)^{W_1}.$$

Now let Λ be the set of lines of $\mathcal{S}$ incident with at least one point of Ω, and let Π be the set of points of $\mathcal{S}$ incident with at least two lines of Λ. Then we now show that $\mathcal{S}^* = (\Pi, \Lambda, \mathrm{I})$ is a subGQ of $\mathcal{S}$. We do that in a number of steps.

STEP 1. If $x_0' \in \{x_2, x_6\}^\perp \setminus \{x_0, x_4\}$, then we claim that $x_0'^{W_1} = x_0'^{W_3}$. Indeed, by the Moufang Property, there exists a unique collineation $u \in G^{[x_2 x_0', x_0', x_0' x_6]}$ mapping x_0 to x_4. Since u fixes all points on the line $x_2 x_0'$, we deduce that $x_0'^{W_1} \subseteq x_0'^{W_3}$. Using u^{-1}, the claim follows.

As a consequence, the groups W_1 and W_3 preserve Ω and hence act on $\mathcal{S}^*$ as collineation groups.

STEP 2. We claim that both U_0 and U_4 stabilize Ω and hence act as collineation groups on $\mathcal{S}^*$. Indeed, this follows readily from $(x_0^{W_3})^{u_4} = (x_0^{u_4})^{W_3}$, for all $u_4 \in U_4$, and the fact that $x_0^{u_4} \in \{x_2, x_6\}^\perp$. Similarly for U_0.

STEP 3. Every line of $\mathcal{S}$ through any element of Π belongs to Λ. Indeed, this is clear for the points of Ω. Now let x be the intersection of two elements L, M of Λ. We may assume that L, M are not incident with x_2, and that they are concurrent with the distinct lines L', M', respectively, incident with x_2. Using the action of U_0 and U_4, we can map $\{L', M'\}$ onto $\{x_1, x_3\}$. Subsequently, we can combine this with the actions of W_1 and W_3 to eventually map the intersection of L and L' onto x_4, and the intersection of M and M' onto x_0. Hence we may assume that x belongs to $\{x_0, x_4\}^\perp$. Let Y be any line through x, and let z be the projection of x_2 onto Y. It is necessary and sufficient to prove that z belongs to Ω. Clearly we may assume that $z \notin \{x_0, x_4\}$. Let $u_2 \in U_2$ map x_6 onto x and let $u_4 \in U_4$ map x_1 onto $x_2 z$. Set $z' := \text{proj}_{x_2 z} x_6$. Then

$$z'^{[u_4, u_2]} = z'^{u_4^{-1} u_2^{-1} u_4 u_2} = x_0^{u_2^{-1} u_4 u_2} = x_0^{u_4 u_2} = z'^{u_2} = z.$$

Since $[U_4, U_2] \leq W_3$, the proof of STEP 3 is complete.

STEP 4. The incidence structure $\mathcal{S}^*$ is a GQ. Indeed, let x be any point of $\mathcal{S}^*$, and L any line of $\mathcal{S}^*$, not incident with x. By STEP 3, the line $\text{proj}_x L$ belongs to $\mathcal{S}^*$, and if $\text{proj}_L x$ does not belong to $\Omega \cup \{x_2\}$, then by definition of Π, it belongs to Π since it is incident with two elements of Λ. Applying Lemma 11.2.2, the result follows.

Clearly $\mathcal{S}^*$ is ideal, and x_3 is an axis of symmetry. Since $\mathcal{S}^*$ is also a Moufang GQ by Proposition 11.5.2, every line is an axis of symmetry and $\mathcal{S}^*$ is a Moufang TGQ.

The theorem is proved.   ■

We end with another result on Moufang TGQs.

Theorem 11.5.6 *All root groups of a Moufang* TGQ *are abelian.*

Proof. Let S be a Moufang GQ with a translation point. Since the translation group of a TGQ is uniquely determined by the symmetries about the lines through the translation point, the group $U_1 U_2 U_3$ is the translation group with respect to the translation point x_2, using the notation of Theorem 11.5.3. Since the translation group is abelian, both U_1 and U_2, as subgroups of $U_1 U_2 U_3$, are abelian. Hence all root groups are abelian.   ■

11.6 Half 3-Moufang Quadrangles

In this section, our principal goal is to show that a half 3-Moufang GQ is a Moufang GQ. This result is due to Haot and Van Maldeghem [68].

So let $\mathcal{S} = (\mathcal{P}, \mathcal{B}, \mathrm{I})$ be a thick generalized quadrangle with automorphism group G, satisfying the half 3-Moufang Condition. More exactly, we assume that for all dual roots $\{y_0, y_1, y_2, y_3, y_4\}$, with $y_0 \mathrm{I} y_1 \mathrm{I} \ldots \mathrm{I} y_4$, and with $y_0, y_2, y_4 \in \mathcal{P}$, the group $G_{y_0}^{[y_2, y_3]}$ acts transitively on the apartments containing $y_0, \ldots, y_4$.

Our first aim is to show that $\mathcal{S}$ is half Moufang — more exactly, that all panels are Moufang.

As in the previous section, we fix some apartment

$$\Sigma := \{x_0, x_1, \ldots, x_7\}, \quad x_0 \mathrm{I} x_1 \mathrm{I} \cdots \mathrm{I} x_7 \mathrm{I} x_0,$$

where $x_0 \in \mathcal{P}$, and where we read the subscripts modulo 8. We prove some lemmas, under the assumptions just stated.

Lemma 11.6.1 *All sequences* (x, y, y', z), *with* $x \mathrm{I} y \mathrm{I} y' \mathrm{I} z$, $x \in \mathcal{B}$, $x \neq y'$, *and* $y \neq z$, *form a single orbit under* G. *In particular, all groups* $G_z^{[x,y]}$ *are conjugate.*

Proof. For any given $(x, y) \in \mathcal{B} \times \mathcal{P}$, with $x \mathrm{I} y$, it suffices to prove that the group $G_{x,y}$ acts transitively on the set of flags $\{y', z\}$, with $y \mathrm{I} y' \mathrm{I} z$, $y \neq z$ and $y' \neq x$. So let, with obvious notation, $\{y_1', z_1\}$ and $\{y_2', z_2\}$ be two such pairs. First suppose that $y_1' \neq y_2'$. Then we choose arbitrarily some element z' in $\{z_1, z_2\}^\perp \setminus \{y\}$. Set $x_1 := \mathrm{proj}_x z'$ and $x_2 := \mathrm{proj}_{z'} x$. There is a collineation $u \in G_y^{[x_1, x_2]}$ mapping z_1 to z_2, and hence y_1' to y_2'. If $y_1' = y_2'$, then we consider any pair $\{y_3', z_3\}$, with $y \mathrm{I} y_3' \mathrm{I} z_3$, $y_1' \neq y_3' \neq x$, and $z_3 \neq y$, apply twice the previous paragraph and get the result. ∎

Lemma 11.6.2 *If in* $\mathcal{S}$ *the span* $\{x, y\}^{\perp\perp}$ *of some opposite points* x, y *contains at least 3 elements, then all dual panels of* $\mathcal{S}$ *are Moufang.*

Proof. We may assume without loss of generality that $\{x, y\} = \{x_2, x_6\}$. Let $x_6' \in \{x_2, x_6\}^{\perp\perp}$, with $x_2 \neq x_6' \neq x_6$. Let x_5' denote the line incident with x_4 and x_6'. As $G_{x_6}^{[x_3, x_4]}$ fixes $\{x_2, x_6\}$, the span $\{x_2, x_6\}^{\perp\perp}$ has to be stabilized as a set, but as the lines through x_4 are fixed as well, this implies that the span is fixed pointwise, and hence in particular x_6' is fixed. Consider an arbitrary element $g \in G^{[x_3,x_4]}_{x_6,x'_6}$ and choose an element $h \in G^{[x'_5,x'_6]}_{x_4}$ mapping x_2 to x_6 (h exists by the half 3-Moufang assumption on the dual root $\{x_0, x_0 x'_6, x'_6, x'_5, x_4\}$). The commutator $[g, h]$ clearly belongs to $G^{[x_4,x'_5,x'_6]}_{x_6}$ and hence is trivial. Consequently $g = g^h \in G^{[x_3,x_4,x_5]}$. ∎

Let Ω denote the set of lines incident with x_0.

Lemma 11.6.3 *Let x range over the set of points incident with x_1, $x \neq x_0$, and let y range over the set of points not on x_1 but collinear with x. If the action of $G^{[x_1,x]}_y$ on Ω is independent of x and y, then all panels of $\mathcal{S}$ are Moufang.*

Proof. It suffices to show that there is an element $g \in G^{[x_0,x_1,x_2]}$ mapping x_6 to an arbitrary point $z \neq x_0$ on x_7. Let us start with an arbitrary nontrivial collineation $u \in G^{[x_1,x_2]}_{x_4}$. Then there is a unique point z' on x_5^u collinear with z. Hence, if we denote by x'_2 the unique point on x_1 collinear with z', then the collineation $u' \in G^{[x_1,x'_2]}_{z'}$ mapping x_7^u to x_7 maps x_6^u to z. The composition uu' fixes all points on x_1 and — by assumption — it also fixes all lines incident with x_0, since the action of u on Ω must be the inverse of the action of u' on Ω. Moreover, uu' maps x_6 to z. Also, the action of uu' on the set of lines through x_2 is the same as the action of u' on that set (since u fixes every line through x_2). Interchanging now the roles of x_0 and x_2, we see that the collineation $u'' \in G^{[x_0,x_1]}_z$ mapping $x_3^{u'}$ back to x_3 has an action on the lines through x_2 that is inverse to the action of uu' on that set. This implies that $uu'u'' \in G^{[x_0,x_1,x_2]}$. Since $uu'u''$ maps x_6 to z, the assertion follows. ∎

So, with the notation of the previous lemma, we will first fix x, vary y, and prove independence of the appropriate action; afterwards, we vary x and make particular choices for the points y.

Lemma 11.6.4 *Let y be any point not on x_1 collinear with x_2. Then the action of $G^{[x_1,x_2]}_y$ on Ω is independent of y.*

Proof. First we note that we may assume y to be incident with x_3. Indeed, this follows immediately from the fact that the group $G^{[x_0,x_1]}_{x_6}$ acts transitively on the lines through x_2 distinct from x_1, and so any group $G^{[x_1,x_2]}_z$, with z any point not on x_1 but collinear with x_2, can thus be seen as a conjugate of some $G_y^{[x_1,x_2]}$, with $y\mathrm{I}x_3$, under a collineation which fixes all lines through x_0.

Now, if the action of $G_y^{[x_1,x_2]}$ on Ω were not independent of the choice of y, with y incident with x_3, then we may assume that for some $y\mathrm{I}x_3$, $y \neq x_2$, the action of the group $G_1 := G_y^{[x_1,x_2]}$ on Ω differs from the action of the group $G_2 := G_{x_0}^{[x_2,x_3]}$ on Ω. Since this statement does not involve x_4, x_5, x_6 and x_7, we may rename x_4 as y and hence assume $G_1 = G_{x_4}^{[x_1,x_2]}$. Note that both G_1 and G_2 act faithfully on Ω.

Suppose first that there is an element $u \in G_1 \cup G_2$ such that u commutes with every element of $G_1 \cup G_2$. We claim that G_1 and G_2 must have the same action on Ω. Indeed, if not, then there is a collineation $g_1 \in G_1$ such that its action on Ω is not induced by any element of G_2. Let $g_2 \in G_2$ be such that g_2 maps $x_7^{g_1}$ back to x_7. Noting that $(x_6^u)^{g_1 g_2} = (x_6^{g_1 g_2})^u = x_6^u$, we see that $g_1 g_2 \in G_{x_6,x_6^u}^{[x_2]}$. If x_6^u were not contained in $\{x_2, x_6\}^{\perp\perp}$, then $g_1 g_2$ would fix at least three points on some line through x_2, implying that $g_1 g_2$ would fix an ideal subGQ, by Lemmas 11.2.2, 11.2.3 and Proposition 11.2.4. This contradicts the fact that $g_1 g_2$ does not fix all lines through x_0. Hence we have a point span of at least three elements. Lemma 11.6.2 now implies that S is half Moufang with respect to dual roots, i.e., $G_1 = G_2$.

Hence we may assume that the centralizer of $G_1 \cup G_2$ in $G_1 \cup G_2$ is trivial. Note that G_1 and G_2 normalize each other. We claim that G_1 cannot have a commutative action on Ω. Indeed, if G_1 were commutative, then also G_2 would be commutative (as by Lemma 11.6.1 the groups G_1 and G_2 are conjugate, and they both act faithfully on Ω). If only the identity in G_1 has the same action on Ω as some element of G_2, then G_1 and G_2 centralize each other, contradicting the assumption that the centralizer of $G_1 \cup G_2$ in $G_1 \cup G_2$ is trivial (alternatively, by Proposition 9.1.2(iii) two abelian groups acting regularly on a set Ω and centralizing each other must have the same action on Ω). Hence there is some nontrivial element c_1 in G_1 having the same action on Ω as an element c_2 in G_2. Now c_1 centralizes G_1 since G_1 is commutative. Since c_1 and c_2 have the same action on Ω, this implies that $[c_2, G_1]$ acts trivially on Ω. Since $[c_2, G_1] \leq G_1 \cap G_2$, this implies $[c_2, G_1] = \{\mathrm{id}\}$. Consequently both c_1, c_2 centralize $G_1 \cup G_2$, again a contradiction with our assumptions. The claim is proved.

Next we claim that only the identity in G_1 has the same action on Ω as some element of G_2. Indeed, suppose by way of contradiction that there is a $u_1 \in G_1^\times$ inducing the same action on Ω as some $u_2 \in G_2$. Since u_1 cannot lie in the center of $G_1 \cup G_2$, we may suppose there is a $g \in G_1 \cup G_2$ such that the commutator $[u_1, g] \neq \mathrm{id}$ (and this is equivalent to the assumption that the action on Ω of that commutator be nontrivial). Suppose $g \in G_2$ — the case $g \in G_1$ is similar, if one interchanges the roles of x_0 and x_4, noting that the action of G_1 and G_2 on Ω is permutation equivalent with their action on the set of lines through x_4. Consider an arbitrary $h \in G_{x_6}^{[x_0, x_1]}$; then g^h induces the same action on Ω as g. It is clear that all the commutators $[u_1, g]$, $[u_2, g]$ and $[u_2, g^h]$ induce the same action on Ω, and each of them fixes all points of x_3. This easily implies $[u_1, g] = [u_2, g] = [u_2, g^h] =: u$ (use Corollary 11.2.5). Since $[u_2, g^h]$ fixes the line x_3^h pointwise and since h is arbitrary, we see that u, which is not trivial, fixes all points collinear with x_2. So, the image of x_6 under u must lie in the span of x_2 and x_6 which forces the generalized quadrangle to be half Moufang with respect to dual roots by Lemma 11.6.2. But then again $G_1 = G_2$, a contradiction.

Hence the regular actions of G_1 and G_2 on Ω normalize each other and share only the identity. This easily implies that they centralize each other, and the actions on Ω are opposite, see Proposition 9.1.2(iii) and (iv).

We conclude that, for arbitrary yIx_3, $y \neq x_2$, the action of $G_y^{[x_1, x_2]}$ on Ω either is the same as the action of G_2 on Ω, or it is opposite.

Suppose both really occur. So for some yIx_3, $y \neq x_2$, the action of $G_1 = G_y^{[x_1, x_2]}$ on Ω is opposite the action of G_2 on Ω, and for some zIx_3, $z \neq x_2$, the action of $G_3 := G_z^{[x_1, x_2]}$ on Ω is the same as the action of G_2 on Ω. Since $G_1 \cap G_2$ is trivial, no nontrivial element of G_2 can fix all points on x_1. This implies that $G_2 \cap G_3$ is trivial. But G_2 and G_3 normalize each other, hence they centralize each other. This means that the action of G_3 on Ω — which is the same as the action of G_2 on Ω — centralizes the action of G_2 on Ω, hence this action is commutative! This contradicts a previous claim.

We conclude that all actions of $G_y^{[x_1, x_2]}$ on Ω, yIx_3, $y \neq x_2$, either are the same as the action of G_2 on Ω, or are opposite. In particular, the action is independent of y.

The lemma is proved.   ∎

Lemma 11.6.5 *If x_2' is an arbitrary point on x_1, $x_2' \neq x_0$, and x_4' is the unique point on x_5 collinear with x_2', then the action of $G_{x_4}^{[x_1, x_2]}$ on Ω coincides with the action of $G_{x_4'}^{[x_1, x_2]}$ on Ω.*

Proof. Let U_2 be the permutation group acting on Ω given by the action of $G_{x_4}^{[x_1, x_2]}$. Define U_2' as the permutation group on Ω given by the action of $G_{x_4'}^{[x_1, x_2]}$. We assume that $U_2 \neq U_2'$ and seek a contradiction.

Let U_6 be the permutation group acting on Ω defined by $G_{x_4}^{[x_6,x_7]}$, and note that by Lemma 11.6.4, this action is the same as the one induced by $G_{x_4'}^{[x_6,x_7]}$ on Ω. Now notice that $G_{x_4}^{[x_1,x_2]}$ is normal in G_{x_0,x_2,x_4}, that $G_{x_4}^{[x_6,x_7]}$ is normal in G_{x_0,x_6,x_4}, and that $G_{x_4}^{[x_1,x_2]}$ is conjugate to $G_{x_4}^{[x_6,x_7]}$ in the group H generated by both (indeed, if g is nontrivial in $G_{x_4}^{[x_6,x_7]}$, then there is a unique element $h \in (G_{x_4}^{[x_1,x_2]})^g$ mapping x_3 to x_5 and hence x_2 to x_6 and x_1 to x_7; then $(G_{x_4}^{[x_1,x_2]})^h = G_{x_4}^{[x_6,x_7]}$). Consequently, U_2 and U_6 are conjugate in $K := \langle U_2, U_6 \rangle$, $U_2 \trianglelefteq K_{x_1}$ and $U_6 \trianglelefteq K_{x_7}$ (although K is defined as a permutation group acting (only) on Ω, it is clear what we mean with K_{x_1} and K_{x_7}). By Lemma 9.1.3, (K, Ω) defines a Moufang set with root groups U_2 and U_6. Likewise, (K', Ω), with $K' = \langle U_2', U_6 \rangle$, defines a Moufang set with root groups U_2' and U_6. We now want to apply Theorem 9.1.5 to obtain a contradiction. So we show that U_2 and U_2' normalize each other.

Choose arbitrary nontrivial $u_2 \in U_2$ and $u_2' \in U_2'$, and let them be induced by the collineations $g \in G_{x_4}^{[x_1,x_2]}$ and $g' \in G_{x_4'}^{[x_1,x_2]}$, respectively. Then $g^{g'}$ belongs to $G_{x_4^{g'}}^{[x_1,x_2]}$, which has the same action on Ω as $G_{x_4}^{[x_1,x_2]}$ by Lemma 11.6.4. Hence $u_2^{u_2'} \in U_2$ and U_2' normalizes U_2. Similarly, U_2 normalizes U_2'.

Now Lemma 9.1.5 leads to a contradiction.   ∎

Lemmas 11.6.3, 11.6.4 and 11.6.5 complete the proof of the following theorem.

Theorem 11.6.6 *If every flag of a GQ is half 3-Moufang at its point, then all panels are Moufang.*   ∎

We have the following easy corollary, see also Tent [151] and Haot and Van Maldeghem [67].

Corollary 11.6.7 *Every half Moufang GQ is a Moufang GQ. Hence every half 3-Moufang GQ is a Moufang GQ.*

Proof.   Assume all dual panels are Moufang. Then obviously all flags are half 3-Moufang at their points. According to Theorem 11.6.6, every panel is Moufang.   ∎

There is an interesting consequence for TGQs.

Corollary 11.6.8 *If every point of a GQ is a translation point, then the GQ is Moufang. Equivalently, if every line is an axis of symmetry, then the GQ is Moufang.* ∎

11.7 2-Moufang Quadrangles and Fong-Seitz Quadrangles

We apply the results of the previous section to the 2-Moufang quadrangles and Fong-Seitz quadrangles. We will show that the 2-Moufang Condition implies the 3-Moufang Condition, and that any Fong-Seitz quadrangle is half Moufang. Hence in both cases, the quadrangle is a Moufang GQ. The next result was first observed by Van Maldeghem [231].

Lemma 11.7.1 *If the thick GQ $S = (\mathcal{P}, \mathcal{B}, \mathbf{I})$ has a center x of transitivity incident with an axis L of transitivity, then the group of whorls about both x and L acts transitively on $x'^{\perp} \setminus x^{\perp}$, for each $x'\mathbf{I}L$, $x' \neq x$.*

Proof. Given the flag $\{x, L\}$, $x \in \mathcal{P}$, $L \in \mathcal{B}$, $x\mathbf{I}L$, and given two points y, z opposite x with $\mathrm{proj}_L y = \mathrm{proj}_L z$, we must show that there exists a collineation of S fixing all lines through x, fixing all points on L, and mapping y to z. We prove this first in a special case, namely, when there are lines M, N opposite L, and a point w opposite x with $y\mathbf{I}M\mathbf{I}w\mathbf{I}N\mathbf{I}z$.

By the 2-Moufang Condition, there exists a collineation u_x fixing all lines through x and mapping y to w. Also, there exists a collineation u_L fixing all points on L and mapping M to N. Note that u_L fixes w, as $\mathrm{proj}_w L$ is fixed by u_L (otherwise a triangle arises). Also, u_L maps y onto z. Hence $[u_x^{-1}, u_L]$ maps y to z and belongs to $G^{[x,L]}$.

As the unique GQ of order $(2, 2)$ is Moufang, we may assume that the order of S is not $(2, 2)$. Let (Γ, E) be the graph with set of vertices the points of S opposite x and the lines of S opposite L, where adjacency is just incidence. Suppose that S is not a 3-Moufang GQ, and let y, y' be two points at minimal distance, say $2n$, in (Γ, E) such that $\mathrm{proj}_L y = \mathrm{proj}_L y'$ and such that there does not exist a collineation of $G^{[x,L]}$ mapping y to y' (this distance is well-defined by the connectivity of (Γ, E), see Lemma 11.2.7). Hence $2n > 4$. Let $(y = y_0, Y_1, y_2, Y_3, \ldots, y_{2n} = y')$ be a minimal path in (Γ, E) connecting y and y'. Then $z := \mathrm{proj}_{Y_3}(\mathrm{proj}_L y)$ and y satisfy the assumptions of our special case above, and so there exists an $u \in G^{[x,L]}$ mapping y to z. Now z and y' are at distance at most $2n - 2$, and hence there exists an $u' \in G^{[x,L]}$ mapping z to y'. Consequently $uu' \in G^{[x,L]}$ maps y to y', a contradiction.

The proof of Lemma 11.7.1 is complete. ∎

There are two interesting corollaries.

Theorem 11.7.2 *Every thick 2-Moufang generalized quadrangle $S = (\mathcal{P}, \mathcal{B}, \mathrm{I})$ is a 3-Moufang generalized quadrangle.* ∎

Corollary 11.7.3 *If in a GQ every point is an elation point and every line is an elation line, then the GQ is Moufang.* ∎

Now we turn to Fong-Seitz GQs. The next theorem was proved by Tent and Van Maldeghem [153].

Theorem 11.7.4 *Every thick Fong-Seitz generalized quadrangle $S = (\mathcal{P}, \mathcal{B}, \mathrm{I})$ is a half Moufang generalized quadrangle.*

Proof. We use the same notation as before. In particular, we have an apartment Σ consisting of the elements $x_1, x_2, \ldots, x_8$. Let B be the stabilizer in $G = \mathrm{Aut}(S)$ of the flag $\{x_1, x_8\}$, and let $N = G_\Sigma$. The fact that S is a Fong-Seitz GQ means that we have a nilpotent subgroup $U \trianglelefteq B$ such that $UH = B$, for $H = B \cap N$. Note that H is the stabilizer in B of the flag $\{x_4, x_5\}$. Furthermore, we know that B acts transitively on the set of flags opposite $\{x_1, x_8\}$ (since S is in particular a Tits GQ). Let F be an arbitrary flag opposite $\{x_1, x_8\}$. Let $b \in B$ map F to $\{x_4, x_5\}$. Let $u \in U$ and $h \in H$ be such that $uh = b$. Then $F^u = F^{bh^{-1}} = \{x_4, x_5\}^{h^{-1}} = \{x_4, x_5\}$. Hence U acts transitively on the set of flags opposite $\{x_1, x_8\}$. We claim that $UH' = B$, for every conjugate H' of H in B. Indeed, let $H' = H^g$, $g \in B$. For arbitrary $b \in B$, there is a unique element $u \in U$ mapping $\{x_4, x_5\}^{gb^{-1}}$ to $\{x_4, x_5\}^g$. Hence $(b^{-1}u)^{g^{-1}}$ fixes $\{x_4, x_5\}$ and so belongs to H, implying that $b^{-1}u \in H'$, or $b \in uH'$. The claim follows. Consequently we may vary the apartment Σ through $\{x_1, x_8\}$ in our arguments.

Since U is nilpotent, it has a nontrivial center $Z(U)$. We claim that every element of $Z(U)$ belongs to $G^{[x_1, x_8]}$. Indeed, suppose some element $v \in Z(U)$ does not fix the point x_2 (which can be considered as an arbitrary point on x_1). Then there is a unique line x_3' through x_4 meeting the line x_3. By the transitivity property of U, there exists an $u \in U$ mapping $\{x_4, x_5'\}$ to $\{x_4, x_5\}$. Since $v \in Z(U)$, $[u, v] = \mathrm{id}$, so $x_2 = x_2^{[u, v^{-1}]} = (x_2^v)^{uv^{-1}}$, hence

x_2^v is fixed under u. Similarly, x_3^v is fixed under u. Since also x_4 is fixed under u, this implies that x_5' is fixed under u, so $u = $ id and $v = $ id, a contradiction. We conclude that $Z(U)$ fixes x_1 pointwise. Together with the dual, the claim follows.

We now show the following assertion. This is the only place where we use the full strength of the hypothesis that U is nilpotent.

CLAIM 1. *The group $G^{[x_1, x_8]}$ acts transitively either on the set of lines through x_2 distinct from x_1, or on the set of points on x_7 distinct from x_8. Also, $G^{[x_1]}$ acts transitively on the set of lines through x_2 distinct from x_1, and $G^{[x_8]}$ acts transitively on the set of points incident with x_7 but distinct from x_8.*

Let U be nilpotent of class n and let $i > 0$ be maximal with respect to the property that $[U, U]_{[i-1]}$ does not fix all elements incident with x_1 and those incident with x_8. Since $[U, U]_{[n-1]} \leq Z(U)$, and all elements of the center fix all elements incident with x_1 and those incident with x_8, we have $i < n$. Without loss of generality, we may assume that $v \in [U, U]_{[i-1]}$ does not fix all elements incident with x_8. If v fixes x_7, then consider $u \in U$ mapping x_7 onto some element x_7' not fixed by v. Then clearly, $[v, u^{-1}]$ would not fix x_7. But is has to because $[v, u^{-1}] \in [U, U]_{[i]}$ and i was chosen to be maximal. Varying x_7 (as is allowed), we conclude that v does not fix any line through x_8, except for x_1, of course.

Choose x_6' arbitrary on x_7 with $x_6' \neq x_8$. We may rechoose x_2 such that x_4 is collinear with x_6^v, and put $x_5' = \text{proj}_{x_6'} x_3$. Let x_4' be the unique point on x_5' collinear with x_6^v. Clearly the flags $\{x_4, x_5\}$ and $\{x_4', x_5'\}$ are opposite $\{x_8, x_1\}$. Hence there exists an $u \in U$ mapping $\{x_4, x_5\}$ onto $\{x_4', x_5'\}$. As x_7 is concurrent with both x_5 and x_5', we see that u fixes x_7, and hence maps x_6 to x_6'. Since the commutator $[u, v^{-1}]$ belongs to $[U, U]_{[i]}$, it fixes all lines through x_8. This is impossible if u did not fix x_7^v. Hence u fixes x_7^v. Since x_6^v is collinear with both x_4 and x_4', it must also be fixed under u.

Hence $x_6^{[v^{-1}, u]} = x_6^{vu^{-1}v^{-1}u} = x_6'$. But $[v^{-1}, u] \in [U, U]_{[i]}$, implying that it belongs to $G^{[x_8, x_1]}$. So the first part of the claim is proved (since x_6' was arbitrary).

The second part is now proved similarly (considering the dual, and choosing i now to be maximal with respect to the property that $[U, U]_{[i-1]}$ does not fix all elements incident with x_1).

CLAIM 1 is proved.

By CLAIM 1, we may assume, up to duality, that $G^{[x_1,x_8]}$ acts transitively on the set of lines through x_2 distinct from x_1. Of course, by transitivity of B, the same is true for any point x_2' on x_1, $x_2' \neq x_8$. The same claim implies that, interchanging the roles of x_1 and x_7, the group $G^{[x_8]}$ acts transitively on the points of x_1 distinct from x_8. Hence $G^{[x_8]}$ acts transitively on the set of lines concurrent with x_1 and not incident with x_8.

We now show that the group $G^{[x_7,x_8,x_1]}$ is nontrivial. As this will be an important step, we call it CLAIM 2.

CLAIM 2. *The group $G^{[x_7,x_8,x_1]}$ is nontrivial — in other words, there is some nontrivial dual root elation.*

Indeed, first we prove that this is true if $Z(U)$ contains some nontrivial root elation α. Since $\alpha \in G^{[x_1,x_8]}$, we necessarily have that α fixes all lines through some point (distinct from x_8) on x_1; without loss of generality we may take this point to be x_2. Considering an arbitrary $u \in U$, it follows easily from $[\alpha, u] = \mathrm{id}$ that α fixes all lines through x_2^u. By the transitivity of U, we see that α is an axial elation with axis x_1. Let h be some nontrivial element of $G^{[x_3,x_4]}$ not fixing x_1 (this exists by CLAIM 1 above, interchanging the roles of x_8 and x_4 and of x_1 and x_3). Then, again by CLAIM 1, there is some $u \in G^{[x_1]}$ with $x_1^{hu} = x_3$. We now see that both $[\alpha, h]$ and α^{hu} belong to $G^{[x_2,x_3]}$, and that they have the same action on the points incident with x_1 (this action is the one of α^h, because u acts trivially on the points of x_1). The latter implies that $v := [\alpha, h]^{-1}\alpha^{hu} \in G^{[x_1,x_2,x_3]}$. There remains to show that v is nontrivial. We first show that the image of x_5 under α cannot be equal to the image of x_5 under α^h. Indeed, suppose by way of contradiction that $x_5^{\alpha} = x_5^{h^{-1}\alpha h}$. Then h fixes x_5^{α}. Since it also fixes x_5, it stabilizes $\{x_5, x_5^{\alpha}\}$. But α is axial, hence $\{x_5, x_5^{\alpha}\}^{\perp} = \{x_5, x_1\}^{\perp}$, and h maps $\{x_5, x_1\}^{\perp}$ to $\{x_5, x_1^h\}$. The latter cannot be equal to $\{x_5, x_1\}^{\perp}$. We conclude that $x_5^{\alpha} \neq x_5^{\alpha^h}$. Since α^{hu} fixes all lines through x_4, we deduce that v indeed does not act trivially and CLAIM 2 is proved in this case.

So we may assume that $Z(U)$ does not contain nontrivial root elations. We remark that, in particular, this implies that $Z(U)$ does not fix any line of $\mathcal{S}$ through x_2 except for x_1.

Next we show CLAIM 2 in the case that there is a nontrivial root elation. So let α now be a nontrivial element of $G^{[x_8,x_1,x_2]}$. Let w be any element of G fixing x_1 and mapping x_8 to x_2; we denote $U^* := U^w$. By our previous remark, there exists $v \in Z(U^*)$ with $x_7' := x_7^v \neq x_7$. By CLAIM 1, there exists $u \in G^{[x_7']}$ with $x_1^u = x_7$. Without loss of generality, we may also assume that $x_2^u = x_6$. Put $x_6' := x_6^v$.

Now consider the product of three root elations $\beta := \alpha \alpha^{-u} \alpha^{uv}$ and set $x'_8 = x_2^{\alpha^{-u}}$. We verify easily that $\beta \in G^{[x_7, x_8, x_1]}$. Moreover, if β were trivial, then v would belong to $G^{[x'_8, x_1, x_2]}$, contradicting our hypothesis that $Z(U^*)$ does not contain any nontrivial root elation.

Thus we may now assume that $G^{[x_8, x_1]}$ does not contain any nontrivial root elation. This means that $G^{[x_8, x_1]}$ must act faithfully on the set of elements incident with x_7 and on the set of elements incident with x_2.

We claim that $G^{[x_8, x_1]} = Z(U)$. Indeed, let U^* be as above, and let α and β be two arbitrary nontrivial elements of $G^{[x_7, x_8]}$ and $Z(U^*) \le G^{[x_1, x_2]}$, respectively, with the only restriction that α does not fix x_2. Clearly, $[\alpha, \beta] \in G^{[x_8, x_1]}$. Our hypotheses imply the existence of some $u \in G^{[x_7]}$ mapping x_7^β (which is different from x_7 since otherwise β would be a root elation inside $Z(U^*)$) to x_1. It is easy to see that $[\alpha, \beta]$ and $\alpha^{\beta u}$ induce the same action on the set of elements incident with x_7. But since they both belong to $G^{[x_8, x_1]}$, and since the latter acts faithfully on the elements incident with x_7, we conclude that $[\alpha, \beta] = \alpha^{\beta u}$. On the other hand, our arguments do not really use the fact that $\beta \in Z(U^*)$, except for the fact that $\beta \in G^{[x_1, x_2]}$ and that β does not fix x_7. Also, we only used Part 2 of CLAIM 1, which is independent of the choice of the duality class of S made after the proof of CLAIM 1. Hence our assumption that α does not fix x_2 permits to interchange the roles of α and β in the above argument (interchanging the roles of x_7, x_8 with those of x_2, x_1, respectively) and so $[\alpha, \beta]$ is conjugate to both α and β. Hence $\alpha \in G^{[x_7, x_8]} \setminus G_{x_2}$ and $\beta \in Z(U^*)$ are mutually conjugate. Consequently, α is conjugate to some element of $Z(U)$. Since the latter is defined independently of x_2, we see that every element of $G^{[x_7, x_8]}$ is conjugate to some element of $Z(U)$ (taking into account that $G^{[x_7, x_8]}$ acts faithfully on the set of points incident with x_1 and hence every nontrivial element of it maps some point of x_2 onto a different one). It follows that each nontrivial element w of $G^{[x_8, x_1]}$ is conjugate to some element w' of $Z(U)$, clearly under an element of B (since the flag $\{x_8, x_1\}$ is the unique flag of S with the property that w and w' fix all elements incident with at least one of the elements of the flag). Since $Z(U)$ is normal in B, we conclude that $G^{[x_8, x_1]} = Z(U)$, showing our claim.

It now follows that $Z(U)$ acts transitively on the set of lines incident with x_2, distinct from x_1, and thus every element in U fixing some line through x_2 distinct from x_1 must fix every line through x_2. In particular, the stabilizer of x_3 in U, which acts transitively on the set of points collinear with x_8 and opposite x_2, belongs to $G^{[x_2]}$. Since CLAIM 1 implies that $G^{[x_2]}$ also acts transitively on the set of points incident with x_1, but different from x_2, we conclude that $G^{[x_2]}$ acts transitively on the set of points opposite x_2.

Since $Z(U)$ acts faithfully and nontrivially on both the set of elements incident with x_2 and the set of those incident with x_7, we may assume, by changing Σ if necessary, that some element $z \in Z(U)$ does neither fix x_4 nor x_6. Then the line $x_5' := x_5^z$ is opposite x_5 and the point $x_6' := \text{proj}_{x_5'} x_4$ is opposite x_2. Let $g \in G^{[x_2]}$ map x_6 to x_6'. Then g fixes x_4 (since g fixes x_3 and $\text{proj}_{x_3} x_6 = \text{proj}_{x_3} x_6'$). Set $x_4' := \text{proj}_{x_3^z} x_6^z = \text{proj}_{x_3^z} x_6' = x_4^z$. If $x_4'^g = x_4'$, then $\text{proj}_{x_3^z} x_6' = \text{proj}_{x_3^z} x_6$, implying the existence of the triangle $\{x_6, \text{proj}_{x_3^z} x_6^z, x_6^z\}$. Hence g does not fix x_4^z and consequently $x_4^{[g,u^{-1}]} = x_4^{g^{-1} u g u^{-1}} = x_4'^{g u^{-1}} \neq x_4$. So $[g, u^{-1}]$ is a nontrivial root elation, contradicting the previous part of the proof of CLAIM 2.

CLAIM 2 is proved.

We now finish the proof of Theorem 11.7.4.

First we show that there are fixed elements $x_2' \text{I} x_1$, $x_2' \neq x_8$, and $x_7' \text{I} x_8$, $x_7' \neq x_1$, such that for each nontrivial $h \in G^{[x_8,x_1]}$, there exists a collineation $\alpha \in G^{[x_8,x_1,x_2']}$ inducing the same action as h on the set of points incident with x_7'.

We choose some nontrivial element β in $G^{[x_5,x_6,x_7]}$ (this exists by CLAIM 2), we put $x_7' = x_1^{\beta^{-1}}$, and we let $u \in G^{[x_1]}$ be such that $x_1^{\beta u} = x_7$. Define $x_2' \text{I} x_1$ as $x_2'^{\beta u} = x_6$. It is easily seen that $h' := [h, \beta]^{u^{-1}\beta^{-1}}$ and h induce the same action on the points of x_7'. There are two possibilities. Firstly, if h fixes x_6, then $[h, \beta]$ is clearly a nontrivial root elation related to the panel $\{x_6, x_7, x_8\}$, hence we may put $\alpha = h'$. Secondly, suppose that h does not fix x_6. By CLAIM 1 and our assumptions preceding CLAIM 2, there is some $v \in G^{[x_6]}$ mapping x_5^h to x_1. Clearly, $[h, \beta]$ and β^{-hv} induce the same action on the set of lines through x_6, hence $[h, \beta]\beta^{hv} \in G^{[x_6]}$. Since β^{-hv} belongs to $G^{[x_1]}$, it immediately follows that $\gamma = [h, \beta]\beta^{hv}$ is a root elation with respect to the panel $\{x_6, x_7, x_8\}$ inducing the same action on the set of points incident with x_1 as $[h, \beta]$, and hence also as $h^{\beta u}$. Consequently we may now put $\alpha = \gamma^{u^{-1}\beta^{-1}}$.

Now let $g \in G^{[x_8,x_1]}$ be arbitrary, and let $\alpha \in G^{[x_8,x_1,x_2']}$ induce the same action on the points of x_7' as g and guaranteed to exist by the previous arguments. Then $\theta = g\alpha^{-1} \in G^{[x_7',x_8,x_1]}$ and has the same action on the set of lines incident with x_2 as g. Since by CLAIM 1 and the assumptions made preceding CLAIM 2, $G^{[x_8,x_1]}$ acts transitively on the set of lines through x_2 distinct from x_1, we conclude that S is half Moufang and the theorem is proved. ∎

As a corollary, we obtain a characterization of groups acting on Moufang quadrangles and containing the little projective group. Recall the definition of a split Tits system on page 236.

Corollary 11.7.5 *Let* (G, B, N) *be a split Tits system. Denoting* $K :=$ $\bigcap_{g \in G} B^g$, *then the GQ* $\mathcal{S}_{G,B,N}$ *is Moufang,* G/K *acts faithfully on it and contains the little projective group.*

Proof. It is easy to see that $(G/K, B/K, N/K)$ is also split. Moreover, if $U \trianglelefteq B$ is nilpotent with $B = UH$, $H = B \cap N$, then $B = UHS$, with $S = \bigcap_{n \in N} B^n$. Hence the corresponding saturated group with BN-pair is also split. The assertion now follows from Theorem 11.4.3 and Theorem 11.7.4. ∎

Remark 11.7.6 In the finite case, Fong and Seitz [59, 60] show in a completely group theoretical way that a split Tits system is related to one of the classical quadrangles (but they talk about their groups rather than about the geometries). The way the proof works is as follows. They observe that a split Tits system contains two kinds of Moufang sets, and then they use the classification of finite Moufang sets to get control over the Tits system as a group. The observation just mentioned is the group theoretical analogue of the next proposition.

Proposition 11.7.7 *Let* $\mathcal{S} = (\mathcal{P}, \mathcal{B}, \mathbf{I})$ *be a Moufang GQ. Using the same notation as before, the pairs*

$$\mathcal{M}_2 := (\{x_0, x_4\}^{\perp}, (G^{[x_0 x, x, x x_4]})_{x \in \{x_0, x_4\}^{\perp}})$$

and

$$\mathcal{M}_3 := (\{x_1, x_5\}^{\perp}, (G^{[x_1 \cap x, x, x \cap x_5]})_{x \in \{x_1, x_5\}^{\perp}})$$

are Moufang sets.

Proof. We prove the assertion for $\mathcal{M}_2$. Since, for every $x \in \{x_0, x_4\}^{\perp}$, every element of $G^{[x_0 x, x, x x_4]}$ fixes both x_0 and x_4, the groups $G^{[x_0 x, x, x x_4]}$ act on $\{x_0, x_4\}^{\perp}$. Moreover, the group $G^{[x_0 x, x, x x_4]}$ fixes x and acts sharply transitively on $\{x_0, x_4\}^{\perp} \setminus \{x\}$. Finally, the groups $G^{[x_0 x, x, x x_4]}$ and $G^{[x_0 y, y, y x_4]}$ are conjugate, for all $x, y \in \{x_0, x_4\}^{\perp}$, and $G^{[x_0 x, x, x x_4]}$ is normal in G_{x_0, x, x_4}. ∎

For the finite Moufang GQs, the Moufang sets arising are always related to $\mathbf{PSL}_2(q)$, except for $\mathbf{H}(4,q^2)$, where $\mathcal{M}(\mathbf{H}(2,q^2))$ turns up for $\mathcal{M}_2$. Indeed, in all cases but this one, the root groups are abelian and act on a projective line in the appropriate ambient projective spaces of the natural embeddings of the classical GQs. However, in $\mathbf{H}(4,q^2)$, the perp of a pair of noncollinear points is a Hermitian unital, and hence geometrically isomorphic to $\mathbf{H}(2,q^2)$. Our assertion above is now clear.

11.8 Conclusion

From the previous sections we conclude:

Theorem 11.8.1 *Let $\mathcal{S}$ be a thick generalized quadrangle. Then $\mathcal{S}$ is a Moufang GQ if and only if it is a Fong-Seitz GQ if and only if it is a half Moufang GQ if and only if it is a 3-Moufang GQ if and only if it is a half 3-Moufang GQ if and only if it is a 2-Moufang GQ.* ∎

A large class of examples of Moufang quadrangles is given by the so-called *classical* and *dual classical quadrangles*. These are the geometries related to classical groups of Witt index 2. In the finite case, all Moufang quadrangles are classical or dual classical, and they are precisely the classical and dual classical examples $\mathbf{W}(q)$, $\mathbf{Q}(4,q)$, $\mathbf{Q}(5,q)$, $\mathbf{H}(3,q^2)$, $\mathbf{H}(4,q^2)$ and $\mathbf{H}(4,q^2)^D$ of Section 1.4. Since the first four classes are TGQs or dual TGQs with respect to every point or line, it follows from Corollary 11.6.8 that these are Moufang. For $\mathbf{H}(4,q^2)$, one needs to identify the root elations and dual root elations with certain collineations of the surrounding ambient projective space $\mathbf{PG}(4,q^2)$. We leave this to the reader. An explicit proof can be found in Chapter 4 of Van Maldeghem [232].

In fact, in the finite case, the group $U_1U_2U_3U_4$ of the previous sections is exactly the unique Sylow p-subgroup in the stabilizer B of a flag in the quadrangle, in the collineation group induced by the linear automorphisms of the projective space, where q is a power of the prime p.

Chapter 12

Translation Ovoids in Translation Quadrangles

In general, ovoids in EGQs and TGQs do not have special properties inherited from the elation or translation group of the GQ they are living in. Even in the case where the ovoid contains the elation or the translation point, this does not seem to put additional structure on the ovoid. For instance, it is known that it is hopeless to try to classify the ovoids of $\mathbf{H}(3, q^2)$ (which is an EGQ with respect to every point), because there are too many (wild) examples. Although the situation is much better in $\mathbf{Q}(4, q)$, there is yet no hope to see a general pattern for the ovoids in this TGQ.

However, when an ovoid arises from two particular general constructions (namely, from polarities, or subtended in a subGQ), then the fact that the ambient quadrangle is an EGQ or TGQ does put additional structure on the ovoid. In particular we will show that in these cases the ovoid is an elation or translation ovoid. Applied to the classical GQs, this gives rise to the classical and *semiclassical ovoids* (the latter being defined as the ovoids arising from polarities in classical GQs; in the finite case these are the the Suzuki-Tits ovoids which we will construct in Section 12.3 below), and the corresponding Moufang sets naturally arise. It is worthwhile to note that in this way we can geometrically describe three of the four infinite classes of finite proper Moufang sets. Among these three, we will take a closer look at the ones related to the Suzuki groups.

In this chapter, we assume the GQs to be thick, unless explicitly mentioned otherwise.

12.1   Ovoids, Elation or Translation with respect to a Flag or a Point

In this section we define elation ovoids and translation ovoids, with respect to flags and points in arbitrary GQs. Our definition will generalize the definition of translation ovoid with respect to a point given in Section 7.2.

Let $\mathcal{S}$ be a GQ and $\mathcal{O}$ an ovoid of $\mathcal{S}$. Let $x \in \mathcal{O}$ be arbitrary, and let L be a line incident with x. Let G be the full automorphism group of $\mathcal{S}$ and let H be a subgroup of $G_{\mathcal{O}}$. Then we say that $\mathcal{O}$ is *elation with respect to the flag* $\{x, L\}$ *and with respect to* H if there is a normal subgroup $E \trianglelefteq H_L$, called *elation group of* $\mathcal{O}$, acting sharply transitively on $\mathcal{O} \setminus \{x\}$. We usually assume that H is given and omit the phrase "with respect to H".

If $\mathcal{O}$ is elation with respect to every flag containing the point $x \in \mathcal{O}$, such that all corresponding groups E coincide, and such that $E \trianglelefteq H_x$, then we say that $\mathcal{O}$ is *elation with respect to the point* x *(and with respect to H)*. In this case E is a group of elations of $\mathcal{S}$ about x. For, if some element $e \in E$ fixed a point z opposite x, then e would fix all the points of $\mathcal{O}$ collinear with z.

If in the previous paragraph, the group E is abelian, then we say that $\mathcal{O}$ is *translation with respect to* x *(and H)*, and we call E a *translation group of* $\mathcal{O}$.

Now suppose $\mathcal{S} \cong \mathbf{Q}(4, q)$ and let $\mathcal{O}$ be a translation ovoid with respect to a point x and with respect to the stabilizer of $\mathcal{O}$ inside the little projective group of $\mathcal{S}$, in the above sense, and denote by T the corresponding translation group. Let L be any line incident with x and let y and y' be two points of $\mathcal{O}$ with $\mathrm{proj}_L y = \mathrm{proj}_L y'$. Note first that T is a p-group, for the unique prime dividing q. Hence there exists an element $\theta \in T$ of order p^n mapping y to y'. Since T is a subgroup of the little projective group of $\mathcal{S}$, it follows easily that θ fixes L pointwise. Hence $\mathcal{O}$ is a translation ovoid with respect to x in the sense of Section 7.2. The converse is easy. Hence the definitions of the present chapter generalize those of Section 7.2.

We also have the following immediate result.

Lemma 12.1.1 *Suppose an ovoid $\mathcal{O}$ of the generalized quadrangle $\mathcal{S}$ is elation with respect to some group H and with respect to two flags $\{x, L\}$ and $\{x', L'\}$, with $x \neq x'$, and with corresponding groups E and E'. Suppose also that $\langle E, E' \rangle_x$ fixes L and $\langle E, E' \rangle_{x'}$ fixes L'. Then we can choose E' such that $(\mathcal{O}, E, E')$ is a Moufang triple.*

Proof. The group $\langle E, E' \rangle$ acts doubly transitively on $\mathcal{O}$, hence we may assume that E' is conjugate to E (the condition that $\langle E, E' \rangle_{x'}$ fixes L' implies that the conjugate E^g, with $g \in \langle E, E' \rangle$ mapping x to x', also fixes L'). Now the condition that $\langle E, E' \rangle_x$ (which is a subgroup of H_x) fixes L combined with $E \trianglelefteq H_x$ implies that $E \trianglelefteq \langle E, E' \rangle_L = \langle E, E' \rangle_x$. Similarly $E' \trianglelefteq \langle E, E' \rangle_{L'} = \langle E, E' \rangle_{x'}$ and the result follows. ∎

In the finite case, being elation with respect to two different flags containing different points is a rather strong restrictive property, in view of the classification of Moufang sets. If an ovoid is translation with respect to two different points, then either the full stabilizer of the ovoid contains a sharply 2-transitive group, or it contains $\mathbf{PSL}_2(q)$ in its natural action. This follows from Theorem 9.2.1.

Example 12.1.2 The above definitions can best be illustrated with the smallest example. Indeed, let $\mathcal{O}$ be an ovoid in $\mathbf{W}(2)$. We may use the description of $\mathbf{W}(2)$ with as point set the set of pairs of $\{1, 2, 3, 4, 5, 6\}$, and with as line set the set of partitions of $\{1, 2, 3, 4, 5, 6\}$ into three pairs (with natural incidence relation). We can take for $\mathcal{O}$ the set of pairs containing the element 6, and we can put $x = \{5, 6\}$. Then Klein's fourgroup $\{\mathrm{id}, (1\ 2)(3\ 4), (1\ 3)(2\ 4), (1\ 4)(2\ 3)\}$ acts sharply transitively on $\mathcal{O} \setminus \{x\}$ and fixes all lines through x. It is obviously a normal subgroup of the stabilizer of x and $\mathcal{O}$ in the full automorphism group of $\mathbf{W}(2)$.

Now let L be an arbitrary line incident with x. We may take $L = \{\{1, 3\}, \{2, 4\}, \{5, 6\}\}$. The group generated by the 4-cycle $(1\ 2\ 3\ 4)$ acts sharply transitively on $\mathcal{O} \setminus \{x\}$, and it is a normal subgroup of the stabilizer H of the flag $\{x, L\}$ in the full collineation group of $\mathbf{W}(2)$ stabilizing $\mathcal{O}$. Indeed, one checks easily that H has order 8 (as there are 45 flags in $\mathbf{W}(2)$ and 2 ovoids through x) and is generated by $(1\ 2)(3\ 4)$ and $(1\ 2\ 3\ 4)$.

So we see that $\mathcal{O}$ is elation with respect to every flag containing a point of $\mathcal{O}$, but there are very different elation groups, and there are elation groups with respect to a flag that are not elation groups with respect to any point. This hybrid behavior can be explained by the fact that an ovoid of $\mathbf{W}(2)$ is at the same time a classical and a semiclassical ovoid. Indeed, in the standard embedding of $\mathbf{W}(2)$ in $\mathbf{PG}(4, 2)$, every ovoid spans a hyperplane, and every ovoid also arises from a polarity.

12.2   Self-Polar Elation Generalized Quadrangles

Let $S = (\mathcal{P}, \mathcal{B}, \mathrm{I})$ be a (finite) EGQ with elation point x and elation group G. In this section, we investigate the case where S is self-polar. It follows that S has order s, for some $s \geq 2$.

Let ρ be a polarity of S. An *absolute point* for ρ is a point a which is incident with its image a^ρ. Similarly one defines an *absolute line* for ρ. It is well-known that the set of absolute points of ρ is an ovoid of S, see 1.8.2 in Payne and Thas [128]. Indeed, let L be any line of S. If L is absolute, then L^ρ is an absolute point incident with L. If L is not absolute, then $\mathrm{proj}_L L^\rho$ is an absolute point with image $\mathrm{proj}_{L^\rho} L$ under ρ. Also, if a and b were two collinear absolute points, then $(ab)^\rho$ is the intersection point c of a^ρ and b^ρ. Since there are no triangles in S, we may assume without loss of generality that $a^\rho = ab$, But then $c = b$, contradicting $b^\rho \neq ab$ (by the bijectivity of ρ).

So let $\mathcal{O}$ be the ovoid of absolute points of ρ. We claim that we may assume that $x \in \mathcal{O}$. Indeed, suppose $x \notin \mathcal{O}$. Since x^ρ is an elation line, all points in $(\mathrm{proj}_{x^\rho} x)^\perp$ not on x^ρ are elation points, and this easily implies that also $\mathrm{proj}_{x^\rho} x$ is an elation point. But we have remarked in the previous paragraph that this is an absolute point for ρ. So we may rename this point by x and the claim is proved.

Since x^ρ is an elation line incident with x, we first study this situation in the next proposition.

Proposition 12.2.1 *Let S be a GQ of order s with an elation point x and associated elation group G and an elation line L incident with x and associated elation group H. Let a be an arbitrary point opposite x and R an arbitrary line opposite L but incident with a. Put $b := \mathrm{proj}_R x$ and $S = \mathrm{proj}_a L$. Then we have*

  (i) *The group $K := G \cap H$ has order s^2.*

 (ii) *The group $\langle G, H \rangle$ is equal to $G_R K_b K_S H_a = G_R K H_a = G_R H = G H_a$ and has order s^4. It acts sharply transitively on the set of flags opposite $\{x, L\}$.*

(iii) *The group G is the only "full" elation group about x and the group H is the only "full" elation group about L. Consequently, G is normal in $\mathrm{Aut}(S)_x$ and H is normal in $\mathrm{Aut}(S)_L$.*

Proof. The proposition is certainly true for $s = 2$, as there are unique elation groups in this case, and they arise from the root groups when we view the GQ as a Moufang GQ. So we may assume that $s > 2$.

Since x is a center of transitivity and L is an axis of transitivity, Lemma 11.7.1 tells us now that the group $J \leq \mathrm{Aut}(\mathcal{S})$ of whorls about both x and L has order s^2 and acts sharply transitively on the set of flags $\{y, M\}$, with $y \sim \mathrm{proj}_L a$, $y \mathrel{\text{I}} L$ and $M \sim \mathrm{proj}_x R$. We now show that $J \leq K$. By symmetry, it suffices to show that $J \leq G$.

Let p be any prime dividing s and let $g \in G_S$ have order p^n, for some positive integer n. Let u be the unique element of J mapping a to a^g. Since every nontrivial orbit of g on the set of $s - 1$ points on L distinct from x and $\mathrm{proj}_L a$ has size a multiple of p, and $s - 1$ is not divisible by p, there must be at least one trivial orbit, say $\{z\}$. The collineation gu^{-1} fixes the quadrangle containing the elements a, R, x, L, all lines through x, and the additional point z on L, hence by Proposition 11.2.4 it fixes a thick subGQ of order (s', s). Lemma 1.1.1 implies $s' = s$ and hence $g = u$. Since the Sylow p-subgroups of G_S, for p ranging over all primes dividing s, generate G_S, we have shown that $G_S \leq J$, and clearly $G_S = J_S$. Hence $J_S \leq G$.

The very same argument shows that $G_{b, \mathrm{proj}_L a} \leq J$, implying $J_b \leq G$. Since $J_b \cap J_S$ is trivial, we see that $|J_b J_S| = |J_b| \cdot |J_S| = s^2$, hence $J_b J_S = J$ and $J \leq G$.

So $J \leq K$. Since K clearly acts freely on the set of flags $\{y, M\}$, with $y \sim \mathrm{proj}_L a \neq y$ and $M \sim \mathrm{proj}_x R$, this immediately implies $J = K$. This proves (i).

From the foregoing follows that $K_b K_S = K$. Now note that $[K, H_a] \leq K$, so that $K H_a$ is a group. Since $K \cap H_a = \{\mathrm{id}\}$, we have $|K H_a| = s^3$, implying $K H_a = H$. We now show that $\langle G, H \rangle = G H_a$. First note that $[G, H_a] \leq J = K \leq G$ so that $G H_a$ is indeed a group. But $G \leq G H_a$ and also $H = K H_a \leq G H_a$, so that $\langle G, H \rangle = G H_a$. The symmetric arguments show $G_R K = G$ and $\langle G, H \rangle = G_R H$. Noting $G \cap H_a = \{\mathrm{id}\}$, we deduce $|G H_a| = s^4$ and (ii) now follows, noting that the number of flags opposite $\{x, L\}$ is also s^4 and the fact that the stabilizer in $G_R H$ of the flag $\{a, R\}$ is obviously trivial.

In order to prove (iii), assume that e belongs to an elation group with center x, and that $e \notin G$. We first claim that e acts fixed point freely on the set V of points of L distinct from x. Indeed, suppose by way of contradiction that e fixes the point $c \in V$. Assume first that all orbits of e in $V \setminus \{c\}$ have equal size s^*. Then s^* divides the order of e, which divides s^3 (since $\langle e \rangle$ is an elation group and hence acts freely on the s^3 points of S opposite x). But s^* also divides $|V \setminus \{c\}| = s - 1$; this implies $s^* = 1$ and the claim follows in this case. Assume now that not all orbits of e in $V \setminus \{c\}$ have the same size. Then some multiple e^n of e fixes at least two, but not all elements of V. By composing with a suitable element of K (namely the element bringing R^{e^n} back to R), we see that we obtain a nontrivial collineation fixing the quadrangle $\{c, x, b, \mathrm{proj}_R c\}$, all lines through x and at least three points on L. This collineation is, by Proposition 11.2.4, the identity, a contradiction. The claim is proved.

We now compose e with a suitable element of K to obtain a collineation, which we denote again by e, fixing R and acting fixed point freely on the set V. Note that e might now not be an elation, but this is not important for the rest of the proof. Since we still have $e \notin G$, the action of e on V does not coincide with the action of any element of G_R. If W is the group of all whorls about x, then we consider W_R and we note that the foregoing implies that W_R acts transitively but not sharply transitively on V. Also, if some element $w \in W_R$ fixes at least two points of V, then a thick subGQ of order (s', s) is fixed pointwise, and as before this implies $w = \mathrm{id}$. So W_R acts faithfully as a Frobenius group on V. The Frobenius kernel necessarily coincides with G_R, contradicting the fact that also e belongs to it. Hence e must already belong to G_R. Together with the dual arguments, this proves (iii).   ∎

We now go back to our polarity ρ and the ovoid of absolute points $\mathcal{O}$. We denote the spread of absolute lines by $\mathcal{T}$. It will also be useful to denote the set of flags consisting of an absolute point and an absolute line by $\mathcal{F}$. Hence $\mathcal{O}^\rho = \mathcal{T}$, $\mathcal{T}^\rho = \mathcal{O}$ and $\mathcal{F}^\rho = \mathcal{F}$. We may assume $x \in \mathcal{O}$, as we showed above, and we put $L = x^\rho$. From the previous proposition follows that the elation groups G and H with respect to x and L, respectively, are uniquely determined, and hence $G^\rho = H$ and $H^\rho = G$. Also, denoting $K = G \cap H$, we have $K^\rho = K$.

Proposition 12.2.2 *If S is an elation generalized quadrangle with elation point x and if ρ is a polarity with absolute point x and corresponding set of absolute points $\mathcal{O}$, then $\mathcal{O}$ is an elation ovoid with respect to the flag $\{x, x^\rho\}$ and with respect to the stabilizer of $\mathcal{O}$ in the full collineation group of S.*

Proof.   With the previous notation, we consider an arbitrary element u of the group $D := \langle G, H \rangle$ mapping an element $F \in \mathcal{F}$ to another element $F^u \in \mathcal{F}$. We prove that u stabilizes $\mathcal{F}$. As ρ interchanges G and

H by conjugation, it normalizes D. We claim that u and ρ commute. Indeed, $F^{\rho u \rho} = (F^u)^\rho = F^u$, and hence by Proposition 12.2.1(ii), u coincides with u^ρ, proving our claim. So for any flag $F' \in \mathcal{F}$, we now have $(F'^u)^\rho = (F'^\rho)^u = F'^u$, which means that $F'^u \in \mathcal{F}$. So u stabilizes $\mathcal{F}$. By Proposition 12.2.1(ii), u acts freely on $\mathcal{F} \setminus \{\{x, L\}\}$, and $E := D_{\mathcal{F}}$ acts sharply transitively on $\mathcal{O} \setminus \{x\}$. Since $D \trianglelefteq \mathrm{Aut}(\mathcal{S})_{\{x,L\}}$, we also have $E \trianglelefteq \mathrm{Aut}(\mathcal{S})_{\{\mathcal{O},L\}}$. ∎

We now combine (the proofs of) Lemma 12.1.1 and Proposition 12.2.2.

Proposition 12.2.3 *Suppose that $\mathcal{S}$ is an elation generalized quadrangle with elation points x and y, $x \neq y$, and that ρ is a polarity with absolute points x and y, and with corresponding set of absolute points $\mathcal{O}$. Let U_x and U_y be the elation groups of the elation ovoid $\mathcal{O}$ with respect to the flags $\{x, x^\rho\}$ and $\{y, y^\rho\}$, respectively. Then $(\mathcal{O}, U_x, U_y)$ is a Moufang triple.*

Proof. By the proof of Proposition 12.2.2 the elation groups U_x and U_y are well-defined and it reveals that these groups act on the set $\{\{z, z^\rho\} \parallel z \in \mathcal{O}\}$ of fixed flags of ρ. Also, since the elation groups of $\mathcal{S}^{(x)}$ and $\mathcal{S}^{(y)}$ are unique, and hence conjugate (even in the stabilizer of $\mathcal{O}$), and since U_x and U_y are uniquely defined only using these elation groups and the ovoid $\mathcal{O}$, we conclude that U_x and U_y are conjugate in the group $\langle U_x, U_y \rangle$. Now the proof of Proposition 12.1.1 immediately implies that $(\mathcal{O}, U_x, U_y)$ is a Moufang triple. ∎

12.3 Suzuki-Tits Moufang Sets

12.3.1 Polarities in the Symplectic Quadrangle $\mathbf{W}(q)$

We specialize the situation of Proposition 12.2.2 to the case of $\mathbf{W}(q)$. Since for q odd, all lines of $\mathbf{W}(q)$ are translation lines, but no point is a translation point, $\mathbf{W}(q)$ cannot be self-polar in this case. For q even, $\mathbf{W}(q)$ is self-dual (see Theorem 1.4.1(i)); this can be seen by projecting $\mathbf{Q}(4, q)$ from its nucleus n in $\mathbf{PG}(4, q)$ onto a 3-dimensional subspace not containing n. We now determine precisely when $\mathbf{W}(q)$ is self-polar. To do this, we need to represent $\mathbf{W}(q)$ with intrinsic coordinates.

There is a general theory of coordinatization of an arbitrary generalized quadrangle, and the next proposition is a result of this. But we will not need to introduce the general theory here.

Proposition 12.3.1 *The symplectic quadrangle* $\mathbf{W}(q)$, *for any power q of the prime* 2, *is isomorphic to the following geometry* $(\mathcal{P}, \mathcal{L}, \mathtt{I})$. *Let ∞ be, as usual, a symbol, not belonging to* $\mathbf{GF}(q)$.

• *The* POINT SET *is equal to*

$$\mathcal{P} = \{(\infty)\} \cup \{(a) \parallel a \in \mathbf{GF}(q)\} \cup$$

$$\{(k,b) \parallel b,k \in \mathbf{GF}(q)\} \cup \{(a,\ell,a') \parallel a,a',\ell \in \mathbf{GF}(q)\},$$

• *the* LINE SET *is*

$$\mathcal{L} = \{[\infty]\} \cup \{[k] \parallel k \in \mathbf{GF}(q)\} \cup$$

$$\{[a,\ell] \parallel a,\ell \in \mathbf{GF}(q)\} \cup \{[k,b,k'] \parallel b,k,k' \in \mathbf{GF}(q)\},$$

and

• INCIDENCE *is given by*

$$(a,\ell,a')\mathtt{I}[a,\ell]\mathtt{I}(a)\mathtt{I}[\infty]\mathtt{I}(\infty)\mathtt{I}[k]\mathtt{I}(k,b)\mathtt{I}[k,b,k'],$$

for all $a,a',b,k,k',\ell \in \mathbf{GF}(q)$, *and*

$$(a,\ell,a')\mathtt{I}[k,b,k'] \iff \begin{cases} b+a' = ak \\ \ell+k' = a^2k \end{cases},$$

and no other incidences occur.

Proof. To describe $\mathbf{W}(q)$, q even, we consider $\mathbf{PG}(3,q)$ and the alternating form $X_0Y_1 + X_1Y_0 + X_2Y_3 + X_3Y_2$ with respect to some given basis. We define the following bijection β from the points of $\mathbf{PG}(3,q)$ to $\mathcal{P}$. Let (x_0,x_1,x_2,x_3) be the coordinates of an arbitrary point of $\mathbf{PG}(3,q)$. If $x_1 \neq 0$, then we define

$$(x_0,x_1,x_2,x_3)^\beta = \left(\frac{x_3}{x_1}, \frac{x_0}{x_1} + \frac{x_2x_3}{x_1^2}, \frac{x_2}{x_1}\right).$$

Now suppose that $x_1 = 0$. If $x_3 \neq 0$, then

$$(x_0,0,x_2,x_3)^\beta = \left(\frac{x_2}{x_3}, \frac{x_0}{x_3}\right).$$

If $x_1 = x_3 = 0$ and $x_2 \neq 0$, then

$$(x_0, 0, x_2, 0)^\beta = \left(\frac{x_0}{x_2} \right).$$

Finally,

$$(1, 0, 0, 0)^\beta = (\infty).$$

We also define β from the set of lines of $\mathbf{W}(q)$ to $\mathcal{L}$ as follows. Note that a line of $\mathbf{PG}(3, q)$ with Plücker coordinates $(p_{01}, p_{02}, p_{03}, p_{12}, p_{13}, p_{23})$ is a line of $\mathbf{W}(q)$ if and only if $p_{01} = p_{23}$ and $p_{01}^2 = p_{02}p_{13} + p_{03}p_{12}$. Hence the lines of $\mathbf{W}(q)$ are in bijective correspondence with the 4-tuples $[p_{02}, p_{13}, p_{03}, p_{12}]$, up to a nonzero (multiplicative) constant. So let $[p_{02}, p_{13}, p_{03}, p_{12}]$ be the 4-tuple of Plücker coordinates of an arbitrary line of $\mathbf{W}(q)$. If $p_{13} \neq 0$, then we define

$$[p_{02}, p_{13}, p_{03}, p_{12}]^\beta = \left[\frac{p_{12}}{p_{13}}, \left(\frac{p_{02}}{p_{13}} + \frac{p_{03}p_{12}}{p_{13}^2} \right)^{1/2}, \frac{p_{03}}{p_{13}} \right].$$

Now we suppose $p_{13} = 0$. If $p_{12} \neq 0$, then

$$[p_{02}, 0, p_{03}, p_{12}]^\beta = \left[\left(\frac{p_{03}}{p_{12}} \right)^{1/2}, \frac{p_{02}}{p_{12}} \right].$$

If $p_{13} = p_{12} = 0$, and $p_{03} \neq 0$, then

$$[p_{02}, 0, p_{03}, 0]^\beta = \left[\frac{p_{02}}{p_{03}} \right].$$

Finally

$$[1, 0, 0, 0]^\beta = [\infty].$$

Some elementary observations show that

$$(a, \ell, a')\mathbf{I}[a, \ell]\mathbf{I}(a)\mathbf{I}[\infty]\mathbf{I}(\infty)\mathbf{I}[k]\mathbf{I}(k, b)\mathbf{I}[k, b, k'],$$

for all $a, a', b, k, k', \ell \in \mathbf{GF}(q)$.

Let k, b, k' be arbitrary in $\mathbf{GF}(q)$. Then the line $[k, b, k']$ of $\mathbf{W}(q)$ has as inverse image under β the line with Plücker coordinates $p_{02} = b^2 + kk'$, $p_{13} = 1$, $p_{03} = k'$ and $p_{12} = k$ (and then $p_{01} = p_{23} = b$). One verifies easily that this line contains the points of $\mathbf{PG}(3, q)$ with coordinates $(b, 0, k, 1)$ and $(k', 1, b, 0)$. Now, the inverse image under β of the point $(a, \ell, a') \in \mathcal{P}$ is clearly the point of $\mathbf{PG}(3, q)$ with coordinates $(\ell + aa', 1, a', a)$. This point belongs to the line generated by $(b, 0, k, 1)$ and $(k', 1, b, 0)$ if and only if $ab + k' = \ell + aa'$ and $ak + b = a'$. Substituting in the former the value of a' obtained by the latter, the assertion follows and the proposition is proved. ∎

We can now determine when precisely $\mathbf{W}(q)$ is self-polar. This is due to Tits [219], but we provide a different proof. Recall that the corresponding ovoids are semiclassical ovoids.

Proposition 12.3.2 $\mathbf{W}(q)$ *is self-polar if and only if* $q = 2^{2e+1}$, *for some nonnegative integer* e.

Proof. If q is odd, then $\mathbf{W}(q)$ is not self-dual. So we may suppose that q is even.

Suppose first that $\mathbf{W}(q)$ is self-polar and let ρ be a polarity. We use the description of $\mathbf{W}(q)$ given in Proposition 12.3.1. By conjugating with suitable elements of the full collineation group of $\mathbf{W}(q)$, one verifies that we may assume that $(\infty)^\rho = [\infty]$, that $(0,0,0)^\rho = [0,0,0]$, and that $(1)^\rho = [1]$. It follows that $(a)^\rho = [k]$ induces a permutation $a \mapsto k := a^\theta$ of $\mathbf{GF}(q)$ with $0^\theta = 0$ and $1^\theta = 1$. Also, we have $(0,0)^\rho = [0,0]$. Clearly the chain

$$(a) \sim (a,0,0)\mathtt{I}[1, a, a^2]\mathtt{I}(1, a) \sim (0,0,a) \sim (0,a)\mathtt{I}[0]$$

must be mapped onto

$$[a^\theta] \sim [a^\theta, 0, 0]\mathtt{I}(1, a^\theta, a^{2\theta})\mathtt{I}[1, a^\theta] \sim [0,0,a^\theta] \sim [0, a^\theta]\mathtt{I}(0).$$

This implies that $[0,0,\ell]^\rho = (0,0,\ell^{\theta^{-1}})$ and hence $[a,\ell]$, which is concurrent with $[0,0,\ell]$, is mapped onto $(a^\theta, \ell^{\theta^{-1}})$. Since (a, ℓ, a') is the unique point incident with $[a, \ell]$ en collinear with $(0, 0, a')$, the line $(a, \ell, a')^\rho$ is the unique line incident with $[a, \ell]^\rho = (a^\theta, \ell^{\theta^{-1}})$ and concurrent with the line $(0, 0, a')^\rho = [0, 0, a'^\theta]$. Consequently

$$(a, \ell, a')^\rho = [a^\theta, \ell^{\theta^{-1}}, a'^\theta],$$

and dually

$$[k, b, k']^\rho = (k^{\theta^{-1}}, b^\theta, k'^{\theta^{-1}}),$$

for all $a, a', b, k, k', \ell \in \mathbf{GF}(q)$.

Expressing that $(a, \ell, a')\mathtt{I}[k, b, k']$ if and only if $(a, \ell, a')^\rho\mathtt{I}[k, b, k']^\rho$, we obtain the identities

$$(ak + b)^\theta + b^\theta = a^\theta(k^{\theta^{-1}})^2, \tag{12.1}$$

$$(a^2 k + \ell)^{\theta^{-1}} + \ell^{\theta^{-1}} = a^\theta k^{\theta^{-1}}. \tag{12.2}$$

Putting $k = 1$ in Equation (12.1), we deduce that θ is additive. Putting $a = 1$ and $b = 0$ in Equation (12.1) we see that $(k^{\theta^{-1}})^2 = k^\theta$. Equation (12.1) now reduces to $(ak)^\theta = a^\theta k^\theta$, implying that θ is a field automorphism satisfying $(k^\theta)^\theta = k^2$. In this case also Equation (12.2) is satisfied.

Conversely, if we have a field automorphism θ satisfying $(k^\theta)^\theta = k^2$, for all $k \in \mathbf{GF}(q)$, then we may define a duality of $\mathbf{W}(q)$ as the action of ρ on $\mathcal{P}$ and $\mathcal{L}$ above and verify easily that this is a polarity.

We conclude that $\mathbf{W}(q)$ is self-polar if and only if $\mathbf{GF}(q)$ admits an automorphism θ with $(k^\theta)^\theta = k^2$, for all $k \in \mathbf{GF}(q)$. Putting $\theta = 2^{e+1}$, for some nonnegative integer e, we see that this is equivalent to requiring that $2q$ is a perfect square (and in this case $q = 2^{2e+1}$).

The proposition is proved. ∎

Remark 12.3.3 The above proof is, without much change, also valid in the infinite case, in which case the conclusion is that the field must be perfect and admit an automorphism whose square is the Frobenius map. Such automorphisms have been called *Tits automorphisms* in [232].

Remark 12.3.4 The proof of the above theorem can be shortened considerably if one uses the general result of Payne (see 1.8.2 of Payne and Thas [128]) stating that a finite GQ of order s admits a polarity only if $2s$ is a perfect square.

12.3.2 The Suzuki-Tits Ovoids

Since $\mathbf{W}(q)$ is elation with respect to all of its points and lines, we can apply Proposition 12.2.3 in the case $q = 2^{2e+1}$ and obtain a Moufang set. We now show that this Moufang set is nothing else than the Suzuki-Tits Moufang set defined in Subsection 9.2.3.

We first give an explicit description of the set $\mathcal{O}$ of absolute points of the polarity ρ above of $\mathbf{W}(q)$ in terms of coordinates in $\mathbf{PG}(3, q)$.

So let $\theta : \mathbf{GF}(q) \to \mathbf{GF}(q) : x \mapsto x^{2^{e+1}}$ be the Tits automorphism of $\mathbf{GF}(q)$, and let ρ be the associated polarity of $\mathbf{W}(q)$, given by the action on $\mathcal{P}$ as follows:

$$(\infty)^\rho = [\infty], \quad (a)^\rho = [a^\theta], \quad (k,b)^\rho = [k^{\theta^{-1}}, b^\theta], \quad (a,\ell,a')^\rho = [a^\theta, \ell^{\theta^{-1}}, a'^\theta],$$

for all $a, a', b, k, \ell \in \mathbf{GF}(q)$. By the calculations in the proof of Proposition 12.3.2, ρ defined in such a way induces indeed a polarity, and the action of ρ on the lines is given by

$$[\infty]^\rho = (\infty), \quad [k]^\rho = [k^{\theta^{-1}}], \quad [a, \ell]^\rho = (a^\theta, \ell^{\theta^{-1}}),$$

$$[k, b, k']^\rho = (k^{\theta^{-1}}, b^\theta, k'^{\theta^{-1}}),$$

for all $a, b, k, k', \ell \in \mathbf{GF}(q)$. Since (∞) is an absolute point, all other absolute points are not collinear with (∞), and hence they are labelled with three coordinates. One now computes that the point (a, ℓ, a') is incident with its image $[a^\theta, \ell^{\theta^{-1}}, a'^\theta]$ if and only if $\ell = a'^\theta + a^{2+\theta}$. Hence

$$\mathcal{O} = \{(\infty)\} \cup \{(a, a^{2+\theta} + a'^\theta, a') \parallel a, a' \in \mathbf{GF}(q)\}.$$

Using the formulae in the proof of Proposition 12.3.1, we can translate this into projective coordinates of $\mathbf{PG}(3, q)$ and obtain a set X, with $X = \{(1, 0, 0, 0)\} \cup \{(a^{2+\theta} + aa' + a'^\theta, 1, a', a) \parallel a, a' \in \mathbf{GF}(q)\}$. This is exactly the set X of Subsection 9.2.3, used to define the Suzuki-Tits Moufang sets. In order to completely see the equivalence between the Moufang set arising from ρ and the Suzuki-Tits Moufang set of Subsection 9.2.3, we have to prove that the respective root groups coincide on X. We show this for the root group U_∞ related to (∞).

One can easily check that the following transformation $u_{x,x'}$ of $\mathcal{P} \cup \mathcal{L}$, defined as

$$(\infty) \mapsto (\infty),$$
$$[\infty] \mapsto [\infty],$$
$$(a) \mapsto (a + x),$$
$$[k] \mapsto [k + x^\theta],$$
$$(k, b) \mapsto (k + x^\theta, b + kx + x' + x^{1+\theta}),$$
$$[a, l] \mapsto [a + x, \ell + a^2 x^\theta + x'^\theta + x^{2+\theta}],$$
$$(a, \ell, a') \mapsto (a + x, \ell + a^2 x^\theta + x'^\theta + x^{2+\theta}, a' + x' + ax^\theta),$$
$$[k, b, k'] \mapsto [k + x^\theta, b + kx + x' + x^{1+\theta}, k' + x'^\theta + kx^2],$$

for all $a, a', b, k, k', \ell \in \mathbf{GF}(q)$, is a collineation of $\mathbf{W}(q)$. It preserves the ovoid $\mathcal{O}$ (in fact, it even commutes with the polarity ρ) and it maps the point $(a, a^{2+\theta} + a'^\theta, a')$ of $\mathcal{O}$ onto to point $(a + x, (a + x)^{2+\theta} +$

$(a'+x'+ax^\theta)^\theta, a'+x'+ax^\theta)$ of $\mathcal{O}$. In Subsection 9.2.3 we have written these points as (a, a') and $(a+x, a'+x'+ax^\theta)$, respectively, and hence we see that $u_{x,x'}$ has exactly the same action as $(x, x')_\infty$ of Subsection 9.2.3. Together with a similar calculation for U_O, which will coincide with the root group related to $(0, 0, 0) \in \mathcal{P}$ (corresponding to the point $(0, 1, 0, 0)$ of $\mathbf{PG}(3, q)$), we conclude that the finite Suzuki-Tits Moufang sets are precisely the Moufang sets arising from polarities of the finite symplectic quadrangles.

Remark 12.3.5 The calculations as performed above are also valid in the infinite case, and without much change also in the case when the field $\mathbb{K}$ is not perfect and the quadrangle is defined over a subspace L of the vector space $\mathbb{K}$ over the field $\mathbb{K}^\theta$. For more information about this situation we refer to Theorem 7.3.2 of [232] and [234].

Remark 12.3.6 Above we mentioned that $u_{x,x'}$ preserves the ovoid and even commutes with the polarity. Obviously, when an automorphism commutes with ρ, it stabilizes $\mathcal{O}$ (since it stabilizes the set of fixed flags of ρ). Reciprocally, every automorphism of $\mathbf{W}(q)$ stabilizing $\mathcal{O}$ commutes with ρ, as shown by Tits [219]. The latter result was generalized to the imperfect case by Van Maldeghem, see 7.6.10 of [232].

12.4 Subtended Elation Ovoids

In this section, we give ourselves a thick elation generalized quadrangle $\mathcal{S}$ of order (s, t) and a proper full subquadrangle $\mathcal{S}^*$ of order (s, t^*) containing the elation point p. In the examples, three cases will occur: $(s, t, t^*) = (q, q, 1)$, $(s, t, t^*) = (q, q^2, q)$ and $(s, t, t^*) = (q^2, q^3, q)$. But the next results are also valid for infinite s, t, t^*.

Take an arbitrary point x in $\mathcal{S}$ collinear with p, and not contained in $\mathcal{S}^*$. Let $\mathcal{O}_x$ be the ovoid of $\mathcal{S}^*$ subtended by x and let H be the stabilizer of x and $\mathcal{S}^*$ in the full automorphism group of $\mathcal{S}$. Also, let E be the group of elations about p (in the given elation group of $\mathcal{S}^{(p)}$) fixing the point x.

Lemma 12.4.1 *With the above notation, $\mathcal{O}_x$ is an elation ovoid of $\mathcal{S}^*$ with respect to p and E.*

Proof. Let $x_1, x_2 \in \mathcal{O}_x$ be arbitrary but distinct from p. Let $\theta \in E$ be the elation of $\mathcal{S}$ mapping x_1 to x_2. Then θ fixes x, and $\mathcal{S}^* \cap \mathcal{S}^{*\theta}$ is an ideal (lines of $\mathcal{S}^*$ through p) and full (points of $\mathcal{S}^*$ collinear with p) subquadrangle (contains x_2) of both $\mathcal{S}^*$ and $\mathcal{S}^{*\theta}$. Proposition 11.2.3 implies that $\mathcal{S}^{*\theta} = \mathcal{S}^*$. ∎

In many situations, the elation group of an EGQ has the property that, whenever an elation fixes a point $y \neq p$ collinear with the elation point p, then it fixes all points on the line py. This is in particular the case for all elation groups of Moufang quadrangles arising naturally from the (dual) root elations. Hence the following lemma does not describe an exceptional situation.

Lemma 12.4.2 *With the above notation, and under the assumption that E fixes all points on the line px, the ovoid $\mathcal{O}_x$ of $\mathcal{S}^*$ is an elation ovoid with respect to p and H.*

Proof. Let $h \in H_p$ and $\theta \in E$ be arbitrary. Choose $z \in \mathcal{O}_x$ distinct from p. Let $\theta' \in E$ be such that it maps $z^{h^{-1}\theta h}$ to z. Then $h^{-1}\theta h \theta'$ fixes all lines of $\mathcal{S}$ through p, all points on px, and the point z, and so is the identity by Corollary 11.2.5. Consequently H_p normalizes E. ∎

As a corollary we obtain the next proposition.

Proposition 12.4.3 *With the above notation, suppose that $\mathcal{S}$ is a Moufang quadrangle. Suppose $p \neq p' \in \mathcal{O}$ and let E and E' be the collineation groups stabilizing $\mathcal{S}^*$, fixing all lines through p and p', respectively, and fixing all points on px and $p'x$, respectively. Then $(\mathcal{O}, E, E')$ is a Moufang triple.*

Proof. By Lemma 12.4.2 all that remains to be shown is that E and E' are conjugate in H. But this follows immediately from the 2-transitive action of $\langle E, E' \rangle \leq H$ on $\mathcal{O}$ and the uniqueness of both E and E' with respect to the sets of their fixed elements. ∎

We now investigate when exactly the previous proposition generates translation Moufang sets (i.e. E and E' are abelian). Of course, if $\mathcal{S}^*$ is rather small (for instance, not thick), then this situation is more likely to occur. We will present some examples below reinforcing the impression that it will be very difficult to state a general sufficient condition for E to be abelian. However, if one restricts to the "maximal" situation, i.e., the situation in which $\mathcal{S}^*$ is as large as theoretically possible, then there is a clear connection between the commutativity of E and properties of the standard elation group of $\mathcal{S}$ with elation point p.

So, what do we mean by "as large as theoretically possible"? Notice first that the larger $\mathcal{O}$ is, the larger S^* is. Clearly, with the above notation, the size of $\mathcal{O}$ is equal to the number of lines through x that meet S^*. So we say that S^* is an *extreme* full subquadrangle if every line of S is incident with at least one point of S^*.

Proposition 12.4.4 *We use the notation of Proposition 12.4.3, and assume that S^* is extreme full in the finite Moufang quadrangle S. Then the following are equivalent.*

(i) *The groups E and E' are abelian.*

(ii) *The Moufang triple $(\mathcal{O}, E, E')$ gives rise to a translation Moufang set.*

(iii) *All root groups of S are abelian.*

(iv) *All points of S are translation points of S.*

(v) *All points of S^* are translation points of S^*.*

In each case, the group U_0 of dual root elations corresponding to the dual panel (L, p, px), with L any line through p different from px, is isomorphic to the group E.

Proof. Note first that E has the same faithful action on the lines of S through x as U_0. Consequently these two groups are isomorphic as permutation groups, and hence also as abstract groups.

We now show that (i) implies (ii), that (ii) implies (iii), that (iii) implies (iv), that (iv) implies (v) and that (v) implies (i).

By definition, (ii) and (i) are equivalent.

Now assume (ii) (and hence also (i) holds). Let U_1 be the group of root elations corresponding to the panel (p, px, x). Since there are only two conjugacy classes of root groups in S (conjugate in the little projective group of S), and since U_0, U_1 belong to different classes, it suffices, in view of the isomorphism $E \cong U_0$, to show that U_1 is abelian.

Suppose by way of contradiction that U_1 is not abelian. By Theorem 11.5.5 we know that there is a thick full or ideal translation subquadrangle S^{**} (with a translation line) of S. In view of the fact that U_1 is nonabelian, it cannot belong to a translation group in any translation subquadrangle and so S^{**} is ideal.

Hence, if S has order (s, t), S^* has order (s, t^*) and S^{**} has order (s^{**}, t), then by Lemma 1.1.1 we have (since S^* is not necessarily thick) $t^* \leq s \leq t$ and at the same time (since S^{**} is thick) $s > t \geq s^{**}$, a contradiction.

Now assume (iii). The finite Moufang quadrangles with abelian root groups have either regular lines or regular points (this is easy to check given that fact that S is a classical quadrangle). Since S has a full subquadrangle, Lemma 1.1.1 implies that $s \leq t$. If $s < t$, then no point can be regular, hence all lines are regular and all points of S are translation points. If $s = t$, then S^* is a grid and hence all lines are regular. The result follows.

The implications (iv) $\Rightarrow$ (v) $\Rightarrow$ (i) are obvious.   ∎

Remark 12.4.5 The foregoing proposition remains valid in the infinite case, when the quadrangle arises from algebraic or classical groups. Indeed, in the former case, the root groups have a certain dimension (as algebraic groups). Just like in the previous proof, the assumption that the root group U_1 is nonabelian leads to a subquadrangle S^{**}. Let the dimensions of the root groups of S, S^* and S^{**} be given by (s, t), (s, t^*) and (s^{**}, t), respectively, where $(s, t) = (\dim(U_1), \dim(U_0))$, and the others similarly. Then, by [233], we know that $s^{**} \leq t < s$, and also $t = s + t^* \geq s$, a contradiction. Also (without going into details here), these quadrangles are related to ordinary root systems of Type $\mathbf{B}_2$, which implies that certain root groups commute. This, in turn, implies that either points or lines are regular (for details, see Chapter 5 of [232]). If the lines are regular then the assertions follow. If the points are regular, then the existence of a non-thick ideal subquadrangle implies $s \leq t$ (by [233] again). The existence of the full GQ S^* implies $s \geq t$, so $s = t$ and $s^* = 0$. This now implies, as in the above proof of Proposition 12.4.4, that all lines of S are regular, and hence that all points of S are translation points. If the GQ S arises from a classical group that is not an algebraic group, then one has to perform a direct calculation, which we shall not do here. Finally, if the GQ S arises from a group of mixed Type $\mathbf{F}_4$ (see [104]), then all root groups are abelian, but the GQ does not have any regular point or line. This is the only situation in which counter examples to Proposition 12.4.4 can exist in the infinite case (but we do not know whether they really occur).

Remark 12.4.6 With each point of a Moufang quadrangle corresponds a Moufang set in a canonical way. For, let y be a point of the Moufang quadrangle S and choose an arbitrary point y' opposite y. For every line LIy, we consider the group U_L of all (dual) root elations with respect to the dual panel (L, z, L'), where $y' \text{I} L' \text{I} z \text{I} L$. If X is the set of lines incident with y, then $(X, (U_L)_{L \text{I} y})$ is a Moufang set (and clearly, the isomorphism type of this Moufang set is independent of the choice of the points y and y'). If S' is a subquadrangle, then, putting $x = y$, the Moufang set found in Proposition 12.4.3 can be viewed as a sub Moufang set of this one, and in case the subquadrangle is extreme full they are even isomorphic. In the latter case, this shows that the isomorphism class of the Moufang set found in Proposition 12.4.3 is independent of the chosen point x.

So Proposition 12.4.3 implies that every full subquadrangle of a Moufang quadrangle S gives rise to a Moufang set in this subquadrangle. We will call this Moufang set *induced* by S. We now apply the situation of Propositions 12.4.3 and 12.4.4 to three concrete examples.

Example 12.4.7 Let S be isomorphic to $\mathbf{Q}(4, q)$ and let S^* be a subquadrangle of order $(q, 1)$ (a grid). Then S induces a translation Moufang set in S^*, the little projective group of which is isomorphic to $\mathbf{PSL}_2(q)$. This follows immediately from Remark 12.4.6. If we view S^* as a ruled nondegenerate quadric $\mathbf{Q}$ in $\mathbf{PG}(3, q)$, then the corresponding subtended ovoid $\mathcal{O}$ is a conic contained in $\mathbf{Q}$.

Example 12.4.8 Let S be isomorphic to $\mathbf{Q}(5, q)$ and let S^* be a subquadrangle of order (q, q) (isomorphic to $\mathbf{Q}(4, q)$). Then S induces a translation Moufang set in S^*, the little projective group of which is isomorphic to $\mathbf{PSL}_2(q^2)$. This also follows immediately from Remark 12.4.6. If we view S^* as a nonsingular parabolic quadric $\mathbf{Q}$ in $\mathbf{PG}(4, q)$, then the corresponding subtended ovoid $\mathcal{O}$ is an elliptic quadric contained in some 3-dimensional subspace of $\mathbf{PG}(4, q)$ as the intersection of that subspace with $\mathbf{Q}$.

Example 12.4.9 Let S be isomorphic to $\mathbf{H}(4, q^2)$ and let S^* be a subquadrangle of order (q^2, q) (isomorphic to $\mathbf{H}(3, q^2)$). Then S induces a Moufang set in S^*, the little projective group of which is isomorphic to $\mathbf{PSU}_3(q)$. This also follows immediately from Remark 12.4.6. If we view S^* as a nonsingular Hermitian variety $\mathbf{H}$ in $\mathbf{PG}(3, q^2)$, then the corresponding subtended ovoid $\mathcal{O}$ is a Hermitian curve in some plane of $\mathbf{PG}(3, q^2)$ (the intersection of that plane with $\mathbf{H}$).

In conclusion to Sections 12.3 and 12.4, we can say that three of the four infinite classes of finite proper Moufang sets arise from ovoids in appropriate Moufang quadrangles. The remaining class (related to the Ree groups) arises from ovoids in finite generalized hexagons, but a description of how this happens is beyond the scope of this book. See, however, Chapter 7 of [232] for some details.

Chapter 13

Translation Generalized Quadrangles in Projective Space

We now consider and aim to classify all embeddings of TGQs in finite projective spaces Σ with the property that the translation group is induced by automorphisms of Σ stabilizing the TGQ.

Hence, in this chapter, we will have to deal with explicit collineations of (Pappian) projective spaces, and we will use the following terminology and notation. After introducing homogeneous coordinates for the points in the projective space $\mathbf{PG}(d, \mathbb{K})$, $d > 1$ and $\mathbb{K}$ a (commutative) field, in the usual way, every point is represented by a nonzero $(d + 1)$-tuple $(x_0, x_1, \ldots, x_d)$ (determined up to a nonzero scalar multiple) that we also shall consider as a $1 \times (d + 1)$-matrix. The Fundamental Theorem of Projective Geometry then states that every collineation $\varphi : (x_0, x_1, \ldots, x_d) \mapsto (x'_0, x'_1, \ldots, x'_d)$ can be written as

$$(x'_0, x'_1, \ldots, x'_d) = (x_0, x_1, \ldots, x_d)^\theta M,$$

where M is a nonsingular $(d + 1) \times (d + 1)$-matrix with entries in the field $\mathbb{K}$ and θ is some field automorphism, which we call the *companion field automorphism of φ*. In general, we say that φ is a *semilinear collineation*. If θ is the trivial automorphism, then we say that the collineation φ is *linear*; if moreover $\det M$ is a $(d+1)$st power in $\mathbb{K}$ (and note that M is determined up to a scalar multiple), then we say that φ is *special linear*. The set of special linear collineations forms a normal subgroup of the group of all collineations, called the *little projective group of $\mathbf{PG}(d, \mathbb{K})$*. We will use without further notice the well-known property that, if a linear collineation fixes at least three points on a line, then it fixes all points on that line.

13.1   Generalities about Lax Embeddings

In this chapter we give ourselves a GQ $S = (\mathcal{P}, \mathcal{B}, I)$ laxly embedded in some projective space $\Sigma := \mathbf{PG}(d, \mathbb{K})$, for some field $\mathbb{K}$. "Laxly embedded" means that we can identify the point set $\mathcal{P}$ with a generating set of points of Σ, and the line set $\mathcal{B}$ with a certain set of lines of Σ in such a way that, if a point $x \in \mathcal{P}$ and a line $L \in \mathcal{B}$ are incident in S, then they are also incident in Σ. Note that we do not require the converse of this property (but it will nevertheless follow, see below). We neither require that lines in $\mathcal{B}$ which do not meet in S have no intersection point in Σ, and this can actually happen. To illustrate this, we present the following well-known example.

Example 13.1.1 Let $\mathcal{H}$ be a hyperoval in $\mathbf{PG}(2, 4)$, i.e., a set of six points no three of which are collinear. The 15 points not in $\mathcal{H}$, together with the 15 lines meeting $\mathcal{H}$ in exactly two points, constitute a lax embedding of the symplectic quadrangle $\mathbf{W}(2)$ of order 2. We will call this example the *hyperoval embedding* of $\mathbf{W}(2)$. Observe that every line external to $\mathcal{H}$ contains the five points of an ovoid of $\mathbf{W}(2)$, and since there are six such lines, every ovoid arises in this way. These six lines define a dual hyperoval $\mathcal{H}^*$ that we call *dual to* $\mathcal{H}$. Also, the full group of collineations and *correlations* (these are the isomorphisms of $\mathbf{PG}(2, 4)$ to its dual) stabilizing $\mathcal{H} \cup \mathcal{H}^*$ is a group of order $2 \cdot 6!$, the latter being exactly the order of the full group of automorphisms and *anti-automorphisms* of $\mathbf{W}(2)$ (an anti-automorphism being an isomorphism from $\mathbf{W}(2)$ to its dual). Hence every automorphism and every anti-automorphism of $\mathbf{W}(2)$ is induced by a collineation or correlation of $\mathbf{PG}(2, 4)$ stabilizing $\mathcal{H}$ or mapping it to its dual, respectively. Note that the little projective group of $\mathbf{W}(2)$ is induced by the little projective group of $\mathbf{PG}(2, 4)$, i.e., by special linear collineations.

We say that S is *polarized at some point* $x \in \mathcal{P}$ if the set of lines of S through x does not span Σ. If S is polarized at every point in $\mathcal{P}$, then we briefly say that S is *polarized*. Also, if for the line $L \in \mathcal{B}$, each point of Σ on L belongs to S, then we say that L is a *full line* of S. If all lines of S are full, then we briefly say that S is *full in* Σ, or is *fully embedded (in Σ)*.

There exist GQs which are not fully embedded, but with full lines.

Example 13.1.2 Let $\mathbb{K}$ be an infinite field with a countable number of elements, let $d = 4$, consider five points p_1, p_2, p_3, p_4, p_5 in Σ generating Σ, and let L_i be the line generated by p_{i-2} and p_{i+2}, considering the subscripts modulo five. We now define a *free 4-gonal closure* of this pentagon $\mathcal{S}_0$ as follows. We number the points of the line L_4 using the nonzero natural numbers, assigning 1 to p_1 and 2 to p_2. Furthermore, we select a nonempty finite subset F of points of the line L_5 neither containing p_2 nor p_3. Suppose we arrive in Step n of the construction at some finite set of points and lines in Σ, forming a connected geometry $\mathcal{S}_n$ without triangles, such that

(C0) there is a natural number k such that precisely the first k points of L_4 are contained in $\mathcal{S}_n$,

(C1) no point is contained in F,

(C2) incidence in $\mathcal{S}_n$ is inherited from Σ,

(C3) for every pair of points a, b of $\mathcal{S}_n$ not incident with L_4, the planes $\langle a, L_4 \rangle$ and $\langle b, L_4 \rangle$ coincide (and note that these planes are always well-defined since we assume that incidence in $\mathcal{S}_n$ is induced by the incidence relation in Σ) if and only if the line ab of Σ also belongs to $\mathcal{S}_n$ and so does the intersection point $ab \cap L_4$.

We execute Step $n + 1$ in the following manner. Whenever a point x and a line L of $\mathcal{S}_n$ are at distance 5 in the incidence graph of $\mathcal{S}_n$, then we will choose an appropriate "new" point y on L (so y belongs to Σ, but not to $\mathcal{S}_n$) and declare y and xy to be elements of $\mathcal{S}_{n+1}$. It is obvious that we never introduce triangles, and that the infinite union of all our geometries is a generalized quadrangle. We must now show that we can make our choices such that $\mathcal{S}_{n+1}$ satisfies (C0), (C1), (C2) and (C3).

Therefore, let Ψ be the set of points of $\mathcal{S}_n$, let k be the number of elements of Ψ incident with L_4, and let Φ be the set of planes of Σ containing L_4 and at least one point of $\mathcal{S}_n$ not incident with L_4. Since there are only finitely many points and lines in $\mathcal{S}_n$, there are only finitely many pairs (x, L) as above to consider. We number them and treat them in the given order. Each time we treated one pair, we check the validity of the conditions (C0), (C1), (C2) and (C3). Hence, we may assume that these conditions hold for all points and lines so far constructed, and we consider the new current value of k, add the new points to Ψ and the new planes, if any, containing L_4 and a newly constructed point not on L_4, to the set Φ.

Suppose first that $L \neq L_4$ and x is not incident with L_4. If L is concurrent with L_4, then the point x is not contained in the plane $\langle L, L_4 \rangle$ by (C3). Also, whichever choice for y we make, (C0), with the same k, and (C3) will always still hold. Hence we only have to avoid that the line xy contains some point of Ψ different from x, and we must also make sure that $y \notin F$. But these conditions only rule out a finite number of possibilities for y; as there are infinitely many points on L, we can make a legible choice for y. Note that we do not require that xy does not meet a line of $\mathcal{S}_n$ (or another "already defined line" of $\mathcal{S}_{n+1}$), and we indicate below why we cannot build in this extra condition, which we do not need for our purposes here anyway. If L is not concurrent with L_4, then, since L is incident — in our extension of $\mathcal{S}_n$ sofar obtained — with at least 2 points by our construction, Condition (C3) assures that $\langle L, L_4 \rangle$ is 3-dimensional. Hence there are only a finite number of points of Σ on L that belong to some member of Φ. Also, the plane $\langle x, L \rangle$ meets L_4 in at most one point, it contains only a finite number of elements of Ψ, and we must avoid that these points are on xy. But clearly there is a choice for y meeting these conditions, and then (C0) and (C1) hold trivially, and (C2) and (C3) by construction.

Suppose $L = L_4$. Then the plane $\langle x, L_4 \rangle$ does not contain any element of Ψ besides x and the points of L_4, by (C3). We now choose for y the $(k + 1)$st point of L_4. Then (C0) is satisfied, and trivially also (C1), (C2) and (C3) hold. Note that we have no control over the fact that xy meets some line of $\mathcal{S}_n$, hence we do not necessarily obtain a lax embedding with the extra property that two lines of the quadrangle meet if and only if they meet in the projective space.

Suppose finally $x \,\mathrm{I}\, L_4$. Then, similarly as above, L does not meet L_4 and so only finitely many points of Σ on L belong to some member of Φ. If we avoid those points for y, then (C3) holds. The other conditions hold trivially.

Now the union of all $\mathcal{S}_n$ over $\mathbb{N}$ is a (thick) generalized quadrangle $\mathcal{S}$ laxly embedded in Σ. Moreover, the line L_4 is full because of Condition (C0) and the fact that $\mathcal{S}$ has infinitely many points on every line (this is easy to see and left for the reader to prove). But the line L_5 is not full because the points of F belong to L_5 in Σ, but they do not belong to $\mathcal{S}$.

Lemma 13.1.3 *Let the GQ $\mathcal{S}$ be laxly embedded in Σ. Then the points of $\mathcal{S}$ incident in Σ with some fixed line L of $\mathcal{S}$ are precisely the points of $\mathcal{S}$ incident with L in $\mathcal{S}$.*

Proof. Suppose x is a point of $\mathcal{S}$ not incident with the line L of $\mathcal{S}$. Assume by way of contradiction that x and L are incident in Σ. In $\mathcal{S}$, there is a unique line M meeting L and incident with x. Clearly, in Σ, this line has to coincide with L, a contradiction.   ∎

A celebrated result of Buekenhout and Lefèvre [32] classifies all fully embedded generalized quadrangles in finite projective spaces. We mention it
without proof.

Theorem 13.1.4 *Every fully embedded generalized quadrangle in a finite
projective space is a classical* GQ *in its standard embedding as described in
Section* 1.4.   ∎

Dienst [51, 52] classifies all fully embedded generalized quadrangles in arbitrary projective space (and only *classical* GQs turn up, i.e., GQs associated
to classical groups; in any case all examples are Moufang).

A crucial step in the above classifications is the fact that every full embedding is automatically polarized. We record a proof of this fact here, not
using finite dimensionality. Hence, in the following lemma, we emphasize
that the dimension d of Σ can be any cardinal number.

Lemma 13.1.5 *If S is fully embedded in Σ, then the embedding is polarized.*

Proof.   Let, by way of contradiction, p be a point of S at which the embedding is not polarized. Hence every line px of Σ, with x a point of S opposite
p, is contained in a finite dimensional space Σ_x spanned by some lines of
S through p. We may select x and Σ_x is such a way that the dimension
of Σ_x, say m, is minimal. Clearly, $m > 1$ as otherwise x is on some line
L_1 of S incident with p and hence not opposite p. So there are m lines
$L_1, L_2, \ldots, L_m$ of S through p generating Σ_x. Denote by Σ'_x the space generated by $L_2, L_3, \ldots, L_m$. Note that $m > 1$ implies that Σ'_x is well-defined
and has dimension at least 1. The line M of S through x meeting L_1 is not
contained in Σ'_x (by minimality of m), but it is contained in Σ_x. Hence M
meets Σ'_x in a unique point x', which is a point of S belonging to a space
Σ'_x of dimension $m - 1$ spanned by some lines of S through p. This contradicts the minimality of m. Hence such m cannot exist (since $m \neq 1$), and
consequently every point opposite p must lie outside the space spanned by
the lines through p. Hence the embedding is polarized at p after all.   ∎

Polarization of an embedding at every point is a useful property. For instance, in 3-dimensional space, it immediately implies that all points are
regular.

Proposition 13.1.6 *Let the not necessarily finite GQ S be laxly embedded in* **PG**$(3, \mathbb{K})$, *for some field* $\mathbb{K}$. *If the embedding is polarized at every point, then every point is regular.*

Proof. If S is a grid, then the result is trivial. So we may assume that there are at least three lines incident with every point.

For any point x of S, we denote by ξ_x the plane of **PG**$(3, \mathbb{K})$ spanned by $x^{\perp}$ (and this is indeed a plane by Lemma 13.1.3). If $\xi_x = \xi_y$ for two distinct points x, y, then the plane ξ_x would induce a subquadrangle which is clearly full and ideal in S, hence coincides with S, a contradiction.

Now let $\{x, y\}$ be an opposite pair of points of S. The intersection of ξ_u and ξ_w, for arbitrary $u, w \in \{x, y\}^{\perp}$, $u \neq w$, is precisely the line xy of **PG**$(3, \mathbb{K})$. Hence $\{x, y\}^{\perp\perp} \subseteq xy$. Conversely, let z be a point of S on the line xy of **PG**$(3, \mathbb{K})$. Since $\xi_w \neq \xi_z \neq \xi_u$, and since S is not a grid, there is a line L of S through z not contained in ξ_u and not in ξ_w either. But then $\text{proj}_u L$ is the intersection of ξ_u and $\langle u, L \rangle$, and hence passes through z. Similarly for $\text{proj}_w L$. This means that $z \in \{u, w\}^{\perp}$. Since u, w were arbitrary, we conclude that $z \in \{x, y\}^{\perp\perp}$.   ∎

A natural question to ask is whether the property of being polarized is already enough to allow a classification. Thas and Van Maldeghem [194], in the finite case, and Steinbach and Van Maldeghem [146, 147], in the general case, answer affirmatively this question and the latter classify all polarized embedded quadrangles in arbitrary projective space. They are all Moufang GQs and in each case, the little projective group of the GQ is induced by the projective group of the projective space. In the finite case, a polarized embedding is almost always a full embedding in some subspace of Σ, obtained by restricting the (possibly infinite) field $\mathbb{K}$ to a finite subfield. There is only one counter example to this phenomenon, and it is given by an exceptional embedding of **W**(2) in **PG**$(4, \mathbb{K})$. We describe it later.

Here, we want to prove a kind of converse to the classification result of Steinbach and Van Maldeghem, namely, roughly speaking, we aim to show that polarization of an embedding follows from an assumption on the collineation group of the GQ in question. In particular, if the finite GQ is Moufang and all (dual) root elations are induced by the little projective group of the projective space Σ, then the embedding is polarized, except in some small well-understood cases.

But first we give some examples of nonpolarized embeddings.

Example 13.1.7 Let $S := \mathbf{W}(q)$ be embedded in $\Sigma := \mathbf{PG}(3,q)$ (in the standard way). Let p be a point of S, and let π be the plane of Σ containing all lines of S through p (the plane π exists since the embedding is polarized at p; it is the image of p under the symplectic polarity corresponding to S). Then we define the quadrangle S' as follows.

- The POINT SET is the set of points of $\Sigma \setminus \pi$.

- The LINES are all lines of S not through p, plus all lines of Σ through p, but not in π.

It is easy to see that S' is isomorphic to the Ahrens-Szekeres quadrangle $\mathbf{AS}(q)$ of order $(q-1, q+1)$, see 3.2.6 of Payne and Thas [128]. The definition given here entails a lax nonpolarized embedding, and the lines of S' incident with p in Σ are examples of lines not meeting in the quadrangle, but intersecting in the projective space.

Example 13.1.8 A second class of (nonpolarized) examples is given by the $\mathbf{T}_2^*(\mathcal{O})$, with $\mathcal{O}$ a hyperoval in some plane π of the 3-dimensional space $\mathbf{PG}(3,q)$, with q even. Here, every line meets another line in $\mathbf{PG}(3,q)$ that is not concurrent in the quadrangle.

The two previous examples show that, even if one assumes that every line of the embedded GQ misses only one point of the projective space, the embedding does not have to be polarized.

Before proceeding, we remark that all the finite examples we have given so far (including the natural embeddings of the classical quadrangles as quadrics and Hermitian varieties in finite projective spaces) have an ambient projective space — that is, the space generated by the points of the quadrangle — of dimension at most 5. This is not a coincidence and can be proved quite easily.

Proposition 13.1.9 *If a GQ S of order (s,t), $t > 1$, is laxly embedded in $\mathbf{PG}(d,q)$, then $d \leq 5$.*

Proof. Suppose the GQ S of order (s,t), $t > 1$, is laxly embedded in $\mathbf{PG}(d,q)$, with $d \geq 5$. Consider two opposite lines L, M of S. The intersection of S with the projective space $\langle L, M \rangle$ is a full subquadrangle S' of order (s,t'), with $t' < t$. Considering a line N of S concurrent with L but not belonging to S', we intersect S with the 4-space $\langle L, M, N \rangle$ and obtain a subquadrangle $\mathcal{S}''$ of order (s, t'') with $t' < t'' < t$. Considering a line K of $\mathcal{S}$ concurrent with L but not belonging to $\mathcal{S}''$, we intersect $\mathcal{S}$ with the 5-space $\langle K, L, M, N \rangle$ and obtain a subquadrangle $\mathcal{S}'''$ of order (s, t''') with $t' < t'' < t''' \leq t$. Now Lemma 1.1.1 implies $t' = 1$, $t'' = s$ and $t''' = s^2 \geq t$, hence $t = t'''$ and $\mathcal{S}''' = \mathcal{S}$. The assertion is now clear.   ∎

Other examples of nonpolarized embeddings are given by first extending the ground field of the projective space in which a classical GQ is naturally embedded, and then projecting from a suitable point onto some hyperplane (and one may repeat this procedure a number of times). A remarkable result of Thas and Van Maldeghem [195] now shows that every lax embedding of the classical quadrangles $\mathbf{Q}(4, s)$, $\mathbf{Q}(5, s)$, $\mathbf{H}(3, s)$ and $\mathbf{H}(4, s)$ in $\mathbf{PG}(d, q)$, $d \geq 3$, $\gcd(s, q) \neq 1$, arises from the standard embedding as described above. Since we will need this result for $\mathbf{Q}(4, s)$ and $d \in \{3, 4\}$, and for $\mathbf{Q}(5, s)$ and $d = 5$, we sketch a proof of it now.

Lemma 13.1.10 *Every lax embedding of* $\mathbf{Q}(4, s)$ *and* $\mathbf{Q}(5, s)$ *in* $\mathbf{PG}(4, q)$ *and* $\mathbf{PG}(5, q)$, *respectively, with* $\gcd(s, q) \neq 1$, *arises from the standard embedding by extending the field* $\mathbf{GF}(s)$ *over which the projective space is defined. Every lax embedding of* $\mathbf{Q}(4, s)$ *in* $\mathbf{PG}(3, q)$, *with* $\gcd(s, q) \neq 1$, *arises from the standard embedding in* $\mathbf{PG}(4, s)$ *by extending the field over which the projective space is defined and then projecting from a well-chosen point onto a hyperplane.*

Sketch of Proof. First consider $\mathcal{S} \cong \mathbf{Q}(4, s)$, laxly embedded in $\mathbf{PG}(4, q)$, with $\gcd(s, q) \neq 1$. Let L be a line of $\mathcal{S}$ and let π be a plane of $\mathbf{PG}(4, q)$ skew to L. We project the set of lines of $\mathcal{S}$ concurrent with L from L onto π and obtain an embedding of the dual affine plane $\mathcal{A}$ corresponding with the regular line L into π. The condition $\gcd(s, q) \neq 1$ implies, by Limbos [99], that the "parallel points" of this dual affine plane are collinear in π, i.e., the lines of $\mathcal{S}$ through a fixed point x of $\mathcal{S}$ on L are contained in a 3-space. This is exactly polarization at x. Moreover, [99] also implies that, under our assumptions, $\mathcal{A}$ extends uniquely to a projective subplane $\mathcal{P}$ of π. Hence $\mathbf{GF}(s)$ is a subfield of $\mathbf{GF}(q)$, and the points of any line M of $\mathcal{S}$ opposite L form a projective subline over $\mathbf{GF}(s)$ of M. Now it is quite easy to show that every grid of $\mathcal{S}$ of order $(s, 1)$ generates a 3-dimensional subspace (if such a grid were contained in a plane, then by considering a subquadrangle in an appropriate 3-dimensional space we obtain a contradiction to Lemma 1.1.1), and is fully embedded in some 3-dimensional subspace over

GF(s) (and this subspace is uniquely defined by four points of the grid lying in a quadrangle and a fifth point of the grid not collinear with any of the four previous points). We fix one such subspace **PG**(3, s) and note that any line M opposite L, but not belonging to the grid in **PG**(3, s), together with **PG**(3, s) generate (over **GF**(q)) the whole space **PG**(4, q) as otherwise the space over **GF**(q) generated by **PG**(3, s) would already contain S — this follows from Lemma 1.1.1. Now we add the points of S incident with M to obtain a unique subspace **PG**(4, s) of **PG**(4, q), containing **PG**(3, s) and the points of S on M. It is then easily seen that all the points of the grids of S containing M and sharing two (intersecting) lines with **PG**(3, s) are contained in **PG**(4, s). Repeating this argument for M substituted by other lines of S not in **PG**(3, s), but whose points (of S) have already been proven to be in **PG**(4, s), we eventually obtain that all points of S are contained in **PG**(4, s), i.e., the embedding is full in a subspace over **GF**(s), and hence it is isomorphic to the appropriate natural embedding therein.

Now consider $S \cong \mathbf{Q}(5, s)$ laxly embedded in **PG**(5, q). By the previous paragraph, every subquadrangle $S' \cong \mathbf{Q}(4, s)$ of S is fully embedded in the standard way in some subspace **PG**(4, s) (one uses again Lemma 1.1.1 to show that such a subGQ is never contained in a 3-dimensional subspace). For a point x of S outside S', the subtended ovoid $\mathcal{O}_x$ of S' is an elliptic quadric in some 3-dimensional subspace **PG**(3, s) of **PG**(4, s). Hence the lines of S through x are contained in the 4-dimensional space spanned by $\mathcal{O}_x$ and x. Hence the embedding is polarized at x, and so at every point. The fact that the embedding is full in some subspace **PG**(5, s) over **GF**(s) is proven in a similar way as above for $\mathbf{Q}(4, s)$.

Finally, let $S \cong \mathbf{Q}(4, s)$ be laxly embedded in **PG**(3, q), with $\gcd(s, q) \neq 1$. We will be very sketchy here and leave some nontrivial details to the reader. We can again show that the points of S on any line L of **PG**(5, q) containing two collinear points of S form a projective subline of L over the field **GF**(s). Next we construct a **PG**(3, s) containing all points of S of a grid $\mathcal{G}$ of order (s, 1) (here one needs to prove first that there exists a grid $\mathcal{G}$ which generates **PG**(3, q)). We then embed **PG**(3, q) into a **PG**(4, q) as a hyperplane, choose a point c in **PG**(4, q) outside **PG**(3, q), and consider a line L of S not contained in the grid $\mathcal{G}$. The line L has some point x in common with $\mathcal{G}$, and we choose a line L' in the plane $\langle c, L \rangle$ through x, and project the points of S on L from x onto L'. We call this "lifting". We can now lift uniquely all points of every grid $\mathcal{G}'$ of S containing L and sharing two intersecting lines L_1, L_2 with $\mathcal{G}$, where the lift of a point y is the unique point in the intersection of $\langle c, y \rangle$ and $\langle L_1, L_2, L' \rangle$. Doing this a number of times, substituting L with appropriate other already lifted lines, the lemma eventually follows.   ∎

For finite lax embeddings, dimension five plays a special role, as in this dimension all lax embeddings can be classified. Since we will use this result of Thas and Van Maldeghem [195], we sketch a proof of it here.

Lemma 13.1.11 *If the GQ S of order (s, t) is laxly embedded in $\mathbf{PG}(5, q)$, then $S \cong \mathbf{Q}(5, s)$. If $\gcd(s, q) = 1$, then $s = 2$ and the embedding is as described in Example 13.3.4 below. Otherwise the embedding arises from the standard embedding by extending the field $\mathbf{GF}(s)$.*

Sketch of Proof. If $\gcd(s, q) > 1$, then, in view of the previous lemma, we only have to show that $S \cong \mathbf{Q}(5, s)$. Let L, M be two opposite lines of S. Then the 3-space $\langle L, M \rangle$ induces a subGQ S' of order (s, t') of S. Choose a line N in S meeting S' in a unique point. Then the subspace $\langle L, M, N \rangle$ induces a subGQ S'' of order (s, t'') of S containing S'. By Lemma 1.1.1, this implies $t = s^2$, $t'' = s$ and $t' = 1$. Moreover, the abundant number of ways to choose S'' can be used to conclude, by 5.3.5 of Payne and Thas [128], that $S \cong \mathbf{Q}(5, s)$.

If $\gcd(s, q) = 1$, then the previous argument still holds to conclude that $S \cong \mathbf{Q}(5, s)$. But now we cannot appeal to Lemma 13.1.10 anymore. Nevertheless, the assumption that allowed us to use the results of Limbos [99] in the proof of Lemma 13.1.10 can be replaced by the assumption $s > 3$. The case $s = 2$ is now handled by directly assigning coordinates to all points, in a systematic way, along the lines of the proofs of Theorems 13.2.2 and 13.4.1. The case $s = 3$ cannot occur since in this case any subquadrangle S' isomorphic to $\mathbf{Q}(4, 3)$ is embedded as in Example 13.3.3 below (this requires some calculations with coordinates), and from that description one deduces that any classical ovoid of S' spans the whole 4-dimensional space H spanned by S'. But there are at least two subquadrangles isomorphic to $\mathbf{Q}(4, s)$ containing the points of the same ovoid, and this implies readily that S would be contained in the hyperplane H, a contradiction.   ∎

13.2 Planar Translation-Homogeneous Embeddings of Translation Generalized Quadrangles

The extreme case of a nonpolarized embedding occurs in projective planes (and we speak of *planar embeddings*: no planar embedding of any GQ can ever be polarized in any point). All classification results in the literature concerning embeddings of GQs in projective spaces exclude the planar case, basically because there are no subspaces to put structure on the embedded GQ. However, we can put some structure on the embedding by hypothesizing that the GQ is a TGQ and that the translation group is induced by collineations of the projective plane. That is exactly what we are going to do, and we will do this in arbitrary dimensions. If the dimension is equal to 2, then we will always assume that the projective plane is Desarguesian.

First we give precise definitions.

Let $\mathcal{S} = (\mathcal{P}, \mathcal{B}, \mathrm{I})$ be a translation generalized quadrangle embedded in $\mathbf{PG}(d, \mathbb{K})$, for some field $\mathbb{K}$ and some cardinal number d, and let x be the corresponding translation point. If every translation with respect to x is induced by a collineation of $\mathbf{PG}(d, \mathbb{K})$, then we say that the embedding is *translation-homogeneous (with respect to x)*. If all these collineations of $\mathbf{PG}(d, \mathbb{K})$ are induced by linear transformations or special linear transformations of the underlying vector space, then we say that the translation-homogeneous embedding is *linear* or *special*, respectively.

If $\mathbb{K}$ is an infinite field, then it may have an essentially arbitrary Galois group, which may contain, in principle, the whole translation group. This is a serious obstruction for our methods, and hence we will only consider the finite case when dealing with translation-homogeneous embeddings in general. Likewise, in the linear case, if $\mathbb{K}$ is a nonabelian skewfield, then we have no control over the group of inner automorphisms (which induces a linear group), and so we discard this case.

In the present section, we will deal with planar translation-homogeneous embeddings of TGQs in finite Desarguesian projective planes. Such embeddings can never be linear since every nontrivial symmetry about a line L would be a collineation with an axis L having $s + 1$ centers (the points of the quadrangle on the line L, viewed as a line of the projective plane). We first present some examples. They are all related to $\mathbf{W}(2)$.

Example 13.2.1 Let q be a power of 2, and let ℓ_1, ℓ_2 be two distinct elements of $\mathbf{GF}(q^2)$ such that $\ell_1 + \ell_1^q = 1$, $\ell_1 \neq \ell_2^q$ and $\ell_1 \ell_1^q = \ell_2 \ell_2^q$. Put $k = \ell_1 + \ell_2$. Identify the point set of $\mathbf{W}(2)$ with the pairs of $\{1,2,3,4,5,6\}$ and the lines with the partitions of $\{1,2,3,4,5,6\}$ into pairs. Then the following identification defines a lax embedding of $\mathbf{W}(2)$ in $\mathbf{PG}(2, q^2)$.

$$
\begin{array}{lll}
\{1,2\} \leftrightarrow (0,1,0) & \{5,6\} \leftrightarrow (1,0,0) & \{4,6\} \leftrightarrow (0,0,1) \\
\{3,4\} \leftrightarrow (1,1,0) & \{3,6\} \leftrightarrow (1,0,1) & \{3,5\} \leftrightarrow (0,1,1) \\
\{4,5\} \leftrightarrow (1,1,1) & \{1,3\} \leftrightarrow (k^q, \ell_1^q, 1) & \{1,4\} \leftrightarrow (k^q, \ell_1, 1) \\
\{2,3\} \leftrightarrow (k, \ell_1, 1) & \{1,5\} \leftrightarrow (k^q, \ell_2^q, 1) & \{1,6\} \leftrightarrow (k^q, \ell_2^q + 1, 1) \\
\{2,4\} \leftrightarrow (k, \ell_1^q, 1) & \{2,5\} \leftrightarrow (k, \ell_2, 1) & \{2,6\} \leftrightarrow (k, \ell_2 + 1, 1)
\end{array}
$$

In order to check that the above really defines a lax embedding, one only has to verify that approximately half of the lines of $\mathbf{W}(2)$ are induced by lines of $\mathbf{PG}(2, q^2)$ taking into account that the collineation $(a, b, c) \mapsto (a^q, b^q, c^q)$ of $\mathbf{PG}(2, q^2)$, $a, b, c \in \mathbf{GF}(q^2)$, stabilizes the above set of points. We leave the easy detailed computations to the reader.

Let us consider the point $\{1, 2\}$ as a translation point of $\mathbf{W}(2)$. The corresponding translation group is generated by the symmetries $(1\ 2)(3\ 4)(5\ 6)$, $(1\ 2)(3\ 5)(4\ 6)$ and $(1\ 2)(3\ 6)(4\ 5)$. Hence it is also generated by the automorphisms $(1\ 2)$, $(3\ 4)(5\ 6)$ and $(3\ 5)(4\ 6)$. The latter three are induced by the semilinear collineations of $\mathbf{PG}(2, q^2)$ with companion field automorphism $x \mapsto x^q$ and matrices

$$
\begin{pmatrix} 1 & 0 & 0 \\ 0 & 1 & 0 \\ 0 & 0 & 1 \end{pmatrix}, \begin{pmatrix} 1 & 0 & 0 \\ 0 & 1 & 0 \\ 0 & 1 & 1 \end{pmatrix}, \begin{pmatrix} 1 & 1 & 0 \\ 0 & 1 & 0 \\ 0 & 0 & 1 \end{pmatrix},
$$

as one can easily check. Hence we have a translation-homogeneous embedding with respect to $\{1, 2\}$ of $\mathbf{W}(2)$ in $\mathbf{PG}(2, q^2)$, for an arbitrary even prime power $q \neq 2$ (and it is easy to generalize this embedding to infinite projective planes over fields in even characteristic admitting an involutory automorphism). We call this embedding the *generalized hyperoval embedding with parameters* (ℓ_1, ℓ_2).

We say that a translation-homogeneous embedding of a TGQ $\mathcal{S}$ in $\mathbf{PG}(d, \mathbb{K})$ is *faithful* if the identity is the only collineation of the projective space that fixes all points of the TGQ. In the finite case, this is equivalent with saying that the embedding is not contained in a $\mathbf{PG}(d, \mathbb{K}')$, with $\mathbb{K}'$ a proper subfield of $\mathbb{K}$ (after remarking that thickness and connectivity of $\mathcal{S}$ imply that the point set of $\mathcal{S}$ is not contained in the union of a nontrivial subspace and a complement). In general, it is equivalent with saying that the embedding is not contained in a $\mathbf{PG}(d, \mathbb{K}')$, with $\mathbb{K}'$ a subfield of $\mathbb{K}$, that is fixed pointwise under a nontrivial field automorphism of $\mathbb{K}$.

We now have the following characterization result.

Theorem 13.2.2 *Every faithful translation-homogeneous embedding of a translation generalized quadrangle in a finite Desarguesian projective plane is isomorphic to a generalized hyperoval embedding of* $\mathbf{W}(2)$ *with some parameters* (ℓ_1, ℓ_2). *The faithful embedding is translation-homogeneous with respect to two different translation points if and only if it is isomorphic to the hyperoval embedding of* $\mathbf{W}(2)$ *in* $\mathbf{PG}(2,4)$.

Proof. Let $\mathcal{S} = (\mathcal{P}, \mathcal{B}, \mathbf{I})$ be a TGQ of order (s,t) with translation point x, embedded in the plane $\mathbf{PG}(2,q)$. Since we assume that the embedding is faithful, $\mathcal{P}$ is not contained in any subplane over any subfield $\mathbf{GF}(q')$ of $\mathbf{GF}(q)$. This is equivalent with saying that no nontrivial collineation of $\mathbf{PG}(2,q)$ fixes $\mathcal{P}$ pointwise (since we are dealing with finite fields). A direct consequence of this is that the companion field automorphism of any collineation stabilizing $\mathcal{P}$ is uniquely determined by the action on $\mathcal{P}$ and so the map that assigns the companion field automorphism to a semilinear collineation of $\mathbf{PG}(2,q)$ that induces an element of the translation group T of $\mathcal{S}^{(x)}$ is a group morphism Γ from T to the automorphism group of $\mathbf{GF}(q)$.

The translation group T is an elementary abelian p-group for some prime p. So T has exponent p, and s and t are powers of p. Noting that $\mathrm{Aut}(\mathbf{GF}(q))$ is a cyclic group and hence has at most one subgroup of exponent p, we see that the mapping $\Gamma : T \rightarrow \mathrm{Aut}(\mathbf{GF}(q))$ assigning to any translation of $\mathcal{S}^{(x)}$ the companion field automorphism (see above) of the corresponding collineation of $\mathbf{PG}(2,q)$ has a kernel of size at least $s^2 t/p$. Hence the intersection U of T (the latter viewed as collineation group of $\mathbf{PG}(2,q)$) with $\mathbf{PGL}_3(q)$ is nontrivial and has size at least $s^2 t/p$. As each element of U fixes at least three lines of $\mathbf{PG}(3,q)$ through x, each such element is a central collineation of $\mathbf{PG}(2,q)$ with center x, and hence it is also an axial collineation with some axis A. It is well-known (see e.g. Hughes and Piper [75]) that x is incident with A in $\mathbf{PG}(2,q)$ if and only if the central collineation has order p', where p' is the unique prime dividing q. In this case, the collineation is called an *elation* of $\mathbf{PG}(2,q)$. We now claim that all elements of U are elations of $\mathbf{PG}(2,q)$, i.e., p divides q. By the previous, this is equivalent with saying that U contains at least one nontrivial elation of $\mathbf{PG}(2,q)$.

Indeed, since the order of U exceeds s, the group U cannot have a free action on the s points, different from x, of an arbitrary line L of $\mathcal{S}$ through x. Let $\theta \in U$ be a nontrivial collineation fixing some point y on L, with $y \neq x$. Since the order of θ is p, it must fix a third point of $\mathcal{S}$ on L. Then θ

fixes all points of $\mathbf{PG}(2, q)$ on L and hence is an elation of $\mathbf{PG}(2, q)$ (since it also fixes all lines through x). The claim is proved.

Let L be any line of S through x. The group S of symmetries about L does not contain nontrivial elements of U, hence Γ is injective on S and so S is a cyclic group of order s. This easily implies $s = p$. Hence $t \in \{p, p^2\}$.

We claim that $t = p$. For, suppose by way of contradiction that $t = p^2$. Let L_1 and L_2 be different lines of S through x and let θ_1 and θ_2 be nontrivial symmetries about L_1 and L_2, respectively, chosen such that they have the same image under Γ. Set $\theta = \theta_1 \theta_2^{-1}$. Then θ maps a point u, with $u \not\sim x$, to a point $u' \not\sim u$. Also, θ fixes every element of $\{x, u, u'\}^\perp$, and the latter contains $p + 1$ elements $\{u_0, u_1, \ldots, u_p\}$. Since θ has order p and belongs to U, it fixes all points on the lines xu_i, $i = 0, 1, \ldots, p$, of $\mathbf{PG}(2, q)$, clearly a contradiction. The claim follows.

Since $s = t = p$, Corollary 3.5.6 implies that $S \cong \mathbf{Q}(4, p)$. We now claim that $p = 2$. Suppose by way of contradiction that p is odd. Then there exists a collineation $\sigma \in T$ fixing every point of S on the lines L_1 and L_2 (with the above notation), and fixing every line of S through x. Clearly $\sigma \notin U$ and σ^Γ belongs to the unique subgroup of order p of $\operatorname{Aut}(\mathbf{GF}(q))$. Now we may rechoose $\theta_1 \in S_1$, with S_1 the group of symmetries about L_1, so that $\theta_1^\Gamma = \sigma^\Gamma$. The composition $\sigma \theta_1^{-1}$ belongs to U, and hence is an elation in $\mathbf{PG}(2, q)$ with axis L_1 and center x. Now let a be any point not collinear with x, and let b be its image under $\sigma \theta_1^{-1}$. Since a and b are not collinear, since a and b are collinear with a common point y on L_1, and since p is odd, there is a second point $y' \neq y$ in $\{x, a, b\}^\perp$. But y' is fixed under $\sigma \theta_1^{-1}$, a contradiction. The claim is proved.

Hence S is isomorphic to $\mathbf{W}(2)$. We now give explicit coordinates to the points of the embedded $\mathbf{W}(2)$. We assume that the point set of $\mathbf{W}(2)$ is abstractly given by the pairs of $\{1, 2, 3, 4, 5, 6\}$ and the line set by the set of partitions of $\{1, 2, 3, 4, 5, 6\}$ in pairs. Without loss of generality we may assign the following coordinates (for some $a, b, b', k_1, k_2, \ell_1, \ell_2 \in \mathbf{GF}(q)$):

$$\begin{array}{lll} \{1, 2\} \rightsquigarrow (0, 1, 0), & \{3, 4\} \rightsquigarrow (1, 1, 0), & \{5, 6\} \rightsquigarrow (1, 0, 0), \\ \{3, 5\} \rightsquigarrow (0, 1, 1), & \{4, 6\} \rightsquigarrow (0, 0, 1), & \{3, 6\} \rightsquigarrow (a, b, 1), \\ \{4, 5\} \rightsquigarrow (a, b', 1), & \{2, 3\} \rightsquigarrow (k_1, \ell_1, 1), & \{2, 5\} \rightsquigarrow (k_2, \ell_2, 1). \end{array}$$

Denote the image of any element $k \in \mathbf{GF}(q)$ under the involutory automorphism of $\mathbf{GF}(q)$ by $\bar{k}$. We may assume without loss of generality that $x = \{1, 2\} = (0, 1, 0)$. Set $y = \{5, 6\} = (1, 0, 0)$. The stabilizer in T of y has order 4. If the two elements θ_1, θ_2 of that stabilizer not fixing the point $\{4,6\}$ both have nontrivial companion field automorphism, then the composition θ is a linear collineation and fixes the base triangle. Since q is even, and the order of θ is equal to 2, this leads to a contradiction. Hence we may assume that θ_1 is linear. But it fixes three lines through x, so it is central with center x; it fixes three points on the line xy, so the axis is xy. This easily implies that the image of $(a,b,1)$ under θ_1 is $(a,b+1,1)$. Hence $b' = b+1$. Moreover, interchanging the roles of the first and third coordinates, we also see that $\frac{b+1}{a} = \frac{b}{a}+1$, implying $a=1$.

We can also take the image of $(k_1, \ell_1, 1)$ under θ_1. Note first that θ_1 does not fix the lines through y, as it would otherwise have two different centers. So θ_1 corresponds to the permutation $(3\ 4)(5\ 6)$ of $\{1,2,3,4,5,6\}$. Thus we obtain

$$\{2,4\} = \{2,3\}^{\theta_1} = (k_1, \ell_1, 1)^{\theta_1} = (k_1, \ell_1+1, 1).$$

Likewise

$$\{2,6\} = (k_2, \ell_2+1, 1).$$

Now the (semilinear) collineation θ corresponding with the permutation $(1\ 2)$ fixes the points $(1,0,0)$, $(1,1,0)$, $(0,1,0)$, $(0,0,1)$ and $(0,1,1)$, hence has identity matrix. So it follows that

$$\{1,3\} = \{2,3\}^{\theta} = (k_1, \ell_1, 1)^{\theta} = (\overline{k}_1, \overline{\ell}_1, 1).$$

Likewise

$$\{1,4\} = (\overline{k}_1, \overline{\ell}_1+1, 1), \quad \{1,5\} = (\overline{k}_2, \overline{\ell}_2, 1), \quad \{1,6\} = (\overline{k}_2, \overline{\ell}_2+1, 1).$$

Now we express that collinearity in the quadrangle implies collinearity in the projective plane.

Expressing that the points $\{5,6\}, \{2,3\}, \{1,4\}$ are collinear, we see that $\overline{\ell}_1 + \ell_1 + 1 = 0$. Doing the same for $\{3,4\}, \{2,5\}, \{1,6\}$, we obtain $\overline{k}_2 + k_2 + \overline{\ell}_2 + \ell_2 + 1 = 0$.

One checks easily that the points on an arbitrary line of $\mathbf{W}(2)$ incident with $\{4,6\}$ or $\{3,5\}$, but not with both, are collinear in $\mathbf{PG}(2,q)$ if and only if $k_1\overline{\ell}_2 = \overline{k}_2\ell_1$, and then automatically the points on every line of $\mathbf{W}(2)$ incident with $\{4,6\}$ or $\{3,5\}$ (or both) are collinear in $\mathbf{PG}(2,q)$.

The points on an arbitrary line of $\mathbf{W}(2)$ incident with $\{3,6\}$ or $\{4,5\}$, but not with both, are collinear in $\mathbf{PG}(2,q)$ if and only if

$$b = \frac{\ell_1 + \ell_2 + k_2}{\overline{k}_1 + k_2} = \frac{\overline{\ell}_1 + \overline{\ell}_2 + \overline{k}_2}{k_1 + \overline{k}_2}.$$

In particular, if follows that $b = \bar{b}$.

Let θ_3 be the (semilinear) collineation corresponding to the permutation $(1\ 2)(3\ 5)(4\ 6)$. Then, since $(0,1,0)$, $(0,0,1)$ and $(0,1,1)$ are fixed, and since $(1,0,0)$ and $(1,1,0)$ are permuted, θ_3 maps $(x_1, x_2, 1)$ onto $(\bar{x}_1, \bar{x}_1 + \bar{x}_2, 1)$. Applying this to the points $\{2,3\}$ and $\{1,5\}$, we see that $k_1 = k_2$ and $\ell_1 + \ell_2 + k_2 = 0$. Hence $b = 0$, and we can put $k = k_1 = k_2 = \ell_1 + \ell_2$.

Noting that from $k_1 \bar{\ell}_2 = \bar{k}_2 \ell_1$ follows that $\ell_1 \bar{\ell}_1 = \ell_2 \bar{\ell}_2$, and that $\ell_1 = \bar{\ell}_2$ implies $k = \bar{k}$ and hence $\{1,4\} = \{2,3\}$, we conclude that we have a generalized hyperoval embedding of $\mathbf{W}(2)$.

We now note that in the generalized hyperoval embedding above, the points $\{1,2\}$, $\{1,3\}$, $\{1,4\}$, $\{1,5\}$, $\{1,6\}$ are incident with one line in $\mathbf{PG}(2,q)$, and similarly for the points $\{1,2\}$, $\{2,3\}$, $\{2,4\}$, $\{2,5\}$, $\{2,6\}$. We may conclude that whenever a planar embedding of $\mathbf{W}(2)$ is translation-homogeneous with respect to some point x, then every ovoid of $\mathbf{W}(2)$ containing x is contained in a line of $\mathbf{PG}(2,q)$.

If the faithful embedding is translation-homogeneous with respect to two distinct collinear translation points, then we may assume without loss of generality that, for the previous embedding, in addition to $\{1,2\}$, also $\{5,6\}$ is one of these two points. By the previous paragraph, the point $\{1,6\}$ is incident with the line of $\mathbf{PG}(2,q)$ through $(1,0,0)$ and $(0,0,1)$. This easily implies that $\ell_2 = 1$. But then $\ell_1 \bar{\ell}_1 = 1$ and so ℓ_1 satisfies the quadratic equation $X^2 + X + 1 = 0$, which implies that ℓ_1, and hence also k, are contained in $\mathbf{GF}(4)$. By faithfulness, $q = 4$, and it is easy to see that we now have the hyperoval embedding.

Finally, if the embedding is translation-homogeneous with respect to two noncollinear points, then it is translation-homogeneous with respect to every (translation) point and the result follows from the previous paragraph. The theorem is proved.   ∎

Remark 13.2.3 The second part of the previous theorem has a more geometric proof using a result of Temmermans (unpublished) that characterizes the hyperoval embedding of $\mathbf{W}(2)$ as the unique (up to field extension) planar embedding of $\mathbf{W}(2)$ with the property that the points of each ovoid of $\mathbf{W}(2)$ lie on a line of the plane.

13.3 Exceptional Non-Planar Translation-Homogeneous Embeddings

In this section we classify the translation-homogeneous embeddings of some small quadrangles. These will give rise to exceptional embeddings. The reason to treat them separately is because our general classification technique will fail for these small parameters. It turns out, however, that precisely in these cases, exceptional embeddings exist.

We start with the case $s = t = 2$, i.e., the symplectic quadrangle $\mathbf{W}(2)$. We first describe some embeddings.

Example 13.3.1 Some Embeddings of $\mathbf{W}(2)$ in $\mathbf{PG}(3, q)$.

Suppose that the prime power q is an odd perfect square and let ℓ, $\ell \neq 0$, be an element of $\mathbf{GF}(q)$ that is mapped to $-\ell$ under the unique involutory field automorphism. As before, we denote that automorphism by $k \mapsto \bar{k}$. We define the following embedding of $\mathbf{W}(2)$ in $\mathbf{PG}(3, q)$.

$$
\begin{array}{ll}
\{1,2\} \rightsquigarrow (0,0,0,1) & \{5,6\} \rightsquigarrow (1,0,0,0) \\
\{3,4\} \rightsquigarrow (1,0,0,1) & \{3,6\} \rightsquigarrow (0,1,0,1) \\
\{4,5\} \rightsquigarrow (0,1,0,0) & \{1,3\} \rightsquigarrow (1,1,1,\ell+1) \\
\{2,3\} \rightsquigarrow (-1,-1,-1,\ell-1) & \{1,5\} \rightsquigarrow (-1,-1,1,\ell-1) \\
\{2,4\} \rightsquigarrow (-1,1,1,\ell+1) & \{2,5\} \rightsquigarrow (1,1,-1,\ell+1)
\end{array}
$$

$$
\begin{array}{l}
\{4,6\} \rightsquigarrow (0,0,1,0) \\
\{3,5\} \rightsquigarrow (0,0,1,1) \\
\{1,4\} \rightsquigarrow (1,-1,-1,\ell-1) \\
\{1,6\} \rightsquigarrow (-1,1,-1,\ell-1) \\
\{2,6\} \rightsquigarrow (1,-1,1,\ell+1)
\end{array}
$$

One checks that the linear collineations with matrices

$$
\begin{pmatrix}
1 & 0 & 0 & 0 \\
0 & -1 & 0 & -1 \\
0 & 0 & -1 & -1 \\
0 & 0 & 0 & 1
\end{pmatrix}
\text{ and }
\begin{pmatrix}
-1 & 0 & 0 & -1 \\
0 & -1 & 0 & -1 \\
0 & 0 & 1 & 0 \\
0 & 0 & 0 & 1
\end{pmatrix},
$$

respectively, together with the unique involutory semilinear collineation having identity matrix, generate a group of order eight acting on $\mathbf{W}(2)$ as the translation group with respect to the translation point $\{1, 2\}$.

The same group acts similarly on the following embedding of $\mathbf{W}(2)$.

$\{1,2\} \rightsquigarrow (0,0,0,1)$ $\{5,6\} \rightsquigarrow (1,0,0,0)$
$\{3,4\} \rightsquigarrow (1,0,0,1)$ $\{3,6\} \rightsquigarrow (0,1,0,1)$
$\{4,5\} \rightsquigarrow (0,1,0,0)$ $\{1,3\} \rightsquigarrow (2\ell+1,1,1,\ell+1)$
$\{2,3\} \rightsquigarrow (2\ell-1,-1,-1,\ell-1)$ $\{1,5\} \rightsquigarrow (2\ell-1,-1,1,\ell-1)$
$\{2,4\} \rightsquigarrow (2\ell-1,1,1,\ell+1)$ $\{2,5\} \rightsquigarrow (2\ell+1,1,-1,\ell+1)$

$\{4,6\} \rightsquigarrow (0,0,1,0)$
$\{3,5\} \rightsquigarrow (0,0,1,1)$
$\{1,4\} \rightsquigarrow (2\ell+1,-1,-1,\ell-1)$
$\{1,6\} \rightsquigarrow (2\ell-1,1,-1,\ell-1)$
$\{2,6\} \rightsquigarrow (2\ell+1,-1,1,\ell+1)$

If we call two embeddings *equivalent* whenever there exists a collineation of the ambient projective space taking the points and lines of the first embedded GQ to the points and lines, respectively, of the second embedded GQ, then the two embeddings defined above are not equivalent. Indeed, in the first embedding every nontrivial axial root elation with axis through $\{1,2\}$ has nontrivial companion field automorphism, while in the latter example this is true for only one nontrivial axial root elation, namely the one with axis $\{\{1,2\},\{3,4\},\{5,6\}\}$.

Example 13.3.2 A Class of Non-Polarized Translation-Homogeneous Embeddings.

Consider the standard embedding of the generalized quadrangle $\mathbf{Q}(4,q)$ in $\mathbf{PG}(4,q)$, q even, and extend the groundfield $\mathbf{GF}(q)$ of the projective space to a field $\mathbf{GF}(q')$, keeping the quadrangle unchanged. Select a point x of $\mathbf{Q}(4,q)$ and let n be the nucleus of $\mathbf{Q}(4,q)$ in $\mathbf{PG}(4,q)$. Select a point y of $\mathbf{PG}(4,q')$ on the line xn such that y does not belong to $\mathbf{PG}(4,q)$, except if $y = n$, and project $\mathbf{Q}(4,q)$ from y onto a hyperplane H of $\mathbf{PG}(4,q')$ not incident with y to obtain an embedding of $\mathbf{Q}(4,q')$ in $\mathbf{PG}(3,q')$. We can identify the translation group T of $\mathbf{Q}(4,q)^{(x)}$ with the corresponding linear projective group of $\mathbf{PG}(4,q')$. Then T fixes the points x and n, and since every element of T is linear and has order 2, the group T fixes y and so it also "projects" from y. Hence the projected embedding is translation-homogeneous. It is polarized at every point z that has the following property:

the hyperplane, tangent to the quadric $\mathbf{Q}(4,q)$, at the pre-image z' of z under the projection, contains y.

In case $y \neq n$, this is equivalent to saying that this hyperplane contains x; hence, if $y \neq n$, then the embedding is polarized exactly at the projections of the points of $x^{\perp}$. If $y = n$, then the embedding is polarized at every point and it is the standard embedding of $\mathbf{W}(q)$ in a subspace $\mathbf{PG}(3, q)$ of H.

Example 13.3.3 Small Quadrangles in 4-Dimensional Projective Space.

The quadrangles $\mathbf{W}(2)$ and $\mathbf{Q}(4, 3)$ admit embeddings in $\mathbf{PG}(4, q)$ for various q. These embeddings are translation-homogeneous with respect to every (translation) point. We start with $\mathbf{W}(2)$.

The Universal Embeddings of $\mathbf{W}(2)$.

Let $\mathbb{K}$ be an arbitrary field of characteristic different from 3 and identify the basis vectors of a 6-dimensional vector space V over $\mathbb{K}$ with the six ovoids of $\mathbf{W}(2)$. The coordinate 6-tuple associated with an arbitrary point x of $\mathbf{W}(2)$ has 1 in every position of an ovoid in which it is not contained, and -2 in the positions of ovoids it belongs to. Since every point belongs to exactly two ovoids, the points are contained in the hyperplane with equation $\sum_{i=1}^{6} X_i = 0$, and since every line of $\mathbf{W}(2)$ is incident with exactly one point of each ovoid, the sum of the coordinates of the three points incident with an arbitrary line of $\mathbf{W}(2)$ equals zero. In other words, collinear points belong to the same vector plane. Consequently we obtain a lax embedding of $\mathbf{W}(2)$ in a hyperplane H of $\mathbf{PG}(V)$.

More concrete, and using the model for $\mathbf{W}(2)$ of pairs and partitions into pairs of the set $\{1, 2, 3, 4, 5, 6\}$, we may identify the ovoids of $\mathbf{W}(2)$ with the numbers $1, 2, \ldots, 6$ (and the ovoid i contains the points $\{i, j\}, j \neq i$), and then the number i is identified with the basis vector $\vec{e}_i$. The point $\{i, j\}$ now is identified with the vector $\sum_{k=1}^{6} a_k \vec{e}_k$, with $a_i = a_j = -2$ and $a_k = 1$ for $k \notin \{i, j\}$.

The transposition $(1\ 2)$ induces a symmetry (with center) in $\mathbf{W}(2)$, and it is easy to check that this symmetry is also induced by the linear transformation of V with matrix

$$
\begin{pmatrix}
\frac{1}{3} & -\frac{2}{3} & \frac{1}{3} & \frac{1}{3} & \frac{1}{3} & \frac{1}{3} \\
-\frac{2}{3} & \frac{1}{3} & \frac{1}{3} & \frac{1}{3} & \frac{1}{3} & \frac{1}{3} \\
\frac{1}{3} & \frac{1}{3} & -\frac{2}{3} & \frac{1}{3} & \frac{1}{3} & \frac{1}{3} \\
\frac{1}{3} & \frac{1}{3} & \frac{1}{3} & -\frac{2}{3} & \frac{1}{3} & \frac{1}{3} \\
\frac{1}{3} & \frac{1}{3} & \frac{1}{3} & \frac{1}{3} & -\frac{2}{3} & \frac{1}{3} \\
\frac{1}{3} & \frac{1}{3} & \frac{1}{3} & \frac{1}{3} & \frac{1}{3} & -\frac{2}{3}
\end{pmatrix}.
$$

Since this linear map fixes all vectors in the 4-dimensional space with equations

$$\begin{cases} X_1 + X_2 + X_3 + X_4 + X_5 + X_6 = 0, \\ \qquad\qquad\;\; X_3 + X_4 + X_5 + X_6 = 0, \end{cases}$$

it induces a special linear collineation in H. We can do this for every transposition, and thus prove that the embedding is a special translation-homogeneous embedding with respect to every (translation) point.

This description is also valid in characteristic 2. In this case, the coordinates of all the points of $\mathbf{W}(2)$ satisfy the quadratic equation $\sum_{i=2}^{6} \sum_{j=1}^{i-1} X_i X_j = 0$ which represents a nonsingular parabolic quadric in H isomorphic to $\mathbf{Q}(4,2)$ in the 4-dimensional subspace H' of H over the subfield $\mathbf{GF}(2)$ of $\mathbb{K}$.

If $\mathbb{K}$ has characteristic 3, then we consider $\mathbf{PG}(6, \mathbb{K})$ and associate with the point $\{i, j\}$ of $\mathbf{W}(2)$ the point with coordinates all 1, except in the positions i and j, where we write -1. This defines a lax embedding of $\mathbf{W}(2)$ in the 4-dimensional projective space obtained by projecting the hyperplane H'' with equation $X_1 + X_2 + \cdots + X_7 = 0$ from the point $c := (1,1,1,1,1,1,0)$ onto an arbitrary 4-dimensional space in H'' not containing c. In fact, it is desirable to leave that 4-space unspecified so that it is easier to see the automorphisms (the symmetric group acting on the first six coordinates in a natural way). As in the previous case, one can show that the embedding is a special translation-homogeneous one.

In all cases, the embeddings described here are polarized. This is obvious since there are only three lines through each point in $\mathbf{W}(2)$, and these cannot generate a 4-dimensional space.

These embeddings are called *universal* because *every* lax embedding of $\mathbf{W}(2)$ in $\mathbf{PG}(3, \mathbb{K})$ arises from a projection of the universal one from a well-chosen point in $\mathbf{PG}(4, \mathbb{K})$ after embedding $\mathbf{PG}(3, \mathbb{K})$ in $\mathbf{PG}(4, \mathbb{K})$. This is proved in [195] and we do not repeat the proof here (it goes along the lines of the proof of Theorem 13.4.1 below).

We remark that the hyperoval embedding of $\mathbf{W}(2)$ is *not* a projection of the universal lax embedding.

The Universal Embeddings of $\mathbf{Q}(4, 3)$.

We label the base vectors of a six-dimensional vector space V_6 over the field $\mathbf{GF}(3)$ with the ordered pairs $\{(01), (02), (03), (12), (13), (23)\}$ of the set $\{0, 1, 2, 3\}$ and write coordinates in lexicographic order. In the projective space $\mathbf{PG}(V_6)$ we consider the Klein quadric with equation $X_{01}X_{23} -$

$X_{02}X_{13} + X_{03}X_{12} = 0$. In $\mathbf{PG}(3,3)$, we consider the symplectic bilinear form β given in ordinary coordinates by

$$\begin{vmatrix} x_0 & x_1 \\ y_0 & y_1 \end{vmatrix} + \begin{vmatrix} x_2 & x_3 \\ y_2 & y_3 \end{vmatrix} - \begin{vmatrix} x_0 & x_3 \\ y_0 & y_3 \end{vmatrix},$$

and let the image of the line of $\mathbf{PG}(3,3)$ incident with the points (x_0, x_1, x_2, x_3) and (y_0, y_1, y_2, y_3) under the Klein correspondence be given by the line in $\mathbf{PG}(V_6)$ with coordinates $x_{ij} = x_i y_j - x_j y_i$. The totally isotropic lines with respect to β are mapped under this correspondence onto the set of points of the quadric $\mathbf{Q}(4,3)$ given by the equations

$$\begin{cases} X_{02}X_{13} = X_{01}X_{12} + X_{01}X_{23} + X_{12}X_{23}, \\ X_{03} = X_{01} + X_{23}. \end{cases}$$

An arbitrary line — which we can label with a homogeneous 4-tuple (a_0, a_1, a_2, a_3) that represents a point of $\mathbf{PG}(3,3)$ — of $\mathbf{Q}(4,3)$ is obtained by intersecting this quadric with the plane spanned by the points

$$(a_1, a_2, a_3, 0, 0, 0), \ (a_0, 0, 0, -a_2, -a_3, 0), \ (0, a_0, 0, a_1, 0, -a_3),$$
$$(0, 0, a_0, 0, a_1, a_2).$$

We now write down all the points of $\mathbf{Q}(4,3)$ in coordinates. But we substitute sometimes -1 by the letter k, and consequently we sometimes write 2 as $1 - k$, 0 as $2 - k$ or as $1 + k$ and 1 as k^2. The reason to do so will become clear below.

$(0,1,0,0,0,0)$	$(0,1,1,0,0,1)$	$(1,k,1,k,1,0)$
$(1,0,1,0,0,0)$	$(0,0,0,0,1,0)$	$(k,k^2,k,k,1,0)$
$(1,1,1,0,0,0)$	$(1,0,1,0,1,0)$	$(0,k,1,0,0,1)$
$(1,k,1,0,0,0)$	$(k,0,k,0,1,0)$	$(k,k,1+k,0,1,1)$
$(0,0,0,1,0,0)$	$(0,0,0,1,1,0)$	$(0,0,1,0,1,1)$
$(0,1,0,1,0,0)$	$(1,1,1,1,1,0)$	$(1,1,2,0,1,1)$
$(0,k,0,1,0,0)$	$(k,k,k,1,1,0)$	$(1,k,1+k,0,1,k)$
$(0,0,1,0,0,1)$	$(0,0,0,k,1,0)$	$(k,k^2,2k,0,1,k)$

$(0,0,k,0,1,k)$	$(1,1,2,1-k,1-k,1)$
$(1,1+k,2,1,2-k,1)$	$(0,1,1,1-k,1-k,1)$
$(1,k,2,1,2-k,1)$	$(1-k,1,2-k,1,1-k,1)$
$(1,1,2,1,2-k,1)$	$(1,1+k,2,1,1-k,1)$
$(k,k,1+k,k,1,1)$	$(0,k,1,1,1-k,1)$
$(0,k,1,k,1,1)$	$(k,k,1+k,1,1,1)$
$(1,k,2,k,1,1)$	$(0,1,1,1,1,1)$
$(1-k,1,2-k,1-k,1-k,1)$	$(1,1+k,2,1,1,1)$

Now let $\mathbb{K}$ be a field admitting an element r satisfying $r^2 - r + 1 = 0$. Then it is a tedious exercise to check that, with $k = r$, the above 40 points define a lax embedding of $\mathbf{Q}(4,3)$ in the hyperplane H of $\mathbf{PG}(5,\mathbb{K})$ with equation $X_{03} = X_{01} + X_{23}$.

One can now check that the little projective group, which contains the translation group with respect to any (translation) point, of this lax embedded GQ $\mathcal{S}$ is induced by the projective special linear group $\mathbf{PSL}_5(\mathbb{K})$ of H. In any case, no other collineation than those of the little projective group are induced by $\mathbf{PGL}_5(\mathbb{K})$ unless $\operatorname{char}(\mathbb{K}) = 3$. For $\mathbb{K} \cong \mathbf{GF}(q)$, with q not divisible by 3, the group $\mathbf{P\Gamma L}_5(q)$ induces the full automorphism group of $\mathcal{S}$ if and only if q is a perfect square and $\sqrt{q} \equiv -1 \mod 3$. These are tedious exercises, and expressions with matrices can be found in [195], to which we refer for more details. Hence the embedding is a special translation-homogeneous embedding with respect to any point.

This embedding is called *universal* because every embedding of $\mathbf{Q}(4,3)$ in a 3-dimensional projective space over $\mathbb{K}$ is a projection of this one, for all fields $\mathbb{K}$, see [195] (where the proofs are only carried out for finite fields, but they hold without much change for arbitrary fields).

Example 13.3.4 The Universal Embeddings of $\mathbf{Q}(5,2)$ in $\mathbf{PG}(5,\mathbb{K})$, with $\mathbb{K}$ Any Field.

Let $\mathcal{S} \cong \mathbf{Q}(5,2)$ and let $\mathcal{S}'$ be a subquadrangle of order 2. We embed $\mathcal{S}'$ as in Example 13.3.3. Recall that $\mathcal{S}$ can be constructed as follows from $\mathbf{W}(2)$. Add to the point set of $\mathbf{W}(2)$ two copies of the set of ovoids — since we can identify each ovoid with a number $i \in \{1,2,3,4,5,6\}$, we may write these twelve additional points as $\mathbf{i}$ and $\mathbf{i}'$, with $i \in \{1,2,3,4,5,6\}$. The thirty extra lines are formed by the triples $\{\mathbf{i},\{i,j\},\mathbf{j}'\}$, for $i,j \in \{1,2,3,4,5,6\}$, $i \neq j$. In other words, the ovoid $\mathbf{i}$, the ovoid $\mathbf{j}'$ and the unique point in the intersection of these ovoids form a line. So it suffices to attach coordinates to these 12 additional points in such a way that the points of each of the additional thirty lines are collinear. This can be done in the following way. First let the characteristic of $\mathbb{K}$ be different from 3. Assign to the point $\mathbf{i}$ the 6-tuple with 0 in every position, except for position i, where we put 1. Assign to the point $\mathbf{i}'$ the 6-tuple with 1 in every position, except in position i, where we put -2. This defines a lax embedding of $\mathbf{Q}(5,2)$ in $\mathbf{PG}(5,\mathbb{K})$. If the characteristic of $\mathbb{K}$ is equal to 2, then the embedding is full in the subspace $\mathbf{PG}(5,2)$ over the subfield $\mathbf{GF}(2)$, and the points of $\mathcal{S}$ are precisely those points of $\mathbf{PG}(5,2)$ whose coordinates satisfy the quadratic equation $\sum_{i=2}^{6}\sum_{j=1}^{i-1} X_i X_j = 0$ above.

If the characteristic of $\mathbb{K}$ is equal to 3, then give the point $\mathbf{i}$ in $\mathbf{PG}(6, \mathbb{K})$ coordinates $(0, \ldots, 0, 1, 0, \ldots, 0)$, where the 1 is in the position i, and the point $\mathbf{i}'$ gets the coordinates $(0, \ldots, 0, 1, 0, \ldots, 0, 1)$, where the first 1 is in the position i. We now still project from the point $c = (1, 1, 1, 1, 1, 1, 0)$, but do not restrict to points of H''. We obtain a lax embedding in the 5-dimensional projective space over $\mathbb{K}$ obtained from $\mathbf{PG}(6, \mathbb{K})$ by considering the subspaces through c. Note that no element of this embedding lies in the hyperplane with equation $X_1 + \cdots + X_6 = 0$. This means we always have an embedding in an affine space, and, in particular, for $|\mathbb{K}| = 3$, the embedding is full in that affine space.

For every field $\mathbb{K}$, the above lax embedding is translation-homogeneous and linear, even special. This follows easily from the following two observations.

(i) All collineations of $\mathcal{S}'$ extend uniquely to linear collineations of $\mathbf{PG}(5, \mathbb{K})$ preserving $\mathcal{S}$ (these are just the permutations of the coordinates).

(ii) By the uniqueness of this embedding, i.e., the embedding is uniquely determined by $\mathcal{S}'$ and one additional line (the two points not in $\mathcal{S}'$ of which may be chosen freely; we will prove this below in Theorem 13.4.4), the embedding has an automorphism group acting transitively on the subGQs of order 2.

(iii) The little projective group of $\mathcal{S}$ is a simple group, and if the embedding were not special, then the intersection with the special linear group of $\mathbf{PG}(5, \mathbb{K})$ would be a nontrivial normal subgroup of the little projective group of $\mathcal{S}$.

Again, this embedding is *universal* in the sense that every embedding of $\mathcal{S}$ in $\mathbf{PG}(d, \mathbb{K})$, $d \in \{3, 4\}$, can be considered as projection of this universal one (and for $d = 5$ every embedding is equivalent to the universal one).

13.4 Non-Planarly Embedded Small Translation Generalized Quadrangles

We start with classifying the translation-homogeneous embeddings of $\mathbf{W}(2)$.

Theorem 13.4.1 *If* $\mathbf{W}(2)$ *is faithfully and translation-homogeneously embedded in* $\mathbf{PG}(3,q)$ *(with respect to a single translation point), then either* q *is an odd perfect square and the embedding is as described in Example 13.3.1, or* q *is even and the embedding is as described in Example 13.3.2.*

Proof. Let $\mathbf{W}(2)$ be embedded in $\mathbf{PG}(3,q)$. We assume that the embedding is translation-homogeneous with respect to the point $x = \{1,2\}$.

We first treat the case where the embedding is not polarized at x.

Then, without loss of generality, we may assign coordinates as follows:

$$\begin{array}{ll}
\{1,2\} \rightsquigarrow (0,0,0,1) & \{5,6\} \rightsquigarrow (1,0,0,0) \\
\{4,6\} \rightsquigarrow (0,0,1,0) & \{3,4\} \rightsquigarrow (1,0,0,1) \\
\{3,6\} \rightsquigarrow (0,1,0,1) & \{3,5\} \rightsquigarrow (0,0,1,1) \\
\{4,5\} \rightsquigarrow (0,1,0,0) & \{1,3\} \rightsquigarrow (a,b,c,d),
\end{array}$$

for some $a,b,c,d \in \mathbf{GF}(q)$. Note that none of a,b,c,d is zero since otherwise the point $\{1,3\}$ is contained in a plane π together with two lines through x. Applying the translation group with respect to x, this implies that all points opposite x are contained in π, and hence, since the intersection of $\mathbf{W}(2)$ with π is a subquadrangle of $\mathbf{W}(2)$, this would mean that the embedding is planar, a contradiction.

Taking the lines through $\{1,3\}$ into account, there must be elements $a',b',c' \in \mathbf{GF}(q)$ such that

$$\{2,4\} \rightsquigarrow (a',b,c,d) \Big| \{2,5\} \rightsquigarrow (a,b,c',d) \Big| \{2,6\} \rightsquigarrow (a,b',c,d).$$

The line $\{\{3,5\},\{2,6\},\{1,4\}\}$ forces $\{1,4\}$ to have coordinates $(a,b',c+k,d+k)$, for some $k \in \mathbf{GF}(q)$. The line $\{\{5,6\},\{1,4\},\{2,3\}\}$ implies $\{2,3\} = (a'',b',c+k,d+k)$, $a'' \in \mathbf{GF}(q)$. Considering the lines through $\{2,3\}$, we then have $\{1,5\} = (a'',b',c'',d+k)$ and $\{1,6\} = (a'',b'',c+k,d+k)$, for some $b'',c'' \in \mathbf{GF}(q)$. Expressing that the respective points of the lines through the points $\{3,5\}$, $\{3,6\}$ and $\{3,4\}$ are collinear in $\mathbf{PG}(3,q)$, an elementary calculation reveals that

$$\begin{cases}
a' = a'' = a + k, \\
b' = b'' + k = b + k, \\
c' = c'' + k = c + k.
\end{cases}$$

Note that the unique nontrivial central root elation θ with center $\{1,2\}$ has nontrivial companion field automorphism. This implies that either the axial root elation θ' with axis $\{\{1,2\},\{3,4\},\{5,6\}\}$ has trivial companion field automorphism or the composition $\theta\theta'$ does. Since the matrix of that linear collineation is determined by the image of the points in $x^{\perp}$, this matrix is equal to

$$\begin{pmatrix} 1 & 0 & 0 & 0 \\ 0 & -1 & 0 & -1 \\ 0 & 0 & -1 & -1 \\ 0 & 0 & 0 & 1 \end{pmatrix}.$$

Since $\{1,3\}^{\theta'} = \{2,4\}$ and $\{1,3\}^{\theta\theta'} = \{1,4\}$, we either have $(a,-b,-c,d-b-c) = (-a-k,-b,-c,-d)$, or $(a,-b,-c,d-b-c) = (a,b+k,c+k,d+k)$. Since clearly $k \neq 0$, both cases imply that q is odd. Also, q must be a square. We denote the unique involutory field automorphism as usual by $\ell \mapsto \overline{\ell}$.

CASE I: $(a,-b,-c,d-b-c) = (a,b+k,c+k,d+k)$.   Then $b = c = -k/2$ and the point $(\overline{a},-\overline{b},-\overline{c},\overline{d}-\overline{b}-\overline{c})$, which is just equal to $\{1,3\}^{\theta'}$, coincides with $(a+k,b,c,d)$, which is $\{2,4\}$. One checks that this implies that

$$\frac{\overline{a}}{\overline{k}} = \frac{a+k}{-k} \quad \text{and} \quad \frac{\overline{a}}{\overline{d}+\overline{k}} = \frac{a+k}{d}.$$

Similarly, the linear collineation θ'' with matrix

$$\begin{pmatrix} -1 & 0 & 0 & -1 \\ 0 & -1 & 0 & -1 \\ 0 & 0 & 1 & 0 \\ 0 & 0 & 0 & 1 \end{pmatrix}$$

maps the point $\{1,3\}$ either to $\{1,5\}$ or to $\{2,5\}$, and then $\theta\theta''$ maps $\{1,3\}$ to either $\{2,5\}$ or $\{1,5\}$, respectively. So we have again two cases.

First suppose that $\{1,3\}^{\theta''} = \{1,5\}$.   Then $2a = -k$. Putting $\ell = -1-2d/k$, we obtain as embedding the first example of Example 13.3.1.

Next suppose that $\{1,3\}^{\theta''} = \{2,5\}$.   Then $d = \frac{a-k/2}{2}$. Putting $\ell = -\frac{2a+k}{2k}$, we obtain the second example of Example 13.3.1.

CASE II: $(a,-b,-c,d-b-c) = (-a-k,-b,-c,-d)$.   In this case, the non-trivial axial root elation θ' with axis $\{\{1,2\},\{3,4\},\{5,6\}\}$ has trivial companion field automorphism. Then, if θ'' is an axial root elation, the composition $\theta\theta'\theta''$ is an axial root elation with nontrivial companion field automorphism, and this is equivalent to the second example of Example 13.3.1.

But, if θ'' is not an axial root elation, then $\theta\theta''$ is, and we again obtain an embedding equivalent to the second example of Example 13.3.1.

Now suppose that the embedding is polarized at x.

Let L be any line of $\mathbf{W}(2)$ incident with x. The pointwise stabilizer of L in the translation group T has order 4, and hence there is some element θ of the translation group T induced by a linear collineation of $\mathbf{PG}(3, q)$, fixing all points of $\mathbf{W}(2)$ — and consequently also of $\mathbf{PG}(3, q)$ — on L, and not fixing $x^{\perp}$ pointwise. In the plane of $\mathbf{PG}(3, q)$ spanned by $x^{\perp}$ the collineation θ is induced by an axial collineation with axis L. the center must necessarily be x, as θ fixes at least three lines of that plane through x (namely, the three lines of $\mathbf{W}(2)$ through x). So q is even and without loss of generality we may assign coordinates as follows:

$$\{1, 2\} \rightsquigarrow (1, 0, 0, 0) \mid \{5, 6\} \rightsquigarrow (0, 0, 1, 0)$$
$$\{4, 6\} \rightsquigarrow (x, 1, 1, 0) \mid \{3, 4\} \rightsquigarrow (1, 0, 1, 0)$$
$$\{3, 6\} \rightsquigarrow (1, 1, 0, 0) \mid \{3, 5\} \rightsquigarrow (x + 1, 1, 1, 0)$$
$$\{4, 5\} \rightsquigarrow (0, 1, 0, 0) \mid \{1, 3\} \rightsquigarrow (0, 0, 0, 1),$$

and we may put $\{2, 3\} = (a, b, c, 1)$, for some $a, b, c \in \mathbf{GF}(q)$. But now the coordinates of all other points are uniquely determined, in fact in two different ways. For instance, $\{1, 5\}$ is the unique point on the line $\langle (x, 1, 1, 0), (a, b, c, 1) \rangle$ and in the plane $\langle (1, 0, 1, 0), (0, 1, 0, 0), (0, 0, 0, 1) \rangle$ (this plane contains the point $\{3, 4\}$ and the line $\{\{1, 3\}, \{2, 6\}, \{4, 5\}\}$ and hence also the line $\{\{1, 5\}, \{2, 6\}, \{3, 4\}\}$), which implies that it has coordinates $(a+cx, a+b+c+bx, a+cx, x+1)$; also it is the unique point on the line $\langle (x, 1, 1, 0), (a, b, c, 1) \rangle$ and in the plane $\langle (1, 1, 0, 0), (0, 0, 1, 0), (0, 0, 0, 1) \rangle$, which implies that it has coordinates $(a + bx, a + bx, a + b + c + cx, x + 1)$. Hence we already deduce $b = c$. Considering similarly the point $\{1, 6\}$, it follows that $b = c = 0$ and we obtain the following coordinates.

$$\{1, 3\} \rightsquigarrow (0, 0, 0, 1) \quad \mid \{2, 4\} \rightsquigarrow (0, 0, a, x + 1)$$
$$\{2, 3\} \rightsquigarrow (a, 0, 0, 1) \quad \mid \{1, 4\} \rightsquigarrow (a(x + 1), 0, a, x + 1)$$
$$\{1, 5\} \rightsquigarrow (a, a, a, x + 1) \mid \{2, 6\} \rightsquigarrow (0, a, 0, x + 1)$$
$$\{2, 5\} \rightsquigarrow (ax, a, a, x + 1) \mid \{1, 6\} \rightsquigarrow (a(x + 1), a, 0, x + 1)$$

Applying the recoordinatization $(X_0, X_1, X_2, X_3) \mapsto (X_0, X_1, X_2, aX_3)$, we may assume that $a = 1$. Now we embed $\mathbf{PG}(3, q)$ in $\mathbf{PG}(4, q)$ and see that $\mathcal{S}$ is the projection from the point $(x, 1, 1, 0, 1)$ onto $\mathbf{PG}(3, q)$ (which has equation $X_4 = 0$) of the following embedding of $\mathbf{W}(2)$.

$$
\begin{array}{lll}
\{1,2\} \rightsquigarrow (1,0,0,0,0) & \{5,6\} \rightsquigarrow (0,0,1,0,0) & \{4,6\} \rightsquigarrow (0,0,0,0,1) \\
\{3,4\} \rightsquigarrow (1,0,1,0,0) & \{3,6\} \rightsquigarrow (1,1,0,0,0) & \{3,5\} \rightsquigarrow (1,0,0,0,1) \\
\{4,5\} \rightsquigarrow (0,1,0,0,0) & \{1,3\} \rightsquigarrow (0,0,0,1,0) & \{1,4\} \rightsquigarrow (1,1,0,x+1,1) \\
\{2,3\} \rightsquigarrow (1,1,1,x+1,1) & \{1,5\} \rightsquigarrow (1,1,1,x+1,0) & \{1,6\} \rightsquigarrow (1,0,1,x+1,1) \\
\{2,4\} \rightsquigarrow (0,0,1,x+1,0) & \{2,5\} \rightsquigarrow (0,0,0,x+1,1) & \{2,6\} \rightsquigarrow (0,1,0,x+1,0)
\end{array}
$$

Applying the recoordinatization $(X_0, X_1, X_2, X_3, X_4) \mapsto (X_0, X_1, X_2, X_3/(x+1), X_4)$, we now easily see that all these points belong the the intersection of the nondegenerate quadric with equation $X_0 X_3 + X_1 X_2 + X_1 X_4 + X_2 X_4 = 0$ with the projective 5-space over the field $\mathbf{GF}(2)$ obtained by restricting the coordinates to $\mathbf{GF}(2)$. The proof of the theorem is now complete. ∎

By Theorem 1.4(ii) of [195], every embedding of $\mathbf{W}(2)$ in $\mathbf{PG}(4, q)$ is isomorphic to the universal one, and hence independent of being translation-homogeneous. So we may state the following result, and we do so for general fields $\mathbb{K}$.

Theorem 13.4.2 *Every lax embedding of* $\mathbf{W}(2)$ *in* $\mathbf{PG}(4, \mathbb{K})$ *is isomorphic to the universal embedding of* $\mathbf{W}(2)$ *in* $\mathbf{PG}(4, \mathbb{K})$ *and hence polarized at every point and translation-homogeneous with respect to every point (and the embedding is linear).*

Proof. The proof is almost identical to the proof of Theorem 13.4.1, except that we can now choose freely the coordinates of all points of $x^{\perp}$ in some 3-dimensional subspace, and additionally two points opposite x outside that 3-space. The coordinates of all other points are now determined and uniqueness follows similarly as in the proof of Theorem 13.4.1, except that we do not have to identify the embedding with a projection. We leave the details to the reader. They can be found in [195]. ∎

Now we come to $\mathbf{Q}(4, 3)$ in $\mathbf{PG}(4, q)$. Also here, one can classify without the hypothesis of being translation-homogeneous and the result is that every lax embedding is translation-homogeneous with respect to every point.

Theorem 13.4.3 *Every lax embedding of* $\mathbf{Q}(4, 3)$ *in* $\mathbf{PG}(4, q)$, *with q not necessarily divisible by 3, is isomorphic to the embedding described in Example 13.3.3. In particular, $q \equiv 0$ or $1 \mod 3$.*

Proof. The proof of this lemma is a straightforward calculation assigning well-chosen but essentially arbitrary coordinates to the points, and then expressing that the points collinear in $\mathbf{Q}(4,3)$ are also collinear in $\mathbf{PG}(4,q)$. We will not perform this exercise here, but refer the reader to [195] for more details.

The fact that q must be congruent to 0 or 1 modulo 3 is because there must exist an element r of the field satisfying $r^2 - r + 1 = 0$, implying $r^3 = -1$, with $r \neq -1$ if the characteristic is different from 3. ∎

Finally we treat $\mathbf{Q}(5,2)$ in $\mathbf{PG}(5,q)$. As in the previous case, one can classify without the hypothesis of being translation-homogeneous and the result is that every lax embedding is translation-homogeneous with respect to every point. Since we already dealt with the case q even in Lemma 13.1.10, we could concentrate on the case q odd, but we can in fact do the general case.

Theorem 13.4.4 *Every lax embedding of* $\mathbf{Q}(5,2)$ *in* $\mathbf{PG}(5,\mathbb{K})$, *with* $\mathbb{K}$ *any field, is isomorphic to the embedding described in Example* 13.3.4.

Proof. Let $\mathcal{S} \cong \mathbf{Q}(5,2)$ and let $\mathcal{S}'$ be a subquadrangle of order 2. We treat the case where the characteristic of the underlying field $\mathbb{K}$ is different from 3 (otherwise the arguments are completely similar, but the calculations differ). Without loss of generality, we may embed $\mathcal{S}'$ as in Example 13.3.3. Also, we may choose the coordinates $(1,0,0,0,0,0)$ for the point $\mathbf{1}$ and $(1,-2,1,1,1,1)$ for the point $\mathbf{2}'$. For $i \in \{4,5,6\}$, the points $\mathbf{2}', \mathbf{i}', \{3,i\}$ and $\{3,2\}$ are coplanar (since they lie on two lines through $\mathbf{3}$) — note that we know the exact coordinates of three points of that plane π, the points $\mathbf{1}, \mathbf{i}'$ and $\{1,i\}$ are collinear, and we know the coordinates of two points of this line K. It follows that we can calculate the coordinates of the intersection $\mathbf{i}'$ of π and K. By interchanging the roles of 3 and 4 we obtain the coordinates of $\mathbf{3}'$. Now, for $i \in \{1,4,5,6\}$, the coordinates of the point $\mathbf{i}$ are calculated by considering the intersection of the lines through $\mathbf{2}', \{2,i\}$ and $\mathbf{3}', \{3,i\}$, respectively. Interchanging the roles of 2 and 4 and of 3 and 5, the coordinates of $\mathbf{2}, \mathbf{3}$ follow. Finally, we can now also easily calculate the coordinates of $\mathbf{1}'$.

The proof of the theorem is complete. ∎

13.5 Non-Planarly Embedded Translation Generalized Quadrangles

We now treat the general case of a TGQ of order (s,t) translation-homogeneously embedded in $\mathbf{PG}(d,q)$, $d \geq 3$, that is, $d \in \{3,4,5\}$ (cf. Proposition 13.1.9). By Lemma 13.1.11, we may assume $d < 5$. Also, the previous section allows us to assume that $t > 2$ for $d = 3$ and $t > 3$ for $d = 4$. In other words, we may assume $t \geq d$, with $d \in \{3,4\}$.

Lemma 13.5.1 *If the TGQ $\mathcal{S}$ of order (s,t) is translation-homogeneously embedded in $\mathbf{PG}(d,q)$, $d \in \{3,4\}$, $t \geq d$, with respect to the translation point x, then the embedding is polarized at x.*

Proof. Without loss of generality, we may assume that the embedding is faithful. Suppose by way of contradiction that the embedding is not polarized at x. Let T be the corresponding translation group (an elementary abelian p-group, for some prime p), and let Γ be the map that assigns to every semilinear collineation the companion field automorphism. Since T is not cyclic, Γ cannot be one-to-one if restricted to T, hence there is some nontrivial linear collineation θ in T.

We first claim that p divides q. Since $t \geq d$, we can choose d lines $L_1, L_2, \ldots,$ L_d through x in $\mathcal{S}$ that generate $\mathbf{PG}(d,q)$. Let L be a line of $\mathcal{S}$ through x different from L_i, for all $i \in \{1, 2, \ldots, d\}$. Let $I \subseteq \{1, 2, \ldots, d\}$ be (set-theoretically) minimal with respect to the property that L is contained in the space H generated by $\{L_i \parallel i \in I\}$. Let $y \neq x$ be a point of $\mathcal{S}$ on L. The stabilizer T_y has order st and is not cyclic. As above, this means that there exists a nontrivial linear collineation θ' in T_y. But since p divides s, θ' fixes at least three points on L, and hence fixes all points of $\mathbf{PG}(d,q)$ on L. Now, the condition on L and I implies easily that θ' fixes H and induces a central collineation in H with center x. But since all points on L are fixed, the axis contains L and hence also x. So, if θ' is not the identity in H, then it is an elation in the projective space H and the claim follows. Now assume that θ' is the identity on H. Then θ' is central in $\mathbf{PG}(d,q)$. Also, it is certainly not a symmetry about xy. Suppose that $s > p$. Then, since $\mathrm{Aut}(\mathbf{GF}(q))$ contains at most $p-1$ elements of order p, there are two different symmetries about xy with same companion field automorphism. Taking the "difference" of these, this implies the existence of a "linear" symmetry about xy. Hence $s = p$. Now Corollary 3.5.6 implies that $\mathcal{S}$ is classical and isomorphic to either $\mathbf{Q}(4,p)$, $p > 2$, or $\mathbf{Q}(5,p)$. In the first case, no collineation of $\mathcal{S}$

fixes all the points of three lines incident with x (by the antiregularity of the point x), in the second case we argue as follows. If at least two lines incident with x are not contained in H, then the center of θ' is x, since the center must lie on every such line. But in that case, the claim is proved again. Hence all lines through x but one are contained in H, and θ' fixes all points on these lines. Again, there is no such collineation in $\mathbf{Q}(5, p)$ but the identity. Hence the claim is proved.

So, in the previous paragraph, we found a linear collineation $\theta' \in T_y$ of order p, and clearly it cannot fix every point of $x^\perp$. So there is a point $z \sim x$ not fixed under θ'. Since p divides q, and since θ' fixes all lines of $\mathcal{S}$ through x, the collineation θ' is central in $\mathbf{PG}(d, q)$ with center x. Let M be a line of $\mathcal{S}$ incident with z but not with x. Then $M^{\theta'}$ is contained in the plane $\Pi = \langle x, M \rangle$ and the latter induces a subquadrangle $\mathcal{S}'$ of order (s, t'), $t' \geq 1$, entirely contained in Π. Since T fixes the lines of $\mathcal{S}$ through x in Π, and these lines span Π, every element of T maps u, with $u \not\sim x$ a point of $\mathcal{S}'$ in Π, to a point of Π. By the transitivity of T on points opposite x, we see that this implies that $\mathcal{S} = \mathcal{S}'$, a contradiction.

The lemma is proved.   ∎

We continue with a general lemma for TGQs of order s for even s.

Lemma 13.5.2 *Let $\mathcal{S}^{(x)}$ be a translation generalized quadrangle of order s, with s even. Let L_1, L_2, L_3 be three arbitrary lines incident with x, and let y be an arbitrary point opposite x. Also, let z be any point on L_1 distinct from x and distinct from $\mathrm{proj}_{L_1} y$. Then there exists a subquadrangle of order 2 containing x, y, z and L_1, L_2, L_3.*

Proof. Let $M_2 = \mathrm{proj}_y L_2$ and let $z^* = \mathrm{proj}_{M_2} z$. Let $s = 2^n$ and let $\mathcal{O}$ be a pseudo-oval in $\mathbf{PG}(3n - 1, 2)$ corresponding with $\mathcal{S}$. Let $\pi_i \in \mathcal{O}$ correspond with L_i, $i = 1, 2, 3$. Let π be the nucleus of $\mathcal{O}$ (see Theorem 3.9.1). Embed $\mathbf{PG}(3n - 1, 2)$ as a hyperplane in $\mathbf{PG}(3n, 2)$ and let y' and z' be the points of $\mathbf{PG}(3n, 2) \setminus \mathbf{PG}(3n - 1, 2)$ corresponding with y and z^*, respectively. Then $y'z'$ meets π_2 in a unique point a_2. The $2n$-space generated by a_2, π_1, π_3 meets π in a point a (this follows immediately from the definitions of a pseudo-oval and the nucleus). The line aa_2 meets the space $\langle \pi_1, \pi_3 \rangle$ in a point a', and there is a unique line $a_1 a_3$, with $a_1 \in \pi_1$ and $a_3 \in \pi_3$, containing a'. Hence we have found a unique plane $\xi := \langle a_1, a_2, a_3, a \rangle$ meeting all of π_1, π_2, π_3, π in a single point, and containing the given point a_2. Now the projective space spanned by ξ and y' defines the subquadrangle of order 2 with the required properties.   ∎

It is easy to see that the previous lemma is equivalent to the following corollary.

Corollary 13.5.3 *Let $\mathcal{S}^{(x)}$ be a translation generalized quadrangle of order s, with s even. Then every ordinary pentagon through x is contained in a subquadrangle of order 2.* ∎

We now come to the first main result of this section.

Theorem 13.5.4 *Let $\mathcal{S}^{(x)}$ be a translation generalized quadrangle of order (s,t), $t > 2$, laxly but translation-homogeneously embedded in $\mathbf{PG}(3,q)$, with respect to x. Then $\mathcal{S} \cong \mathbf{W}(s)$, with s even, and the embedding is as described in Example 13.3.2 above.*

Proof. Again, we may assume that the embedding is faithful. By Lemma 13.5.1, we know that the embedding is polarized at x. Let p be the unique prime dividing s. We claim that p divides q as well. Let $y \sim x$, $z \sim x$, $y \neq x$ and $z \not\sim y$. Let T be the translation group with respect to x. Suppose first that all linear collineations of T_y fix z. Since each member of $T_{y,z}$ fixes at least three points of $\mathcal{S}$ on xy and at least three points of $\mathcal{S}$ on xz, every such linear collineation fixes the plane $H := \langle x, y, z \rangle$ pointwise. Then, as in the proof of Lemma 13.5.1, $s = p$ and so $\mathcal{S} \cong \mathbf{Q}(4,p)$, $p > 2$, or $\mathcal{S} \cong \mathbf{Q}(5,p)$. In both cases $\mathcal{S}$ does not admit central root elations. Hence there is a linear collineation $\theta \in T_y$ not fixing z. It is easily seen that the restriction of θ to H is a nontrivial elation with center x and axis xy. The claim follows.

Suppose now, by way of contradiction, that the embedding is not polarized at y. Then applying T, we see that the embedding is not polarized at any point of xy, excluding x. An arbitrary linear symmetry about xy then has s centers on xy, implying it is necessarily the identity, a contradiction. Hence no nontrivial symmetry about xy is linear. If $s > p$, then, since $\mathbf{GF}(q)$ contains at most $p-1$ automorphisms of order p, there are two such symmetries with the same companion field automorphism, implying that there is such a linear symmetry. Hence $s = p$. If $t = p^2$, then the same argument shows that we again have linear central collineations with center x, contradicting the fact that in this case $\mathcal{S} \cong \mathbf{Q}(5,p)$. If $t = p$, and hence $\mathcal{S} \cong \mathbf{Q}(4,p)$, then as in the proof of Theorem 13.2.2, there is a linear collineation fixing all points on two lines through x and not fixing any other point collinear with x, a contradiction. Hence the embedding is polarized at every point collinear with x. For such a point u, we denote by ξ_u the plane containing $u^\perp$.

Consider again y, z as above. The planes ξ_y and ξ_z meet in a line L, which contains $\{y, z\}^\perp$. Let u be an arbitrary point of $\{y, z\}^\perp \setminus \{x\}$, and let v be arbitrary in $\{u, x\}^\perp \setminus \{y, z\}$. Since ξ_v contains u and x, it contains L and hence $v^\perp$ must necessarily contain $\{y, z\}^\perp$. Consequently $v \in \{y, z\}^{\perp\perp}$ and we conclude that $\{x, u\}$ is a regular pair of points. Varying y, z, we see that x is a regular point. Hence, by Theorems 2.1.3 and 2.1.5, $s = t$ and s is even.

If we can show that $\mathcal{S} \cong \mathbf{W}(s)$, then knowing that the embedding is polarized at every point collinear with x, we see that, using Lemma 13.1.10 and the construction of Example 13.3.2, the embedding is necessarily isomorphic to the latter. Hence we show $\mathcal{S} \cong \mathbf{W}(s)$. To achieve this, it suffices to prove that all lines are regular. Since all lines through x are already regular by the fact that $\mathcal{S}$ is a TGQ, we only have to show that two nonconcurrent lines L, M of $\mathcal{S}$ not incident with x and not meeting the same line of $\mathcal{S}$ through x form a regular pair of lines.

So let L and M be two opposite lines of $\mathcal{S}$ not incident with x and not meeting the same line of $\mathcal{S}$ incident with x. The lines L and M are concurrent with some lines L' and M' of $\mathcal{S}$, respectively, which are incident with x. Let y and z be the respective intersections (so $L\mathrm{I}y\mathrm{I}L'\mathrm{I}x\mathrm{I}M'\mathrm{I}z\mathrm{I}M$). We first claim that the point set of $\mathcal{S}$ on L (and also on M) is projectively equivalent with the point set of a projective line over the subfield $\mathbf{GF}(s)$ of $\mathbf{GF}(q)$ (and hence that the former is indeed a subfield of the latter!).

Note that the lines L and M' are not contained in a plane of $\mathbf{PG}(3, q)$, since otherwise, by the action of T, all points of $\mathcal{S}$ opposite x would be contained in that plane, and so the embedding would be planar.

So L and M' are contained in a nonsingular hyperbolic quadric (by the regularity of M') containing all the lines of $\mathcal{S}$ in $\{L, M'\}^\perp \cup \{L, M'\}^{\perp\perp}$, and we see that the point sets of L and M' are projectively equivalent. So, to prove our claim, it suffices to show that the points of $\mathcal{S}$ incident with M' form a projective line over $\mathbf{GF}(s)$. In $\mathbf{PG}(3, q)$, we may coordinatize the line M' in such a way that x is assigned the inhomogeneous coordinate (∞), and two arbitrary but fixed points $z_0 \neq x$ and $z_1 \neq z_0, z_1 \neq x$, of $\mathcal{S}$ on M' get the coordinates (0) and (1), respectively. Choose a fixed point y_0 on L', with $y_0 \neq x$.

Let $K'\mathrm{I}x$, $M' \neq K' \neq L'$, be a variable line through x, and let $u\mathrm{I}K'$, $u \neq x$, be the point on K' that belongs to $\{y_0, z_0\}^{\perp\perp}$. Let K be any line incident with u but different from K'. The hyperbolic quadrics containing $\{M', K\}^{\perp} \cup \{M', K\}^{\perp\perp}$ and $\{L', K\}^{\perp} \cup \{L', K\}^{\perp\perp}$ induce a projectivity $\rho_{K'}$ from the set of points $S(M')$ of $\mathcal{S}$ on M' to the set of points of $\mathcal{S}$ on L', fixing x and mapping z_0 onto y_0. Since x is fixed, this projectivity is a projection from some point c of $\mathbf{PG}(3,q)$ on the line $y_0 z_0$ of $\mathbf{PG}(3,q)$. Since $t > 2$, and since $\mathrm{Aut}(\mathbf{GF}(q))$ contains at most one involution, there exists a linear symmetry θ with axis K'. Obviously, θ defines in the projective plane ξ_x a central collineation with axis K'. Since θ fixes L', M', u and K, it maps z_0^{θ} to y_0^{θ}. It follows that c is the intersection of the lines $y_0 z_0$ and $y_0^{\theta} z_0^{\theta}$. Since θ is an involution of $\mathcal{S}$, and can also be viewed as an involution of $\mathbf{PG}(3,q)$ by faithfulness, the point c is a fixpoint of θ, and hence is contained in the axis, which is K'. Considering two different lines K_1' and K_2' playing the role of K', we see that $\rho := \rho_{K_1'} \rho_{K_2'}^{-1}$ is a nontrivial projectivity preserving $S(M')$, and fixing x and z_0 (the projectivity is indeed nontrivial since the centers c_1 and c_2, respectively — playing the role of c above — are distinct!). In coordinates, if ρ maps (1) to (a), then it maps an arbitrary point (b) of M' to (ab). Since there are $s - 2$ choices for $K_2' \neq K_1'$, given K_1', and since no two choices define the same projectivity $(\rho_{K_1'} \rho_{K_2'}^{-1} = \rho_{K_1'} \rho_{K_3'}^{-1}, K_3' \mathrm{I} x, M' \neq K_3' \neq L'$, implies that $\rho_{K_2'} \rho_{K_3'}^{-1}$ is trivial, hence $K_2' = K_3'$), we obtain, adding the identity, a set of projectivities fixing (∞) and (0) and acting sharply transitively on $S(M') \setminus \{x, z_0\}$. Since the projectivity group of M' acts sharply 3-transitively, we deduce that this set of projectivities is actually a group. Varying z_0 on M' in $\mathcal{S}$, we obtain a group of projectivities acting sharply 2-transitively on $S(M') \setminus \{x\}$. The Frobenius kernel of this action is necessarily a subgroup of the additive group of $\mathbf{GF}(q)$ (identified via the inhomogeneous coordinates on M'). So we see that the inhomogeneous coordinates of the points of $S(M') \setminus \{x\}$ form an additive subgroup of $\mathbf{GF}(q)$, and removing (0), they form a multiplicative subgroup of $\mathbf{GF}(q) \setminus \{0\}$. We conclude that they form a subfield $\mathbf{GF}(s)$ of $\mathbf{GF}(q)$. Our claim is proved.

Now let $L_0 \sim M_0 \sim N_0$ be three arbitrary lines of $\mathcal{S}$ with $L_0 \not\sim N_0$ and such that x is not incident with either. We also assume that L_0 and N_0 do not meet the same line incident with x. We will show that, considering $x, \mathrm{proj}_x L_0$ and $\mathrm{proj}_x N_0$, Corollary 13.5.3 implies that $\langle L_0, M_0, N_0 \rangle$ is $\mathbf{PG}(3,q)$. For, otherwise there is a subGQ $\mathcal{S}^*$ of order 2 containing x, L_0, M_0, N_0 and embedded in a plane, say π. Considering a point u_0 of $\mathcal{S}^*$ opposite x, we see that $x^{\perp}$ and u_0 are contained in π, and so $\mathcal{S}$ is contained in π, a contradiction. Hence the points of $\mathcal{S}$ on the lines L_0, M_0, N_0 define a unique projective subspace $\mathbf{PG}(3,s)$ of $\mathbf{PG}(3,q)$. We claim that, for every point $w_0 I L_0$, the points of $\mathcal{S}$ on the line $\text{proj}_{w_0} N_0$ belong to $\mathbf{PG}(3, s)$. Indeed, it is clear that $|\{\text{proj}_x L_0, \text{proj}_x M_0, \text{proj}_x N_0\}| \in \{2, 3\}$, and either $w_0 \sim x$ or $w_0 \not\sim x$. Hence there are four cases to consider; we will consider the generic case and leave the others to the reader (in fact they will follow immediately from the arguments of the generic case, permuting the notation). Hence we assume that $L_0' := \text{proj}_x L_0$, $M_0' := \text{proj}_x M_0$ and $N_0' := \text{proj}_x N_0$ are all distinct, and that $x \not\sim w_0$. Then the quadrangle $\mathcal{S}^*$ above is uniquely defined. Put $y_0 := \text{proj}_{L_0} x$, $z_0 := \text{proj}_{M_0} x$ and $u_0 := \text{proj}_{N_0} x$. Inside $\mathcal{S}^*$, there is a grid containing L_0, M_0, N_0; two other lines of that grid are $\text{proj}_{N_0} y_0$ and $\text{proj}_{L_0} u_0$, and, viewed as lines of $\mathbf{PG}(3, q)$, they also belong to $\mathbf{PG}(3, s)$. Now the point z_0 belongs to $\mathbf{PG}(3, s)$ and hence so does the unique transversal P_0 through z_0 meeting both $\text{proj}_{N_0} y_0$ and $\text{proj}_{L_0} u_0$, and its intersection points with the latter two. Since a projective subline over $\mathbf{GF}(s)$ is determined by three points, we see that all the points of the grid in $\mathcal{S}$ belong to $\mathbf{PG}(3, s)$. Now, considering the subGQ of order 2 containing $L_0, \text{proj}_{w_0} N_0, N_0$ and x and repeating this argument, the claim follows.

The same claim can be proved for the cases where L_0 and N_0 do meet the same line incident with x, or when one of the lines L_0, M_0, N_0 is incident with x, and the point w_0 is on a line through x or is x, or not. Indeed, this is easy since every line through x is regular, and every grid of $\mathcal{S}$ through such a line is easily shown to be contained in a subspace over $\mathbf{GF}(s)$, noting that each such grid spans $\mathbf{PG}(3, q)$.

Now we finally prove that the pair $\{L, M\}$, with L and M two opposite lines of $\mathcal{S}$ not incident with x and not meeting the same line through x, is regular. Let N be any line of $\mathcal{S}$ meeting both L and M. Then there is a unique projective subspace, which we may again denote by $\mathbf{PG}(3, s)$, over $\mathbf{GF}(s)$, containing the points of $\mathcal{S}$ on the lines L, M, N. Now we build the subquadrangle $\mathcal{S}_0$ generated by the points of $\mathcal{S}$ on L, M, N. This subquadrangle can be constructed as follows. Each step consists of looking at already constructed points and lines at distance 5 in the incidence graph of the thus far constructed subgeometry, and adding the unique line of $\mathcal{S}$ passing through the point and concurrent with the line (and not belonging to the so far constructed subgeometry). It is easy to see that this construction indeed yields a subGQ. But by our above claims, $\mathcal{S}_0$ is entirely contained, and hence fully embedded, in $\mathbf{PG}(3, s)$. Now there are two possibilities (since $\mathcal{S}$ has order s).

(1) For some choice of L and M, the order of $\mathcal{S}_0$ is s, and then $\mathcal{S} = \mathcal{S}_0$

is fully embedded in $\mathbf{PG}(3, s)$, which implies that the embedding is polarized by Lemma 13.1.5. So every point is regular by Proposition 13.1.6, $\mathcal{S} \cong \mathbf{W}(s)$, and we are done.

(2) For all pairs $\{L, M\}$, the subGQ $\mathcal{S}_0$ has order $(s, 1)$, which means that all lines are regular, and we are done again.

∎

Our second main result treats the 4-dimensional case.

Theorem 13.5.5 *Let the TGQ $\mathcal{S}^{(x)} = \mathcal{S}$ of order (s, t) be translation-homogeneously embedded in $\mathbf{PG}(4, q)$ with respect to x, and suppose $t > 3$. Then $\mathcal{S} \cong \mathbf{Q}(4, s)$ and the embedding arises from the natural embedding in $\mathbf{PG}(4, s)$ by field extension.*

Proof. We may assume that the embedding is faithful. By Lemma 13.5.1, we know that the embedding is polarized at x. We first claim that no three lines of $\mathcal{S}$ through x are coplanar. Indeed, suppose by way of contradiction that the three lines L_1, L_2, L_3 through x are coplanar, say, they belong to the plane π. Choose any point y opposite x, not contained in π. Then $H := \langle \pi, y \rangle$ intersects $\mathcal{S}$ in a translation subquadrangle $\mathcal{S}' = (\mathcal{P}', \mathcal{B}', \mathbf{I})$, translation-homogeneously embedded in the 3-dimensional projective space H, because every translation of $\mathcal{S}^{(x)}$ mapping y to a point of $\mathcal{S}$ in H, and viewed as a collineation of $\mathbf{PG}(4, q)$, clearly preserves H. Hence $\mathcal{S}' \cong \mathbf{W}(s)$ by Theorem 13.5.4 and $t = s^2$ by Lemma 1.1.1. We also know from Theorem 13.5.4 that the embedding of $\mathcal{S}'$ in H is polarized at each point $z \sim x$. Consider such a point $z \neq x$. Let π_z be the plane containing $z^{\perp} \cap \mathcal{P}'$, and choose any line $L\mathbf{I}x$ in $\mathcal{S}$, with $z \not\mathbf{I} L$. Then $H' := \langle \pi_z, L \rangle$ intersects $\mathcal{S}$ again in a translation-homogeneously embedded subquadrangle (with respect to x) $\mathcal{S}''$ of order s, which is, by Lemma 13.5.1, polarized at x. Hence the plane $\langle xz, L \rangle$ contains exactly $s + 1$ lines of $\mathcal{S}$. Continuing like this, we see that, for every pair of lines L, M through x, the plane $\langle L, M \rangle$ intersects $\mathcal{S}$ in $s + 1$ lines through x. In this way we obtain a linear space with $s^2 + 1$ points (the lines through x) and every block containing $s + 1$ points (consequently every point is contained in s blocks). Counting the number of incident point-block pairs in two ways, and putting b equal to the total number of blocks, we arrive at the contradiction $(s + 1)b = s(s^2 + 1)$. Our first claim is proved.

Our first claim implies that every linear element θ of the translation group T fixes all lines through x of $\mathbf{PG}(4, q)$ which belong to the 3-space ξ_x generated by the lines of $\mathcal{S}$ through x.

Let p be the unique prime dividing s. We claim that p divides q as well. Let $y \sim x$, $z \sim x$, $y \neq x$ and $z \not\sim y$. Let T be the translation group with respect to x. Suppose first that all linear collineations of T_y fix z. Hence every such linear collineation fixes the plane $\pi' = \langle x, y, z \rangle$ pointwise. Then, as before in the proof of Lemma 13.5.1, $s = p$ and so $\mathcal{S} \cong \mathbf{Q}(4, p)$, $p > 2$, or $\mathcal{S} \cong \mathbf{Q}(5, p)$. In both cases $\mathcal{S}$ does not admit central root elations. Hence there is a linear collineation $\theta \in T_{y,z}$ not acting trivially on ξ_x. Then it induces an elation of ξ_x with center x and axis π', and the claim follows. If, on the other hand, T_y contains a linear collineation that does not fix z, then this collineation is again central in ξ_x with center x, and the axis contains xy, hence the claim again follows.

We now claim that $t < s^2$. For, suppose by way of contradiction that $t = s^2$, and let θ be a nontrivial linear collineation in $T_{y,z}$, with y, z two opposite points collinear with x (note that θ exists since $|T_{y,z}| = t > p$). Let u be a point opposite x and collinear with both y and z. Then $\{x, u, u^\theta\}^\perp$ contains $s + 1$ elements (see Section 2.5) and θ fixes all points on every line xx', with $x' \in \{x, u, u^\theta\}^\perp$. This now easily implies that θ fixes ξ_x pointwise, a contradiction, since this would mean that $|\{x, u, u^\theta\}^\perp| = s^2 + 1$. Our claim is proved.

Next we claim that $t \in \{s, sp\}$. Suppose by way of contradiction that $sp < t < s^2$. By Theorem 3.10.1(ii), $p > 2$. Since both s and t are powers of the same prime p, we have $t \geq p^2 s$. Let y and z be as in the previous paragraph, and let M be an arbitrary line of $\mathcal{S}$ with $x \mathrm{I} M$ and $y \not{\mathrm{I}} M \not{\mathrm{I}} z$. Let $\Gamma : T_{y,z} \to \mathrm{Aut}(\mathbf{GF}(q))$ be the group morphism that assigns to every element of $T_{y,z}$, viewed as a collineation of $\mathbf{PG}(4, q)$, its companion field automorphism. Since $\mathrm{Aut}(\mathbf{GF}(q))$ contains at most $p - 1$ elements of order p, the kernel S of Γ has order at least $t/p \geq ps$. Hence S cannot act fixpoint freely on the set of points of $\mathcal{S}$ on M, without x. Hence there is at least one nontrivial linear element $\theta \in T_{y,z}$ fixing some point of $\mathcal{S}$ on M distinct from x. Since the lines xy, xz and M span ξ_x, and since the order p of θ divides q, this implies that θ fixes all points of ξ_x, and so θ has some center c in $\mathbf{PG}(4, q)$. So θ is a symmetry with center x, while by the dual of Theorem 3.3.2 a TGQ of order (s, t) can only have symmetries about a point if $st(t + 1) \equiv 0 \mod s + t$, and this, together with Theorem 3.8.1(ii), implies $s = t$.

Next we claim that all lines of $\mathcal{S}$ are regular. Indeed, consider two arbitrary opposite lines L, M of $\mathcal{S}$ and consider the subGQ induced by $\langle L, M \rangle$, which has order (s, t''), for some positive integer t''. Assume first that $t'' > 1$. Then $t''^2 \geq s$ and $t \geq st''$, implying $t^2 \geq s^3$. Clearly $s \neq t$, so $t = ps$ and, as

$t \neq s^2$, we have $p^2 = s$ and $p^3 = t$, contradicting Theorem 3.8.1(ii). Hence $t'' = 1$ and the pair $\{L, M\}$ is regular. Our claim is proved.

Corollary 3.8.3 now implies that $\mathcal{S} \cong \mathbf{Q}(4, s)$, since we already proved $t \neq s^2$. Lemma 13.1.10 concludes the proof of the theorem. ∎

Finally, in finite 5-dimensional space, all embeddings are translation-homogeneous, as follows from combining Lemmas 13.1.10 and 13.1.11.

Theorem 13.5.6 *If the generalized quadrangle $\mathcal{S}$ of order (s, t) is laxly embedded in $\mathbf{PG}(5, q)$, then $\mathcal{S} \cong \mathbf{Q}(5, s)$. If $\gcd(s, q) = 1$, then $s = 2$ and the embedding is as described in Example 13.3.4. If $\gcd(s, q) \neq 1$, then the embedding arises from the standard embedding in $\mathbf{PG}(5, s)$ by including the latter in $\mathbf{PG}(5, q)$.* ∎

We now have the following corollary.

Corollary 13.5.7 *If the translation Moufang generalized quadrangle $\mathcal{S}$ of order (s, t) is embedded in $\mathbf{PG}(d, q)$, $d \geq 2$, and all symmetries are induced by collineations of $\mathbf{PG}(d, q)$, then either the embedding is full in a subspace over $\mathbf{GF}(s)$ (in which case this full embedding is the appropriate natural one), or one of the following holds.*

(i) *$d = 2$, $\mathcal{S} \cong \mathbf{W}(2)$ and the embedding is the hyperoval embedding in some subplane $\mathbf{PG}(2, 4)$ (and $q = 4$ corresponds with the only faithful case).*

(ii) *$d = 4$, $\mathcal{S} \cong \mathbf{W}(2)$ and the embedding is the universal embedding in $\mathbf{PG}(4, q)$, as described in Example 13.3.3. It is automatically special.*

(iii) *$d = 4$, $\mathcal{S} \cong \mathbf{Q}(4, 3)$ and the embedding is the universal embedding in $\mathbf{PG}(4, q)$, as described in Example 13.3.3. It is automatically special.*

(iv) *$d = 5$, $\mathcal{S} \cong \mathbf{Q}(5, 2)$ and the embedding is the universal one in $\mathbf{PG}(5, q)$, with q odd, as described in Example 13.3.4. It is automatically special.* ∎

Appendix A

Open Problems

The problems mentioned here are mainly concerned with translation generalized quadrangles, but several other interesting open problems will also be stated.

A.1   Chapter 1

Problem A.1.1 *Are* $\mathbf{Q}(4,q)$ *and* $\mathbf{W}(q)$ *the only GQs of order* q, q *odd and* $q \geq 5$?

Problem A.1.2 *Is there just one GQ of order* $(q-1, q+1)$, *for any odd* q, $q \geq 5$?

Problem A.1.3 *Is* $\mathbf{H}(4, q^2)$ *the only GQ of order* (q^2, q^3), $q \geq 2$?

Problem A.1.4 *Is there a unique GQ of order* $(4, t)$, $t \in \{6, 8, 16\}$?

Problem A.1.5 *Does there exist a GQ of order* $(4, t)$, $t \in \{11, 12\}$?

Problem A.1.6 *Does there exist a GQ of order* 6?

Problem A.1.7 *Classify all ovoids of* $\mathbf{Q}(4, q)$.

All known ovoids of $\mathbf{Q}(4, q)$ are listed in Section 7.4.

Problem A.1.8 *Classify all spreads of* $\mathbf{Q}(5, q)$.

Many spreads of $\mathbf{Q}(5,q)$ are known, see e.g. Payne and Thas [128], Thas [170], and Thas and Payne [190].

Problem A.1.9 *Does* $\mathbf{H}(4,q^2)$, $q > 2$, *have a spread?*

By Brouwer [20], $\mathbf{H}(4,4)$ has no spread.

A.2   Chapter 2

Problem A.2.1 *Does there exist a GQ of order* (s,t), *with* $s < t < s^2$, *for which all lines are regular?*

Problem A.2.2 *Is every GQ of order* (s,t), $1 < s < t$, *with all lines regular, isomorphic to* $\mathbf{Q}(5,s)$?

Thas and Van Maldeghem [195] proved that all lines of a GQ $\mathcal{S}$ of order $(s, s+2)$, $s \geq 2$, are regular if and only if $s = 2$ and $\mathcal{S} \cong \mathbf{Q}(5,2)$.

Problem A.2.3 *Does there exist a GQ of order* s, $s \neq 1$, *with regular point* x, *for which the corresponding projective plane* π_x *is not Desarguesian?*

Problem A.2.4 *Is every GQ of order* s, s *odd and* $s > 1$, *for which each point is antiregular, isomorphic to* $\mathbf{Q}(4,s)$?

Problem A.2.5 *Give a purely geometrical proof of the theorem of Bagchi, Brouwer and Wilbrink [6] on antiregular points in a GQ of order* s, $s > 1$ *and odd; see Theorem 2.4.6.*

A.3   Chapter 3

Problem A.3.1 *Does there exist a TGQ of order* (q^a, q^{a+1}), q *odd and* $a > 1$?

Problem A.3.2 *Is every TGQ of order* q, $q \neq 1$, *isomorphic to a* $\mathbf{T}_2(\mathcal{O})$ *of Tits, that is, is any generalized oval* $\mathcal{O}(n,n,q)$ *regular?*

Problem A.3.3 *Is* $\mathcal{O}(n,n,q)$, q *odd, always isomorphic to its translation dual* $\mathcal{O}^*(n,n,q)$?

Problem A.3.4 *What can be said about $\mathcal{O}(n, n, q)$ if q is even and the $(n-1)$-spread $\mathcal{T}$, respectively $\mathcal{T}_i$, defined in Section 3.9, is regular?*

In the odd case it follows that $\mathcal{O}(n, n, q)$ is classical, see Theorem 3.9.6.

Problem A.3.5 *Is every TGQ of order (q, q^2), $q \neq 1$, with all lines regular isomorphic to $\mathbf{Q}(5, q)$?*

Problem A.3.6 *Does there exist a TGQ of order (q, q^2), with q even, not isomorphic to a $\mathbf{T}_3(\mathcal{O})$ of Tits, that is, is any generalized ovoid $\mathcal{O}(n, 2n, q)$ with q even, regular?*

Problem A.3.7 *Is $\mathcal{O}(n, 2n, q)$, with q even, always isomorphic to its translation dual?*

Problem A.3.8 *A weak generalized ovoid is a set of $q^{2n} + 1$ $(n-1)$-dimensional subspaces of $\mathbf{PG}(4n-1, q)$, every three of which generate a $(3n-1)$-dimensional space. Is every weak generalized ovoid a generalized ovoid?*

Problem A.3.9 *Classify all thick TGQs of order (q^2, q^4), where $\mathbf{GF}(q)$ is the kernel of the TGQ.*

In [97], the authors classify all TGQs of order $(4, 16)$ (they are all isomorphic to $\mathbf{Q}(5, 4)$).

Problem A.3.10 *Classify all EGQs with an axis of symmetry incident with the elation point.*

Problem A.3.11 *Let $(\mathcal{S}^{(\infty)}, G)$ be a thick EGQ. Is G always a p-group for some prime?*

By work of Frohardt [62], the next problem is equivalent.

Problem A.3.12 *Let $(\mathcal{S}^{(\infty)}, G)$ be a thick EGQ. Is G always nilpotent?*

Problem A.3.13 *Classify EGQs $(\mathcal{S}^{(\infty)}, G)$ of order (s, p), with p a prime.*

EGQs of order (p, t) with p a prime were classified by Bloemen, Thas and Van Maldeghem; see [13].

A.4   Chapter 4

Problem A.4.1 *Does there exist a semifield flock of the quadratic cone $\mathcal{K}$ of* $\mathbf{PG}(3, q)$, *q odd and $q \neq 3^h$, which is not a Kantor-Knuth flock?*

Problem A.4.2 *Does there exist a GQ of order (q^2, q), $q \neq 1$, which is neither the point-line dual of a TGQ nor of a flock GQ?*

Problem A.4.3 *Classify all herds of hyperovals of* $\mathbf{PG}(2, q)$, *q even, arising from flocks of the quadratic cone in* $\mathbf{PG}(3, q)$.

Interesting results on this topic can be found in O' Keefe and Penttila [91].

Problem A.4.4 *Is any GQ of order (q, q^2), q even, which satisfies Property (G) at a pair $\{x, y\}$, with x and y distinct collinear points, isomorphic either to a $\mathbf{T}_3(\mathcal{O})$ of Tits or to the point-line dual of a flock GQ?*

In the odd case the GQ is always the point-line dual of a flock GQ, see Theorem 4.10.7.

A.5   Chapter 5

Problem A.5.1 *Is one of $\mathcal{O}(n, 2n, q)$ and $\mathcal{O}^*(n, 2n, q)$ always good?*

Problem A.5.2 *Is $\mathcal{O}(n, 2n, q)$, with q even, always good?*

Problem A.5.3 *Is $\mathcal{O}(n, 2n, q)$, with $\mathcal{O}(n, 2n, q)$ good and q even, either the point-line dual of a flock GQ and hence classical by Theorem 4.5.1, or isomorphic to a $\mathbf{T}_3(\mathcal{O}')$ of Tits?*

Problem A.5.4 *If $\mathcal{O}(n, 2n, q)$, q odd, is not a Kantor-Knuth egg, is necessarily $q = 3^h$?*

Problem A.5.5 *If a good $\mathcal{O}(n, 2n, q)$, q odd, is not a Kantor-Knuth egg, is necessarily $q = 3^h$?*

This is equivalent to Problem A.4.1, by Corollary 5.1.4 and Section 4.7.

Problem A.5.6 *Is an egg $\mathcal{O}(n, 2n, q)$, q even, with at least two good elements regular?*

A.6   Chapter 6

Problem A.6.1 *Let x be a regular point of the GQ $\mathcal{S}$ of order (s,t), $s \geq t >$ 1, and let $\mathcal{N}_x^D$ be the corresponding dual net. Under which conditions do particular automorphisms of $\mathcal{N}_x^D$ extend to automorphisms of $\mathcal{S}$?*

A.7   Chapter 7

Problem A.7.1 *If $\mathcal{O}$ is an ovoid of $\mathbf{Q}(4,q)$, q odd, which is not a Kantor-Knuth ovoid, is necessarily $q = 3^h$?*

Problem A.7.2 *Let $\mathcal{S}$ be a GQ of order (s^2, s^3), having a subGQ $\mathcal{S}'$ isomorphic to $\mathbf{H}(3, s^2)$. Assume that in $\mathcal{S}'$ each ovoid $\mathcal{O}_x$ subtended by a point x of $\mathcal{S} \setminus \mathcal{S}'$ is a Hermitian curve. Prove that $\mathcal{S} \cong \mathbf{H}(4, s^2)$.*

Note that Theorem 7.5.4 is the corresponding theorem for quadrics in even characteristic; see Remark 7.5.5 when the characteristic is odd.

Problem A.7.3 *Let $\mathcal{S}$ be a TGQ of order (s, s^2), s even, with $\mathcal{O} = \mathcal{O}(n, 2n, q)$, and assume that $\mathcal{S}'$ is a subGQ of order s containing the point (∞). Is $\mathcal{S}$ necessarily isomorphic to a $\mathbf{T}_3(\mathcal{O}')$ of Tits?*

Problem A.7.4 *Prove Proposition 7.9.2 without using Theorem 3.3.1.*

Problem A.7.5 *Let $\mathcal{S} = \mathbf{T}(\mathcal{O})$, with $\mathcal{O} = \mathcal{O}(n, 2n, q)$ and q even, be a TGQ such that $\mathcal{O}$ is good at its element π and assume that $\mathcal{S}'$ is a subGQ of order $s = q^n$ of $\mathcal{S}$ which contains the point (∞). By Theorem 7.7.10, $\mathcal{S}' = \mathbf{T}(\mathcal{O}')$, with $\mathcal{O}' = \mathcal{O}'(n, n, q)$ a pseudo-oval on $\mathcal{O}$. Does $\mathcal{O}'$ necessarily contain π?*

Problem A.7.6 *Work under the same hypotheses as in the previous question. Is $\mathcal{O}$ necessarily regular, that is, is $\mathcal{S} \cong \mathbf{T}_3(\overline{\mathcal{O}})$ with $\overline{\mathcal{O}}$ some ovoid of $\mathbf{PG}(3, s)$?*

Problem A.7.7 *Classify all GQs of order (s, s^2) with a thick subGQ $\mathcal{S}'$ of order s fixed pointwise by an involution.*

If s is even, then $\mathcal{S}' \cong \mathbf{Q}(4, s)$, and if s is odd, then each point of the subGQ is antiregular, see [182].

A.8   Chapter 8

Problem A.8.1 *Is any translation generalized oval regular?*

Problem A.8.2 *Prove, without using the classification of finite simple groups, that Theorem 8.5.1 also holds for q odd.*

Problem A.8.3 *Generalize Theorem 8.5.1 by means of other group actions.*

Problem A.8.4 *Classify generalized ovals admitting a nontrivial involution.*

Problem A.8.5 *Classify generalized ovoids admitting a nontrivial involution.*

Problem A.8.6 *Suppose $\mathcal{O}$ is a generalized oval in $\mathbf{PG}(3n-1,q)$, let $\pi \in \mathcal{O}$, and let τ be the tangent space of $\mathcal{O}$ at π. Suppose Θ is the set of involutions of $\mathbf{PGL}_{3n}(q)$ that fix τ pointwise and stabilize $\mathcal{O}$. How big should $|\Theta|$ be at least to conclude that $\mathcal{O}$ is a translation generalized oval?*

Problem A.8.7 *Same question for ovals.*

A.9   Chapter 9

Problem A.9.1 *Are the root groups of any proper Moufang set necessarily nilpotent?*

Problem A.9.2 *Is there an absolute constant $N \in \mathbb{N}$ such that, if the root groups of a proper Moufang set are nilpotent, then the nilpotency class is at most N?*

Problem A.9.3 *Can one classify all proper translation Moufang sets?*

Problem A.9.4 *Are the root groups uniquely and unambiguously determined by the permutation group defined by the (proper) Moufang set?*

A.10   Chapter 10

Problem A.10.1 *Classify all span-symmetric generalized quadrangles of order (s,s^2), $s \neq 1$.*

Problem A.10.2 *Classify all span-symmetric generalized quadrangles of order* (s, s^2), $s \neq 1$, *with a regular line not contained in the base-grid.*

One may want to consult Chapter 7 of the monograph [206] for the known results on both problems.

Problem A.10.3 *Classify EGQs with a line of elation points.*

When the elation group is abelian (that is, the EGQ is a TGQ), this problem has been solved by K. Thas — see Chapter 10.

A.11 Chapter 11

Problem A.11.1 *Prove, without the classification of finite simple groups, that a finite GQ admitting a group acting transitively on pairs of opposite points and on pairs of opposite lines, is classical or dual classical.*

A major step towards the solution of this problem is contained in the papers [209, 210, 211], where the author classifies the finite Tits GQs.

Problem A.11.2 *Prove, without finiteness condition, that half 2-Moufang GQs are necessarily Moufang (or provide a counter example).*

Problem A.11.3 *Is every GQ all points and lines of which are regular, necessarily a Moufang GQ?*

A.12 Chapter 12

Problem A.12.1 *If a not necessarily finite EGQ is self-polar (or self-dual), then are the elation groups unique?*

For the finite case, the answer is given in Proposition 12.2.1.

Problem A.12.2 *Is every polarity in an arbitrary or finite GQ uniquely determined by the set of absolute points?*

Problem A.12.3 *Do there exist a GQ* $\mathcal{S}$*, a proper full subGQ* $\mathcal{S}'$*, and two points of* $\mathcal{S}$ *not in* $\mathcal{S}'$ *subtending nonisomorphic ovoids of* $\mathcal{S}'$*?*

A.13   Chapter 13

Problem A.13.1 *Can one classify all linear translation-homogeneously embedded TGQs $\mathcal{S}^{(x)}$ in Pappian projective spaces?*

Problem A.13.2 *Can one classify all lax embeddings of $\mathbf{W}(2)$ in Pappian or Desarguesian projective planes?*

Problem A.13.3 *Can one classify translation-homogeneously embedded TGQs $\mathcal{S}^{(L)}$, with L a line, in finite projective spaces?*

Problem A.13.4 *Can one classify all (linear) "elation-homogeneously" embedded EGQs in finite projective spaces?*

Problem A.13.5 *Can one classify all (lax) embeddings of the symplectic quadrangle $\mathbf{W}(s)$ in the 3-dimensional projective space $\mathbf{PG}(3,q)$, $q > s$?*

Problem A.13.6 *Can one classify all lax embeddings of GQs in finite projective planes?*

Bibliography

[1] L. BADER. Some new examples of flocks of $Q^+(3, q)$, *Geom. Dedicata* **27** (1988), 213–218.

[2] L. BADER & G. LUNARDON. On the flocks of $Q^+(3, q)$, *Geom. Dedicata* **29** (1989), 177–183.

[3] L. BADER, G. LUNARDON & S. E. PAYNE. On q-clan geometry, $q = 2^e$, *Bull. Belgian Math. Soc. — Simon Stevin* **1** (1994), 301-328.

[4] L. BADER, G. LUNARDON & I. PINNERI. A new semifield flock, *J. Combin. Theory Ser. A* **86** (1999), 49–62.

[5] L. BADER, G. LUNARDON & J. A. THAS. Derivation of flocks of quadratic cones, *Forum Math.* **2** (1990), 163–174.

[6] B. BAGCHI, A. E. BROUWER & H. A. WILBRINK. Notes on binary codes related to the $O(5, q)$ generalized quadrangle for odd q, *Geom. Dedicata* **39** (1991), 339–355.

[7] R. D. BAKER & G. L. EBERT. A nonlinear flock in the Minkowski plane of order 11, *Eighteenth Southeastern International Conference on Combinatorics, Graph Theory, and Computing (Boca Raton, Fla., 1987), Congr. Numer.* **58** (1987), 75–81.

[8] A. BARLOTTI. Un 'estensione del teorema di Segre-Kustaanheimo, *Boll. Un. Mat. Ital. (3)* **10** (1955), 498–506.

[9] S. G. BARWICK, M. R. BROWN & T. PENTTILA. Flock generalized quadrangles and tetradic sets of elliptic quadrics of PG(3,q), *J. Combin. Theory Ser. A* **113** (2006), 273-290.

[10] C. T. BENSON. On the structure of generalized quadrangles, *J. Algebra* **15** (1970), 443–454.

[11] T. BETH, D. JUNGNICKEL, & H. LENZ. *Design Theory*, Bibliograph. Inst. Mannheim, and Cambridge Univ. Press, Cambridge, 1985.

[12] M. BILIOTTI & G. LUNARDON. Derivation sets and Baer subplanes in a translation plane, *Atti Accad. Naz. Lincei, VIII. Ser., Rend., Cl. Sci. Fis. Mat. Nat.* **69** (1980), 135-141.

[13] I. BLOEMEN, J. A. THAS & H. VAN MALDEGHEM. Elation generalized quadrangles of order (p, t), p prime, are classical, *J. Stat. Plann. Inference* **56** (1996), 49-55.

[14] I. BLOEMEN, J. A. THAS & H. VAN MALDEGHEM. Translation ovoids of generalized quadrangles and hexagons, *Geom. Dedicata* **72** (1998), 19-62.

[15] A. BLOKHUIS. On subsets of $GF(q^2)$ with square differences, *Ned. Akad. Wetensch. Proc. Ser. A* **87** (1984), 369–372.

[16]   A. BLOKHUIS, M. LAVRAUW & S. BALL. On the classification of semifield flocks, *Adv. Math.* **180** (2004), 104–111.

[17]   R. C. BOSE. Mathematical theory of the symmetric factorial design, *Sankhyā* **8** (1947), 107–166.

[18]   R. C. BOSE & S. S. SHRIKHANDE. Geometric and pseudo-geometric graphs $(q^2 + 1, q + 1, 1)$, *J. Geom.* **2** (1972), 75–94.

[19]   L. BROUNS, J. A. THAS & H. VAN MALDEGHEM. A characterization of $Q(5, q)$ using one subquadrangle $Q(4, q)$, *European J. Combin.* **23** (2002), 163–177.

[20]   A. E. BROUWER. $\mathbf{H}(4, 4)$ has no spreads, *Private Communication*, 1981.

[21]   A. E. BROUWER. The complement of a geometric hyperplane in a generalized polygon is usually connected, in *Finite Geometry and Combinatorics*, Proceedings Deinze 1992 (ed. F. De Clerck *et al.*), Cambridge University Press, *London Math. Soc. Lecture Note Ser.* **191** (1993), 53 – 57.

[22]   M. R. BROWN. *Generalised Quadrangles and Associated Structures*, Ph.D. Thesis, University of Adelaide, Adelaide, 1999, vii+141 pp.

[23]   M. R. BROWN. Ovoids of $\mathrm{PG}(3, q)$, q even, with a conic section, *J. London Math. Soc.* **62** (2000), 569–582.

[24]   M. R. BROWN. A characterisation of the generalized quadrangle $Q(5, q)$ using cohomology, *J. Algebraic Combin.* **15** (2002), 107–125.

[25]   M. R. BROWN. Projective ovoids and generalized quadrangles, Preprint.

[26]   M. R. BROWN & M. LAVRAUW. Eggs in $PG(4n - 1, q)$, q even, containing a pseudo-conic, *Bull. London Math. Soc.* **36** (2004), 633–639.

[27]   M. R. BROWN & M. LAVRAUW. Eggs in $PG(4n - 1, q)$, q even, containing a pseudo-pointed conic, *European J. Combin.* **26** (2005), 117–128.

[28]   M. R. BROWN & J. A. THAS. Subquadrangles of order s of generalized quadrangles of order (s, s^2). I, *J. Combin. Theory Ser. A* **106** (2004), 15–32.

[29]   M. R. BROWN & J. A. THAS. Subquadrangles of order s of generalized quadrangles of order (s, s^2). II, *J. Combin. Theory Ser. A* **106** (2004), 33-48.

[30]   F. BUEKENHOUT. Une caractérisation des espaces affins basée sur la notion de droite, *Math. Z.* **111** (1969), 367–371.

[31]   F. BUEKENHOUT. *Handbook of Incidence Geometry*, Edited by F. Buekenhout, North-Holland, Amsterdam, 1995.

[32]   F. BUEKENHOUT & C. LEFÈVRE. Generalized quadrangles in projective spaces, *Arch. Math.* **25** (1974), 540–552.

[33]   F. BUEKENHOUT & H. VAN MALDEGHEM. Finite distance-transitive generalized polygons, *Geom. Dedicata* **52** (1994), 41 – 51.

[34]   R. CALDERBANK & W. M. KANTOR. The geometry of two-weight codes, *Bull. London Math. Soc.* **18** (1986), 97–122.

[35]   I. CARDINALI & S. E. PAYNE. *The q-Clan Geometries with $q = 2^e$*, Monograph, Manuscript, 2003.

[36]   L. CARLITZ. A theorem on permutations in a finite field, *Proc. Amer. Math. Soc.* **11** (1960), 456–459.

[37] L. R. A. CASSE, J. A. THAS & P. R. WILD. $(q^n + 1)$-sets of $PG(3n - 1, q)$, generalized quadrangles and Laguerre planes, *Simon Stevin* **59** (1985), 21–42.

[38] X. M. CHEN & D. FROHARDT. Normality in a Kantor family, *J. Combin. Theory Ser. A* **64** (1993), 130–136.

[39] Y. CHEN & G. KAERLEIN. Eine Bemerkung über endliche Laguerre- und Minkowski-Ebenen, *Geom. Dedicata* **2** (1973), 193–194.

[40] W. CHEROWITZO, T. PENTTILA, I. PINNERI & G. F. ROYLE. Flocks and ovals, *Geom. Dedicata* **60** (1996), 17–37.

[41] S. D. COHEN & M. J. GANLEY. Commutative semifields, two-dimensional over their middle nuclei, *J. Algebra* **75** (1982), 373–385.

[42] T. CZERWINSKI. Finite translation planes with collineation groups doubly transitive on the points at infinity, *J. Algebra* **22** (1972), 428–441.

[43] F. DE CLERCK & N. L. JOHNSON. Subplane covered nets and semipartial geometries, *Discrete Math.* **106/107** (1992), 127–134.

[44] P. DEMBOWSKI. Möbiusebenen gerader Ordnung, *Math. Ann.* **157** (1964), 179–205.

[45] P. DEMBOWSKI. *Finite Geometries*, Berlin/Heidelberg/New York, Springer, 1968.

[46] T. DE MEDTS. *Moufang Quadrangles: A Unifying Algebraic Structure, and Some Results on Exceptional Quadrangles*, Ph.D. Thesis, Ghent University, Ghent, 2003, 186 pp.

[47] T. DE MEDTS. An algebraic structure for Moufang quadrangles, *Mem. Amer. Math. Soc.* **173** (2005), 99 pp.

[48] T. DE MEDTS, F. HAOT, K. TENT & H. VAN MALDEGHEM. Split BN-pairs of rank at least 2 and the uniqueness of splittings, *J. Group Theory* **8** (2005), 1–10.

[49] M. DE SOETE & J. A. THAS. A coordinatization of generalized quadrangles of order $(s, s + 2)$, *J. Combin. Theory Ser. A* **46** (1988), 1-11.

[50] S. DE WINTER & K. THAS. Automorphisms of generalized quadrangles of order s with a regular point and Payne derivatives, Preprint.

[51] K. J. DIENST. Verallgemeinerte Vierecke in Pappusschen projektiven Räumen, *Geom. Dedicata* **9** (1980), 199–206.

[52] K. J. DIENST. Verallgemeinerte Vierecke in projektiven Räumen, *Arch. Math.* **35** (1980), 177–186.

[53] S. DIXMIER & F. ZARA. Etude d'un quadrangle généralisé autour de deux de ses points non liés, Preprint, 1976.

[54] J. D. DIXON & B. MORTIMER. *Permutation Groups*, Graduate Texts in Mathematics **163**, Springer-Verlag, New York/Berlin/Heidelberg, 1996.

[55] J. R. FAULKNER. Groups with Steinberg relations and coordinatization of polygonal geometries, *Mem. Amer. Math. Soc.* **10** (1977), 135 pp.

[56] W. FEIT & G. HIGMAN. The nonexistence of certain generalized polygons, *J. Algebra* **1** (1964), 114–131.

[57] G. FELLEGARA. Gli ovaloidi in uno spazio tridimensionale di Galois di ordine 8, *Atti Accad. Naz. Lincei Rend. Cl. Sci. Fis. Mat. Natur.* **32** (1962), 170–176.

[58] J. C. FISHER & J. A. THAS. Flocks in $PG(3, q)$, *Math. Z.* **169** (1979), 1–11.

[59] P. FONG & G. M. SEITZ. Groups with a (B,N)-pair of rank 2, I, *Invent. Math.* **21** (1973), 1 – 57.

[60] P. FONG & G. M. SEITZ. Groups with a (B,N)-pair of rank 2, II, *Invent. Math.* **24** (1974), 191 – 239.

[61] H. FREUDENTHAL. Une étude de quelques quadrangles généralisés, *Ann. Mat. Pura Appl.* **102** (1975), 109–133.

[62] D. FROHARDT. Groups which produce generalized quadrangles, *J. Combin. Theory Ser. A* **48** (1988), 139 – 145.

[63] D. GLYNN. The Hering classification for inversive planes of even order, *Simon Stevin* **58** (1984), 319–353.

[64] D. GORENSTEIN. *Finite Simple Groups. An Introduction To Their Classification*, University Series in Mathematics, Plenum Publishing Corp., New York, 1982.

[65] D. HACHENBERGER. On finite elation generalized quadrangles with symmetries, *J. London Math. Soc.* **53** (1996), 397–406.

[66] M. HALL JR. Affine generalized quadrilaterals, *Studies in Pure Math.* (ed. L. Mirsky), Academic Press (1971), 113–116.

[67] F. HAOT & H. VAN MALDEGHEM. Some characterizations of Moufang generalized quadrangles, *Glasgow Math. J.* **46** (2004), 335 – 343.

[68] F. HAOT & H. VAN MALDEGHEM. A half 3-Moufang quadrangle is Moufang, *Bull. Belg. Math. Soc. — Simon Stevin* **12** (2006), 805–811.

[69] C. HERING, W. M. KANTOR & G. M. SEITZ. Finite groups with a split BN-pair of rank 1, *J. Algebra* **20** (1972), 435 – 475.

[70] D. G. HIGMAN. Partial geometries, generalized quadrangles and strongly regular graphs, in *Atti Convegno di Geometriae Combinatorica e Sue Applicazioni* (Univ. Perugia, Perugia, 1970) Ist. Mat., Univ. Perugia, Perugia (1971), 263–293.

[71] J. W. P. HIRSCHFELD. *Projective Geometries over Finite Fields*, Oxford Mathematical Monographs, The Clarendon Press, Oxford University Press, New York, 1979.

[72] J. W. P. HIRSCHFELD. *Finite Projective Spaces of Three Dimensions*, Oxford Mathematical Monographs, The Clarendon Press, Oxford University Press, New York, 1985.

[73] J. W. P. HIRSCHFELD. *Projective Geometries over Finite Fields, Second Edition*, Oxford Mathematical Monographs, The Clarendon Press, Oxford University Press, New York, 1998.

[74] J. W. P. HIRSCHFELD & J. A. THAS. *General Galois Geometries*, Oxford Mathematical Monographs. Oxford Science Publications, The Clarendon Press, Oxford University Press, New York, 1991.

[75] D. R. HUGHES & F. C. PIPER. *Projective Planes*, Springer-Verlag, New York/Heidelberg/Berlin, 1973.

[76] N. L. JOHNSON. Semifield flocks of quadratic cones, *Simon Stevin* **61** (1987), 313–326.

[77] N. L. JOHNSON. Flocks of hyperbolic quadrics and translation planes admitting affine homologies, *J. Geom.* **34** (1989), 50–73.

[78] N. L. JOHNSON. Homology groups of translation planes and flocks of quadratic cones, I. The structure, *Bull. Belgian Math. Soc. — Simon Stevin* **12** (2006), 827–844.

[79] N. L. JOHNSON & S. E. PAYNE. Flocks of Laguerre planes and associated geometries, in *Mostly Finite Geometries* (Iowa City, IA, 1996), *Lecture Notes in Pure and Appl. Math.* **190**, Dekker, New York (1997), 51–122.

[80] M. KALLAHER. *Translation Planes*, Chapter 5 of *Handbook of Incidence Geometry*, Edited by F. Buekenhout, North-Holland, Amsterdam (1995), 137–192.

[81] W. M. KANTOR. Generalized quadrangles associated with $G_2(q)$, *J. Combin. Theory Ser. A* **29** (1980), 212–219.

[82] W. M. KANTOR. Ovoids and translation planes, *Canad. J. Math.* **34** (1982), 1195–1207.

[83] W. M. KANTOR. Some generalized quadrangles with parameters q^2, q, *Math. Z.* **192** (1986), 45–50.

[84] W. M. KANTOR. Generalized quadrangles, flocks, and BLT sets, *J. Combin. Theory Ser. A* **58** (1991), 153–157.

[85] W. M. KANTOR. Automorphism groups of some generalized quadrangles, in *Adv. Finite Geom. and Designs*, Proceedings Third Isle of Thorns Conference on Finite Geometries and Designs, Brighton 1990, Edited by J. W. P. Hirschfeld et al., Oxford University Press, Oxford (1991), 251–256.

[86] W. M. KANTOR. Note on span-symmetric generalized quadrangles, *Adv. Geom.* **2** (2002), 197–200.

[87] W. M. KANTOR. Grid-symmetric generalized quadrangles, *Bull. Belgian Math. Soc. — Simon Stevin* **12** (2005), 137–139.

[88] N. KNARR. The nonexistence of certain topological polygons, *Forum Math.* **2** (1990), 603–612

[89] N. KNARR. A geometric construction of generalized quadrangles from polar spaces of rank three, *Results Math.* **21** (1992), 332–344.

[90] C. M. O'KEEFE & T. PENTTILA. Elation generalized quadrangles of order (q^2, q), in *Geometry, Combinatorial Designs and Related Structures* (Spetses, 1996), *London Math. Soc. Lecture Note Ser.* **245**, Cambridge Univ. Press, Cambridge (1997), 181–192.

[91] C. M. O'KEEFE & T. PENTTILA. Characterizations of flock quadrangles, *Geom. Dedicata* **42** (2000), 171–191.

[92] C. M. O'KEEFE & T. PENTTILA. Subquadrangles of generalized quadrangles of order (q^2, q), q even, *J. Combin. Theory Ser. A* **94** (2001), 218–229.

[93] M. LAVRAUW. *Scattered Spaces with respect to Spreads, and Eggs in Finite Projective Spaces*, Ph.D. Thesis, Eindhoven University of Technology, Eindhoven, 2001, viii+115 pp.

[94] M. LAVRAUW. Characterisations and properties of good eggs in $PG(4n - 1, q)$, q odd, *Discrete Math.* **301** (2005), 106–116.

[95] M. LAVRAUW. Semifield flocks, eggs, and ovoids of $Q(4, q)$, *Adv. Geom.* **5** (2005), 333–345.

[96] M. LAVRAUW & G. LUNARDON. Good eggs and Veronese varieties, *Discrete Math.* **294** (2005), 119–122.

[97] M. LAVRAUW & T. PENTTILA. On eggs and translation generalised quadrangles, *J. Combin. Theory Ser. A* **96** (2001), 303–315.

[98] H. LENZ. Zur Begründung der analytischen Geometrie, *S.-B. Math.-Nat. Kl. Bayer. Akad. Wiss.* (1954), 17–72.

[99] M. LIMBOS. *Plongement et arcs projectifs*, Ph.D.-thesis Université Libre de Bruxelles, 1981.

[100] G. LUNARDON. Flocks, ovoids of $Q(4, q)$ and designs, *Geom. Dedicata* **66** (1997), 163–173.

[101] H. LÜNEBURG. *Translation Planes*, Springer-Verlag, Berlin-New York, 1980.

[102] F. MAZZOCCA. Immergibilità in un $S_{4,q}$ di certi sistemi rigati di seconda specie, *Atti Accad. Naz. Lincei Rend. Cl. Sci. Fis Mat. Natur.* (8) **56** (1974), 189–196.

[103] F. MAZZOCCA. Caratterizzazione dei sistemi rigati isomorfi ad una quadrica ellittica dello $S_{5,q}$, con q dispari, *Atti Accad. Naz. Lincei Rend. Cl. Sci. Fis. Mat. Natur.* (8) **57** (1974), 360–368 (1975).

[104] B. MÜHLHERR & H. VAN MALDEGHEM. Exceptional Moufang quadrangles of type F_4, *Canad. J. Math.* **51** (1999), 347 – 371.

[105] W. F. ORR. *The Miquelian Inversive Plane $IP(q)$ and the Associated Projective Planes*, Ph.D. Thesis, University of Wisconsin, Madison WI, 1973.

[106] G. PANELLA. Caratterizzazione delle quadriche di uno spazio (tridimensionale) lineare sopra un corpo finito, *Boll. Un. Mat. Ital.* (3) **10** (1955), 507–513.

[107] S. E. PAYNE. A complete determination of translation ovoids in finite Desarguian planes, *Atti Accad. Naz. Lincei Rend. Cl. Sci. Fis. Mat. Natur.* (8) **51** (1971), 328–331.

[108] S. E. PAYNE. All generalized quadrangles of order 3 are known, *J. Combin. Theory Ser. A* **18** (1975), 203–206.

[109] S. E. PAYNE. Generalized quadrangles with symmetry. II, *Simon Stevin* **50** (1976/77), 209–245.

[110] S. E. PAYNE. Generalized quadrangles of order 4, I, *J. Combin. Theory Ser. A* **22** (1977), 267–279.

[111] S. E. PAYNE. Generalized quadrangles of order 4, II, *J. Combin. Theory Ser. A* **22** (1977), 280–288.

[112] S. E. PAYNE. Generalized quadrangles as group coset geometries, *Congr. Numer.* **29** (1980), 717–734.

[113] S. E. PAYNE. Generalized quadrangles as group coset geometries, *Proceedings of the Eleventh Southeastern Conference on Combinatorics, Graph Theory and Computing (Florida Atlantic Univ., Boca Raton, Fla., 1980), Vol. II, Congr. Numer.* **29** (1980), 717–734.

[114] S. E. PAYNE. Span-symmetric generalized quadrangles, in *The Geometric Vein*, Springer, New York/Berlin (1981), 231 – 242.

[115] S. E. PAYNE. Collineations of finite generalized quadrangles, in *Finite Geometries, Lecture Notes in Pure and Appl. Math.* **82**, Dekker, New York (1983), 361–390.

[116] S. E. PAYNE. A garden of generalized quadrangles, *Algebras, Groups and Geometries* **3** (1985), 323–354.

[117] S. E. PAYNE. A new infinite family of generalized quadrangles, *Proceedings of the Sixteenth Southeastern International Conference on Combinatorics, Graph Theory and Computing (Boca Raton, Fla., 1985), Congr. Numer.* **49** (1985), 115–128.

[118] S. E. PAYNE. An essay on skew translation generalized quadrangles, *Geom. Dedicata* **32** (1989), 93–118.

[119] S. E. PAYNE. Collineations of the generalized quadrangles associated with q-clans, in *Combinatorics '90* (Gaeta, 1990), North-Holland, Amsterdam, *Ann. Discrete Math.* **52** (1992), 449–461.

[120] S. E. PAYNE. A tensor product action on q-clan generalized quadrangles with $q = 2^e$, *Linear Algebra Appl.* **226/228** (1995), 115–137.

[121] S. E. PAYNE. The fundamental theorem of q-clan geometry, *Des. Codes Cryptogr.* **8** (1996), 181–202.

[122] S. E. PAYNE. Flocks and generalized quadrangles: an update, in *Generalized Polygons*, Proceedings of the Academy Contact Forum 'Generalized Polygons' (October 2000), Palace of the Academies, Brussels, Belgium, Edited by F. De Clerck et al., Universa Press (2001), 61–98.

[123] S. E. PAYNE. *Private Communication*, 2004.

[124] S. E. PAYNE & C. C. MANERI. A family of skew-translation generalized quadrangles of even order, *Proceedings of the thirteenth Southeastern conference on Combinatorics, Graph Theory and Computing (Boca Raton, Fla., 1982), Congr. Numer.* **36** (1982), 127–135.

[125] S. E. PAYNE & L. A. ROGERS. Local group actions on generalized quadrangles, *Simon Stevin* **64** (1990), 249–284.

[126] S. E. PAYNE & J. A. THAS. Generalized quadrangles with symmetry, Part II, *Simon Stevin* **49** (1975/76), 81–103.

[127] S. E. PAYNE & J. A. THAS. Moufang conditions for finite generalized quadrangles, in *Finite Geometries and Designs*, Proceedings Second Isle of Thorns Conference 1980, Lond. Math. Soc. Lect. Note Ser. **49**, Cambridge University Press, Cambridge (1981), 275-303.

[128] S. E. PAYNE & J. A. THAS. *Finite Generalized Quadrangles*, Research Notes in Mathematics **110**, Pitman Advanced Publishing Program, Boston/London/Melbourne, 1984.

[129] S. E. PAYNE & J. A. THAS. Generalized quadrangles, BLT-sets, and Fisher flocks, *Proceedings of the Twenty-second Southeastern Conference on Combinatorics, Graph Theory, and Computing (Baton Rouge, LA, 1991), Congr. Numer.* **84** (1991), 161–192.

[130] S. E. PAYNE & K. THAS. Notes on elation generalized quadrangles, *European J. Combin.* **24** (2003), 969–981.

[131] T. PENTTILA & B. WILLIAMS. Ovoids of parabolic spaces, *Geom. Dedicata* **82** (2000), 1–19.

[132] B. QVIST. Some remarks concerning curves of the second degree in a finite plane, *Ann. Acad. Sci. Fennicae Ser. A I. Math.-Phys.* **134** (1952), 1–27.

[133] L. A. ROGERS. Characterization of the kernel of certain translation generalized quadrangles, *Simon Stevin* **64** (1990), 319–328.

[134] R. ROSTERMUNDT. Some new elation groups of $H(3, q^2)$, Preprint, 2004.

[135] W. M. SCHMIDT. *Equations over Finite Fields*, Lecture Notes in Mathematics **536**, Springer, 1976.

[136] R. -H. SCHULZ. Über Translationsebenen mit Kollineationsgruppen, die die Punkte der ausgezeichneten Geraden zweifach transitiv permutieren, *Math. Z.* **122** (1971), 246–266.

[137] B. SEGRE. Sulle ovali nei piani lineari finiti, *Atti Accad. Naz. Lincei Rend. Cl. Sci. Fis. Mat. Natur.* **17** (1954), 141–142.

[138] B. SEGRE. On complete caps and ovaloids in three-dimensional Galois spaces of characteristic two, *Acta Arith.* **5** (1959), 315–332.

[139] B. SEGRE. *Lectures on Modern Geometry,* Consiglio Nazionale delle Ricerche Monografie Matematiche **7**, Edizioni Cremonese, Rome, 1961.

[140] B. SEGRE. Teoria di Galois, fibrazioni proiettive e geometrie non desarguesiane, *Ann. Mat. Pura Appl. (4)* **64** (1964), 1–76.

[141] J. J. SEIDEL. Strongly regular graphs with $(-1, 1, 0)$ adjacency matrix having eigenvalue 3, *Linear Algebra Appl.* **1** (1968), 281–298.

[142] E. E. SHULT. Characterizations of certain classes of graphs, *J. Combin. Theory Ser. B.* **13** (1972), 142–167.

[143] E. SHULT. On a class of doubly transitive groups, *Illinois J. Math.* **16** (1972), 434–455.

[144] R. R. SINGLETON. On minimal graphs of maximum even girth, *J. Combin. Theory* **1** (1966), 306–332.

[145] J. C. D. SPENCER. On the Lenz-Barlotti classification of projective planes, *Quart. J. Math. Oxford Ser. (2)* **11** (1960), 241–257.

[146] STEINBACH & H. VAN MALDEGHEM. Generalized quadrangles weakly embedded of degree > 2 in projective space, *Forum Math.* **11** (1999), 139–176.

[147] A. STEINBACH & H. VAN MALDEGHEM. Generalized quadrangles weakly embedded of degree 2 in projective space, *Pacific J. Math.* **193** (2000), 227–248.

[148] L. STORME & J. A. THAS. k-Arcs and partial flocks, *Linear Algebra Appl.* **226/228** (1995), 33–45.

[149] G. TALLINI. Ruled graphic systems, in *Atti del Convegno di Geometria Combinatoria e sue Applicazioni* (Univ. Perugia, Perugia, 1970), Ist. Mat., Univ. Perugia, Perugia (1971), 403–411.

[150] K. TENT. Very homogeneous generalized polygons of finite Morley rank, *J. London Math. Soc.* **62** (2000), 1–15.

[151] K. TENT. Half Moufang implies Moufang for generalized quadrangles, *J. Reine Angew. Math.* **566** (2004), 231–236.

[152] K. TENT. Split BN-pairs of rank 2: the octagons, *Adv. Math.* **181** (2004), 308–320.

[153] K. TENT & H. VAN MALDEGHEM. Split BN-pairs of rank 2 and Moufang polygons, I., *Adv. Math.* **174** (2003), 254–265.

[154] J. A. THAS. *Eindige Meetkunden (Finite Geometries),* Master Thesis, Ghent University, Ghent, 1966.

[155] J. A. THAS. *Een Studie Betreffende de Projectieve Rechte over de Totale Matrix Algebra $M_3(K)$ der 3×3-Matrices met Elementen in een Algebraïsch Afgesloten Veld K,* Verh. Kon. Vl. Acad. Wet. Lett. Sch. K. van België, Kl. der Wet. **112**, 1969.

[156] J. A. THAS. The m-dimensional projective space $S_m(M_n(GF(q)))$ over the total matrix algebra $M_n(GF(q))$ of the $n \times n$-matrices with elements in the Galois field $GF(q)$, *Rend. Mat. (6)* **4** (1971), 459–532.

[157] J. A. THAS. 4-gonal subconfigurations of a given 4-gonal configuration, *Atti Accad. Naz. Lincei Rend. Cl. Sci. Fis. Mat. Natur.* **52** (1972), 520–530.

[158] J. A. THAS. Ovoidal translation planes, *Arch. Math.* **23** (1972), 110–112.

[159] J. A. THAS. Flocks of finite egglike inversive planes, in *Finite Geometric Structures and their Applications* (C.I.M.E., II Ciclo, Bressanone, 1972), Edizioni Cremonese, Rome (1973), 189–191.

[160] J. A. THAS. 4-gonal configurations, in *Finite Geometric Structures and Their Applications*, (ed. A. Barlotti), Ed. Cremonese Roma (1973), 251–263.

[161] J. A. THAS. On 4-gonal configurations, *Geom. Dedicata* **2** (1973), 317–326.

[162] J. A. THAS. Geometric characterization of the $[n-1]$-ovaloids of the projective space $PG(4n-1, q)$, *Simon Stevin* **47** (1974), 365–375.

[163] J. A. THAS. Translation 4-gonal configurations, *Atti Accad. Naz. Lincei Rend. Cl. Sci. Fis. Mat. Natur.* **56** (1974), 303–314.

[164] J. A. THAS. On 4-gonal configurations with parameters $r = q^2 + 1$ and $k = q + 1$, *Geom. Dedicata* **3** (1974), 365–375.

[165] J. A. THAS. 4-gonal configurations with parameters $r = q^2 + 1$ and $k = q + 1$. II, *Geom. Dedicata* **4** (1975), 51–59.

[166] J. A. THAS. Flocks of non-singular ruled quadrics in $\mathrm{PG}(3, q)$, *Atti Accad. Naz. Lincei Rend. Cl. Sci. Fis. Mat. Natur. (8)* **59** (1975), 83–85.

[167] J. A. THAS. Series of Lectures at Ghent University, Unpublished manuscript, 1976.

[168] J. A. THAS. Combinatorial characterizations of generalized quadrangles with parameters $s = q$ and $t = q^2$, *Geom. Dedicata* **7** (1978), 223–232.

[169] J. A. THAS. Some results on quadrics and a new class of partial geometries, *Simon Stevin* **55** (1981), 129–139.

[170] J. A. THAS. Semi-partial geometries and spreads of classical polar spaces, *J. Combin. Theory Ser. A* **35** (1983), 58–66.

[171] J. A. THAS. 3-Regularity in generalized quadrangles of order (s, s^2), *Geom. Dedicata* **17** (1984), 33–36.

[172] J. A. THAS. Generalized quadrangles and flocks of cones, *European J. Combin.* **8** (1987), 441–452.

[173] J. A. THAS. Flocks, maximal exterior sets, and inversive planes, in *Finite Geometries and Combinatorial Designs* (Lincoln, NE, 1987), *Contemp. Math.* **111**, Amer. Math. Soc., Providence, RI (1990), 187–218.

[174] J. A. THAS. Recent results on flocks, maximal exterior sets and inversive planes, in *Combinatorics '88*, Vol. 1 (Ravello, 1988), *Res. Lecture Notes Math.*, Mediterranean, Rende (1991), 95–108.

[175] J. A. THAS. The affine plane $\mathrm{AG}(2, q)$, q odd, has a unique one point extension, *Invent. Math.* **118** (1994), 133–139.

[176] J. A. THAS. Generalized quadrangles of order (s, s^2), I, *J. Combin. Theory, Ser. A* **67** (1994), 140–160.

[177] J. A. THAS. Recent developments in the theory of finite generalized quadrangles, *Acad. Analecta*, Med. Konink. Acad. Wetensch. Lett. Sch. K. België (1994), 99 – 113.

[178] J. A. THAS. *Projective Geometry over a Finite Field*, Chapter 7 of *Handbook of Incidence Geometry*, Edited by F. Buekenhout, North-Holland, Amsterdam (1995), 295–347.

[179] J. A. THAS. *Generalized Polygons*, Chapter 9 of *Handbook of Incidence Geometry*, Edited by F. Buekenhout, North-Holland, Amsterdam (1995), 383–431.

[180] J. A. THAS. Generalized quadrangles of order (s, s^2), II, *J. Combin. Theory Ser. A* **79** (1997), 223–254.

[181] J. A. THAS. Symplectic spreads in $PG(3, q)$, inversive planes and projective planes, *Discrete Math.* **174** (1997), 329–336.

[182] J. A. THAS. 3-Regularity in generalized quadrangles: a survey, recent results and the solution of a longstanding conjecture, in *Combinatorics '98* (Mondello), *Rend. Circ. Mat. Palermo (2)* Suppl. No. **53** (1998), 199–218.

[183] J. A. THAS. Generalized quadrangles of order (s, s^2), III, *J. Combin. Theory Ser. A* **87** (1999), 247–272.

[184] J. A. THAS. Generalized quadrangles of order (s, s^2): recent results, *Discrete Math.* **208/209** (1999), 577 – 587.

[185] J. A. THAS. Characterizations of translation generalized quadrangles, *Des. Codes Cryptogr.* **23** (2001), 249–257.

[186] J. A. THAS. Geometrical constructions of flock generalized quadrangles, *J. Combin. Theory Ser. A* **94** (2001), 51–62.

[187] J. A. THAS. Flocks and partial flocks: a survey, *J. Statist. Plann. Inference* **94** (2001), 335–348.

[188] J. A. THAS. Moufang generalized quadrangles and the theorem of Fong and Seitz on BN-pairs, Manuscript in preparation.

[189] J. A. THAS & F. DE CLERCK. Partial geometries satisfying the axiom of Pasch, *Simon Stevin* **51** (1977), 123–137.

[190] J. A. THAS & S. E. PAYNE. Spreads and ovoids in finite generalized quadrangles, *Geom. Dedicata* **52** (1994), 227–253.

[191] J. A. THAS & K. THAS. Translation generalized quadrangles and translation duals, part I, *Discrete Math.*, To appear.

[192] J. A. THAS & K. THAS. Translation generalized quadrangles in even characteristic, *Combinatorica*, To appear.

[193] J. A. THAS & H. VAN MALDEGHEM. Generalized quadrangles and the Axiom of Veblen, In *Geometry, Combinatorial Designs and Related Structures*, Edited by J. W. P. Hirschfeld, *London Math. Soc. Lecture Note Ser.* **245**, Cambridge University Press, Cambridge (1997), 241–253.

[194] J. A. THAS & H. VAN MALDEGHEM. Generalized quadrangles weakly embedded in finite projective space, *J. Statist. Plann. Inference* **73** (1998), 353–361.

[195] J. A. THAS & H. VAN MALDEGHEM. Lax embeddings of generalized quadrangles in finite projective spaces, *Proc. London Math. Soc.* **82** (2001), 402–440.

[196] J. A. THAS, S. E. PAYNE & H. VAN MALDEGHEM. Half Moufang implies Moufang for finite generalized quadrangles, *Invent. Math.* **105** (1991), 153 – 156.

[197] K. THAS. *Symmetrieën in Eindige Veralgemeende Vierhoeken (Symmetries in Finite Generalized Quadrangles)*, Master Thesis, Ghent University, Ghent, 1999.

[198] K. THAS. On symmetries and translation generalized quadrangles, in *Finite Geometries, Developments in Mathematics* **3**, Proceedings of the Fourth Isle of Thorns Conference 'Finite Geometries', 16–21 July 2000, Edited by A. Blokhuis et al., Kluwer Academic Publishers (2001), 333–345.

[199] K. THAS. Automorphisms and characterizations of finite generalized quadrangles, in *Generalized Polygons*, Proceedings of the Academy Contact Forum 'Generalized Polygons' (October 2000), Palace of the Academies, Brussels, Belgium, Edited by F. De Clerck et al., Universa Press (2001), 111–172.

[200] K. THAS. A theorem concerning nets arising from generalized quadrangles with a regular point, *Des. Codes Cryptogr.* **25** (2002), 247–253.

[201] K. THAS. Classification of span-symmetric generalized quadrangles of order s, *Adv. Geom.* **2** (2002), 189–196.

[202] K. THAS. The classification of generalized quadrangles with two translation points, *Beiträge Algebra Geom.* **43** (2002), 365–398.

[203] K. THAS. On generalized quadrangles with some concurrent axes of symmetry, *Bull. Belg. Math. Soc. — Simon Stevin* **9** (2002), 217–243.

[204] K. THAS. Symmetry in generalized quadrangles, *Des. Codes Cryptogr.* **29** (2003), 227–245.

[205] K. THAS. Translation generalized quadrangles for which the translation dual arises from a flock, *Glasgow Math. J.* **45**(3) (2003), 457–474.

[206] K. THAS. *Symmetry in Finite Generalized Quadrangles, Frontiers in Mathematics* **1**, Birkhäuser Verlag, Basel/Boston/Berlin, 2004.

[207] K. THAS. Some basic questions and conjectures on elation generalized quadrangles, and their solutions, *Bull. Belgian Math. Soc. — Simon Stevin* **12** (2006), 909–918.

[208] K. THAS. Elation generalized quadrangles of order (q, q^2), q even, with a classical subGQ of order q, *Adv. Geom.*, To appear.

[209] K. THAS. Finite BN-pairs and the Tits conjecture, I. Local analysis in even characteristic, Preprint.

[210] K. THAS. Finite BN-pairs and the Tits conjecture, II. Local analysis in odd characteristic, Preprint.

[211] K. THAS. Finite BN-pairs and the Tits conjecture, III. Generalized Baer quadrangles, Preprint.

[212] K. THAS. *Lectures on Elation Quadrangles*, Monograph, Preprint.

[213] K. THAS. Property (F) and Kantor's conjecture, Manuscript in preparation.

[214] K. THAS & S. E. PAYNE. Foundations of elation generalized quadrangles, *European J. Math.* **27** (2006), 51–62.

[215] K. THAS & H. VAN MALDEGHEM. Geometric characterizations of Chevalley groups of Type B_2, *Trans. Amer. Math. Soc.*, To appear.

[216] F. G. TIMMESFELD. *Abstract Root Subgroups and Simple Groups of Lie Type, Monographs in Mathematics* **95**, Birkhäuser Verlag, Basel/Boston/Berlin, 2001.

[217] J. TITS. Sur la trialité et certains groupes qui s'en déduisent, *Inst. Hautes Etudes Sci. Publ. Math.* **2** (1959), 13–60.

[218] J. TITS. Géométries polyédriques et groupes simples, *Deuxième Réunion du Groupement de Mathématiciens d'Expression Latine*, Florence, sept. 1961, 66 – 88.

[219] J.TITS. Ovoïdes et groupes de Suzuki, *Arch. Math.* **13** (1962), 187–198.

[220] J. TITS. Théorème de Bruhat et sous-groupes paraboliques, *C. R. Acad. Sci. Paris Ser. I Math.* **254** (1962), 2910 – 2912.

[221] J. TITS. Ovoïdes à translations, *Rend. Mat. e Appl. (5)* **21** (1962), 37–59.

[222] J. TITS. *Buildings of Spherical Type and Finite BN-Pairs*, Lecture Notes in Mathematics **386**, Springer, Berlin, 1974.

[223] J. TITS. Non-existence de certains polygones généralisés, I, *Invent. Math.* **36** (1976), 275 – 284.

[224] J. TITS. Classification of buildings of spherical type and Moufang polygons: a survey, in *Teorie Combinatoria, Vol. I*, Accad. Naz. dei Lincei, Roma **17** (1976), 229 – 246.

[225] J. TITS. Non-existence de certains polygones généralisés, II, *Invent. Math.* **51** (1979), 267 – 269.

[226] J. TITS. Moufang octagons and the Ree groups of type $^2F_4$, *Amer. J. Math.* **105** (1983), 539 – 594.

[227] J. TITS. Twin buildings and groups of Kac-Moody type, *London Math. Soc. Lecture Note Ser.* **165**, Proceedings of a conference on *Groups, Combinatorics and Geometry*, Edited by M. Liebeck and J. Saxl, Durham 1990, Cambridge University Press, Cambridge (1992), 249 – 286.

[228] J. TITS. Moufang polygons, I. Root data, *Bull. Belg. Math. Soc. — Simon Stevin* **1** (1994), 455 – 468.

[229] J. TITS & R. WEISS. *Moufang Polygons*, Springer Monographs in Mathematics, Springer Verlag, Berlin, 2002.

[230] H. VAN MALDEGHEM. Generalized polygons with valuation, *Arch. Math.* **53** (1989), 513–520.

[231] H. VAN MALDEGHEM. Some consequences of a result of Brouwer, *Ars Combin.* **48** (1998), 185 – 190.

[232] H. VAN MALDEGHEM. *Generalized Polygons*, Monographs in Mathematics **93**, Birkhäuser Verlag, Basel/Boston/Berlin, 1998

[233] H. VAN MALDEGHEM. On the dimensions of the root groups of full subquadrangles of Moufang quadrangles arising from algebraic groups, in *Finite Geometry and Combinatorics* (ed. F. De Clerck *et al.*), *Bull. Belg. Math. Soc. — Simon Stevin* **5** (1998), 469–475.

[234] H. VAN MALDEGHEM. Moufang lines defined by (generalized) Suzuki groups, Preprint.

[235] H. VAN MALDEGHEM & R. WEISS. On finite Moufang polygons, *Israel J. Math.* **79** (1992), 321 – 330.

[236] H. VAN MALDEGHEM, J. A. THAS & S. E. PAYNE. Desarguesian finite generalized quadrangles are classical or dual classical, *Des. Codes Cryptogr.* **1** (1992), 299 – 305.

[237] M. WALKER. A class of translation planes, *Geom. Dedicata* **5** (1976), 135–146.

[238] R. WEISS. The nonexistence of certain Moufang polygons, *Invent. Math.* **51** (1979), 261 – 266.

[239] J. C. D. S. YAQUB. The non-existence of finite projective planes of Lenz-Barlotti class III 2, *Arch. Math. (Basel)* **18** (1967), 308–312.

Index